Encapsulated Corrosion Inhibitors for Eco-Benign Smart Coatings

This book covers all the recent advancements and technologies developed for encapsulated corrosion inhibitors-based coatings by using eco-benign and sustainable encapsulated smart coatings based on self-healing functionality. It includes an overview of properties and applicability of encapsulated inhibitors, methodologies to detect corrosion, and recent developments made in the field of corrosion science to study the inhibition potential of encapsulated corrosion inhibitors for eco-benign smart coatings in several corrosive systems.

Features:

- Covers encapsulation of the corrosion inhibitor which explores the self-healing mechanism of the smart coatings.
- Includes encapsulated corrosion inhibitor fabrication, synthesis, modeling, functionalization, classification, characteristics, and so forth.
- Reviews effectiveness and significant constraints of scale-up engineering.
- Discusses the fundamental characteristics of industrial-scale application research for encapsulated corrosion inhibitors.
- Explores entire aspects of encapsulated corrosion inhibitors-based sustainable smart coatings in one place.

This book is aimed at researchers and graduate students of corrosion engineering, surface engineering, and chemistry engineering .

Encapsulated Corrosion Inhibitors for Eco-Benign Smart Coatings

Edited by
Ashish Kumar and Abhinay Thakur

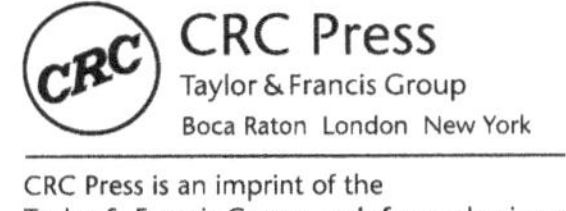

CRC Press is an imprint of the
Taylor & Francis Group, an informa business

Designed cover credit: shutterstock.

First edition published 2025
by CRC Press
2385 NW Executive Center Drive, Suite 320, Boca Raton FL 33431

and by CRC Press
4 Park Square, Milton Park, Abingdon, Oxon, OX14 4RN

CRC Press is an imprint of Taylor & Francis Group, LLC

ISBN: 9781032534770 (hbk)
ISBN: 9781032677231 (pbk)
ISBN: 9781032677255 (ebk)

DOI: 10.1201/9781032677255

Typeset in Times
by Newgen Publishing UK

Contents

Preface

In the realm of materials science and engineering, corrosion stands as an enduring adversary, silently eroding structures, machinery, and infrastructure, siphoning away trillions of dollars annually in repairs and replacements. The battle against corrosion has witnessed numerous skirmishes, with researchers and engineers arming themselves with an array of tools and strategies to fend off this relentless foe. Among the weapons in this arsenal, smart coatings fortified with encapsulated corrosion inhibitors have emerged as a promising frontier in the ongoing war against material degradation. The *Encapsulated Corrosion Inhibitors for Eco-Benign Smart Coatings* embarks on a journey through the landscape of corrosion science, unveiling the recent advancements and technologies that have reshaped our approach to combating corrosion. This book endeavors to illuminate the path from ground zero to the peak of the sky, offering readers a comprehensive platform to comprehend the intricacies of encapsulated corrosion inhibitors and their role in the realm of smart coatings.

While this book is primarily designed for corrosion engineering students, its reach extends far beyond the confines of academia. Corrosion scientists and engineers, battling corrosion in their professional capacities, will find within these pages a wealth of insights and strategies to augment their ongoing efforts. Moreover, individuals with a vested interest in the science of corrosion will discover a treasure trove of knowledge that unveils the very core of this phenomenon and its potential mitigation. Central to the narrative of this book is the exploration of eco-benign and sustainable encapsulated smart coatings, endowed with the remarkable ability to self-heal and fend off corrosion. These coatings, a testament to human ingenuity, represent a convergence of cutting-edge technology and ecological consciousness. As the global community strives to fulfill the Green Earth pledge, these coatings offer a ray of hope, bridging the chasm between technological progress and environmental sustainability.

For researchers and students venturing into the labyrinthine field of corrosion, this book assumes the role of a guiding light. It ushers them through the fundamentals of utilizing encapsulated corrosion inhibitors, unraveling the science behind their function, and unveiling their promise as cornerstones of eco-benign smart coatings. With a pedagogical approach, the book ensures that even those unfamiliar with the nuances of corrosion science can grasp the essence of this intricate field. From understanding the nuanced properties of encapsulated corrosion inhibitors to discerning their applicability across diverse contexts, readers will embark on a comprehensive exploration. Methodologies to detect corrosion, critical for the timely intervention of degradation processes, are scrutinized and demystified, empowering practitioners to take proactive measures against material deterioration. As the panorama of this book unfolds, it will traverse the recent developments that have propelled the use of encapsulated corrosion inhibitors to the forefront of corrosion inhibition strategies. The narrative threads through various corrosive systems, dissecting their complexities and laying bare the potential of these inhibitors in safeguarding against corrosion-related catastrophes.

About the Editors

Ashish Kumar is Professor, Head of the Department, and Dean at Nalanda College of Engineering, Bihar Engineering University, Department of Science, Technology and Technical Education, Government of Bihar, India. He is the author of more than 120 highly reputed research papers and 70 book chapters in world top publication houses. He authored four international books and working on six new books with RSC, ACS, Elsevier, Wiley, Springer, DeGuryter, CRC, etc. His area of research is materials chemistry, corrosion, nano materials, environmental science, surface chemistry, and solution thermodynamics. He is the editor and reviewer of top research journals with all top publication houses. He is an evaluator of Ph.D thesis from India and abroad. Kumar is listed in world top 2% researchers list data by Stanford University and Scopus 2023 and recipient of several research awards including Asia's prominent Scientists Award 2023. His current h-index is 34.

Abhinay Thakur is Assistant Professor at the Division of Research and Development Cell, Lovely Professional University, India. He specializes in physical chemistry, green chemistry, corrosion chemistry, and solution thermodynamics. Abhinay has also edited several books related to corrosion and contributed chapters to numerous book for publishers such as Elsevier, Springer, IGI Global, CRC Press, and De Gruyter. Abhinay has served as a reviewer for several international journals and is a life member of The Indian Science Congress Association. He has presented his research findings at various conferences and workshops, showcasing his commitment to staying at the forefront of advancements in his field. Abhinay's research contributions extend to numerous published papers in reputed journals, covering topics ranging from corrosion inhibition studies to the application of nanomaterials in biofuel production and environmental remediation.

Contributors

Nayad Abdallah
Molecular Chemistry Laboratory
Coordination Chemistry and Catalysis Unit
Faculty of Sciences Semlalia, Cadi Ayyad University
Marrakech, Morocco

Abouelfida Abdesselam
Laboratory of Applied Chemistry and Biomass
Department of Chemistry, Faculty of Science Semlalia
University Cadi Ayyad
Marrakech, Morocco

Athira Ajayan
Department of Chemistry, University of Calicut
Kerala, India

Valentine Chikaodili Anadebe
Department of Chemical Engineering
Alex Ekwueme Federal University Ndufu Alike
Abakakili, Ebonyi State, Nigeria

Ruby Aslam
School of Civil Engineering and Architecture
Chongqing University of Science and Technology
Chongqing, China

Fidya Ayuningtyas
Department of Metallurgical Engineering
University of Sultan Ageng Tirtayasa, Cilegon
Banten, Indonesia

Mahboobeh Azadi
Faculty of Materials and Metallurgical Engineering
Semnan University
Semnan, Iran

Lai Xuan Bach
Future Materials & Devices Lab.
Institute of Fundamental and Applied Sciences, Duy Tan University
Ho Chi Minh City, Viet Nam
The Faculty of Environmental and Chemical Engineering
Duy Tan University, Danang, Viet Nam

Elyor Berdimurodov
Chemical & Materials Engineering, New Uzbekistan University
Tashkent, Uzbekistan
Akfa University, Tashkent, Uzbekistan
Faculty of Chemistry, National University of Uzbekistan
Tashkent, Uzbekistan

Khasan Berdimuradov
Faculty of Industrial Viticulture and Food Production Technology
Shahrisabz Branch of Tashkent Institute of Chemical Technology
Shahrisabz, Uzbekistan

Bakhtiyor Borikhonov
Faculty of Chemistry-Biology, Karshi State University
Karshi, Uzbekistan

El Ibrahimi Brahim
Department of Applied Chemistry
Faculty of Applied Sciences
Ibn Zohr University
Aït Melloul, Morocco

Nkechinyere Amaka Chikaodili
Department of Chemical Engineering
Alex Ekwueme Federal University Ndufu Alike
Abakakili, Ebonyi State, Nigeria
Department of Chemical Engineering
Chukwuemeka Odumegwu Ojukwu Univesity
Anambra State, Nigeria

Carolina Garín Correa
Laboratório de corrosão, Federal University of Rio de Janeiro
Rio de Janeiro, Brazil

Omar Dagdag
Department of Mechanical Engineering
Gachon University, Seongnam
Republic of Korea

Nam Nguyen Dang
Future Materials & Devices Lab.
Institute of Fundamental and Applied Sciences, Duy Tan University
Ho Chi Minh City, Viet Nam
The Faculty of Environmental and Chemical Engineering
Duy Tan University, Danang, Viet Nam

Thi-Bich-Ngoc Dao
Future Materials & Devices Lab.
Institute of Fundamental and Applied Sciences
Duy Tan University
Ho Chi Minh City, Viet Nam
The Faculty of Environmental and Chemical Engineering
Duy Tan University, Danang, Viet Nam

Walid Daoudi
Laboratory of Molecular Chemistry, Materials and Environment (LCM2E)
Department of Chemistry, Multidisciplinary Faculty of Nador
University Mohamed I, Nador, Morocco

Kim Long Duong Ngo
Future Materials & Devices Lab.
Institute of Fundamental and Applied Sciences
Duy Tan University, Ho Chi Minh City, Viet Nam
The Faculty of Environmental and Chemical Engineering, Duy Tan University, Danang, Viet Nam

Abdelmalik El Aatiaoui
Laboratory of Molecular Chemistry, Materials and Environment (LCM2E)
Department of Chemistry, Multidisciplinary Faculty of Nador
University Mohamed I, Nador, Morocco

Abdellah Elyoussfi
Laboratory: Applied Chemistry and Environment (LCAE)
Department of Chemistry, Faculty of Sciences
University Mohammed Premier
Oujda, Morocco

Hicham Es-soufi
Laboratory of Sciences and Professions of the Engineer
Materials and Processes Department ENSAM-Meknes Marjane II
Moulay Ismail University, Meknes, Morocco

Joseph Okechukwu Ezeugo
Department of Chemical Engineering
Chukwuemeka Odumegwu Ojukwu Univesity
Anambra State, Nigeria

Luana B. Furtado
Department of Organic Processes
School of Chemistry
Federal University of Rio de Janeiro
Rio de Janeiro, Brazil

Netravati Gayakwad
Department of Chemistry
KLE Institute of Technology Hubli
Karnataka, India

Reza Goudarzi
Department of Surface Coatings and Corrosion
Institute for Color Science and Technology (ICST)
Tehran, Iran

Shweta Goyal
Civil Engineering Department
Thapar Institute of Engineering and Technology
Patiala, India

Cam Tu Hoang Ngoc
Future Materials & Devices Lab.
Institute of Fundamental and Applied Sciences
Duy Tan University
Ho Chi Minh City, Viet Nam
The Faculty of Environmental and Chemical Engineering
Duy Tan University, Danang, Viet Nam

Abraham Joseph
Department of Chemistry, University of Calicut
Kerala, India

Vismaya Joseph
Department of Chemistry, University of Calicut
Kerala, India

Bening Nurul Hidayah Kambuna
Department of Metallurgical Engineering
University of Sultan Ageng Tirtayasa, Cilegon, Banten, Indonesia

Harpreet Kaur
Department of Chemistry
School of Chemical Engineering and Physical Sciences
Lovely Professional University
Phagwara, Punjab, India

Hedie Kazemi
Department of Surface Coatings and Corrosion
Institute for Color Science and Technology (ICST)
Tehran, Iran

Akshay Kumar
Department of Chemistry
Dyal Singh College, Karnal, Haryana, India

Ashish Kumar
Nalanda College of Engineering, Bihar Engineering University, Department of Science, Technology and Technical Education, Government of Bihar, India

Salma Lamghafri
Laboratory of Applied Sciences
National School of Applied Sciences Al-Hoceima
Abdelmalek Essaâdi University, Tetouan, Morocco

Sadhucharan Mallick
Departments of Chemistry
Indira Gandhi National Tribal University (Central University)
Amarkantak, Madhya Pradesh, India

Lasri Mohammed
Laboratory of Applied Chemistry and Biomass
Department of Chemistry, Faculty of Science Semlalia
University Cadi Ayyad
Marrakech, Morocco

Khadiri Mohy Eddine
Laboratory of Applied Chemistry and Biomass
Department of Chemistry, Faculty of Science Semlalia
University Cadi Ayyad
Marrakech, Morocco

Rafaela C. Nascimento
Department of Chemistry, LAQV-REQUIMTE
Instituto de Investigação e Formação Avançada
Colégio Luís António Verney, Universidade de Évora
Évora, Portugal

Patrick Chukwudi Nnaji
Department of Chemical Engineering
Michael Okpara University of Agriculture
Umudike, Abia State, Nigeria

Okechukwu Dominic Onukwuli
Department of Chemical Engineering
Nnamdi Azikwe University, Awka
Anambra State, Nigeria

Adyl Oussaid
Laboratory of Molecular Chemistry, Materials and Environment (LCM2E)
Department of Chemistry, Multidisciplinary Faculty of Nador
University Mohamed I, Nador, Morocco

Cindy Putri Pancaningtias
Department of Metallurgical Engineering
University of Sultan Ageng Tirtayasa, Cilegon
Banten, Indonesia

Anila Paul
Department of Chemistry, University of Calicut
Kerala, India

Anjali Peter
Apex University, Jaipur

Siska Prifiharni
Research Center for Metallurgy
National Research and Innovation Agency
South Tangerang, Indonesia

Purnima
Civil Engineering Department
Thapar institute of Engineering and Technology
Patiala, India

Idouhli Rachid
Laboratory of Applied Chemistry and Biomass
Department of Chemistry
Faculty of Science Semlalia
University Cadi Ayyad
Marrakech, Morocco

Bahram Ramezanzadeh
Department of Surface Coatings and Corrosion
Institute for Color Science and Technology (ICST)
Tehran, Iran

Ankit Sharma
Department of Pure & Applied Chemistry
University of Kota, Kota
Rajasthan, India

Shobhana Sharma
Department of Chemistry
S.S. Jain Subodh P.G. College, Rambagh
Jaipur, Rajasthan, India

Sushil Kumar Sharma
Department of Pure & Applied Chemistry
University of Kota, Kota
Rajasthan, India

Ambrish Singh
Department of Chemistry
Nagaland University, Nagaland, India

Amita Somya
Department of Applied Chemistry
Amity School of Engineering and Technology
Amity University, Bengaluru
Karnataka, India

Abhinay Thakur
Department of Chemistry
School of Chemical Engineering and Physical Sciences
Lovely Professional University
Punjab, India

Ashish Kumar Tiwari
Civil Engineering Department
Thapar Institute of Engineering and Technology
Patiala, India

Burak Tuzun
Plant and Animal Production Department
Technical Sciences Vocational School of Sivas
Sivas Cumhuriyet University, Sivas, Turkey

Amit Varshney
Tech Lead in IT, Bengaluru, India

Qihui Wang
School of Civil Engineering and Architecture
Chongqing University of Science and Technology
Chongqing, China

Zhitao Yan
School of Civil Engineering and Architecture
Chongqing University of Science and Technology
Chongqing, China

Kaoutar Zaidi
Laboratory: Applied Chemistry and Environment (LCAE)
Department of Chemistry, Faculty of Sciences
University Mohammed Premier
Oujda, Morocco

1 An Overview of Corrosion and Smart Healing Coatings

Netravati Gayakwad

1.1 INTRODUCTION

Metals are found concentrated as ores in the earth's crust. They are extracted by suitable metallurgical processing depending on their electropositive nature and occurrence in nature. There are 90 naturally occurring elements and most of them are metals. These metals have metallic characteristics. The main metallic characteristics are solid structure, lustrous, malleable, ductile, electropositive, and electrical and thermal conductivity [1]. Most of these metals form alloys. These metals are classified as alkali metals, alkaline earth metals, transition metals and inner transition metals. Transition metals are widely studied because they show variable oxidation states due to partially filled (n-1) d orbital and form coordination compounds. The 3d series transition metals are found in large percentages in the earth's crust. Among these metals, iron and aluminum are available in larger quantities. Most of these 3d metals are known as alloy steel, which are mainly used in automobile parts, machinery, cutting tools, utensils, and several industrial infrastructure materials. Iron exposure to air reacts with oxygen to form mainly oxides known as rust. Alloy steels are used widely as engineering materials because they are easily manufactured to desired quality steels with less production cost [2, 3]. Iron has mainly +2 and +3 oxidation states, with the latter being more stable due to a partially filled 3d orbital. Iron doesn't show passivity due to oxide coating over the surface like aluminum. Compounds of iron react with hydroxide solution to form reddish-brown ferric hydroxide and residue on heating forms stoichiometric ferric oxide in the presence of air. The above property is used to estimate iron analytically. A vexing problem associated with every use of the metal object is corrosion. Silver tarnishes, copper develops patina and iron rusts. Corrosion is a surface phenomenon that proceeds at slow rates but accounts for heavy loss [4, 5]. The losses cannot be estimated by considering only the metal loss. The indirect losses are much higher. Corrosion badly damages all equipment, instruments, and chemical unit structures made up of metals. Owing to the corrosion property of metals, infrastructure made from metals/alloys de-creases the life span of the basic framework. Sometimes corrosion damages materials heavily and the cost of repairing the reactor units is beyond affordability. When structures like buildings or bridges collapse due to corrosion, the loss of life is also considered seriously. Costs include alternate arrangements and subsequent repairing of damaged infrastructure [6]. Most of the engineering materials are made up of metals, especially alloy steels. These ferroalloys known for corrosion in all medium damage infrastructure of industries result in heavy loss to the users. "The destruction of metal when it reacts with an environment either by chemical or electrochemical method is called corrosion" [7].

DOI: 10.1201/9781032677255-1

1.2 CORROSION COST

The total cost of corrosion can be estimated by physical damages to infrastructure, parts of machinery, or reaction vessels, due to production losses, or due to factory shutdown. Prevention of corrosion of assets and protection of the environment are most important to industries. It is troublesome to manage the harm to the environment due to erosion. The strategy of estimating adding up to a toll of erosion was evaluated by Uhling in one of his studies that investigated the cost of erosion in the UK and Japan. The erosion control strategies that were considered for the estimation of erosion fetched incorporate defensive adsorption film, corrosion-resistant metals and alloys, erosion inhibitors, polymer, anodic and cathodic security, rusting hindrance services, corrosion mitigation tactics, instruction and giving preparing approximately erosion control. For the year 1998, in common each year's rate of erosion turned into expected being \$121.41 or 1.38% of the \$8.79 trillion gross domestic product (GDP). The assets of misfortune due to rust were \$364 billion in 2004, separated from roundabouts taking a toll and consequent harms. If included it would be around 3.1% of GDP as suggested by George Hays, president of the National Association of Corrosion Engineers (NACE), Houston, in the Mumbai conclave of NACE worldwide India segment. In India alone, this appraisal is around ₹36,000 crore and in the United States (US) it would be over \$360 billion. Agreeing with the Indian chapter of NACE universal, a worldwide organization of over 50,000 engineers, India was paying a charge of around₹1.5 lakh crore each year owing to erosion misfortunes, 16,000 erosion engineers. Concurring with Mr. RajanBahri, trustee of NACE Worldwide India, the fetched is essentially bigger than any of the nation's past common calamities. According to an investigation conducted within the joined-together states, the coordinated costs of the rusting sum to almost 4.9% of an industrialized country's net residential item. This sum surpasses the overall money-related fetches of all of the country's fires, surges, storms, and seismic tremors combined. Of this 4.9%, 1–2% can be maintained at a strategic distance by applying suitable advances that are accessible. Framework costs ₹22,600 crore per year, utility administrations ₹47,100 crore per year, generation and fabricating ₹17,650 crore per year, and defense and atomic squander ₹20,000 crore per year. As a result, the toll of relief has been calculated at 1% of the investment funds. The cost of erosion to the US economy as of 1995 costs is \$300 billion. This could be diminished to one-third of the fetched by utilizing the coatings of erosion-resistant materials and receiving the best erosion diminishment hones as per the report given by Battelle Researchers in 1978. This was created by Ponder as an expound show of around 130 financial areas. The fetched due to erosion in metals was \$92 to joined-together states as of 1975 at 4.9% of GDP. The curious reality was that 60% of this fetched was unavoidable. The yearly fetching of rusting metal to the US industry and the common populace is evaluated to be \$170 billion. Erosion is a deceptive foe that ruins our vehicles, plumbing, huge buildings, connective bridges, engines, and manufacturing plants, indeed in spite of the fact that it is simply nature's strategy of reusing, or returning a metal question to its most reduced vitality frame. In India, there's a shortage of erosion mindfulness. The enormous sum of misfortunes due to erosion, which amounts to over 3% of GDP, can be decreased to a sensible level by simply raising erosion mindfulness. Erosion can be avoided and postponed by utilizing straightforward design methods and material decisions. Erosion encom-passes a critical cost to the industry. However, an impressive division of the taken toll may be minimized and spared through basic designing machines under the application of demonstrated strategies besides improvement of strategies of anticipation. The misfortune of mellow steel in India is assessed to be around ₹25,000 crore per year, with accessible erosion combating advances sparing generally 33% of that sum. Direct losses are those incurred as a result of replacing corroded equipment, components, or structures. The costs of maintaining equipment and structures to prevent or slow down corrosion are also considered, in terms of both labor and materials. This class of labor and materials consists of painting, making use of shielding coatings or linings, and working prices for protecting pipes and structures by sacrificing cathode and habitual inspections or checking out of devices using online corrosion tracking sensors. Domestic losses include the alternative of residential water heaters and the breakdown of home water pipelines and different miscellaneous items. Additional prices are incurred through the use of corrosion-resistant metals or alloys in preference to less-expensive iron that has mechanical properties but inadequate corrosion resistance.

The sources of indirect losses can be listed; however, actual figures on these losses are difficult to estimate. Equipment breakdown due to corrosion causes unplanned shutdown of machinery that causes losses in timely production. Although maintenance work may be inexpensive, the value of loss in production might be substantial. If this occurs frequently, the expense is usually included in the production cost. The fluids being transported, stored, processed, or packaged in a metallic component are contaminated by the metallic compounds formed by the corrosion. These corrosion products can damage the product, resulting in decreased shelf life of the products and even changing the color of the products by affecting the dyes used. In some cases it leads to the inability to finish intermediate product in production stages. Nowadays lead pipes are not used to transport water because of lead poisoning which is caused by the dissolution of lead products in water. Corrosion can be so severe that it may result in product loss due to leakage of the gases or liquid ingredients. If leakage occurs in a pipeline, it may go undetected for a long time, thereby causing continuous product loss. If the leaking substance is corrosive, it will attack the surrounding environment, causing even more damage. Corrosion products cause a thin layer of scale over the surface of the container or in a plumbing system. This scale reduces the heat transfer while simultaneously increasing the energy required to pump the fluid through the system. Corrosion can also have an impact on the efficiency of mechanical devices, which includes higher operating costs, increased fuel consumption, lubricant losses, and decreased work production. Corrosion of equipment used to handle reactant, pipelines, or reactor may result in the leakage of reactant or products from the processing unit that may be toxic and enter the atmosphere. This affects the environment as the reaction vessel is normally designed based on the expected corrosion rate. In considering industrial safety additional thicknesses of reaction vessels are usually included in the vessel design, known as overdesign. The term for this is "corrosion allowance." This thickness is above what is required for the design parameters, hence it costs more. The loss of raw material metallic resources due to corrosion waste is referred to as conservation losses. This waste also includes the additional loss of energy and water reserves resulting from the production and construction of metallic components. Replacement and redesign of corroded equipment and components need more human effort and resources.

1.2.1 Importance of Corrosion Studies

It is currently essential to pay more attention to steel rusting due to:

- Increasing use of metals in all sectors of machinery technology.
- Use of uncommon and costly metals that are less corrosive but not in practice for fabricating infrastructure. As a result demand for ferroalloy was considerably increasing.
- In recent years, there has been a fast development in the manufacturing process. For this reason, modification of earlier equipment or new equipment is required. Frequent changes in these infrastructures need materials that should be easily fabricated and cheap. So materials like ferroalloys are used.
- An enhancement in the level of pollutants and moisture culminates to an excessive corrosive environment.
- Strict protection requirements for working equipment, which might also additionally fail catastrophically because of corrosion.

1.3 CORROSION PREVENTION

Corrosion of steel is a natural spontaneous process, which transforms into a greater strong compound state. Therefore, inhibition of corrosion control is greater than corrosion prevention. The types of corrosion are numerous, the mechanisms of corrosion are very different, and the situations under which corrosion takes place are so numerous that no single approach can be used to control all feasible corrosion cases.

1.3.1 Protective Coatings

Protective coatings protect the metal surface from rusting by acting as a barrier film on the metal surface to the environment. There are many types of coatings, but a few are discussed here.

1.3.1.1 Metal Coating

The coating of a protective metal over the surface of the underlying metal is called metal coating. It is a valuable and well-tried method for improving the corrosion life of the underlying metal [8]. From the corrosion point of view, metal coating is divided into two: anodic and cathodic metal coating.

1.3.1.2 Inorganic Coating

Inorganic coatings are generally chemical conversation coatings in which a metal gets transformed into a compound by chemical or electrochemical reaction that forms a protective barrier between the base metal and the environment. This type of coating is formed by chemical dip, spray, or electrolytic methods to enhance adherence to paints or dyes. There are two types of inorganic coatings: anodizing inorganic coating and phosphate inorganic coating [9, 10].

1.3.1.3 Cathodic Coating

Cathodic coating is a method of protecting a metal or alloy from corrosion by converting it completely into a cathode. Impressed methods of cathodic coating and sacrificial cathodic coating are the two main types of cathodic protection.

1.3.1.4 Physical Barrier Coatings

In general, a physical process can be defined as a process that protects the substrate from contact with the external environment through an inert medium. The protection of the functioning of the system is mainly based on its safety and the protection of the coating material. For both single and double layers, it is necessary to determine the properties of the material and the preparation process of the data layer [11]. In the past, metal layers, metal hydroxides and oxides, and silane sol-gel coatings were frequently utilized as dormant stages for metal security. Polymers and CaP (Calcium Phosphate) compounds are moreover broadly utilized for biomedical applications. Subsequently, later advancements in metal coatings, metal hydroxide and hydrotalcite coatings, metal oxide coatings, silane sol-gel coatings, polymer coatings, and CaeP coatings are common [12].

1.3.1.5 Metal-Based Coatings

As we all know, each metal has its own electrode potential/corrosion resistance, which shows its ability in a corrosive environment. Security of dynamic metal substrates with dormant metal coatings may be a broadly utilized concept in mechanical applications. Cold spraying, plasma showering, electroplating, and electroplating are the most common strategies for planning the metal structure. Cold showering can fix the metal powder on the surface of the magnesium amalgam to make a coating, such as copper and nickel-based coatings [13]. Plasma spraying relies on plasma as heat to melt or semi-melt the metal powder, and then slowly spray it onto the surface of the magnesium alloy to prepare the whole process like a metal structure [14]. Galvanic plating and electroless plating are based on forms that diminish metal particles to their metals in arrangement. The distinction is that galvanic plating depends on an outside current to nourish the fabric, whereas nonelectric plating depends on diminishing specialists such as sodium phosphate, hydrazine hydrate, and sodium borohydride. In past considerations, the electroless coating was favored for the following reasons:

i. Simple equipment and simple preparation process
ii. Easy to control and modify
iii. More importantly, the prepared coating has a uniform, high hardness, strong adhesion, and superior corrosion resistance. Among the various electroless coatings, nickel-based electroless coating is the most researched one.

1.3.1.6 Metal Hydroxides and Hydrotalcite Coatings

Considering that the erosion handle of the magnesium amalgam comprises free $Mg(OH)_2$, the analysts attempted to get ready the colored and thick $Mg(OH)_2$ film of the magnesium amalgam by aqueous treatment. Temperature and time-controlled testimony of $Mg(OH)_2$ thin films with diverse morphologies (e.g., plate, strip, and hexagonal plate) on ZK60 Amalgam [15, 16]. Application of silicate containing $Mg(OH)_2$ film in AZ31 amalgam made strides in erosion resistance and hydrophilicity of $Mg(OH)_2$ film [17]. However, the unsteady appearance of $Mg(OH)_2$ prevents its erosion assurance because when the chloride concentration surpasses 30 mm, $Mg(OH)_2$ starts to change over to dissolvable $MgCl_2$. Hence, hydrophobic or superhydrophobic films based on $Mg(OH)_2$ were arranged by uniting long chain atoms (CeC or CeF) with mono surface vitality such as 1-dodecanthiol and stearic acid [18, 19]. However, there has been a small work centered on developing modern strategies to create $Mg(OH)_2$ films with better ways of erosion resistance, which can be ascribed to the following two reasons. First, $Mg(OH)_2$ contains a basic structure (layered hydroxide, trigonal); hence, exceptionally small constraint can be balanced. $Mg(OH)_2$ thin films were developed in the first decade of the 21st century [20]. But then, the $Mg(OH)_2$ thin film showed low erosion resistance and biocompatibility, so researchers searched for other ways to alter the surface [21]. In these cases, the LDH thin film with a structure comparable to $Mg(OH)_2$ came to the fore [22].

1.3.1.7 Metal Oxide Coatings

Since metal oxides are steadier than metal hydroxides, the metal oxide layer is additionally expanded to adjust the surface of magnesium combinations for erosion resistance. Since the re-layering handle incorporates the layering preparation, the pressed oxide can be partitioned into layer-by-layer handling, and not blended handling. Different strategies have been developed to prepare substrate-inviting magnesium combination layers, such as atomic layer deposition (ALD) [23, 24], electrophoretic deposition (EPD) [25], electro deposition [26], spraying coating (including cold spraying and plasma spraying coatings) [27], sol-gel [28], magnetron sputtering [29], and atmospheric plasma jet [30]. Among them, ALD, EPD, sol-gel, and splashing are the most commonly used strategies. Unlike ALD, EPD, and sputtering strategies, sol-gels are frequently utilized to form two layers, such as to seal the pores of plainly visible micro-arc oxidation (MAO) coatings [31].

1.3.1.8 Silane Sol-Gel Coatings

A silane sol-gel coating could be a coating arranged by filling and thermally hydrolyzing a silane compound forerunner. They have a solid relationship with the Mg substrate due to the arrangement of covalent bonds (MgeOeSi). The structure of the silane sol-gel layer can be depicted in the following preparation: silane compound forerunner arrangement/sol/gel/impregnation (recyclable)/

drying. The arrangement handle is exceptionally straightforward and naturally inviting. However, during drying, handle issues can emerge for two reasons: (1) shrinkage and inside stretch when overwhelming and water vanishes, and (2) The coefficient of warm development of the Mg substrate is related to the coefficient. Warm development does not coordinate. Pre-treatment, doping-added substances (particles and inhibitors), and the development of two-stage coatings can be examined and utilized in past considerations to dodge the peril of handle absconds and move forward the defensive properties of the method[32]. Later, Peres et al. looked at the impact of corrosive-base pretreatment on the defensive properties of silane coatings on AZ91 magnesium alloy and found that antacid pretreatment (Na_3PO_4/NaOH) for silane sol-gel coating was way better than pickling (utilizing CH_3COOH or utilizing hydrogen fluoride [HF]) [33–36].

1.3.1.9 Polymer Coatings

Two types of polymer coating are available: natural and synthetic [37]. Common polymers are composed of chitosan and collagen. The former comprises glucosamine and N-acetyl glucosamine units connected by one to four glycosidic bonds, whereas the latter has three interlaced polypeptide chains in a superhelical design. Later, silk fibroin, a degradable polymer-based protein, has also been assessed as a defensive layer for magnesium alloys [38, 39]. The most common engineered materials are polylactic acid (PLA), polyglycolic acid (PGA), polycaprolactone (PCL), and polylactic-co-glycolic acid (PLGA). Polymer coatings can be formed on magnesium alloys by spinning, electrospinning, dip coating, and EPD methods [40, 41]. The first two strategies are restricted to the substrate layer, and the EPD layer shows the strongest attachment. However, there is a risk that a corrosive element will enter the method, and then the fast advancement of the magnesium substrate will lead to the degradation of the polymer layer. In this manner, pretreatment is critical for Mg substrates. Application of different subcoatings such as silane coatings, MgF_2 thin films, $Mg(OH)_2$ coating, and HA coatings on magnesium amalgams has shown to be effective coatings with polymers [42–45]. Recently, many researchers have used additives such as MgOeAg, graphene oxide-TiO_2, and metal-organic frame-works [46–48] to move forward the properties of the framework. Undefined calcium carbonate (ACC) particles can be included in the PCL layer, and the division of ACC has appeared to provide better erosion resistance by shaping a magnesium phosphate layer on the surface of the structure [49]. The addition of simvastatin-loaded HA nanoparticles to the PCL layer of AZ31 alloy improved the corrosion resistance and osteocompatibility [50].

1.3.1.10 Smart Self-Healing Coatings

In general, plating process can meet the corrosion resistance requirements of steel, but in some tough services, the plating will not be damaged when used. In this case, the coating will be limited. It is therefore important to develop a highly corrosion-resistant process that will start in the damaged area, the service life will be shortened and the use of protection system with suitable self-healing products to provide long-term protection is needed. Conceptually, smart self-healing [51] coatings can completely or partially repair defects or restore coating functionality and itself [52]. This section provides an overview of the various advanced self-healing coatings devolved on Mg alloys in the past decade. There are many reasons as to why a particular method might fail. The failures can be attributed to one of four reasons: wrong application layer, wrong layer, wrong layer selection, or wrong assumption. When coating failure falls into one of the four categories, or perhaps a combination of these, there are several important factors that can cause coating failure [53]. During its service life, the mechanical properties of the film layer change, resulting in microcracks that subsequently expand and expose the substrate to atmospheric moisture and oxygen. This effect causes the paint and flakes to move quickly through the metal coating interface [54]. A new self-healing method has been developed using self-healing polymers for microcracks and all types of damage [55]. Self-healing polymers are a class of smart materials recently developed for complete,

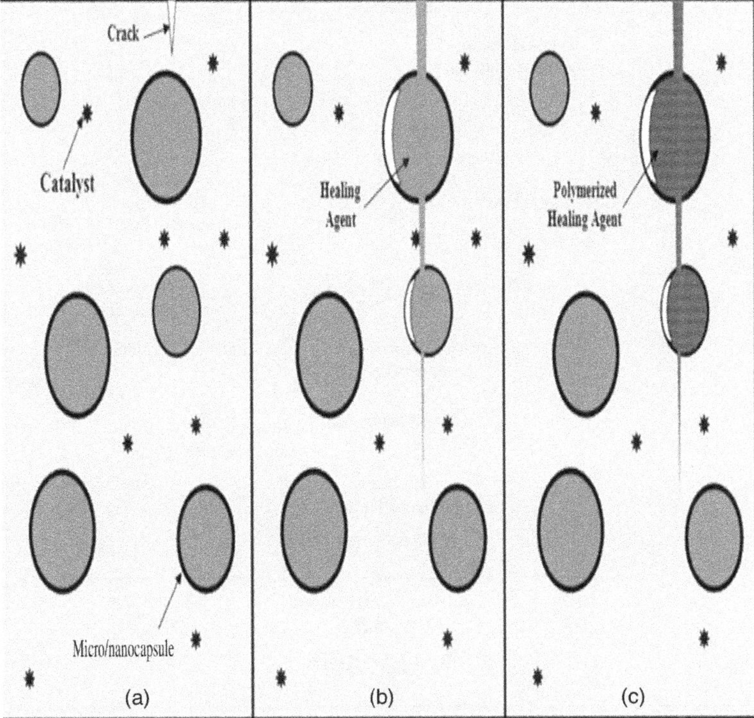

FIGURE 1.1 (a) Cracks form in the matrix wherever damage occurs. (b) The crack ruptures the mi-cro/nanocapsules, releasing the healing agent into the crack plane through capillary action. (c) The healing agent contacts the catalyst, triggering polymerization that bonds the crack faces closed.

minimally invasive therapy without detection or any human intervention [56–60]. The most common techniques used for self-healing are micro/nanocapsule embedding, hollow fiber embedding, and microvessel system [61]. Micro/nanocapsules are evenly dispersed in the passive matrix, making the curing agent "right" and thus avoiding negative interactions between active ingredients and harmful substances that can have negative consequences. According to the scheme shown in Figure 1.1, when the environment is changed or the active surface is externally affected, the micro/nanocapsules respond to the signal and release the encapsulated active species to repair cracks [62]. This survey attempts to review the research activities of micro/nanocapsule-based self-healing coatings and key design concepts for self-healing materials.

1.3.1.11 Other Coatings

The six sorts of physical boundary encapsulation coatings specified have been most considered on Mg amalgams for a long time. All things considered, numerous coatings with other chemical compounds have been investigated. Hydrofluoric acid or fluoride salt treatment is a successful strategy to manufacture MgF_2 passivation layers on Mg combinations. The MgF_2 stage is much more steady than $Mg(OH)_2$ and MgO and hence appears as a promising application. Stanislava et al. [63] applied an MgF_2 coating on AZ61 combination employing an Na [BF4] salt dissolved at 430°C and 450°C, and no signs of erosion splits were observed on the coating after inundation in SBF for 168 hours. In expansion to MgF_2 film, MgH_2 film was moreover gotten centers. Lately, Li et al. [64] applied an MgH_2 film on Mg amalgam and found that it improved the erosion resistance and biocompatibility of Mg combination. Also, there are a few innovations for creating extraordinary

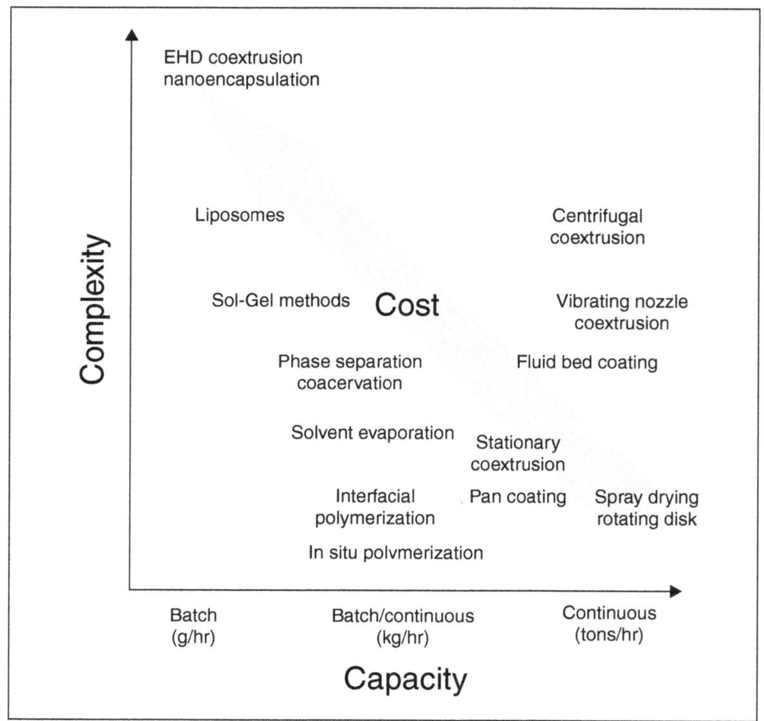

FIGURE 1.2 Comparison between the different encapsulation methods.

coatings on Mg amalgams. Plasma immersion ion implantation (PIII) utilizes high voltage (10e200 keV) to get high-energy plasma particles and embed them into the Mg substrate. Metals (i.e., Sr, Zr, and Fe) and gasses (i.e., N_2, NH_3, and O_2) are embedded or co-implanted into the surface of Mg amal gams to improve their erosion resistance. Compared with single-layer coatings, multi-layer coatings appear more promising since they can effortlessly control the erosion resistance of Mg amalgams and bless the surface with different capacities. Coatings often select films with strong adhesive properties, such as those treated with HF, solvents, and plasma electrolytic oxidation (PEO), to serve as the internal layer in multi-layer coatings. The choice of the external layer depends on the properties of the inward layer and the application prerequisites of the modified Mg combinations. Li et al. arranged a limit inter-rod-dispersed HA nanorod layer on a PEO coating employing an aqueous method and found that the bilayer coating significantly improved osteoblast capacities, counting multiplication, separation, and extracellular framework mineralization and in vivo osteogenesis. Tian et al. [65] combined PEO, PCL, and PDA coatings to manufacture a multifunctional (erosion assurance, cytocompatibility, and antibacterial) surface on AZ31 amalgam and different encapsulation methods shown in Figures 1.2 and 1.3.

1.4 PRINCIPLES

According to the Peeling Bed Worth law, more corrosion will occur if the oxide film on the metal surface is porous, unstable, and discontinuous. If the oxide film is nonporous, stable, and uniform, there will be no further corrosion. The coated surface may or may not be shielded; it depends on the nature of the metal. Alkali metals and alkaline earth metals form a porous oxide film that allows more oxygen to come into contact. Similarly, heavy metals like aluminum, chromium, lead, and tin also contribute to this effect [66–68].

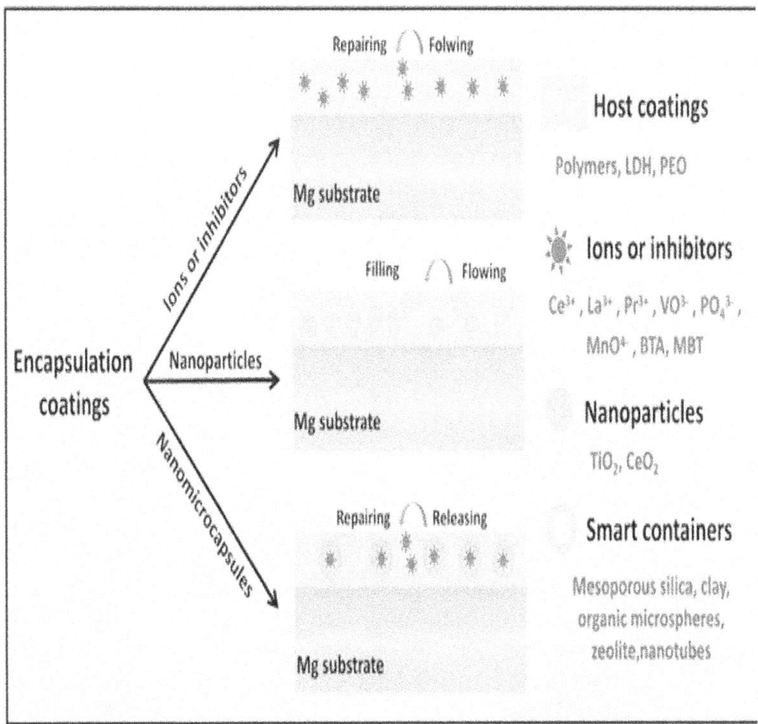

FIGURE 1.3 Diagram of the encapsulation coatings.

1.4.1 Micro/Nanocapsules Synthesis

In general, micro/nanocapsules are small particles surrounded by coatings or shells, containing particles, liquid droplets, or carbon monoxide as the main material. Commercial microcapsules are usually 3–800 μm in diameter and contain 10–90% by weight of the active ingredient. They are used in many engineering applications such as carbonless printing, adhesives, cosmetics, pesticides, and pharmaceuticals [69]. There are many reasons as to why multifunctional core materials are encapsulated, such as improving long-term efficiency, stabilizing environmental degradation, ease of use with liquid core curing, and keeping degradation products nontoxic [70]. Recently, microencapsulation techniques have been adopted in many technological applications, such as electronic paper [71, 72]. The synthesis method of microcapsules includes interface polymerization [73] and coacervation method [74], in situ polymerization [75], extrusion, and sol-gel method. However, among these methods, in situ polymerization is the easiest and best encapsulation process since it does not require high technology. Therefore, almost everyone who studies self-healing processes uses in situ polymerization as the main process to fabricate micro/nanocapsules and the difference of encapsulation. The most common repair materials are liquids as they flow out of the fracture surface. In order for the body to heal itself, many liquid-based treatment agents and their results are being researched. Investigation of the possibility of encapsulating two types of resins and three different solvents as a dual system to see the effect of the restorative agents used in its restorative role and other studies on shell thickness is under way [76]. To thin the shell and prevent the capsule from clumping, the scientists had to modify Brown's method by encapsulating the shell material wall and the aqueous phase in half. Capsules containing resin solvent are also produced by the sonication and stabilization process.

1.4.2 Approval of Self-Healing Ability

After the preparation and encapsulation of the curing agents to achieve self-healing ability, micro/nanocapsules should be included in the coating process. Different mixing and dispersing methods of micro/nanocapsules were used. Finally, the possibility of self-healing has been confirmed by various comparative experiments. Suryanarayana et al. colored linseed oil-filled microcapsules mixed with epoxy resin solution under slow stirring at ambient temperature [77, 78]. Investigation of inconsistency in mixing micro/nanocapsules in the matrix layer is an important issue to consider. In their research, the effects of mixing speed, solution viscosity, and dispersion time on the mechanical stability of microcapsules were also investigated. The notched panels were exposed into a salt spray booth. Unlike the control panel, the capsule-coated sample showed no wear on the labels [79]. Therefore, the microcapsules in the film layer release healing properties, repair cracks during bursting, and have good anti-corrosion properties. Different catalysts for the polymerization of the curing agent were evaluated as a separate study versus the previous system without the polymerization catalyst. Suryanarayana et al. explored two methods for the self-healing process, starting with silicon-based materials (polydimethylsiloxane). In the first method, the catalyst is microencapsulated and the siloxane exists as phase-separated droplets, and in the second method, siloxane is also encapsulated and dispersed in the matrix layer. Encapsulation of the two phases (catalyst and curative) is good when the matrix can interact with the curative. All control structures quickly corroded in a short time and were subject to many forms of rust, mostly seen in the grooves in the scratched area, but also adhered to the entire substrate surface [80–83]. Another interesting and powerful negative that will be important for personal growth is the location of the capsule in the matrix layer. Therefore, two methods of using microcapsules in the primer layer were studied and healing process is illustrated in Figure 1.4 [84].

1.4.3 Micro/Nanocapsule Embedment Disadvantages and Limitations

The self-healing process has many advantages, and the self-healing organization provides a reasonable and reliable answer. However, the negative effects of capsule embedding, which can lead to prevention problems and shorten the life cycle, are not taken into account. In such systems, it is possible to repeat the self-healing process after the first treatment only if the fluid is persistent in the damaged area; and it is impossible to know when the solution is completely cured. Cui et al. [85] studied the effect of adding embedded microcapsules on Young's modulus and the ultimate stress of dog bone epoxy resin samples. The sample was subjected to uniaxial tensile testing and both Young's modulus and ultimate stress decreased with increasing capsule size. Further analysis confirmed that the size of the microcapsule did not have a significant effect on the tensile properties. Kumar et al. studied the effect of microcapsule addition on the adhesion strength of the coating. In the second test, the microcapsules were examined and the same systems were used as in the first test. The authors stated that by grinding the microcapsules in two layers of primer, the microcapsules are held close to the substrate but not touching the substrate [86–88]. In this way, the microcapsules do not reduce the liner's ability to act as it did when stained by the dentist in the first test. They also found that the waiting time for the application of sandwich-format microcapsules had an effect on adhesion. All these tests were repeated after accelerating the medium and showed the same results, but in one case better adhesion to the substrate was achieved with the microcapsule coating containing tung oil [89].

1.5 CONCLUSION

The development of next-generation self-healing coatings with both passive properties derived from the substrate and sensitive responses to changes in the local environment or poor activity

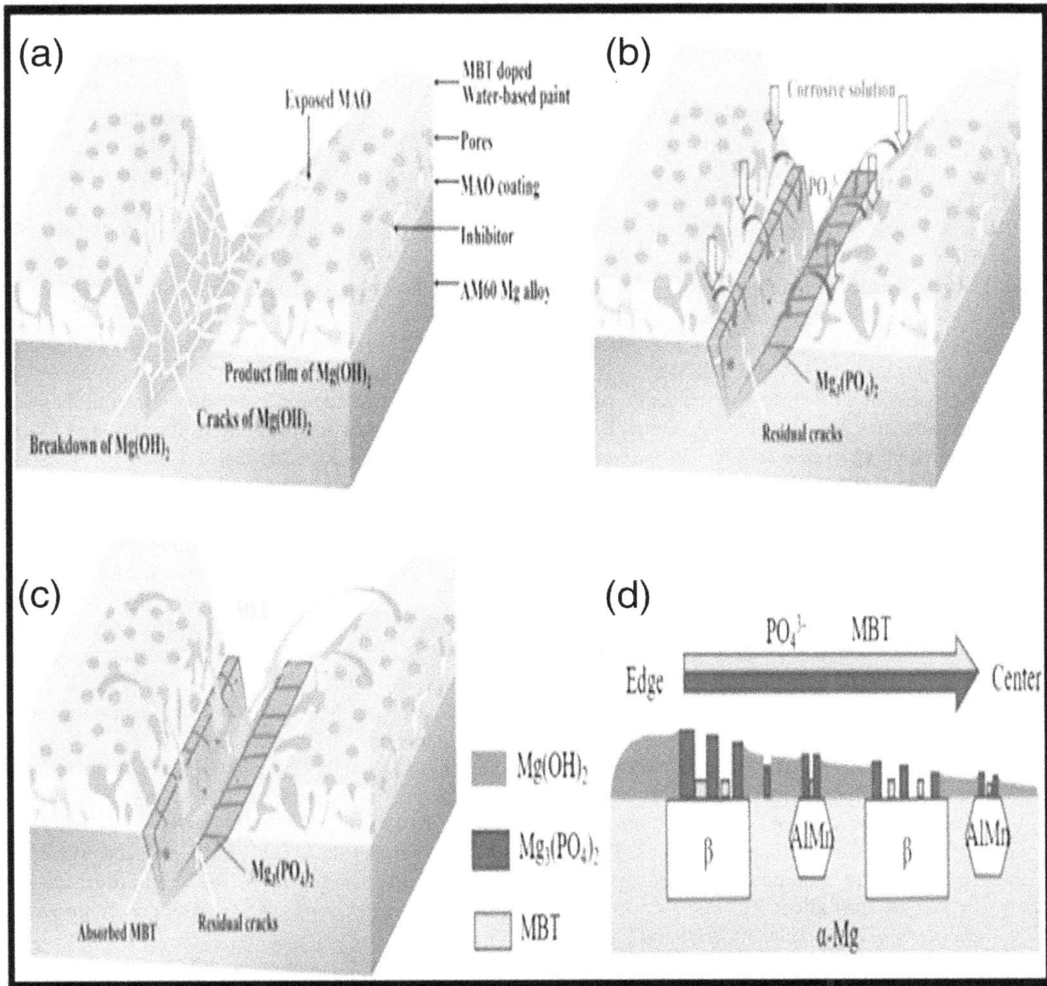

FIGURE 1.4 Schematic illustrations of the healing process of duplex self-healing coating.

performance of the substrate will open a wide window for the creation of truly smart products polymer coating. In this chapter, many parameters such as substrate material, micro/nanocapsule diameter, micro/nanocapsule core and shell, micro/nanocapsule dispersion, catalyst, and application layer are mentioned to improve the self-healing process. Layers must be consumed to account for thickness and layer substrate. Although the embedding of the capsule causes some physical degradation of the system, more research and effort should be devoted to the weakness of the burial site. Implementation of the self-healing process will be the most effective and cost-effective way to ensure corrosion protection and durability of the metal structure. Self-healing systems can protect structures from automobiles to airplanes, chemical plants to household appliances. However, further economic exploration of these layers is recommended for future research. This review only considers micro/nanocapsules that respond to material degradation or microcrack propagation, but the need to connect and use new micro/nanocapsule families that are sensitive to these changes in pH, temperature, environment, and their distribution in the layer should be considered.

REFERENCES

[1] R. Winston Revie, Herbert H. Uhlig, *Corrosion and corrosion control: An introduction to corrosion science and engineering*, Fourth edition, John Wiley & Sons, 2008.

[2] V. Bonu, H.C. Barshilia, High-temperature solid particle erosion of aerospace components: its mitigation using advanced nanostructured coating technologies. *Coatings*, 12 (12), 2022, 1979.

[3] A. Philip, P.E. Schweitzer, *Corrosion engineering handbook*, Second edition, CRC Press Taylor & Francis, 2006.

[4] R. Pierre. *Roberge handbook of corrosion engineering*, Third edition, McGraw-Hill Publishers, New York, 2000.

[5] Smita Verma, G. Mehta, Effect of acid extracts of acacia arabica on acid corrosion of mild steel. *Bulletin of Electrochemistry*, 15 (2), 1999, 67–70.

[6] Raj Narayanan, *An introduction to metallic corrosion and its prevention*, Oxford and IBH Publishing Co. Pvt. Ltd., New Delhi, 1983.

[7] Nestor Perez, *Electrochemistry and corrosion science*, Kluwer Academic Publishers, Boston, 2004.

[8] M. Acharya, J.S. Chouhan, A. Dixit, D.K. Gupta, Green inhibitor for prevention of metal and alloys corrosion: an overview. *Chemistry and Materials Research*, 3 (6), 2013, 16–24.

[9] A. M.Al-Fakih, M. Aziz, H.M. Sirat, Turmeric and ginger as green inhibitors of mild steel corrosion in acidic medium. *Journal of Materials and Environmental Science*, 6 (5), 2015, 1480–1487.

[10] T. Rajesh, A.S. Mideen, J. Karthikeyan et al., Inhibitive effect of N, N-Bis (2-chloro ethyl amino benzaldehyde) ethyl thio semicarbazone on the corrosion of mild steel in 1NH2SO4. *Indian Journal of Science and Technology*, 5 (6), 2012, 2810–2815.

[11] Saman Nikpour, R. Mohammed, B. Ghasem et al., Eriobotryajaponicalindl leaves extract application for effective corrosion mitigation of mildsteel in HCl solution: experimental and computational studies. *Construction and Building Materials,* 220, 2019, 161–176.

[12] Mokhtar Benarioua, Abdelkader Mihi, Nora Bouzeghaia et al., Mild steel corrosion inhibition by Parsley extract in acid media Egyptian. *Journal of Petroleum*, 28, 2019, 155–159.

[13] Xu, Ruizhen, Yi Shen, Jiangshan Zheng, Qiang Wen, Zhi Li, Xiongbo Yang, and Paul K. Chu, Effects of one-step hydrothermal treatment on the surface morphology and corrosion resistance of ZK60 magnesium alloy. *Surface Coating Technology*, 309, 2017, 490–496.

[14] Y.S. Lin, S. Cai, S. Jiang, D. Xie, R. Ling, J.Y. Sun, J.L. Wei, K. Shen, G.H. Xu, Enhanced corrosion resistance and bonding strength of Mg substituted btricalcium phosphate/Mg(OH)2 composite coating on magnesium alloys via one-step hydro thermal method. *Journal of the Mechanical Behavior of Biomedical Materials*, 90, 2019, 547–555.

[15] L.W. Zhu, C. Peng, K. Kuroda, M. Okido, Hydrophilic thin films for-mation on AZ31 alloys by hydrothermal treatment in silicate containing solution and the evaluation of corrosion protection in phosphate buffered saline. *Materials Research Express* 6, 2019, 116–424.

[16] W. Wu, F. Zhang, Y.C. Li, L. Song, D. Jiang, R.C. Zeng, S.C. Tjong, D.C. Chen, Corrosion resistance of dodecanethiol-modified magnesium hydroxide coating onAZ31 magnesium alloy. *Applied Physics*, A126, 2020, 1–11.

[17] Y.J. Zhang, H. Cao, H.X. Huang, Z.P. Wang, Hydrophobic modification of magnesium hydroxide coating deposited cathodically on magnesium alloy and its corrosion protection. *Coatings*, 9, 2019, 477.

[18] X.N. Gu, W. Zheng, Y. Cheng, Y.F. Zheng, A study on alkaline heat treated Mg-Ca alloy for the control of the biocorrosion rate. *Acta Biomaterial,* 5, 2009, 2790–2799.

[19] L. Li, J. Gao, Y. Wang, Evaluation of cyto-toxicity and corrosion behavior of alkali-heat-treated magnesium in simulated body fluid. *Surface Coating Technology*, 185, 2004, 92–98.

[20] L. Guo, W. Wu, Y. Zhou, F. Zhang, R. Zeng, J. Zeng, Layered double hydroxidecoatings on magnesium alloys: a review. *Journal of Material Science and Technology*, 34, 2018, 1455–1466.

[21] M.Q. Zhao, Q. Zhang, J.Q. Huang, F. Wei, Hierarchical nanocomposites derived from nanocarbons and layered double hydroxides – properties, synthesis, and applications. *Advanced Functional Materials*, 22, 2012, 675–694.

[22] J. Yu, Q. Wang, D. O'Hare, L. Sun, Preparation of two dimensional layered double hydroxide nano sheets and their applications. *Chemical Society Review*, 46, 2017, 5950–5974.

[23] T.S.N. Sankara Narayanan, I.S. Park, M.H. Lee, Strategies to improve the corrosion resistance of micro arc oxidation (MAO) coated magnesium alloys for degradable implants: prospects and challenges. *Progress in Materials Science*, 60, 2014, 1–71.

[24] M. Ali, M. Elsherif, A.E. Salih, A. Ul-Hamid, M.A. Hussein, S. Park, A.K. Yetisen, H. Butt, Surface modification and cytotoxicity of Mg-based bio-alloys: an overview of recent advances. *Journal of Alloys Compound*, 825, 2020, 154140.

[25] M. Kalaiyarasan, K. Saranya, N. Rajendran, In-vitro corrosion assessment of silicate-coated AZ31 Mg alloy in Earle's solution. *Journal of Materials Science,* 55, 2020, 3571–3587.

[26] Y.G. Ko, S. Namgung, D.H. Shin, Correlation between KOH concentration and surface properties of AZ91 magnesium alloy coated by plasma electrolytic oxidation. *Surface and Coatings Technology* , 205, 2010, 2525–2531.

[27] J.P. Lu, G.P. Cao, G.F. Quan, C. Wang, J.J. Zhuang, R.G. Song, Effects of voltage on microstructure and corrosion resistance of micro-arc oxidation ceramic coatings formed on KBM10 magnesium alloy. *Journal of Materials Engineering and Performance,* 27, 2018, 147–154.

[28] M. Rahmati, K. Raeissi, M.R. Toroghinejad, A. Hakimizad, M. Santamaria, Effect of pulse current mode on microstructure, composition and corrosion performance of the coatings produced by plasma electrolytic oxidation on AZ31 Mgalloy. *Coatings*, 9, 2019, 688.

[29] L. Toro, A.A. Zuleta, E. Correa, D. Calderon, Y. Galindez, J. Calderon, P. Chacon, A. Valencia-Escobar, E.F. Echeverria, New insights on the influence of low frequency pulsed current on the characteristics of PEO coatings formed onAZ31B. *Materials Research Express*, 7, 2020, 016539.

[30] A. Heydarian, M. Atapour, A. Hakimizad, K. Raeissi, The effects of anodicsion performance of the coatings grown by plasma electrolytic oxidation on AZ91 Mg alloy from an aluminate bath. *Surface and Coatings Technology*, 383, 2020, 125235.

[31] X.T. Shi, Y.Y. Zhu, S.F. Zhang, R.F. Zhao, R.F. Zhang, L. Chen, Y.J. Zhang, Characteristics of selenium-containing coatings on WE43 magnesium alloy by micro-arc oxidation. *Material Letters*, 261, 2020, 126944.

[32] F. Zanotto, V. Grassi, A. Frignani, F. Zucchi, Protection of the AZ31 magnesium alloy with cerium modified silane coatings. *Materials Chemistry and Physics*, 129, 2011, 1–8.

[33] H.A. Sorkhabi, S.M. Alavian, A. Kazempour, Salt-nanoparticle systems incorporated into sol-gel coatings for corrosion protection of AZ91 magnesium alloy. *Progress in Organic Coatings*, 135, 2019, 475–482.

[34] H.A. Sorkhabi, S.M. Alavian, M.D. Esrafili, A. Kazempour, Hybrid sol-gel coatings based on silanes-amino acids for corrosion protection of AZ91 magnesium alloy: electrochemical and DFT insights. *Progress in Organic Coatings*, 131, 2019, 191–202.

[35] R.N. Peres, E.S.F. Cardoso, M.F. Montemor, H.G. de Melo, A.V. Benedetti, P.H. Suegama, Influence of the addition of SiO2 nanoparticles to a hybrid coating applied on an AZ31 alloy for early corrosion protection. *Surface and Coatings Technology*, 303, 2016, 372–384.

[36] J. Li, J. Cui, J. Yang, Y. Ma, H. Qiu, J. Yang, Silanized graphene oxide reinforced organo functional silane composite coatings for corrosion protection. *Progress in Organic Coatings,* 99, 2016, 443–451.

[37] C.X. Wang, H. Fang, X.Y. Qi, C.J. Hang, Y.R. Sun, Z.B. Peng, W. Wei, Y.S. Wang, Silk fibroin film-coated MgZnCa alloy with enhanced in vitro and in vivo performance prepared using surface activation. *Acta Biomaterials*, 91, 2019, 99–111.

[38] P. Wang, P. Xiong, J. Liu, S. Gao, T.F. Xi, Y. Cheng, A silk-based coating containing GREDVY peptide and heparin on Mg-Zn-Y-Nd alloy: improved corrosion resistance, hemo compatibility and endothelialization. *Journal of Materials Chemistry B*, 6, 2018, 966–978.

[39] H.F. Qi, S. Heise, J.C. Zhou, K. Schuhladen, Y.Y. Yang, N. Cui, R.X. Dong, S. Virtanen, Q. Chen, A.R. Boccaccini, T.L. Lu, Electrophoretic deposition of bioadaptive drug delivery coatings on magnesium alloy for bone repair. *ACS Applied Materials Interfaces,* 11, 2019, 8625–8634.

[40] M. Daroonparvar, X.B. Chen, Coating biodegradable magnesium alloys with electrospun poly-L-lactic acid-akermanited oxycycline nanofibers for enhanced biocompatibility, antibacterial activity, and corrosion resistance. *Surface of Coatings Technol*ogy, 377, 2019, 124898

[41] Y.F. Ren, E. Babaie, S.B. Bhaduri, Nanostructured amorphous magnesium phosphate/poly(lactic acid) composite coating for enhanced corrosion resistance and bioactivity of biodegradable AZ31 magnesium alloy. *Progress in Organic Coatings*, 118, 2018, 1–8.

[42] Z. Wei, P. Tian, X. Liu, B. Zhou, In vitro degradation, hemolysis, and cyto-compatibility of PEO/PLLA composite coating on biodegradable AZ31 alloy. *Journal of Biomedical Materials Research Part B: Applied Biomaterials* 103(2), 2015, 342–354.

[43] A.A. Nazeer, E. Al-Hetlani, M.O. Amin, T. Quinones-Ruiz, I.K. Lednev, A poly (butylmethacrylate)/graphene oxide/TiO2 nano composite coating with superior corrosion protection for AZ31 alloy in chloride solution. *Chemical Engineering Journal*, 361, 2019, 485–498.

[44] Q.Y. Zheng, J. Li, W. Yuan, X.M. Liu, L. Tan, Y.F. Zheng, K.W.K. Yeung, S.L. Wu, Metal-organic frameworks incorporated polycaprolactone film for enhanced corrosion resistance and biocompatibility of Mg alloy. *ACS Sustainable Chemical Engineering*, 7, 2019, 18114–18124.

[45] S.Q. Jia, Y.T. Guo, W. Zai, Y.C. Su, S.S. Yuan, X.S. Yu, Y.C. Xu, G.Y. Li, Preparation and characterization of a composite coating composed of polycaprolactone (PCL) and amorphous calcium carbonate (ACC) particles for enhancing corrosion resistance of magnesium implants. *Progress in Organic Coatings,* 136, 2019, 105225.

[46] A.I. Rezk, H.M. Mousa, J. Lee, C.H. Park, C.S. Kim, Composite PCL/HA/simva-statin electrospunnan of ber coating on biodegradable Mg alloy for ortho-pedic implant application. *Journal of Coating Technology Research*, 16, 2019, 477–489.

[47] L. Jia, F. Han, H. Wang, C. Zhu, Q. Guo, J. Li, Z. Zhao, Q. Zhang, X. Zhu, B. Li, Poly dopamine-assisted surface modification for orthopaedic implants. *Journal of Orthopaedic Translation*, 17, 2019, 82–95.

[48] L. Qiao, Y. Yao, Y. Wang, N. Lai, W. Wang, Preparation and characterization of polydopamine films on pure magnesium surface. *Journal of South China University of Technology*, 46, 2018, 16–23.

[49] A. Ghanbari, F. Warchomicka, C. Sommitsch, A. Zamanian, Investigation of the oxidation mechanism of dopamine functionalization in an AZ31 magnesium alloy for biomedical applications. *Coatings*, 9, 2019, 584.

[50] S. Hong, Y.S. Na, S. Choi, I.T. Song, W.Y. Kim, H. Lee, Non-covalent self-assembly and covalent polymerization co-contribute to polydopamine formation. *Advanced Functional Materials*, 22, 2012, 4711–4717.

[51] F. Pearlstein, V.S. Agarwala, Trivalent chromium solutions for applying chemical conversion coatings to aluminum alloys or for sealing anodized aluminum. *Plating Surface Finish*, 81, 1994, 50–55.

[52] J.L. Zhang, C.D. Gu, Y.Y. Tong, W. Yan, J.P. Tu, A smart super hydrophobic coating on AZ31B magnesium alloy with self-healing effect. *Advanced Materials Interfaces*, 3, 2016, 1500694.

[53] A.S. Hamdy, I.D.H. Mohwald, Smart self-healing anti-corrosion vanadia coating for magnesium alloys. *Progress in Organic Coatings*, 72, 2011, 387–393.

[54] A.S. Hamdy, I.D.H. Mo€hwald, Assessment of one-step intelligent self-healing vanadia protective coating for magnesium alloys in corrosive media, *Electrochimica Acta,* 56, 2011, 2493–2502.

[55] K. Li, J.Y. Liu, T. Lei, T. Xiao, Optimization of process factors for self-healing vanadium-based conversion coating on AZ31 magnesium alloy, *Applications of Surface Science,* 353, 2015, 811–819.

[56] A.S. Hamdy, I.D.H. Mo€hwald, Vanadia based coating of self-repairing functionality for advanced magnesium Elektron ZE41 Mg-Zn-rare earth alloy. *Surface and Coatings Technology*, 206, 2012, 3686–3692.

[57] A.S. Hamdy, D.P. Butt, Novel smarts tannate based coatings of self-healing functionality for AZ91D magnesium alloy. *Electrochimica Acta,* 97, 2013, 296–303.

[58] N. Yang, Q. Li, F.N. Chen, P. Cai, C. Tan, Z.X. Xi, A solving reprecipitation theory for self-healing functionality of stannate coating with a high environment stability. *Electrochimica Acta,* 174, 2015, 1192–1202.

[59] S.S. Jamali, S.E. Moulton, D.E. Tallman, Y. Zhao, J. Weber, G.G. Wallace, Self-healing characteristic of praseodymium conversion coating on AZNd Mg alloy studied by scanning electrochemical microscopy. *Electrochemistry Communications,* 76, 2017, 6–9.

[60] X. Jiang, R.G. Guo, S.Q. Jiang, Evaluation of self-healing ability of Ce-V conversion coating on AZ31 magnesium alloy. *Journal of Magnesium and Alloy,* 4, 2016, 230–241.

[61] S.Y. Jian, C.Y. Yang, J.K. Chang, Robust corrosion resistance and self-healing characteristics of a novelCe/Mn conversion coatings on EV31 magnesium alloys. *Applications of Surface Science,* 510, 2020, 145385.

[62] S. Hiromoto, Self-healing phosphate coatings on pure magnesium and magnesium alloy. *Corrosion Science*, 100, 2015, 284–294.

[63] F. Stanislava, D. Julia´na, B. Hadzima, L. Trˇsko, P. Doleˇzal, J. Wasserbauer, Degradation of unconventional fluoride conversion coating on AZ61 magnesium alloy in SBF solution. *Surface and Coatings Technology,* 380, 2019, 125012.
[64] Z.X. Li, L.T. Guo, H. Yao, X.S. Di, K. Xing, J.P. Tu, C.D. Gu, Formation and in vitro evaluation of a deepeutectic solvent conversion film on biodegradable magnesium alloy. *ACS Applied Materials Interfaces,* 12, 2020, e33315–e33324.
[65] T. Jin, Z.L. Xie, D. Fullston, C.J. Huang, R.C. Zeng, R.Q. Bai, Corrosion resistance of copolymerization of acrylamide and acrylic acid grafted graphene oxide composite coating on magnesium alloy. *Progress in Organic Coatings* 136, 2019, 105222.
[66] C.A. Kim, M.K. Kim, M.J. Joung, S.D. Ahn, S.Y. Kang, Y.E. Lee, K.S. Suh, Amino resin microcapsules containing polystyrene-coated electrophoretic titanium oxide particle suspension. *Journal of Industrial Engineering Chemistry,* 9, 2003, 674–678.
[67] A. Shulkin, H.D.H. Stöver, Polymer microcapsules by interfacial polyaddition between styrene–maleic anhydride copolymers and amines. *Journal of Membrane Science*, 209, 2002, 421.
[68] C.Y. Lii, S.C. Liaw, V.M.F. Lai, P. Tomasik, Xanthan gum–gelatin complex. *European Polymer Journal*, 38, 2002, 1377.
[69] E.N. Brown, M.R. Kessler, N.R. Sottos, S.R. White, In situ poly (urea-formaldehyde) microencapsulation of dicyclopentadiene. *Journal of Microencapsulation,* 20, 2003, 719.
[70] B.J. Blaiszik, N.R. Sottos, S.R. White, Nanocapsules for self-healing materials. *Composites Science and Technology,* 68, 2008, 978.
[71] T.J. Manson, J.P. Lorimer, *Applied Sonochemistry*, Wiley-VCH, 2002.
[72] B.J. Blaiszik, M.M. Caruso, D.A. McIlroy, J.S. Moore, S.R. White, N.R. Sottos, Microcapsules filled with reactive solutions for self-healing materials. *Polymer Journal,* 50(4), 2009, 990–997.
[73] C. Suryanarayana, K. Chowdoji Rao, D. Kumar, Preparation and characterization of microcapsules containing linseed oil and its use in self-healing coatings. *Progress in Organic Coatings,* 63(1) , 2008, 72–78.
[74] V. Sauvant-Moynot, S. Gonzalez, J. Kittel, Self-healing coatings: An alternative route for anticorrosion protection. *Progress in Organic Coatings,* 63(3), 2008, 307–315.
[75] S. Kumar, P. Suryanarayana, On the bending of rectangular atomic monolayers along different directions: an ab initio study. *Nanotechnology*, 34(8), 2022, 08570.
[76] C.A. Kim, M.K. Kim, M.J. Joung, S.D. Ahn, S.Y. Kang, Y.E. Lee, K.S. Suh, Amino resin microcapsules containing polystyrene-coated electrophoretic titanium oxide particle suspension. *Journal of Industrial and Engineering Chemistry,* 9(6), 2003, 674–678.
[77] A. Ashok, K. Doriya, D.R. Mohan Rao, and D. Santhosh Kumar. Design of solid state bioreactor for industrial applications: an overview to conventional bioreactors. *Biocatalysis and Agricultural Biotechnology*, 9, 2017, 11–18.
[78] J.J. Guo, X.F. Liu, K.Q. Du, Q.Z. Guo, Y. Wang, Y.Y. Liu, L. Feng, An anti-stripping and self-healing micro-arcoxidation/acrylamide gel composite coating on magnesium alloy AZ31. *Materials Letters,* 260, 2020, 126912.
[79] A. Yabuki, M. Sakai, Self-healing sensitive organic agent. *Corrosion Science*, 53, 2011, 829–833.
[80] Z.J. Jia, P. Xiong, Y.Y. Shi, W.H. Zhou, Y. Cheng, Y.F. Zheng, T.F. Xi, S.C. Wei, Inhibitor encapsulated, self-healable and cytocompatible chitosan multilayer coating on biodegradable Mg alloy: a pH-responsive design. *Journal of Materials Chemistry B,* 4, 2016, 2498–2511.
[81] L.M. Caladoa, M.G. Tarybaa, M.J. Carmezima, M.F. Montemora, Self-healing ceria-modified coating for corrosion protection of AZ31 magnesium alloy. *Corrosion Science*, 142, 2018, 12–21.
[82] A. Kartsonakis, E.P. Koumoulos, C.A. Charitidis, G. Kordas, Hybrid organic-inorganic coatings including nanocontainers for corrosion protection of magnesium alloy ZK30. *Journal of Nanoparticle Research*, 15, 2013, 1871.
[83] M. Saremi, M. Yeganeh, Application of mesoporous silica nanocontainers as smart host of corrosion inhibitor in polypyrrole coatings. *Corrosion Science*, 86, 2014, 159–170.
[84] Z.H. Xie, D. Li, Z.K. Skeete, A.J. Sharma, C.J. Zhong, Nano container-enhanced self-healing for corrosion resistance Ni coating on Mg alloy. *ACS Applied Materials Interfaces,* 9, 2017, 36247–36260.
[85] Z.Q. Cui, Y.K. Zhang, Y.L. Cheng, D.Q. Gong, W.X. Wang, Microstructure, mechanical, corrosion properties and cytotoxicity of beta-calcium poly-phosphate reinforced ZK61 magnesium alloy composite by spark plasmas intering. *Materials Science and Engineering C-Material,* 99, 2019, 1035e1047.

[86] D.H. Yan, Y.L. Wang, J.L. Liu, D.L. Song, T. Zhang, J.Y. Liu, F. He et al., Self-healing system adapted to different pH environments for active corrosion protection of magnesium alloy. *Journal of Alloys and Compounds,* 824, 2020, 153918.

[87] S. Kumar, H. Singh, N. Gaur, S. Patil, D. Kumar, N. Singh, Imparting increased corrosion passive and bio-active character to Al2O3 based ceramic coating on AZ91 alloy. *Surface and Coatings Technology*, 383, 2020, 125231.

[88] J.Y. Chen, X.B. Chen, J.L. Li, B. Tang, N. Birbilis, X.G. Wang, Electrosprayed PLGA smart containers for active anti-corrosion coating on magnesium alloy AM lite. *Journal of Material Chemistry*, 2, 2014, 5738–5743.

[89] Y.B. Zhao, Z. Zhang, L.Q. Shi, F. Zhang, S.Q. Li, R.C. Zhang, Corrosion resistance of a self-healing multilayer film based on SiO2 and CeO2 nanoparticles layer-by-layer assembly on Mg alloys. *Material Letters*, 237, 2019, 14–18.

2 Capsule Synthesis and Fabrication for Encapsulation of Corrosion Inhibitors

Lasri Mohammed, Nayad Abdallah, Idouhli Rachid, El Ibrahimi Brahim, Khadiri Mohy Eddine, and Abouelfida Abdesselam

2.1 INTRODUCTION

Corrosion describes the electrochemical process that occurs when a metallic surface comes into contact with a corrosive chemical environment [1]. One promising approach being explored is the use of encapsulation as a new method for metal protection [2–4]. The use of capsules guarantees long-term and effective protection of metals against corrosion. The encapsulation process typically involves the application of a protective layer or coating over the smart coating, creating a barrier that prevents direct contact with the surrounding environment. This protective layer can be made of various materials, such as polymers, ceramics, and hybrid coatings, depending on the specific requirements of the smart coating and the desired properties of the encapsulation [5]. Therefore, researchers have directed their efforts toward the development of an additive, known as a self-healing agent that possesses the capability to automatically and promptly repair any cracks or imperfections in coatings, eliminating the need for manual intervention. Self-healing agents can be classified into two categories: intrinsic self-healing, where the healing agent is inherent within the polymer itself, and extrinsic self-healing, which involves the integration of a healing agent into the polymer through the encapsulation phenomenon. These healing agents are stored within containers, such as microcapsules, hollow fibers, or vascularized networks, which serve as reservoirs for the self-healing agent [6–11].

A variety of techniques can be utilized to encapsulate an inhibiting substance within a polymer matrix. These techniques involve in situ polymerization, interfacial polymerization, the Pickering emulsion model, solvent evaporation, sol-gel methods, layer-by-layer assembly, and ion exchange. The selection of an appropriate encapsulation method relies on several factors, including the characteristics of the active components, the nature of the incorporated inhibitor, the desired charge density, and the compatibility with the substrate or surface to be protected. Encapsulation plays a pivotal role in preserving the functionality of the inhibitors and ensuring their long-term performance. By encapsulating the inhibitors, their unique properties and responsiveness to external stimuli can be effectively harnessed, enabling their practical application in various industries such as automotive, aerospace, electronics, energy, and healthcare. These smart coatings provide advanced functionality and enhanced performance, catering to the specific requirements of these industries. Loading the inhibitor into the capsule involves the process of incorporating or placing the inhibitor substance within a small container or enclosure intended for storage, transportation, or controlled release purposes. In this context, a "capsule" typically refers to a compact vessel or enclosure, which may consist of organic or inorganic materials such as polyurethane, gelatin, silica-based materials, or other porous substances suitable for safely containing the inhibitor. It is crucial to acknowledge

DOI: 10.1201/9781032677255-2

that the loading process can vary significantly based on the specific requirements, properties, and handling considerations associated with both the inhibitor substance and the capsules employed. Adherence to established guidelines, industry best practices, and regulatory requirements is of utmost importance to ensure the safety, effectiveness, and integrity of the loaded inhibitor capsules.

Conventional organic coatings are susceptible to various types of damage, such as the formation of microcracks, pinholes, cavities, as well as mechanical abrasions and scratches that may arise during transportation and diverse service applications [12]. Furthermore, the concept of self-healing coatings was originally introduced in the 1980s with the objective of reducing the need for frequent repairs and the associated costs [13]. Typically, micro/nanocapsules are particles with a core (solid, liquid droplets, or gas) surrounded by a coating layer or shell (micro/nanocontainers). The terms used here are based on a 2012 International Union of Pure and Applied Chemistry (IUPAC) recommendation. This means that nanoparticles and microparticles are particles with sizes ranging from 1 to 100 nm and 0.01 to 100 μm, respectively [14]. For spherical particles, the term "capsules" refers to hollow particles with a solid shell and inner space, while the term "spheres" refers to particles without differentiation between inner and outer sections. The characteristics of containers are closely related to their responsiveness, release mechanisms, and applications [15]. Polymeric micro/nanocontainers play a crucial role as substrates in self-healing smart coating applications. Several techniques, including interfacial polymerization, coacervation, extrusion, sol-gel methods, and electrospinning, are employed to fabricate these polymeric micro/nanocontainers. Notably, the interfacial polymerization method, utilizing an oil-in-water (O/W) emulsion to encapsulate desired materials, is widely employed for this purpose [16]. This approach provides a commonly adopted and efficient means to produce polymeric micro/nanocontainers, suitable for a wide range of applications in the field of material protection. This chapter aims, firstly, to explore the commonly employed techniques for synthesizing microcapsules and understand their inherent advantages, and, secondly, to their applications in the field of corrosion protection. In addition, the chapter aims to gain a comprehensive understanding of the various synthesis methods, their respective strengths and their effective use in solving corrosion-related problems.

2.2 ROLE AND CHARACTERISTICS OF SMART RELEASE SYSTEMS

The protection of metal surfaces by different inhibitors is crucial to prevent or reduce their degradation by electrochemical corrosion when exposed to aggressive media. The main function of corrosion inhibitors is to adhere to the metal surface by forming a protective thin layer through strong bonds at a very low concentration. It is known that due to different circumstances (chemical, mechanical, or biological issues), traditional coatings mixed with inhibitor species are not able to sustainably protect metal and alloy surfaces because of the formation of microcracks. In order to increase the lifespan of the anticorrosion protection, inhibitors are loaded into nano- or micro-capsule materials and continuously diffuse onto the surface when a corrosive reaction occurs (Figure 2.1). Besides, a polymeric agent can also be added to repair the damaged surface due to corrosion. The corrosion inhibition and healing mechanisms can be activated either by an autonomous response or by an external environmental intervention. Different environmental changes can cause the container to become unstable and, as a response, release the inhibitor or polymeric species in the matrix: chemical, mechanical, electrochemical, and physical triggers. Considering the great importance of inhibitors and containers for the intelligent/controlled protection of steel against deterioration, it is important to overview them along with their characteristics.

2.2.1 NANOCONTAINERS: FACTORS AND CLASSIFICATION

Developing self-healing coating systems is critical to protecting steel surfaces from long-term corrosive degradation. One of the keys is the intelligent capsule material that will store the inhibitor

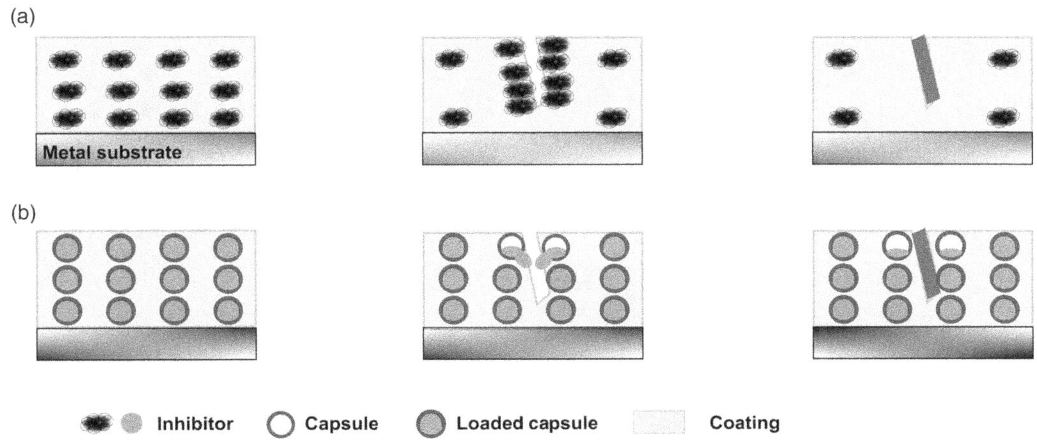

FIGURE 2.1 Illustration of inhibition mechanism by (a) traditional coating system with unloaded inhibitor and (b) modern coating system with encapsulated inhibitor.

or healing polymer within its cavities. To ensure the efficacy, durability, and safety of engineered nano- or microcapsules, different requirements are demanded, such as:

- A well-defined size and distribution in the nanometer to micrometer scale to ensure an efficient controlled release of the corrosion inhibitor.
- A high level of stability (mechanical, electrical, biological, or optical) against corrosive media that avoids any degradation of the smart coating's integrity or leakage of the inhibitor.
- A good protective barrier for preventing corrosive chemicals from coming into direct contact with the metal surface. They should adequately protect the metal from the environment, resulting in increased corrosion resistance.
- A good compatibility with the coating system that ensures effective adhesion on the surface of the metal substrate to avoid its detachment.
- The ability to store and release a high amount of the compatible inhibitor inside the shell in order to provide a high efficiency of corrosion inhibition over time.
- The capsules should be biocompatible for the environment and human safety.

A variety of materials with different structural morphologies respond to the above factors, and some of them are shown in Figure 2.2. Organic polymers and copolymers are among the most used encapsulating materials in the smart coating fields, such as poly(urea-formaldehyde) (PUF) [17], polyurethane (PU) [18], polyaniline (PANI) [19], polyurea (PUa) [20], polystyrene (PS) [21], and polypyrrole (PPy) [22]. They bear astounding mechanical and barrier properties, along with excellent compatibility as guest materials for a wide range of corrosion inhibitors. Moreover, they can easily be integrated in combination with other coating systems and enhance adhesion, thus allowing effective coverage of different types of metal surfaces for prolonged corrosion protection. Besides, many inorganic materials with hollow, spherical, mesoporous, or tubular morphologies have served as nanocontainers for inhibitors. They include mesoporous silica ($mSiO_2$) [23], halloysite nanotubes (HNT) [24], zeolite [25], metal-organic frameworks [26], and cerium oxide (CeO_2) nanoparticles (Figure 2.2) [27]. These materials exhibit great mechanical properties and high compatibility with the coating system. Recently, bidimensional materials such as layered-double hydroxides (LDH) [28], layered clay materials, and graphene oxide (GO) [29] have received much attention as novel nanocontainers in corrosion science [30].

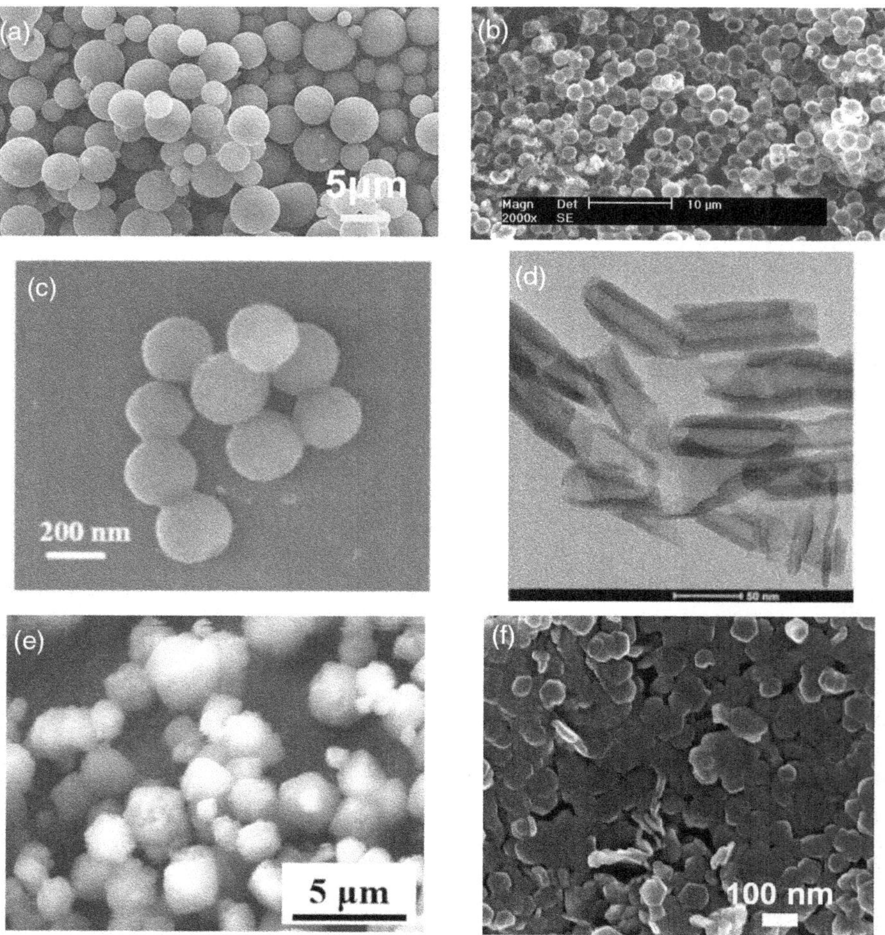

FIGURE 2.2 Illustration of different capsule-like materials for smart corrosion protection. (a) PUF. (Adapted with permission from reference [17], Copyright 2022, Elsevier.) (b) PANI. (Adapted with permission from reference [19], Copyright 2021, Elsevier.) (c) $mSiO_2$. (Adapted with permission from reference [23], Copyright 2023, Elsevier.) (d) HNT. Adapted with permission from reference [24], Copyright 2022, Elsevier.) (e) Zeolite. (Adapted with permission from reference [25], Copyright 2022, Elsevier.) (f) LDH. (Adapted with permission from reference [28], Copyright 2022, Taylor & Francis.)

In addition, chitosan [31], cellulose [32, 33], lignin [34], and β-cyclodextrin [35] biopolymers were investigated as environmentally friendly reservoirs for the loading of inhibitors. Recently, hybrid organic/inorganic composites have been investigated as innovative nanocontainers by combining the advantages of both materials, which lead to a significant improvement of the encapsulation process through bonding with the inhibitor or healing agent, the adhesion of the coating, the barrier property, and a better control release [36–44]. Other complex nanocapsules have been prepared which combine a two-dimensional (2D) material such as GO, graphene, or LDH with $mSiO_2$ to enhance the protective physical barrier of the coating layer [45–47]. On the other hand, the surface modification of the capsule materials with polar chains provides an enhancement of the chemical compatibility, dispersion, hydrophobicity, and stability with the coating system [37, 48–50]. Moreover, these new advanced reservoirs exhibit better autonomous sensing of any damage to the coating matrix and provide a more effective recombination and healing process along with the control of the release of the encapsulated inhibitor or healing agent.

2.2.2 Types of Loaded Inhibitors

A smart corrosion system is composed of a loaded inhibitor in a capsule material that acts as a healing agent against cathodic and anodic corrosion reactions. When facing a corrosion phenomenon, a stimulus response causes the liberation of the inhibitor from the container, which is diffused on the surface of the metal substrate by chemisorption or physisorption. To protect different metal surfaces or alloys such as iron, copper, magnesium, and aluminum, different inhibitors are used. The protective mechanism of the inhibitors is subject to several factors in the environment that can affect its performance, which include the concentration, the pH, the temperature, the interaction with the coating system, and the corrosive medium. Corrosion agents can be organic or inorganic and their loading in the micro/nano-container matrix is achieved in one-step process or after the capsule synthesis. Different bondings ensure the storage of the inhibitors within the capsule material, such as hydrogen bonds, π-bonds, chemical bonds, ion–ion interactions, acid–base interactions, and electrostatic forces.

Heterocyclic compounds are the most studied organic inhibitors in the literature. They bear π-bonding system along with O, S, N, or P atoms, which can easily adhere to the active sites of the metal surface to form a thin, protective layer on the surface [51–55]. Among them, benzotriazole (BTA) and its derivatives have been the most commonly used organic corrosion inhibitor in the industry since the 1950s [56]. It is constituted of a rich electronic benzene system and three N heteroatoms, which can be easily passivated on the metal surfaces by chemisorption. Turano et al. [57] investigated its adsorption mechanism on copper surfaces, and its coverage on the metal surfaces was observed using scanning tunneling microscopy. The results, supported by density functional theory, revealed the deprotonating of BTA molecules that are then coated on the copper substrate, resulting in the formation of Cu(BTA) and $Cu(BTA)_2$ metal-organic complexes. However, BTA is a soluble compound in water, thus limiting its effectiveness in corrosion media due to its rapid dissolution. The incorporation of BTA molecules in encapsulated materials has been extensively reported in recent literature for corrosion protection against different alloy substrates [26, 58, 59–66, 67]. Other organic heterocyclic compounds were also used as inhibitors loaded in nanocontainers such as imidazole [68–70], 2-mercaptobenzothiazole [71–73], 2-mercaptobenzimidazole [74–77], 8-hydroxyquinoline [78–81], 5-aminoindazole [82], and sulfamethazine [83, 84]. Another class of organic compounds that are corrosion inhibitors are amine-based compounds with $-NH_2$ groups, and they have been well studied for the protection of metallic surfaces [85, 86]. The encapsulation of amines such as dodecyl amine and triethanolamine has been recently realized in a smectite clay mineral [87] and a polymeric seed latex [88] for the long-term protection of copper and aluminum alloy 3003 surfaces, respectively.

On the other hand, inorganic corrosion inhibitors are composed of anodic-type and cathodic-type retarders. The most reported ones are anodic-type, such as molybdate, phosphates, vanadates, and cerium nitrates. On the metal surface, they usually form a protective oxide layer, which inhibits oxidation reactions and induces an increase in the corrosion potential and a decrease in the corrosion current density of the metal. For this purpose, several nanomaterials were used as nanocontainers for the loading of sodium molybdate inhibitors such as PPy/TiO_2 [89], $mSiO_2$ [90], zinc-layered hydroxide salt [91], and SnO_2 [92], which resulted in the release of molybdate anion (MoO_4^{2-}) for the protection of different steel substrates. Besides, a significant interest was devoted to the nano-encapsulation of phosphates [93–96] and vanadates [97–100] in 2D materials, for instance, LDHs and GO. Nitrate salts such as cerium nitrate were used as strong anodic-type inhibitor in smart coatings [101–104].

Nowadays, environmental challenges urge scientists to use biodegradable and healthy materials for nature. In this context, an increasing number of articles are oriented toward the use of renewable natural compounds as corrosion inhibitors [105–108]. Plant extracts are among the suitable sustainable natural compounds as corrosion inhibitors and have been extensively studied [109–112]. With regard to their performances, natural oil extracts or compounds have been loaded into various reservoir materials as renewable healing agents in smart coatings [113]. They include tannic acid [114–116], linseed oil [117–119], vanillin [120], caffeine [121, 122], and amino acids [123,1 24].

Finally, the development of sophisticated smart coatings with dual or triple actions led to the integration of two or more inhibitors/healing agents to enhance damage sensing, corrosion protection, and self-reparation. In this regard, the encapsulation of different organic inhibitors [70, 125–131] or organic–inorganic inhibitors [44, 132–135] in smart coating systems has shown an enhancement in the efficiency and durability of the anticorrosion performance. In the next section, we will discuss the different strategies for the synthesis of encapsulated micro/nano-containers for the smart release of corrosion inhibitors.

2.3 DIFFERENT METHODS FOR CAPSULE SYNTHESIS

Several micro and nano-encapsulation techniques can be employed, encompassing both physical and chemical methods. Physical techniques include solvent evaporation, spray drying, centrifugal extrusion, the spinning disk method, and fluid bed coating. Chemical procedures include interfacial polymerization, complex coacervation, sol-gel, phase separation, and electrospinning. This section presents examples of these manufacturing methods [136].

2.3.1 Interfacial Polymerization: Oil in Water (O/W)

When two immiscible liquids are mixed, an intriguing phenomenon occurs, resulting in distinct phases within the system. The first phase, called the dispersed phase, consists of tiny droplets dispersed in the second phase, the continuous phase. The liquid with the lowest volume fraction makes up the composition of the dispersed phase. The composition of the dispersed phase is predominantly comprised of the lowest volume fraction. However, the size of these droplets is not solely determined by chance; rather, it is influenced by various mixing parameters including stirring or sonication conditions, vessel shape, temperature, and even the composition of the medium itself. It is essential to control the emulsion obtained by manipulating the different factors involved; we can influence the result of the emulsion and obtain the desired characteristics and properties. For example, adjusting the stirring speed, sonication intensity, or temperature can produce emulsions with specific droplet sizes, thus ensuring a consistent and predictable result.

Understanding and managing the process of emulsion formation is of great importance in various fields such as pharmaceuticals, and in the smart coatings we deal with, the precise control of emulsion properties is essential as the size and stability of the droplets influence the release and absorption of active compounds like inhibitors.

In the commercial field, interfacial polymerization has gained great popularity as a method of making resin-based capsules, which have applications in the paper industry [137], self-healing materials [138, 139], and thermal energy storage [140]. This encapsulation process starts with an emulsion, where oil droplets are suspended in water, serving as a template. Next, the water-soluble monomers undergo cross-linking, resulting in the formation of precipitates that adhere to the interface between the oil and water. These precipitates eventually transform into polymeric films, effectively enclosing the oil droplets and resulting in the formation of capsules [141–143] as shown in Figure 2.3(a) and (b).

Interfacial polymerization can be performed under a variety of conditions and compositions, providing a wide range of possibilities. Table 2.1 provides relevant examples of these conditions

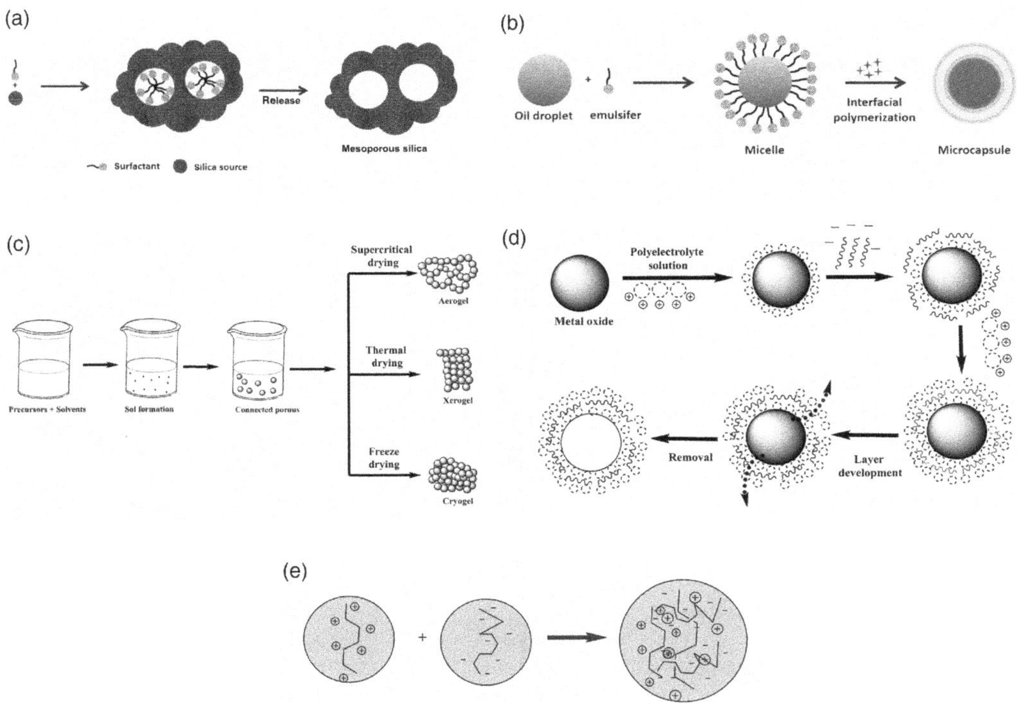

FIGURE 2.3 Illustration of different synthesis strategies of container materials. (a, b) Interfacial polymerization, (c) sol-gel method, (d) layer-by-layer assembly, and (e) coacervation method.

and compositions. Depending on the composition and formation method, the size of the particles formed in the emulsion can vary, ranging from nanometer dimensions of tens of nanometers to larger sizes on the order of hundreds of microns.

The advantages of this technique arise from the simplicity of its application and preparation. In simple terms, the hydrophilic monomers of the shell material are dissolved in the continuous aqueous phase. The polymerization process begins, involving all the reactants responsible for the formation of the shell. As the polymerization progresses, it spreads to the surface of the base material droplets, ultimately resulting in the creation of a robust polymer shell.

2.3.2 Complex Coacervation

Complex coacervation is an interesting technique that occurs when two molecular species, possessing opposite charges, intertwine and undergo a phase separation process. This complex phenomenon usually results in the formation of two distinct phases with contrasting properties. The first phase, called coacervate, is characterized by its dense and concentrated nature. In contrast, the second phase, called supernatant, has a relatively more dilute composition. This complex coacervation phenomenon illustrates the intriguing behavior of charged molecules as they interact and create distinct regions within a system as shown in Figure 2.3(e) [158, 159].

In 1929, Bungenberg de Jong and Kruyt conducted research on encapsulation technology, specifically focusing on the production of gelatin spheres through agglomeration [160]. Later, in the 1950s, Barret Green achieved a significant milestone by inventing the world's first carbonless paper. microparticle research, the process consists of encapsulating and enclosing small droplets or granules containing various active substances such as inhibitors, or species play a protective

TABLE 2.1
Nature and size of capsules synthesized by interfacial polymerization

Type	Nature	Emulsion formation	Size	Ref
O/W	PUa	vigorous agitation	100–1000 nm	[20]
O/W	PUa	stirring	50 μm	[144]
O/W	PUa	stirring	60 μm	[145]
O/W	PUa	stirring	100 nm–2 μm	[146]
O/W	PU	stirring	40–400 μm	[18]
O/W	PU MCs	blade propeller	5–350 μm	[147]
O/W	Melamine formaldehyde	vigorous shaking	40–117 μm	[148]
O/W	Silica	vigorous stirring	100–150 nm	[149]
O/W	Silica	sonication	100–400 nm	[150]
O/W	Silica-imidazoline	vigorous stirring	3; 5 μm	[151]
O/W	Polyaniline	sonication	700±400nm	[152]
	Polypyrrole		100 ± 5 nm	
O/W	PUa/silica hybrid	blade propeller	50–300 μm	[153]
O/EtOh	SiO_2-imidazoline	stirring	50–100 nm	[154]
W/O	PU	sonication	200 nm	[155]
	PUa		100–300 nm	
	Poly (thiourea)		370–420 nm	
W/O	PUa	stirring	10 μm	[156]
W/O	Urea formaldehyde	sonication	7–18 μm	[16]
W/O	PANI	sonication	500 nm–1 μm	[157]

barrier. The main objective of this technology is to ensure the stability of the particles, to protect the active core, and/or to create isolation from the surrounding environment, which allows a better control of their application. Therefore, microencapsulation offers several advantages, including controlled release of encapsulated bioactive molecules, protection against oxidation of encapsulated materials, and better stability against aggressive factors in the environment [161, 162].

This technique is primarily aimed at ensuring particle stability, and protecting and isolating the active core from the surrounding environment. With this approach, materials can be handled more efficiently during application, ensuring optimal performance and functionality. The technique aims to improve overall particle stability, preserve the integrity of the active core, and create a controlled environment for efficient use. By achieving these goals, the technique allows for improved material handling and application, resulting in better results in a variety of application areas. One advantage of these models is that they do not require an exact representation of highly concentrated loads, making them more accessible for practical use. Their proper parameterization generally yields accurate results. However, unlike liquid state theories, these models have limitations when it comes to providing predictive insights into the behavior of local charge correlations; their ability to accurately capture and predict the complex dynamics of charge correlations at a local level is limited. It is important to consider these limitations when using these models and to seek alternative approaches when a detailed understanding of local load correlations is essential [163].

2.3.3 Sol-Gel Methods

Various methods for the preparation and synthesis of micro/nanoparticles are widely used today, one of which is the sol-gel process. It is a solution-based technique that stands out for its prevalence and commercial application. Although other techniques can produce significant amounts of nanomaterials, the sol-gel process has gained greater importance due to its widespread use and extensive commercial applications [164, 165]. The sol-gel process works on the principle of initially creating a uniform soil by combining precursor materials, followed by its conversion into a gel structure. This gel is formed by an interconnected network of particles suspended in a solvent. Subsequently, the solvent in the gel matrix is carefully removed by evaporation or other appropriate methods, which causes a solid to form dried gel. This drying process ensures that the gel structure is preserved, resulting in a solid material with unique properties. The characteristics of the dried gel are highly dependent on the drying method chosen, which is tailored to the intended application of the gel (Figure 2.3(c)). Dried gels find their use in a variety of industries, such as surface coating, building insulation, and specialty textile production. Importantly, nanoparticles can be obtained by grinding or crushing the dried gel using specialized mills. This process generates finely divided particles with the desired properties, which opens up new avenues for synthesizing nanoparticles and applying them in various fields [165].

Sol-gel technique provides an approach with excellent control over chemical composition due to its low reaction temperature. This versatile technique has applications in the manufacture of corrosion-resistant coatings. In addition, the sol-gel technique serves as an intermediate layer between thin metal oxide layers in various applications. By acting as a bridge, it can improve adhesion, surface properties, and functionality in areas such as electronics, optics, and energy storage. The adaptability and versatility of sol-gel technology make it a valuable tool for achieving desired material properties and enabling advances in many industries. The versatility of sol-gel materials and the associated technology make them an essential element in driving progress and innovation in many sectors.

2.3.4 Electrospinning

Since the 1990s, a technique called electrospinning, or electrostatic spinning, has been used to generate micro-nanofibers and particles. This method consists in extracting extremely fine fibers from a liquid by applying an electric charge [166, 167].

Core-shell nanomaterials have revolutionized the field by providing a remarkable approach to encapsulating reactive or beneficial components in tiny capsules. This breakthrough was made possible by the use of coaxial electrospinning, a simple and rapid method of fabricating core-shell nanofibers. Core-shell nanomaterials allow delicate or reactive substances to be encapsulated and protected within a protective shell, shielding them from external influences while maintaining their integrity. This opens up a world of possibilities in various industries, such as the manufacture of smart coatings to protect metals from corrosion.

The coaxial electrospinning technique has emerged as one of the leading methods for producing these complex core-shell nanofibers. Using an electrospinning setup, in which two or more concentrically arranged coaxial nozzles are used, a polymer solution containing the desired core material is extruded from the innermost nozzle, while a sheath polymer is simultaneously extruded from the outer nozzle. Electrostatic forces generated by a high-voltage supply cause the solution to stretch and form ultrafine fibers, resulting in core nanofibers with precise control over their dimensions [168].

The simplicity and speed of the coaxial electrospinning process is one of its main advantages. Unlike more complex methods, it offers a straightforward approach, accessible to researchers and practitioners alike. In addition, the ability to tune the composition and structure of the core-shell nanofibers allows the design of materials with tailored properties and functionalities.

Numerous studies have shown that electrospinning outperforms conventional encapsulation methods when it comes to encapsulating essential oils. This superiority can be attributed to the elimination of heat, which is crucial for preserving heat-sensitive bioactive compounds like essential oils. In addition, electrospinning allows for remarkably efficient encapsulation, ensuring the stability of compounds during processing and storage [168, 169].

2.3.5 Layer-by-Layer Methods

Layer-by-layer self-assembly technology, which emerged in the 1990s, is a simple, rapid, and environmentally friendly method for multifunctional surface modification. The most commonly used principle involves the alternating deposition of polyelectrolyte self-assembly multilayers in oppositely charged polyelectrolyte solutions. The polyelectrolyte multilayers most frequently used are negatively charged polystyrene sulphonic acid (PSS) and positively charged polyetherimide (PEI) [170]. Electrostatic interactions between the layers make the polyelectrolyte layer sensitive to external stimuli such as pH and light. This allows for the controlled and gradual release of active ingredients encapsulated between the layers. In this way, controlled release of corrosion inhibitors can be achieved as the corrosion environment changes. It is for this reason that layer-by-layer self-assembly technology is often used in coatings for the loading of corrosion inhibitors. This method uses a sequential assembly process in which metal oxide materials are dispersed in aqueous media. This results in the sedimentation of polyelectrolytes with positive and negative charges onto the metal oxide particles and the formation of stacked polymer layers [171]. This is achieved by means of electrostatic forces (Figure 2.3(d)). After forming multiple layers, an acid treatment is applied to eliminate the metal oxide core, creating voids within the semi-permeable containers or capsules. This method allows easy incorporation of the trigger mechanism and is suitable for both organic and aqueous cores. However, the relatively low physical integrity of the capsules and the need for strong acids to prepare certain cores are some of the drawbacks of this method.

The simplicity, speed, and environmental impact of this process offer a number of advantages. The technology can be used to create multifunctional surfaces with specific properties, such as corrosion resistance. This is achieved by incorporating controllable corrosion inhibitors. Thus, layer-by-layer self-assembly technology represents a promising approach to the production of effective and durable coatings for corrosion protection.

2.4 MICROENCAPSULATION

Microencapsulation is an innovative technique that consists of enclosing solid particles, liquid droplets, or gases of micrometer size in a protective and inert envelope. This process effectively isolates and protects these entities from the external environment, ensuring their preservation and safeguarding their integrity [172]. The reactivity between the base material and the shell determines the level of inertness of the shell. Once the microencapsulation process is complete, the final product is a microcapsule. This microcapsule consists of two components: the core and the shell. The shape of these components can be round or irregular. Their size can vary from nanoscale to microscale. Microcapsules play a crucial role in the development of self-healing polymer composites, as they incorporate healing agents or catalysts into their structure [173].

Microencapsulation enables a wide range of materials to be encapsulated, giving them enhanced stability and extended shelf life. Solid particles, such as inorganic inhibitors or functional additives, can be effectively encapsulated to prevent the degradation of metals against corrosion. Similarly, liquid droplets, such as essential oils or volatile substances, can be encapsulated to prevent evaporation or oxidation, thus preserving their therapeutic properties. Even micron-sized gases can be encapsulated, offering unique applications in various industries. The inert shell used in microencapsulation acts as a protective barrier, effectively isolating the encapsulated materials from the external environment. This shell can be composed of different materials, such as polymers or lipids, depending on the specific requirements of the encapsulated substances and the desired release profile.

The benefits of microencapsulation go beyond preservation and protection. It also allows for modified release rates, improved solubility, and facilitates the controlled release of encapsulated materials at specific target sites. This versatility and flexibility makes microencapsulation a valuable technique for research, development, and commercial applications.

Microencapsulation of substances has attracted significant research interest and has been extensively studied in universities for several decades [174]. This promising process has also found numerous applications in various industries. Microcapsules, composed of specific base materials and protective shells, provide an effective means of isolating the encapsulated content from its environment. These microcapsules serve as storage units when the shell is impermeable, preventing the contents from escaping, or facilitate controlled release when the shell allows variable permeability.

Microencapsulation in smart coatings refers to the process of protecting and enclosing the active components or functionalities of a smart coating within a protective layer or matrix. Smart coatings are advanced materials that can respond to external stimuli, such as temperature, light, or pH, and exhibit specific properties or behaviors in response. Encapsulation is important in smart coatings to ensure the durability, stability, and functionality of the active components [175]. It helps to protect the sensitive materials from environmental factors like moisture, oxygen, UV radiation, or chemical reactions that could degrade or alter their performance [176, 177].

In response to the increasing demand for microcapsules containing a wide range of base materials for various applications in different industries [178–180], many microencapsulation methods have been developed. In the field of self-healing applications, extensive research has been conducted on different types of microcapsule systems. These systems use different shell materials, including PUF [181–183], poly(melamine-formaldehyde) [184], poly(methyl methacrylate) [185], PU [186], epoxy resins [187], and mesoporous silica particles [188]. Of these options, PUF is the most commonly used shell material due to its exceptional mechanical strength, impermeability, chemical resistance, and cost effectiveness.

2.5 APPLICATIONS OF CAPSULE SYNTHESIS FOR CORROSION PREVENTION

To protect metals from the adverse effects of corrosive environmental conditions, a range of polymer coatings can be used. These coatings act as a protective barrier, shielding metal surfaces from exposure to corrosion-inducing factors [188, 189]. Polymer coatings offer several advantages in corrosion protection. They provide a physical barrier to prevent direct metal contact with corrosives such as moisture, oxygen, acids, or salts. This barrier effectively prevents the initiation and progression of corrosion reactions. Different types of polymers can be used for corrosion protection, depending on the specific requirements of the application. Organic coatings, such as epoxy, PU, acrylics, and polyester, are commonly used because of their excellent adhesion properties, chemical resistance, and durability. These coatings form a strong protective layer on the metal surface, providing long-lasting protection against corrosion.

In addition to providing a physical barrier, polymer coatings can also incorporate microcapsules containing corrosion inhibitors into their formulation. These inhibitors act as active agents that further impede corrosion processes by forming a passivation layer on the metal surface when the coating cracks or by impeding the electrochemical reactions involved in corrosion. The choice of polymer coating depends on several factors, including the type of metal substrate, environmental conditions, desired service life, and specific corrosive agents present. The thickness of the coating, the method of application, and the curing process also play an important role in achieving effective corrosion protection.

Polymer coatings have a wide range of applications in many industries, including automotive, aerospace, marine, infrastructure, and electronics. They are used to protect a wide range of metal structures, components, and surfaces, ensuring their longevity, performance, and aesthetics. Improving the anticorrosion performance of polymer coatings through continued research and development innovations focuses on improving coating adhesion, developing self-healing coatings, incorporating intelligent functionalities, and exploring sustainable, environmentally friendly alternatives.

During their lifetime, these coatings are susceptible to degradation, both at the micro and macro level, which can have serious consequences and lead to premature failure [190]. To address this problem, one potential solution is to implement an intelligent self-healing coating that can react to damage either by external triggers such as temperature, pressure, and pH, or autonomously without external intervention. Much research has focused on incorporating self-healing properties into polymer coatings for corrosion protection [191, 192]. In the following section we present some applications of microcapsules in the protection of metals against corrosion.

Inspired by the self-healing characteristics observed in polymer composites [193], Parsaee et al. developed urea-formaldehyde (UF) and PU-based microcapsules [194]. The simultaneous use of two different types of microcapsules resulted in an epoxy-vinyl ester coating with inherent self-repair capabilities. Other approaches have explored the separate encapsulation of epoxy resin and curing agent [195]. Also Zhang et al. have synthesized PUF-based microcapsules containing linseed oil through in situ polymerization within an oil-in-water emulsion [196]. By in situ polymerization, PUF microcapsules were created using a tall oil fatty acid epoxy ester as a base material. The synthesized microcapsules were thoroughly examined using scanning electron microscopy (SEM) and optical microscopy techniques, the core content of the microcapsules represented 67 wt%, while the average diameter measured 101 ± 48 μm. To facilitate the production of self-healing coatings, polyurea formaldehyde-epoxy ester (PUF-epoxy ester) microcapsules were embedded in an epoxy coating matrix.

The weight percentage of epoxy ester in the synthesized microcapsules was approximately 67%. These microcapsules were incorporated into epoxy coatings to create self-healing coatings. In particular, the self-healing ability was observed when artificial microcracks were repaired sealed and a new layer formed in the cracks (Figure 2.4). The formation of this new layer was due to the auto-oxidation and cross-linking reaction of the epoxy ester, which was released at room temperature. The striped self-healing coating performed better than the pure epoxy coating. These results highlight the significant potential of the tall oil fatty acid epoxy ester as an effective healing agent in smart self-healing anticorrosion coatings.

Da Cunha et al. have synthesized polymers as modifiers for pH-sensitive microcapsule containers which have prevailed in various applications [197]. However, there has been a noticeable shift toward the adoption of biopolymers with less environmental impact. Alginate

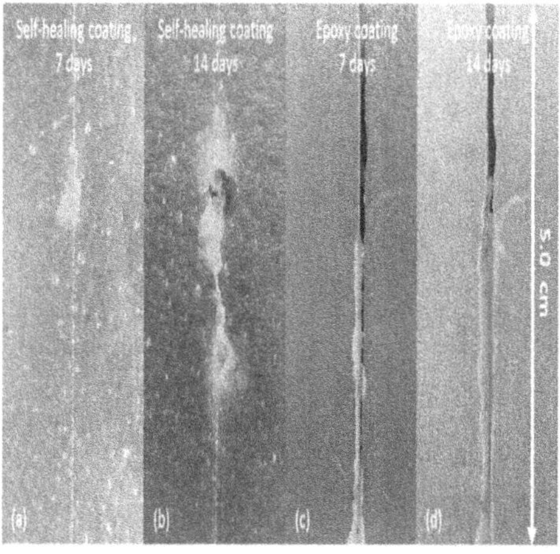

FIGURE 2.4 Images illustrating the coatings after a salt spray test, with a cut length of 5.0 cm, are shown. The self-healing coating after (a) 7 days and (b) 14 days; the pure epoxy coatings after (c) 7 days and (d) 14 days. (Adapted with permission from reference [196], Copyright 2022, Elsevier.)

and chitosan (CS) are widely studied polyelectrolyte pairs, known for their opposite charges and ability to form stable polymer complexes via electrostatic interactions. These complexes are widely used in drug delivery systems and biomaterials. This development is driven by the growing demand for environmentally friendly materials that offer properties and functions comparable to those of their synthetic counterparts; they have been successfully synthesized by bifunctional PUF microcapsules. These microcapsules are composed of linseed oil as core material and BTA as corrosion inhibitor, all enclosed in a multilayer polyelectrolyte shell. In the manufacturing process, sodium alginate and CS, which are oppositely charged biopolymers, were used as polyelectrolytes. The resulting microcapsules were spherical in shape, with an average diameter of 0.99 ± 0.15 µm. To enhance the protective properties of the coatings, an epoxy-based smart coating was developed. This coating demonstrated its reactivity to two distinct stimuli: mechanical stress and pH changes. The inclusion of microcapsules considerably enhanced the protection provided by the coating, as indicated by the average percentage release (PR). The PR value was three times higher than that of the pure resin coating. The results showed an inverse relationship between BTA release rate and pH value. This behavior can be attributed to the response of CS and sodium alginate to pH changes. Under acidic conditions, CS is protonated, resulting in electrostatic repulsion between its chains. As a result, the polymer layer adopts a more open conformation, allowing H^+ ions access to the alginate layer. Although the interaction between alginate and H^+ ions is stronger than that between alginate and BTA, the latter is released from the microcapsules. Specifically, at pH 3.0, the release rate was measured at 0.189% per minute. Furthermore, as pH rises above 6.5, an increased presence of OH^- groups induces deprotonating of the CS amine groups. This in turn leads to shrinkage and bending of the polymer chains. As a result, contact between alginate and H^+ ions is impeded, preventing the release of BTA. These results suggest that core-shell microcapsules can exhibit mechanical and pH-sensitive behavior by taking advantage of the properties of the core material and appropriately modifying the wall material [198].

Another study conducted by Huang et al. successfully created nanocontainers that respond to pH changes, using zeolite imidazolate-8 (ZIF-8) frameworks. These nanocontainers offer a novel approach to developing self-healing sol-gel coatings for corrosion prevention. However, the organic nature of ZIF-8 nanocapsules limits their compatibility with the inorganic sol-gel paint matrix, rendering them ineffective as anticorrosive agents. This study introduces a new pH-sensitive corrosion inhibitor delivery system (BTA-ZIF-8@SiO_2) based on ZIF-8, specifically designed for self-healing sol-gel anticorrosion coatings. The ZIF-8 core was encapsulated within 1H-benzotriazole (BTA) using a simple one-step self-assembly technique. Subsequently, a silica layer was deposited onto the core using a conventional sol-gel method. The silica shell enhances the compatibility of the nanocontainers with the sol-gel matrix, thereby improving the overall corrosion protection [199].

2.6 ENCAPSULATION OF CORROSION INHIBITORS

2.6.1 Importance of Encapsulating Corrosion Inhibitors

The development of the new technology as an effective anticorrosion method of the metallic substrate is an important way to increase the durability of metal structures and components. The encapsulating corrosion inhibitors present these proprieties, which is based on covering or encapsulating the inhibitor particles in a protective shell or matrix as reservoirs for the repairing agent and chemical initiator. This method is frequently employed to improve the efficiency and performance of this technique in different applications. Also it presents various advantages, such as protection against environmental factors (temperature, humidity, chemicals compound, etc.), controlled release of the inhibitor (control the release of the inhibitor over time), increasing the stability of the inhibitor exposed to the aggressive environment (self-reporting, anti-microbial, anti-fouling and self-lubrication functions, etc.), computability with different environments, and their cost-effectiveness for protecting materials against corrosion environment in order to prolong the lifespan [200]. The self-healing function can be attained through either the intrinsic or the extrinsic approach. Intrinsic self-healing coatings rely on reversible physical and chemical interactions among molecules, leading to a permanent self-healing effect. Conversely, the extrinsic method involves embedding micro/nanocontainers within the coatings. These containers can detect changes in the micro-environment and promptly initiate repairs in the micro-cracked areas on the coating surface. However, the self-healing performance in this case is temporary, as it depends on the depletion of the encapsulated active substances [12]. These intelligent behaviors have inspired the development of increasingly versatile coatings with multiple functionalities, which find wide applications across various domains [201].

Generally, encapsulating corrosion inhibitors can improve the effectiveness, stability, and versatility of this new technology, making them an important tool for preventing corrosion in various industries with unique insight into the future development.

2.6.2 Encapsulating for Corrosion Inhibitors

Different methods for capsule synthesis have been discussed which have been used in the fabrication of encapsulated corrosion inhibitor. Rahimi et al. [200] present a study on the development and evaluation of hybrid nanocomposite coatings with encapsulated organic corrosion inhibitors based on [2-mercapto-benzothiazole (MBT) and 2-mercaptobenzimidazole] and b-cyclodextrin. The aim of the study was to investigate the effectiveness of the encapsulated corrosion inhibitors in preventing corrosion on metallic surfaces. The authors then present the method used to prepare the hybrid nanocomposite coatings, which involved the incorporation of both inorganic and organic materials. The coatings were formulated using a combination of epoxy resin, silica, and alumina, and the corrosion inhibitors were encapsulated via a sol-gel technique [200].

As part of their research, Wang et al. have succeeded in developing graphene-polymer coatings with exceptional impermeability and chemical stability. These coatings have been widely regarded as the most optimal solution in this field. However, one of the challenges encountered when incorporating graphene as nanofiller is to achieve miscibility in the coating. To solve this problem, the researchers introduced a bilayer nanocomposite coating that not only overcomes the miscibility problem but also has a remarkable self-healing capacity at the functional level, guaranteeing a strong interfacial interaction. This innovative coating, unlike previous nanofiller coatings, uses benzotriazole- and poly(vinyl butyral-co-vinyl alcohol-co-vinyl acetate)-loaded nano-tanks as the nanocomposite. The manufacturing process involves the use of dip-coating and spin-coating techniques, respectively. A key element of their work is the synthesis of a nanoreservoir, with an effective thickness of 35 nm, which serves as a highly effective vehicle for the delivery of corrosion inhibitors. This nanoreservoir is created using a combination of oil–water superposition and vacuum loading techniques. To better understand the release of corrosion inhibitors, the researchers developed semiempirical models, which are rarely used in this field. These models predict the diffusion behavior of corrosion inhibitors. These graphene-polymer coatings not only exhibit superior barrier properties but also possess a remarkable capacity for self-repair, offering long-term protection against corrosion. This revolutionary strategy represents a promising avenue for the development of corrosion inhibitors based on graphene technology [23].

2.6.3 Mechanism of Loading and Unloading Corrosion Inhibitors

This section focuses mainly on examining the mechanisms involved in regulating the release of corrosion inhibitors from micro/nanocontainers, particularly those with a core-shell or layered structure, encompassing both triggering and response aspects. Understanding how corrosion inhibitors are loaded and unloaded is crucial to corrosion prevention strategies. It involves understanding how these inhibitors are introduced into the system and then released when needed. By studying this process, researchers can develop more effective corrosion inhibition techniques. Discharge of corrosion inhibitors involves the controlled release of these inhibitors from the protection system into the corrosive environment when required. This process can be triggered by a variety of stimuli, including changes in temperature, pH, or exposure to corrosive agents. Release mechanisms can vary depending on the loading techniques used. For example, in the case of physical adsorption, inhibitors can be desorbed from the surface due to changes in environmental conditions. In encapsulation methods, the envelope surrounding the inhibitor may degrade or rupture, allowing the corrosion inhibitor to be released.

The release mechanism of encapsulated inhibitor molecules can be influenced by a number of factors. A common approach is to use stimuli-responsive materials, where release is triggered by specific external stimuli. These stimuli can be changes in temperature, pH, humidity, or exposure to certain chemicals. The reactive nature of the encapsulation material enables selective and timely release of inhibitor molecules when the desired conditions are met.

McCafferty [202] has carried out extensive research into the corrosion process, focusing in particular on local variations in pH at the metal surface. In the cathodic region of the corrosive environment, external oxygen and water molecules, or H^+ ions from the corrosive environment, come into contact with the metal substrate by infiltrating through perforations or cracks in the coating. These molecules and ions acquire electrons from the metal surface, which acts as an anode. Throughout the corrosion process, the alkalinity of this local zone increases progressively due to continuous dynamic changes in the accumulation of OH– ions or the depletion of H+ ions, resulting in an elevated pH level. The specific reduction reactions that occur in the cathode region depend on the acidity of the corrosive medium (Eq. 2.1) and (Eq. 2.2) [203, 204].

$$\boldsymbol{O}_2 + 2\boldsymbol{H}_2\boldsymbol{O} + 4e^- \rightarrow 4\boldsymbol{OH}^- \tag{2.1}$$

$$2\boldsymbol{H}^+ + 2e^- \rightarrow \boldsymbol{H}_2 \tag{2.2}$$

In addition, the anode reactions are briefly described below (Eq. 2.3) and (Eq. 2.4).

$$\boldsymbol{M} \rightarrow \boldsymbol{M}^{n+} + \boldsymbol{ne}^- \tag{2.3}$$

$$\boldsymbol{M}^{n+} + \boldsymbol{nH}_2\boldsymbol{O} \rightarrow \boldsymbol{M}(\boldsymbol{OH})_n + \boldsymbol{nH}^+ \tag{2.4}$$

Therefore, acidic and alkaline stimuli serve as common triggers for the initiation of shell opening [205]. To summarize, an effective approach to mitigating the adverse effects of corrosion is the incorporation of vessels containing pH-sensitive corrosion inhibitors into the coating matrix. The pH value plays a crucial role in controlling the release of corrosion inhibitors inside a polyelectrolyte shell made of oppositely charged polymers. By regulating its permeability, the shell effectively manages the release of these inhibitors [206]. In a study by Shchukin et al. [171], a zirconia (ZrO_2)/SiO_2 hybrid film was introduced onto the surface of an aluminum alloy as an anticorrosion coating. The hybrid film comprises SiO_2 nanoparticles coated with a polyelectrolyte (poly(ethylene imine)/poly(styrene sulfonate), PEI/PSS) and an inhibitor (BTA). Increasing the pH of the surrounding aqueous solution had a significant impact on the structure of the polyelectrolyte layers and led to the decomposition of the PSS/BTA complex. This, in turn, triggered the release of BTA from the shells containing corrosion inhibitors.

For the creation of micro/nanocontainers containing corrosion inhibitors, the core-shell structure provides a reliable framework. This provides an effective approach to the development of these inhibitors, which can either be applied to the outer shell or encapsulated within the inner core. In a study by Jiang et al. [207], the researchers emphasized the redox-reaction properties and the core-shell structure of the SiO_2 nanocapsules. These enable the rapid release of the corrosion inhibitor MBT during chemical reduction, while preventing its release during oxidation. Redox-sensitive permeability properties were obtained by incorporating bis[3-(triethoxysilyl)propyl] tetrasulfide into the SiO_2 shell. In particular, the permeability of the SiO_2 shell is increased when the reducing agent tris(2-carboxyethyl)phosphine hydrochloride triggers the reduction of tetrasulfide bonds.

For the production of anticorrosion films, one suitable method is the exploration of conductive polymers as substitutes for environmentally hazardous and banned heavy metals [208]. Of these options, PANI has attracted considerable research interest. It has unique redox properties that allow it to transition from an emerald to a leuco-emerald state [209]. In a study by Lv et al. [210], corrosion inhibitors, either diglycidyl ether or dicarboxylic acid-terminated polydimethylsiloxane (PDMS-DE or PDMS-DC), were used. A redox-sensitive nanocapsule with a conductive PANI shell encapsulated these inhibitors. The PANI capsules exhibited a delayed release response under oxidative conditions. However, rapid release was observed when exposed to reducing conditions. When PANI was oxidized with hydrogen peroxide, certain benzenoid rings were oxidized. This led to the formation of quinonoid rings. Consequently, intermolecular hydrogen bonding occurred between amines and imines, leading to densification of the polymer chains and preventing inhibitor release; conversely, when PANI was reduced by hydrazine, the strength of the hydrogen bonds decreased, leading to a loosening of the polymer chains in the capsule membrane. As a result, there was an increase in the flexibility of the PANI single bonds and a decrease in the hydrophobicity of the membrane. This led to increased permeability and facilitated the release of the hydrophobic PDMS-DE inhibitor [211].

2.6.4 Industrial Applications of Encapsulation on Corrosion Inhibition

Encapsulation techniques have found valuable applications in industrial environments for corrosion inhibition purposes. Through encapsulation, corrosion inhibitors can be effectively protected and diffused in a controlled manner, enhancing their performance in a variety of industries. This approach offers advantages such as extended inhibitor life, targeted release, and protection against environmental factors. In industrial applications, encapsulation methods are used to encapsulate corrosion inhibitors in coatings or protective materials. In this way, the inhibitors remain confined and isolated until they are needed for corrosion protection. Encapsulated inhibitors can be incorporated into paints, coatings, or composite materials used in industries such as automotive, aerospace, marine, and infrastructure.

Encapsulation techniques play an essential role in industrial corrosion inhibition applications, offering enhanced protection, controlled release, and extended inhibitor life. These advances are helping to improve corrosion prevention strategies, enhancing the durability and longevity of various industrial systems and structures.

2.7 CONCLUSION

In summary, employing diverse polymer coatings presents a dependable and efficient strategy for safeguarding metals against corrosion across a wide range of environmental circumstances. These coatings serve as a vital barrier, upholding the integrity of metallic materials and prolonging their lifespan in crucial applications. The combination of capsules and the incorporation of different inhibitors offer innovative approaches to improving the effectiveness of coatings in protecting against corrosion. The use of different inhibitors in coatings further enhances their corrosion protection capabilities. Different types of inhibitors may target specific corrosion mechanisms or be effective under specific environmental conditions. By incorporating several inhibitors, coatings can provide broader protection against a variety of corrosive agents or environments. The use of capsules and the incorporation of different inhibitors are promising strategies for improving the corrosion protection capabilities of coatings. By using capsules to control inhibitor release and incorporating a variety of inhibitors, coatings can provide targeted and comprehensive corrosion protection across multiple environments and applications.

REFERENCES

1. Izionworu V, Ukpaka C, Oguzie E. Green and eco-benign corrosion inhibition agents: alternatives and options to chemical based toxic corrosion inhibitors. *Chem Int.* 2020;6(4):232–59.
2. Bethencourt M, Botana FJ, Marcos M, Osuna RM, Sánchez-Amaya JM. Inhibitor properties of "green" pigments for paints. *Prog Org Coatings.* 2003;46(4):280–7.
3. de Lima-Neto P, de Araújo AP, Araújo WS, Correia AN. Study of the anticorrosive behaviour of epoxy binders containing non-toxic inorganic corrosion inhibitor pigments. *Prog Org Coatings.* 2008;62(3):344–50.
4. Deyá C, Blustein G, Del Amo B, Romagnoli R. Evaluation of eco-friendly anticorrosive pigments for paints in service conditions. *Prog Org Coatings.* 2010;69(1):1–6.
5. Cao Y, Yuan X, Wang X, Wu H, Wang G, Yang H. Improving the self-healing performance of the water-borne epoxy anticorrosion coating in 3.5 wt% NaCl solution using acid and alkaline dual sensitive organic microspheres. *Prog Org Coatings [Internet].* 2022;167(March):106815. Available from: https://doi.org/10.1016/j.porgcoat.2022.106815
6. White SR, Sottos NR, Geubelle PH, Moore JS, Kessler MR, Sriram SR, et al. Erratum: autonomic healing of polymer composites. *Nature.* 2001;409:794–797.
7. Dry C, McMillan W. Three-part methylmethacrylate adhesive system as an internal delivery system for smart responsive concrete. *Smart Mater Struct.* 1996;5(3):297–300.
8. Dry C. Matrix cracking repair and filling using active and passive modes for smart timed release of chemicals from fibers into cement matrices. *Smart Mater Struct.* 1994;3(2):118–23.
9. Pang JWC, Bond IP. A hollow fibre reinforced polymer composite encompassing self-healing and enhanced damage visibility. *Compos Sci Technol.* 2005;65(11–12):1791–9.
10. Williams G, Trask R, Bond I. A self-healing carbon fibre reinforced polymer for aerospace applications. *Compos Part A Appl Sci Manuf.* 2007;38(6):1525–32.
11. Toohey KS, Sottos NR, Lewis JA, Moore JS, White SR. Self-healing materials with microvascular networks. *Nat Mater.* 2007;6(8):581–5.
12. Liu T, Ma L, Wang X, Wang J, Qian H, Zhang D, et al. Self-healing corrosion protective coatings based on micro/nanocarriers: a review. *Corros Commun.* 2021;1:18–25.
13. Hanson P. Alkyl substituent effects. Part 1: an analysis of alkyl inductive properties in terms of group connectivity. *J Chem Soc Perkin Trans 2.* 1984;16(1):101–8.
14. Yu A, Song X, Lu W, Liang Y, He Z, Sun Y, et al. Facile and safety synthesis of highly loaded phase change microcapsules with paraffin/butyl stearate core and their feasible application in polymer composite. *Sol Energy Mater Sol Cells* [Internet]. 2022;247(August):111955. Available from: https://doi.org/10.1016/j.solmat.2022.111955
15. Cheng J, Kang M, Liu Y, Niu S, Guan Y, Qu W, et al. The preparation and characterization of thermal expansion capric acid microcapsules for controlling temperature. *Energy* [Internet]. 2022;261(PB):125296. Available from: https://doi.org/10.1016/j.energy.2022.125296
16. Matsuda T, Jadhav N, Kashi KB, Jensen M, Suryawanshi A, Gelling VJ. Self-healing ability and particle size effect of encapsulated cerium nitrate into pH sensitive microcapsules. *Prog Org Coatings* [Internet]. 2016;90:425–30. Available from: http://dx.doi.org/10.1016/j.porgcoat.2015.10.021
17. Wang J, Huang Y, Ma L, Guo X, Wu S, Ren C, et al. Corrosion-sensing and self-healing dual-function coating based on 1,10-phenanthroline loaded urea formaldehyde microcapsules for carbon steel protection. *Colloids Surfaces A: Physicochem Eng Asp* [Internet]. 2022;652(August):129855. Available from: https://doi.org/10.1016/j.colsurfa.2022.129855
18. Jinglei Y, Keller MW, Moore JS, White SR, Sottos NR. Microencapsulation of isocyanates for self-healing polymers. *Macromolecules.* 2008;41(24):9650–5.

19. Cao Y, Yuan X, Wang X, Li W, Yang H. Synthesis and controlled release kinetics of pH-sensitive hollow polyaniline microspheres encapsuled with the corrosion inhibitor. *J Mol Liq* [Internet]. 2021;342:117497. Available from: https://doi.org/10.1016/j.molliq.2021.117497
20. Maia F, Tedim J, Bastos AC, Ferreira MGS, Zheludkevich ML. Active sensing coating for early detection of corrosion processes. *RSC Adv*. 2014;4(34):17780–6.
21. Dong Y, Geng C, Wang X, Zhou Q. Porous polystyrene nanoparticles as nanocontainers of inhibitors for corrosion protection of low-alloy steel. *Pigment Resin Technol*. 2020;49(4):305–13.
22. Chen Z, Li X, Gong B, Scharnagl N, Zheludkevich ML, Ying H, et al. Double stimuli-responsive conducting polypyrrole nanocapsules for corrosion-resistant epoxy coatings. *ACS Appl Mater Interfaces* [Internet]. 2023 Jan 11;15(1):2067–76. Available from: https://pubs.acs.org/doi/10.1021/acsami.2c17466
23. Du P, Wang J, Zhao H, Liu G, Wang L. Graphene oxide encapsulated by mesoporous silica for intelligent anticorrosive coating: studies on release models and self-healing ability. *Dalt Trans*. 2019;48(34):13064–73.
24. Ashraf Ismail N, Moussa AM, Kahraman R, Shakoor RA. Study on the corrosion behavior of polymeric nanocomposite coatings containing halloysite nanotubes loaded with multicomponent inhibitor. *Arab J Chem* [Internet]. 2022;15(9):104107. Available from: https://doi.org/10.1016/j.arabjc.2022.104107
25. Chen Q, Lu X, Serdechnova M, Wang C, Lamaka S, Blawert C, et al. Formation of self-healing PEO coatings on AM50 Mg by in-situ incorporation of zeolite micro-container. *Corros Sci* [Internet]. 2022;209(June):110785. Available from: https://doi.org/10.1016/j.corsci.2022.110785
26. Mohammadpour Z, Zare HR. Fabrication of a pH-sensitive epoxy nanocomposite coating based on a Zn-BTC metal–organic framework containing benzotriazole as a smart corrosion inhibitor. *Cryst Growth Des* [Internet]. 2021 Jul 7;21(7):3954–66. Available from: https://pubs.acs.org/doi/10.1021/acs.cgd.1c00284
27. Nawaz M, Shakoor RA, Kahraman R, Montemor MF. Cerium oxide loaded with Gum Arabic as environmentally friendly anti-corrosion additive for protection of coated steel. *Mater Des* [Internet]. 2021;198:109361. Available from: https://doi.org/10.1016/j.matdes.2020.109361
28. Sui Y, Liu X, Bai S, Li X, Sun Z. Phosphate loaded layered double hydroxides for active corrosion protection of carbon steel. *Corros Eng Sci Technol* [Internet]. 2022 Jan 2;57(1):7–14. Available from: www.tandfonline.com/doi/full/10.1080/1478422X.2021.1976086
29. Javidparvar AA, Naderi R, Ramezanzadeh B. Manipulating graphene oxide nanocontainer with benzimidazole and cerium ions: application in epoxy-based nanocomposite for active corrosion protection. *Corros Sci* [Internet]. 2020;165(November 2019):108379. Available from: https://doi.org/10.1016/j.corsci.2019.108379
30. Leal DA, Kuznetsova A, Silva GM, Tedim J, Wypych F, Marino CEB. Layered materials as nanocontainers for active corrosion protection: a brief review. *Appl Clay Sci*. 2022;225(January):106537.
31. Sun A, Cui G, Liu Q. Colloids and surfaces A: physicochemical and engineering aspects capsule corrosion inhibitor loaded with hyperbranched chitosan: carbon dioxide corrosion protection for downhole pipelines in oil fields. *Colloids Surfaces A: Physicochem Eng Asp* [Internet]. 2023;664(February):131106. Available from: https://doi.org/10.1016/j.colsurfa.2023.131106
32. Calegari F, da Silva BC, Tedim J, Ferreira MGS, Berton MAC, Marino CEB. Benzotriazole encapsulation in spray-dried carboxymethylcellulose microspheres for active corrosion protection of carbon steel. *Prog Org Coatings* [Internet]. 2020;138(August 2019):105329. Available from: https://doi.org/10.1016/j.porgcoat.2019.105329
33. Yoshimoto N, Wahyudhin I, Yabuki A. Colloids and surfaces A: physicochemical and engineering aspects self-healing polymer coating with efficient delivery for alginates and calcium nitrite to provide corrosion protection for carbon steel. *Colloids Surfaces A: Physicochem Eng Asp* [Internet]. 2023;662(November 2022):130970. Available from: https://doi.org/10.1016/j.colsurfa.2023.130970
34. Tan Z, Wang S, Hu Z, Chen W, Qu Z, Xu C, et al. pH-responsive self-healing anticorrosion coating based on a lignin microsphere encapsulating inhibitor. *Ind Eng Chem Res* [Internet]. 2020 Feb 19;59(7):2657–66. Available from: https://pubs.acs.org/doi/10.1021/acs.iecr.9b05743
35. Berdimurodov E, Eliboyev I, Berdimuradov K, Kholikov A, Akbarov K, Dagdag O, et al. Green β-cyclodextrin-based corrosion inhibitors: recent developments, innovations and future opportunities.

Carbohydr Polym [Internet]. 2022 Sep;292:119719. Available from: https://linkinghub.elsevier.com/retrieve/pii/S0144861722006245
36. Dehghani A, Bahlakeh G, Ramezanzadeh B, Mofidabadi AHJ. Improvement of the anti-corrosion ability of a silane film with β-cyclodextrin-based nanocontainer loaded with L-histidine: coupled experimental and simulations studies. *Prog Org Coatings* [Internet]. 2021 Aug;157:106288. Available from: https://linkinghub.elsevier.com/retrieve/pii/S0300944021001594
37. Akhondi M, Jamalizadeh E. Fabrication of β-cyclodextrin modified halloysite nanocapsules for controlled release of corrosion inhibitors in self-healing epoxy coatings. *Prog Org Coatings* [Internet]. 2020 Aug;145:105676. Available from: https://linkinghub.elsevier.com/retrieve/pii/S0300944019317151
38. Ma Y, Huang H, Zhou H, Graham M, Smith J, Sheng X, et al. Superior anti-corrosion and self-healing bi-functional polymer composite coatings with polydopamine modified mesoporous silica/graphene oxide. *J Mater Sci Technol* [Internet]. 2021 Dec;95:95–104. Available from: https://linkinghub.elsevier.com/retrieve/pii/S1005030221004382
39. Qian B, Zheng Z, Michailids M, Fleck N, Bilton M, Song Y, et al. Mussel-inspired self-healing coatings based on polydopamine-coated nanocontainers for corrosion protection. *ACS Appl Mater Interfaces* [Internet]. 2019 Mar 13;11(10):10283–91. Available from: https://pubs.acs.org/doi/10.1021/acsami.8b21197
40. Haddadi SA, Hu S, Ghaderi S, Ghanbari A, Ahmadipour M, Pung S-Y, et al. Amino-functionalized MXene nanosheets doped with Ce(III) as potent nanocontainers toward self-healing epoxy nanocomposite coating for corrosion protection of mild steel. *ACS Appl Mater Interfaces* [Internet]. 2021 Sep 8;13(35):42074–93. Available from: https://pubs.acs.org/doi/10.1021/acsami.1c13055
41. Dehghani A, Bahlakeh G, Ramezanzadeh B. Designing a novel targeted-release nano-container based on the silanized graphene oxide decorated with cerium acetylacetonate loaded beta-cyclodextrin (β-CD-CeA-MGO) for epoxy anti-corrosion coating. *Chem Eng J* [Internet]. 2020 Nov;400:125860. Available from: https://linkinghub.elsevier.com/retrieve/pii/S1385894720319884
42. Zhang M, Xu F, Lin D, Peng J, Zhu Y, Wang H. A smart anti-corrosion coating based on triple functional fillers. *Chem Eng J* [Internet]. 2022 Oct;446:137078. Available from: https://linkinghub.elsevier.com/retrieve/pii/S1385894722025700
43. Haddadi SA, Mehmandar E, Jabari H, Ramazani SAA, Mohammadkhani R, Yan N, et al. Zinc-doped silica/polyaniline core/shell nanoparticles towards corrosion protection epoxy nanocomposite coatings. *Compos Part B Eng* [Internet]. 2021 May;212:108713. Available from: https://linkinghub.elsevier.com/retrieve/pii/S1359836821001062
44. Shankar Kashyap S, Basak P, Pendem C, Narayan R, Ahmed M. Role of ordered mesoporous silica loaded with cerium and with an amine-modified surface as an effective binder and nano-container in corrosion resistant hybrid coating. *Chemistry Select* [Internet]. 2023 Apr 5;8(13):e202300657. Available from: https://chemistry-europe.onlinelibrary.wiley.com/doi/10.1002/slct.202300657
45. Xu T, Fang L, Zhao J-P, Zhang J-T, Hu J-M. A novel graphene/silica composite nanocontainer prepared by electrochemically assisted deposition. *Prog Org Coatings* [Internet]. 2023 May;178:107464. Available from: https://linkinghub.elsevier.com/retrieve/pii/S0300944023000607
46. Mohammadi I, Shahrabi T, Mahdavian M, Izadi M. A novel corrosion inhibitive system comprising Zn-Al LDH and hybrid sol-gel silane nanocomposite coating for AA2024-T3. *J Alloys Compd* [Internet]. 2022 Jul;909:164755. Available from: https://linkinghub.elsevier.com/retrieve/pii/S092583882201146X
47. Olya N, Ghasemi E, Mahdavian M, Ramezanzadeh B. Construction of a novel corrosion protective composite film based on a core-shell LDH-Mo@SiO2 inhibitor nanocarrier with both self-healing/barrier functions. *J Taiwan Inst Chem Eng* [Internet]. 2020 Aug;113:406–18. Available from: https://linkinghub.elsevier.com/retrieve/pii/S1876107020302091
48. Su Y, Qiu S, Wei J, Zhu X, Zhao H, Xue Q. Sulfonated polyaniline assisted hierarchical assembly of graphene-LDH nanohybrid for enhanced anticorrosion performance of waterborne epoxy coatings. *Chem Eng J* [Internet]. 2021 Dec;426:131269. Available from: https://linkinghub.elsevier.com/retrieve/pii/S1385894721028503
49. Sun J, Li W, Li N, Zhan Y, Tian L, Wang Y. Effect of surface modified nano-SiO2 particles on properties of TO@CA/SR self-healing anti-corrosion composite coating. *Prog Org Coatings* [Internet].

2022 Mar;164:106689. Available from: https://linkinghub.elsevier.com/retrieve/pii/S030094402 1005609

50. Li D, Peng H, Lin Z, Zhu J, Yu J, Liu J, et al. Corrosion protection coatings embedded with silane-functionalized rGO/SiO2 nanocontainers: enhancing dispersive and corrosion-inhibitor loading capabilities. *Surf Coatings Technol* [Internet]. 2021 Dec;427:127850. Available from: https://linkinghub.elsevier.com/retrieve/pii/S0257897221010240
51. Verma C, Abdellattif MH, Alfantazi A, Quraishi MA. N-heterocycle compounds as aqueous phase corrosion inhibitors: a robust, effective and economic substitute. *J Mol Liq* [Internet]. 2021 Oct;340:117211. Available from: https://linkinghub.elsevier.com/retrieve/pii/S016773222 1019358
52. Quraishi MA, Chauhan DS, Saji VS. Heterocyclic biomolecules as green corrosion inhibitors. *J Mol Liq* [Internet]. 2021 Nov;341:117265. Available from: https://linkinghub.elsevier.com/retrieve/pii/S0167732221019899
53. Louroubi A, Nayad A, Hasnaoui A, Idouhli R, Abouelfida A, El Firdoussi L, et al. Natural hydroxyapatite: green catalyst for the synthesis of pyrroles, inhibitors of corrosion. *J Chem* [Internet]. 2021 Mar 5;2021:1–11. Available from: www.hindawi.com/journals/jchem/2021/6613243/
54. Idouhli R, N'Ait Ousidi A, Koumya Y, Abouelfida A, Benyaich A, Auhmani A, et al. Electrochemical studies of monoterpenic thiosemicarbazones as corrosion inhibitor for steel in 1 M HCl. *Int J Corros* [Internet]. 2018 Mar 21;2018:1–15. Available from: www.hindawi.com/journals/ijc/2018/9212705/
55. Sheetal, Batra R, Singh AK, Singh M, Thakur S, Pani B, et al. Advancement of corrosion inhibitor system through N-heterocyclic compounds: a review. *Corros Eng Sci Technol* [Internet]. 2023 Jan 2;58(1):73–101. Available from: https://www.tandfonline.com/doi/full/10.1080/14784 22X.2022.2137979
56. Hrimla M, Bahsis L, Laamari MR, Julve M, Stiriba S-E. An overview on the performance of 1,2,3-triazole derivatives as corrosion inhibitors for metal surfaces. *Int J Mol Sci* [Internet]. 2021 Dec 21;23(1):16. Available from: www.mdpi.com/1422-0067/23/1/16
57. Turano M, Walker M, Grillo F, Gattinoni C, Edmondson J, Adesida O, et al. Understanding the interaction of organic corrosion inhibitors with copper at the molecular scale: benzotriazole on Cu(110). *Appl Surf Sci* [Internet]. 2021 Dec;570:151206. Available from: https://linkinghub.elsevier.com/retrieve/pii/S0169433221022613
58. Li C, Zhang C, He Y, Li H, Zhao Y, Li Z, et al. Benzotriazole corrosion inhibitor loaded nanocontainer based on g-C3N4 and hollow polyaniline spheres towards enhancing anticorrosion performance of waterborne epoxy coatings. *Prog Org Coatings* [Internet]. 2023 Jan;174:107276. Available from: https://linkinghub.elsevier.com/retrieve/pii/S0300944022005732
59. Liu T, Li W, Zhang C, Wang W, Dou W, Chen S. Preparation of highly efficient self-healing anticorrosion epoxy coating by integration of benzotriazole corrosion inhibitor loaded 2D-COF. *J Ind Eng Chem* [Internet]. 2021 May;97:560–73. Available from: https://linkinghub.elsevier.com/retrieve/pii/S1226086X21001362
60. Rodriguez J, Bollen E, Nguyen TD, Portier A, Paint Y, Olivier M-G. Incorporation of layered double hydroxides modified with benzotriazole into an epoxy resin for the corrosion protection of Zn-Mg coated steel. *Prog Org Coatings* [Internet]. 2020 Dec;149:105894. Available from: https://linkinghub.elsevier.com/retrieve/pii/S030094402031105X
61. Pham TH, Nguyen TC, Le AK, Le PK, Son DN. Insights into effects of metal cations on the adsorption of benzotriazole on halloysite nanotubes: an experimental and DFT study. *J Phys Chem C* [Internet]. 2022 Feb 10;126(5):2920–9. Available from: https://pubs.acs.org/doi/10.1021/acs.jpcc.1c10699
62. Hao Y, Sun W, Jiang L, Cui J, Zhang Y, Song L, et al. Self-healing effect of epoxy coating containing mesoporous polyaniline hollow spheres loaded with benzotriazole. *Prog Org Coatings* [Internet]. 2021 Oct;159:106445. Available from: https://linkinghub.elsevier.com/retrieve/pii/S030094402 1003167
63. Calegari F, da Silva BC, Tedim J, Ferreira MGS, Berton MAC, Marino CEB. Benzotriazole encapsulation in spray-dried carboxymethylcellulose microspheres for active corrosion protection of carbon steel. *Prog Org Coatings* [Internet]. 2020 Jan;138:105329. Available from: https://linkinghub.elsevier.com/retrieve/pii/S0300944018308853

64. Jin Z, Zhao Z, Zhao T, Liu H, Liu H. One-step preparation of inhibitor-loaded nanocontainers and their application in self-healing coatings. *Corros Commun* [Internet]. 2021 Jun;2:63–71. Available from: https://linkinghub.elsevier.com/retrieve/pii/S2667266921000190
65. Chen G, Wen S, Ma J, Sun Z, Lin C, Yue Z, et al. Optimization of intrinsic self-healing silicone coatings by benzotriazole loaded mesoporous silica. *Surf Coatings Technol* [Internet]. 2021 Sep;421:127388. Available from: https://linkinghub.elsevier.com/retrieve/pii/S0257897221005624
66. Zhou S, Zhao Z, Mao H, Wang L, Chen J, Chen J, et al. Bronze preservation by using composite hydrogel coating-loaded corrosion inhibitors. *Herit Sci* [Internet]. 2022 Dec 25;10(1):116. Available from: https://heritagesciencejournal.springeropen.com/articles/10.1186/s40494-022-00747-w
67. Cao J, Guo C, Chen Z. Loading and release mechanisms of MOF-5 @ BTA-X (X = -CH3/-NH2/-CO(CH2)6CH3): experimental and theoretical investigations. *Colloids Surfaces A: Physicochem Eng Asp* [Internet]. 2023 Jun;666:131274. Available from: https://linkinghub.elsevier.com/retrieve/pii/S0927775723003588
68. Matangouo B, Dedzo GK, Dzene L, Nanseu-Njiki CP, Ngameni E. Encapsulation of butylimidazole in smectite and slow release for enhanced copper corrosion inhibition. *Appl Clay Sci* [Internet]. 2021 Nov;213:106266. Available from: https://linkinghub.elsevier.com/retrieve/pii/S0169131721002908
69. Nawaz M, Naeem N, Kahraman R, Montemor MF, Haider W, Shakoor RA. Effectiveness of epoxy coating modified with yttrium oxide loaded with imidazole on the corrosion protection of steel. *Nanomaterials* [Internet]. 2021 Sep 3;11(9):2291. Available from: www.mdpi.com/2079-4991/11/9/2291
70. Khan A, Hassanein A, Habib S, Nawaz M, Shakoor RA, Kahraman R. Hybrid halloysite nanotubes as smart carriers for corrosion protection. *ACS Appl Mater Interfaces* [Internet]. 2020 Aug 19;12(33):37571–84. Available from: https://pubs.acs.org/doi/10.1021/acsami.0c08953
71. Amini M, Naderi R, Mahdavian M, Badiei A. Effect of piperazine functionalization of mesoporous silica type SBA-15 on the loading efficiency of 2-mercaptobenzothiazole corrosion inhibitor. *Ind Eng Chem Res* [Internet]. 2020 Feb 26;59(8):3394–404. Available from: https://pubs.acs.org/doi/10.1021/acs.iecr.9b05261
72. Cao Y, Wu H, Wang X, Wang G, Yang H. Novel long-acting smart anticorrosion coating based on pH-controlled release polyaniline hollow microspheres encapsulating inhibitor. *J Mol Liq* [Internet]. 2022 Aug;359:119341. Available from: https://linkinghub.elsevier.com/retrieve/pii/S0167732222008790
73. Cui M, Njoku DI, Li B, Yang L, Wang Z, Hou B, et al. Corrosion protection of aluminium alloy 2024 through an epoxy coating embedded with smart microcapsules: the responses of smart microcapsules to corrosive entities. *Corros Commun* [Internet]. 2021 Mar;1:1–9. Available from: https://linkinghub.elsevier.com/retrieve/pii/S2667266921000050
74. Yan R, Xue B, Wang C, Zhang H, Yue Y, Luo J. Study on properties of porous SiO2 /GO modified polyurethane coatings loaded with corrosion inhibitor. *J Phys Conf Ser* [Internet]. 2022 Feb 1;2200(1):012013. Available from: https://iopscience.iop.org/article/10.1088/1742-6596/2200/1/012013
75. Li H, Qiang Y, Zhao W, Zhang S. 2-Mercaptobenzimidazole-inbuilt metal-organic-frameworks modified graphene oxide towards intelligent and excellent anti-corrosion coating. *Corros Sci* [Internet]. 2021 Oct;191:109715. Available from: https://linkinghub.elsevier.com/retrieve/pii/S0010938X21004819
76. Haddadi SA, Ramazani SAA, Mahdavian M, Arjmand M. Epoxy nanocomposite coatings with enhanced dual active/barrier behavior containing graphene-based carbon hollow spheres as corrosion inhibitor nanoreservoirs. *Corros Sci* [Internet]. 2021 Jun;185:109428. Available from: https://linkinghub.elsevier.com/retrieve/pii/S0010938X21001943
77. Edraki M, Zaarei D. Azole derivatives embedded in montmorillonite clay nanocarriers as corrosion inhibitors of mild steel. *Int J Miner Metall Mater* [Internet]. 2019 Jan 5;26(1):86–97. Available from: http://link.springer.com/10.1007/s12613-019-1712-1
78. Daradmare S, Pradhan M, Raja VS, Parida S. Epoxy–8-HQ@PS/GO composite coatings: size effect and the loading of fillers on the self-healing and passive protection. *Prog Org Coatings* [Internet]. 2021 Sep;158:106352. Available from: https://linkinghub.elsevier.com/retrieve/pii/S030094402100223X

79. Njoku CN, Bai W, Arukalam IO, Yang L, Hou B, Njoku DI, et al. Epoxy-based smart coating with self-repairing polyurea-formaldehyde microcapsules for anticorrosion protection of aluminum alloy AA2024. *J Coatings Technol Res* [Internet]. 2020 May 24;17(3):797–813. Available from: http://link.springer.com/10.1007/s11998-020-00334-3
80. Siva T, Kumari SS, Sathiyanarayanan S. Dendrimer like mesoporous silica nano container (DMSN) based smart self healing coating for corrosion protection performance. *Prog Org Coatings* [Internet]. 2021 May;154:106201. Available from: https://linkinghub.elsevier.com/retrieve/pii/S0300944021000722
81. Taheri N, Sarabi AA, Roshan S. Investigation of intelligent protection and corrosion detection of epoxy-coated St-12 by redox-responsive microcapsules containing dual-functional 8-hydroxyquinoline. *Prog Org Coatings* [Internet]. 2022 Nov;172:107073. Available from: https://linkinghub.elsevier.com/retrieve/pii/S0300944022003708
82. Ma L, Qiang Y, Zhao W. Designing novel organic inhibitor loaded MgAl-LDHs nanocontainer for enhanced corrosion resistance. *Chem Eng J* [Internet]. 2021 Mar;408:127367. Available from: https://linkinghub.elsevier.com/retrieve/pii/S1385894720334914
83. Yeganeh M, Rabizadeh T, Rabiezadeh MS, Kahvazizadeh M, Ramezanalizadeh H. Corrosion and the antibacterial response of epoxy coating/drug-loaded mesoporous silica. *Polym Bull* [Internet]. 2023 Apr 17;80(4):4209–27. Available from: https://link.springer.com/10.1007/s00289-022-04261-8
84. Yeganeh M, Asadi N, Omidi M, Mahdavian M. An investigation on the corrosion behavior of the epoxy coating embedded with mesoporous silica nanocontainer loaded by sulfamethazine inhibitor. *Prog Org Coatings* [Internet]. 2019 Mar;128:75–81. Available from: https://linkinghub.elsevier.com/retrieve/pii/S0300944018311019
85. Verma C, Quraishi MA, Ebenso EE, Hussain CM. Amines as corrosion inhibitors. In: *Organic Corrosion Inhibitors* [Internet]. Wiley; 2021. p. 75–94. Available from: https://onlinelibrary.wiley.com/doi/10.1002/9781119794516.ch5
86. Verma C, Quraishi MA, Ebenso EE. A review on ammonia derivatives as corrosion inhibitors for metals and alloys. In: *Green Energy and Technology* [Internet]. Springer, Cham; 2020. p. 49–67. Available from: http://link.springer.com/10.1007/978-3-030-35106-9_3
87. Matangouo B, Dedzo GK, Dzene L, Josien L, Nanseu-Njiki CP, Ngameni E. Dodecylamine encapsulation in layered smectite clay mineral: release and application for copper corrosion inhibition in 0.1 M nitric acid solution. *J Phys Chem Solids* [Internet]. 2023 Sep;180:111427. Available from: https://linkinghub.elsevier.com/retrieve/pii/S0022369723002172
88. Kim C, Karayan AI, Milla J, Hassan M, Castaneda H. Smart coating embedded with ph-responsive nanocapsules containing a corrosion inhibiting agent. *ACS Appl Mater Interfaces* [Internet]. 2020 Feb 5;12(5):6451–9. Available from: https://pubs.acs.org/doi/10.1021/acsami.9b20238
89. Chen Z, Yang W, Yin X, Chen Y, Liu Y, Xu B. Corrosion protection of 304 stainless steel from a smart conducting polypyrrole coating doped with pH-sensitive molybdate-loaded TiO2 nanocontainers. *Prog Org Coatings* [Internet]. 2020 Sep;146:105750. Available from: https://linkinghub.elsevier.com/retrieve/pii/S0300944020301144
90. Yeganeh M, Omidi M, Rabizadeh T. Anti-corrosion behavior of epoxy composite coatings containing molybdate-loaded mesoporous silica. *Prog Org Coatings* [Internet]. 2019 Jan;126:18–27. Available from: https://linkinghub.elsevier.com/retrieve/pii/S0300944018301747
91. Abrantes Leal D, Wypych F, Bruno Marino CE. Zinc-layered hydroxide salt intercalated with molybdate anions as a new smart nanocontainer for active corrosion protection of carbon steel. *ACS Appl Mater Interfaces* [Internet]. 2020 Apr 29;12(17):19823–33. Available from: https://pubs.acs.org/doi/10.1021/acsami.0c02378
92. Chen Z, Yang W, Chen Y, Yin X, Liu Y. Smart coatings embedded with polydopamine-decorated layer-by-layer assembled SnO2 nanocontainers for the corrosion protection of 304 stainless steels. *J Colloid Interface Sci* [Internet]. 2020 Nov;579:741–53. Available from: https://linkinghub.elsevier.com/retrieve/pii/S0021979720308699
93. Kong X, Liu J, Li S, Yu M. APTES-modified graphene oxide loaded with cerium dibutylphosphate as two-dimensional nanocomposites for enhancing corrosion protection properties. *Corros Sci* [Internet]. 2023 Apr;213:110966. Available from: https://linkinghub.elsevier.com/retrieve/pii/S0010938X23000082

94. Wu Y. Graphene oxide-loaded zinc phosphate as an anticorrosive reinforcement in waterborne polyurethane resin. *Int J Electrochem Sci* [Internet]. 2019 Jun;5271–86. Available from: www.electrochemsci.org/abstracts/vol14/140605271.pdf
95. Chen Z, He K, Wei R, Lv Y, Liu Z, Han G-C. Synthesis of Mn-MOFs loaded zinc phosphate composite for water-based acrylic coatings with durable anticorrosion performance on mild steel. *RSC Adv* [Internet]. 2021;11(6):3371–9. Available from: http://xlink.rsc.org/?DOI=D0RA09753E
96. Zhang G, Jiang E, Wu L, Tang A, Atrens A, Pan F. Active corrosion protection of phosphate loaded PEO/LDHs composite coatings: SIET study. *J Magnes Alloy* [Internet]. 2022 May;10(5):1351–7. Available from: https://linkinghub.elsevier.com/retrieve/pii/S2213956721000621
97. Cao Y, Zheng D, Luo J, Zhang F, Wang C, Dong S, et al. Enhanced corrosion protection by Al surface immobilization of in-situ grown layered double hydroxide films co-intercalated with inhibitors and low surface energy species. *Corros Sci* [Internet]. 2020 Mar;164:108340. Available from: https://linkinghub.elsevier.com/retrieve/pii/S0010938X19315215
98. Bouali AC, André NM, Silva Campos MR, Serdechnova M, dos Santos JF, Amancio-Filho ST, et al. Influence of LDH conversion coatings on the adhesion and corrosion protection of friction spot-joined AA2024-T3/CF-PPS. *J Mater Sci Technol* [Internet]. 2021 Mar;67:197–210. Available from: https://linkinghub.elsevier.com/retrieve/pii/S1005030220306708
99. Liang Z, Ma Y, Li K, Liao Y, Yang B, Liu L, et al. Formation of layered double hydroxides film on AA2099-T83 Al-Cu-Li alloy and its effect on corrosion resistance. *Surf Coatings Technol* [Internet]. 2019 Nov;378:124967. Available from: https://linkinghub.elsevier.com/retrieve/pii/S0257897219309569
100. Ying L, Wang D, Nie C, Zhu T, Cao F, Liu R, et al. Modification of Mg/Al-LDH by vanadate: effects on tribological properties and corrosion resistance. *RSC Adv* [Internet]. 2023;13(21):14171–80. Available from: http://xlink.rsc.org/?DOI=D3RA01636F
101. Pirhady Tavandashti N, Molana Almas S, Esmaeilzadeh E. Corrosion protection performance of epoxy coating containing alumina/PANI nanoparticles doped with cerium nitrate inhibitor on Al-2024 substrates. *Prog Org Coatings* [Internet]. 2021 Mar;152:106133. Available from: https://linkinghub.elsevier.com/retrieve/pii/S0300944021000035
102. Raj R, Taryba MG, Morozov Y, Kahraman R, Shakoor RA, Montemor MF. On the synergistic corrosion inhibition and polymer healing effects of polyolefin coatings modified with Ce-loaded hydroxyapatite particles applied on steel. *Electrochim Acta* [Internet]. 2021 Aug;388:138648. Available from: https://linkinghub.elsevier.com/retrieve/pii/S0013468621009385
103. Denissen PJ, Homborg AM, Garcia SJ. Requirements for corrosion inhibitor release from damaged primers for stable protection: a simulation and experimental approach using cerium loaded carriers. *Surf Coatings Technol* [Internet]. 2022 Jan;430:127966. Available from: https://linkinghub.elsevier.com/retrieve/pii/S0257897221011403
104. Ghahremani P, Sarabi AA, Roshan S. Cerium containing pH-responsive microcapsule for smart coating application: characterization and corrosion study. *Surf Coatings Technol* [Internet]. 2021 Dec;427:127820. Available from: https://linkinghub.elsevier.com/retrieve/pii/S0257897221009944
105. Wei H, Heidarshenas B, Zhou L, Hussain G, Li Q, Ostrikov K (Ken). Green inhibitors for steel corrosion in acidic environment: state of art. *Mater Today Sustain* [Internet]. 2020 Dec;10:100044. Available from: https://linkinghub.elsevier.com/retrieve/pii/S2589234720300130
106. Gabsi M, Ferkous H, Delimi A, Boublia A, Boulechfar C, Kahlouche A, et al. The curious case of polyphenols as green corrosion inhibitors: a review on their extraction, design, and applications. *Environ Sci Pollut Res* [Internet]. 2023 Apr 5;30(21):59081–105. Available from: https://link.springer.com/10.1007/s11356-023-26753-4
107. Al Jahdaly BA, Maghraby YR, Ibrahim AH, Shouier KR, Alturki AM, El-Shabasy RM. Role of green chemistry in sustainable corrosion inhibition: a review on recent developments. *Mater Today Sustain* [Internet]. 2022 Dec;20:100242. Available from: https://linkinghub.elsevier.com/retrieve/pii/S2589234722001348
108. Yadav M, Goel G, Hatton FL, Bhagat M, Mehta SK, Mishra RK, et al. A review on biomass-derived materials and their applications as corrosion inhibitors, catalysts, food and drug delivery agents. *Curr Res Green Sustain Chem* [Internet]. 2021;4:100153. Available from: https://linkinghub.elsevier.com/retrieve/pii/S2666086521001004

109. Fazal BR, Becker T, Kinsella B, Lepkova K. A review of plant extracts as green corrosion inhibitors for CO2 corrosion of carbon steel. *npj Mater Degrad.* 2022;6(1):5.
110. Salleh SZ, Yusoff AH, Zakaria SK, Taib MAA, Abu Seman A, Masri MN, et al. Plant extracts as green corrosion inhibitor for ferrous metal alloys: a review. *J Clean Prod* [Internet]. 2021 Jul;304:127030. Available from: https://linkinghub.elsevier.com/retrieve/pii/S095965262101249X
111. Alrefaee SH, Rhee KY, Verma C, Quraishi MA, Ebenso EE. Challenges and advantages of using plant extract as inhibitors in modern corrosion inhibition systems: recent advancements. *J Mol Liq* [Internet]. 2021 Jan;321:114666. Available from: https://linkinghub.elsevier.com/retrieve/pii/S0167732220369087
112. Idouhli R, Koumya Y, Khadiri M, Aityoub A, Abouelfida A, Benyaich A. Inhibitory effect of Senecio anteuphorbium as green corrosion inhibitor for S300 steel. *Int J Ind Chem* [Internet]. 2019;10(2):133–43. Available from: https://doi.org/10.1007/s40090-019-0179-2
113. Singh P, Rana A, Karak N, Kumar I, Rana S, Kumar P. Sustainable smart anti-corrosion coating materials derived from vegetable oil derivatives: a review. *RSC Adv* [Internet]. 2023;13(6):3910–41. Available from: http://xlink.rsc.org/?DOI=D2RA07825B
114. Wang J, Tan W, Yang H, Rao X, Luo X, Ma L, et al. Towards weathering and corrosion resistant, self-warning and self-healing epoxy coatings with tannic acid loaded nanocontainers. *npj Mater Degrad.* 2023;7(1):1–12.
115. Raj R, Kahraman R, Shakoor A, Montemor F, Taryba M. Tannic acid-loaded hydroxyapatite carriers for corrosion protection of polyolefin-coated carbon steel. *Appl Sci* [Internet]. 2022 Oct 12;12(20):10263. Available from: www.mdpi.com/2076-3417/12/20/10263
116. Koerner CM, Hopkinson DP, Ziomek-Moroz ME, Rodriguez A, Xiang F. Environmentally friendly tannic acid multilayer coating for reducing corrosion of carbon steel. *Ind Eng Chem Res* [Internet]. 2021 Jan 13;60(1):243–50. Available from: https://pubs.acs.org/doi/10.1021/acs.iecr.0c02925
117. Zhang L, Wu K, Chen Y, Liu R, Luo J. The preparation of linseed oil loaded graphene/polyaniline microcapsule via emulsion template method for self-healing anticorrosion coatings. *Colloids Surfaces A Physicochem Eng Asp* [Internet]. 2022 Oct;651:129771. Available from: https://linkinghub.elsevier.com/retrieve/pii/S0927775722015266
118. Mirabedini SM, Farnood RR, Esfandeh M, Zareanshahraki F, Rajabi P. Nanocomposite coatings comprising APS-treated linseed oil-embedded polyurea-formaldehyde microcapsules and nanoclay, part 2: self-healing and corrosion resistance properties. *Prog Org Coatings* [Internet]. 2020 May;142:105592. Available from: https://linkinghub.elsevier.com/retrieve/pii/S030094401931728X
119. Feng Y, Cui Y, Zhang M, Li M, Li H. Preparation of tung oil-loaded PU/PANI microcapsules and synergetic anti-corrosion properties of self-healing epoxy coatings. *Macromol Mater Eng* [Internet]. 2021 Feb 10;306(2):2000581. Available from: https://onlinelibrary.wiley.com/doi/10.1002/mame.202000581
120. Suresh Kumar S, Kakooei S, Ismail MC, Haris M. Synthesis and characterization of metal ion end capped nanocontainer loaded with duo green corrosion inhibitors. *J Mater Res Technol* [Internet]. 2020 Jul;9(4):8350–4. Available from: https://linkinghub.elsevier.com/retrieve/pii/S2238785420313806
121. Akhavan-Bahabadi Z, Zare HR, Mohammadpour Z. Use of caffeine-containing MIL-100 (Fe) metal organic framework as a high-performance smart anticorrosion coating to protect stainless steel in 3.5 wt% NaCl solution. *J Coatings Technol Res* [Internet]. 2023 May 3;20(3):883–98. Available from: https://link.springer.com/10.1007/s11998-022-00709-8
122. Aguirre-Pulido M, González-Sánchez JA, Dzib-Pérez LR, Soria-Castro M, Ávila-Ortega A, Talavera-Pech WA. Synthesis and electrochemical evaluation of MSNs-PbAE nanocontainers for the controlled release of caffeine as a corrosion inhibitor. *Pharmaceutics* [Internet]. 2022 Nov 30;14(12):2670. Available from: www.mdpi.com/1999-4923/14/12/2670
123. Dehghani A, Bahlakeh G, Ramezanzadeh B, Hossein Jafari Mofidabadi A. Electronic DFT-D modeling of L-citrulline molecules interactions with Beta-CD aligned rGO-APTES multi-functional nano-capsule for anti-corrosion application. *J Mol Liq* [Internet]. 2022 May;354:118814. Available from: https://linkinghub.elsevier.com/retrieve/pii/S0167732222003518

124. Yan D, Wang Y, Liu J, Song D, Zhang T, Liu J, et al. Self-healing system adapted to different pH environments for active corrosion protection of magnesium alloy. *J Alloys Compd* [Internet]. 2020 May;824:153918. Available from: https://linkinghub.elsevier.com/retrieve/pii/S0925838820302814
125. Židov B, Lin Z, Stojanović I, Xu L. Impact of inhibitor loaded mesoporous silica nanoparticles on waterborne coating performance in various corrosive environments. *J Appl Polym Sci* [Internet]. 2021 Jan 5;138(1):49614. Available from: https://onlinelibrary.wiley.com/doi/10.1002/app.49614
126. Habib S, Hassanein A, Kahraman R, Mahdi Ahmed E, Shakoor RA. Self-healing behavior of epoxy-based double-layer nanocomposite coatings modified with Zirconia nanoparticles. *Mater Des* [Internet]. 2021 Sep;207:109839. Available from: https://linkinghub.elsevier.com/retrieve/pii/S0264127521003920
127. Privitera A, Ruggiero L, Venditti I, Pasqual Laverdura U, Tuti S, De Felicis D, et al. One step nanoencapsulation of corrosion inhibitors for gradual release application. *Mater Today Chem* [Internet]. 2022 Jun;24:100851. Available from: https://linkinghub.elsevier.com/retrieve/pii/S2468519422000805
128. Ma L, Wang J, Wang Y, Guo X, Wu S, Fu D, et al. Enhanced active corrosion protection coatings for aluminum alloys with two corrosion inhibitors co-incorporated in nanocontainers. *Corros Sci* [Internet]. 2022 Nov;208:110663. Available from: https://linkinghub.elsevier.com/retrieve/pii/S0010938X22005819
129. Habib S, Fayyad E, Nawaz M, Khan A, Shakoor RA, Kahraman R, et al. Cerium dioxide nanoparticles as smart carriers for self-healing coatings. *Nanomaterials* [Internet]. 2020 Apr 20;10(4):791. Available from: www.mdpi.com/2079-4991/10/4/791
130. Liu T, Zhang D, Zhang R, Wang J, Ma L, Keil P, et al. Self-healing and corrosion-sensing coatings based on pH-sensitive MOF-capped microcontainers for intelligent corrosion control. *Chem Eng J* [Internet]. 2023 Feb;454:140335. Available from: https://linkinghub.elsevier.com/retrieve/pii/S1385894722058156
131. Yin Y, Prabhakar M, Ebbinghaus P, Corrêa da Silva C, Rohwerder M. Neutral inhibitor molecules entrapped into polypyrrole network for corrosion protection. *Chem Eng J* [Internet]. 2022 Jul;440:135739. Available from: https://linkinghub.elsevier.com/retrieve/pii/S1385894722012396
132. Adsul SH, Bagale UD, Sonawane SH, Subasri R. Release rate kinetics of corrosion inhibitor loaded halloysite nanotube-based anticorrosion coatings on magnesium alloy AZ91D. *J Magnes Alloy* [Internet]. 2021 Jan;9(1):202–15. Available from: https://linkinghub.elsevier.com/retrieve/pii/S2213956720301249
133. Ghamsarizade R, Najafi S, Sarabi AA, Roshan S, Eivaz Mohammadloo H. Synthesis of pH sensitive microcapsules containing ZAPP, SAPP, and 8-HQ, and evaluation of their anti-corrosion performance, and mechanical enhancement of epoxy coating. *Prog Org Coatings* [Internet]. 2023 Jan;174:107290. Available from: https://linkinghub.elsevier.com/retrieve/pii/S0300944022005872
134. Eivaz Mohammadloo H, Mirabedini SM, Pezeshk-Fallah H. Microencapsulation of quinoline and cerium based inhibitors for smart coating application: anti-corrosion, morphology and adhesion study. *Prog Org Coatings* [Internet]. 2019 Dec;137:105339. Available from: https://linkinghub.elsevier.com/retrieve/pii/S0300944019308872
135. Guo C, Cao J, Chen Z. Core-shell mesoporous silica–metal–phenolic network microcapsule for the controlled release of corrosion inhibitor. *Appl Surf Sci* [Internet]. 2022 Dec;605:154747. Available from: https://linkinghub.elsevier.com/retrieve/pii/S0169433222022759
136. Zhu DY, Rong MZ, Zhang MQ. Self-healing polymeric materials based on microencapsulated healing agents: From design to preparation. *Prog Polym Sci* [Internet]. 2015;49–50:175–220. Available from: http://dx.doi.org/10.1016/j.progpolymsci.2015.07.002
137. Zhang Y, Mustapha AN, Zhang X, Baiocco D, Wellio G, Davies T, et al. Improved volatile cargo retention and mechanical properties of capsules via sediment-free in situ polymerization with cross-linked poly(vinyl alcohol) as an emulsifier. *J Colloid Interface Sci* [Internet]. 2020;568(June 2019):155–64. Available from: https://doi.org/10.1016/j.jcis.2020.01.115
138. Blaiszik BJ, Caruso MM, McIlroy DA, Moore JS, White SR, Sottos NR. Microcapsules filled with reactive solutions for self-healing materials. *Polymer (Guildf)* [Internet]. 2009;50(4):990–7. Available from: http://dx.doi.org/10.1016/j.polymer.2008.12.040

139. Nesterova T, Dam-Johansen K, Kiil S. Synthesis of durable microcapsules for self-healing anticorrosive coatings: a comparison of selected methods. *Prog Org Coatings* [Internet]. 2011;70(4):342–52. Available from: http://dx.doi.org/10.1016/j.porgcoat.2010.09.032
140. Chen Z, Wang J, Yu F, Zhang Z, Gao X. Preparation and properties of graphene oxide-modified poly(melamine-formaldehyde) microcapsules containing phase change material n-dodecanol for thermal energy storage. *J Mater Chem A*. 2015;3(21):11624–30.
141. Nguon O, Lagugné-Labarthet F, Brandys FA, Li J, Gillies ER. Microencapsulation by in situ polymerization of amino resins. *Polym Rev* [Internet]. 2018;58(2):326–75. Available from: https://doi.org/10.1080/15583724.2017.1364765
142. Desai KGH, Jin Park H. Recent developments in microencapsulation of food ingredients. *Drying Technol.* 2005;*23*(7):1361–94.
143. Loison P, Loison P. Development of a smart coating based on hollow nanoparticles for corrosion detection and protection to cite this version: HAL Id: tel-02475624 développement d ' un revêtement intelligent pour la détection et la protection contre la corrosion. 2020;
144. Li J, Hitchcock AP, Stöver HDH, Shirley I. A new approach to studying microcapsule wall growth mechanisms. *Macromolecules*. 2009;42(7):2428–32.
145. Li G, Feng Y, Gao P, Li X. Preparation of mono-dispersed polyurea-urea formaldehyde double layered microcapsules. *Polym Bull.* 2008;60(5):725–31.
146. Maia F, Yasakau KA, Carneiro J, Kallip S, Tedim J, Henriques T, et al. Corrosion protection of AA2024 by sol-gel coatings modified with MBT-loaded polyurea microcapsules. *Chem Eng J* [Internet]. 2016;283:1108–17. Available from: http://dx.doi.org/10.1016/j.cej.2015.07.087
147. Huang M, Yang J. Salt spray and EIS studies on HDI microcapsule-based self-healing anticorrosive coatings. *Prog Org Coatings* [Internet]. 2014;77(1):168–75. Available from: http://dx.doi.org/10.1016/j.porgcoat.2013.09.002
148. Yi H, Yang Y, Gu X, Huang J, Wang C. Multilayer composite microcapsules synthesized by Pickering emulsion templates and their application in self-healing coating. *J Mater Chem A*. 2015;3(26):13749–57.
149. Galvão TLP, Sousa I, Wilhelm M, Carneiro J, Opršal J, Kukačková H, et al. Improving the functionality and performance of AA2024 corrosion sensing coatings with nanocontainers. *Chem Eng J*. 2018;341(November 2017):526–38.
150. Fickert J, Rupper P, Graf R, Landfester K, Crespy D. Design and characterization of functionalized silica nanocontainers for self-healing materials. *J Mater Chem*. 2012;22(5):2286–91.
151. Liang Y, Guo M, Fan C, Dong H, Ding G, Zhang W, et al. Development of novel urease-responsive pendimethalin microcapsules using silica-IPTS-PEI as controlled release carrier materials. *ACS Sustain Chem Eng*. 2017;5(6):4802–10.
152. Lv LP, Landfester K, Crespy D. Stimuli-selective delivery of two payloads from dual responsive nanocontainers. *Chem Mater*. 2014;26(11):3351–3.
153. Wu G, An J, Sun D, Tang X, Xiang Y, Yang J. Robust microcapsules with polyurea/silica hybrid shell for one-part self-healing anticorrosion coatings. *J Mater Chem A*. 2014;2(30):11614–20.
154. Liang Y, Wang MD, Wang C, Feng J, Li JS, Wang LJ, et al. Facile synthesis of smart nanocontainers as key components for construction of self-healing coating with superhydrophobic surfaces. *Nanoscale Res Lett* [Internet]. 2016;11(1). Available from: http://dx.doi.org/10.1186/s11671-016-1444-3
155. Crespy D, Stark M, Hoffmann-Richter C, Ziener U, Landfester K. Polymeric nanoreactors for hydrophilic reagents synthesized by interfacial polycondensation on miniemulsion droplets. *Macromolecules*. 2007;40(9):3122–35.
156. Kakaroglou A, Domini M, De Graeve I. Encapsulation and incorporation of sodium molybdate in polyurethane coatings and study of its corrosion inhibition on mild steel. *Surf Coatings Technol* [Internet]. 2016;303(Part B):330–41. Available from: http://dx.doi.org/10.1016/j.surfcoat.2016.02.007
157. Vimalanandan A, Lv LP, Tran TH, Landfester K, Crespy D, Rohwerder M. Redox-responsive self-healing for corrosion protection. *Adv Mater*. 2013;25(48):6980–4.
158. Gucht J Van Der, Spruijt E, Lemmers M, Stuart MAC. Polyelectrolyte complexes: bulk phases and colloidal systems. *J Colloid Interface Sci* [Internet]. 2011;361(2):407–22. Available from: http://dx.doi.org/10.1016/j.jcis.2011.05.080
159. Dobrynin AV, Rubinstein M. Counterion condensation and phase separation in solutions of hydrophobic polyelectrolytes. *Macromolecules*. 2001;34(6):1964–72.

160. Sing CE, Perry SL. Recent progress in the science of complex coacervation. *Soft Matter*. 2020;16(12):2885–914.
161. Böger BR, Acre LB, Viegas MC, Kurozawa LE, Benassi MT. Roasted coffee oil microencapsulation by spray drying and complex coacervation techniques: characteristics of the particles and sensory effect. *Innov Food Sci Emerg Technol*. 2021;72(February).
162. Butstraen C, Salaün F. Preparation of microcapsules by complex coacervation of gum Arabic and chitosan. *Carbohydr Polym* [Internet]. 2014;99:608–16. Available from: http://dx.doi.org/10.1016/j.carbpol.2013.09.006
163. Tian Q, Zhou W, Cai Q, Pan X, Ma G, Lian G. In situ complex coacervation supported by self-coated polydopamine interlayer on uniform-sized essential oils droplet. *J Colloid Interface Sci* [Internet]. 2022;623:1027–38. Available from: https://doi.org/10.1016/j.jcis.2022.05.072
164. Pfleging W, Kohler R, Südmeyer I, Rohde M. Laser micro and nano processing of metals, ceramics, and polymers. *Laser-Assisted Fabrication of Materials*. 2013;1:319–74.
165. Bokov D, Jalil AT, Chupradit S, Suksatan W, Ansari MJ, Shewael IH, et al. Nanomaterial by sol-gel method: synthesis and application. *Adv Mat Sci Eng*. 2021;2021: 1–21.
166. Ardekani-zadeh AH, Hosseini SF. Electrospun essential oil-doped chitosan / poly (ε -caprolactone) hybrid nanofibrous mats for antimicrobial food biopackaging exploits. *Carbohydr Polym* [Internet]. 2019;223(March):115108. Available from: https://doi.org/10.1016/j.carbpol.2019.115108
167. Huang ZM, Zhang YZ, Kotaki M, Ramakrishna S. A review on polymer nanofibers by electrospinning and their applications in nanocomposites. *Compos Sci Technol*. 2003;63(15):2223–53.
168. Rostamabadi H, Assadpour E, Tabarestani HS, Falsafi SR. Electrospinning approach for nanoencapsulation of bioactive compounds; recent advances and innovations. *Trends Food Sci Technol* [Internet]. 2020;100(April):190–209. Available from: https://doi.org/10.1016/j.tifs.2020.04.012
169. Lamarra J, Calienni MN, Rivero S, Pinotti A. Electrospun nano fibers of poly (vinyl alcohol) and chitosan-based emulsions functionalized with cabreuva essential oil. *Int J Biol Macromol* [Internet]. 2020;160:307–18. Available from: https://doi.org/10.1016/j.ijbiomac.2020.05.096
170. Raja PB, Assad MA, Ismail M. Inhibitor-encapsulated smart nanocontainers for the controlled release of corrosion inhibitors [Internet]. *Corrosion Protection at the Nanoscale*. Elsevier Inc.; 2020. 91–105 p. Available from: http://dx.doi.org/10.1016/B978-0-12-819359-4.00006-4
171. Shchukin DG, Zheludkevich M, Yasakau K, Lamaka S, Ferreira MGS, Möhwald H. Layer-by-layer assembled nanocontainers for self-healing corrosion protection. *Adv Mater*. 2006;18(13):1672–8.
172. Kessler MR, White SR. Self-activated healing of delamination damage in woven composites. *Compos Part A Appl Sci Manuf*. 2001;32(5):683–99.
173. Drozińska E, Kanclerz A, Kurek MA. Microencapsulation of sea buckthorn oil with β-glucan from barley as coating material. *Int J Biol Macromol* [Internet]. 2019;131:1014–20. Available from: https://doi.org/10.1016/j.ijbiomac.2019.03.150
174. Song X, Yu Y, Yang G, Jiang A, Ruan Y, Fan S. One-step emulsification for controllable preparation of ethyl cellulose microcapsules and their sustained release performance. *Colloids Surfaces B Biointerfaces* [Internet]. 2022;216(May):112560. Available from: https://doi.org/10.1016/j.colsurfb.2022.112560
175. Ouarga A, Lebaz N, Tarhini M, Noukrati H, Barroug A, Elaissari A, et al. Towards smart self-healing coatings: advances in micro/nano-encapsulation processes as carriers for anti-corrosion coatings development. *J Mol Liq* [Internet]. 2022;354:118862. Available from: https://doi.org/10.1016/j.molliq.2022.118862
176. Boh B, Sumiga B. Microencapsulation technology and its applications in building construction materials. *RMZ – Mater Geoenvironment*. 2008;55(3):329–44.
177. Nelson G. Application of microencapsulation in textiles. *Int J Pharm*. 2002;242(1–2):55–62.
178. Zhang H, Zhang X, Bao C, Li X, Sun D, Duan F, et al. Direct microencapsulation of pure polyamine by integrating micro fluidic emulsion and interfacial polymerization for practical self-healing materials. *J Mat Chem*. 2018;24092–9.
179. Zhang H, Zhang X, Chong YB, Peng J, Fang X, Yan Z, Liu B, Yang J. Shell formation mechanism for direct microencapsulation of nonequilibrium pure polyamine droplet. *J Phys Chem C*. 2019; 123(36):22413–23.
180. Yang Z, Fang X, Peng J, Cao X, Liao Z, Yan Z. Versatility of the microencapsulation technique via integrating micro fluidic T-Junction and interfacial polymerization in encapsulating different

polyamines. *Colloids Surfaces A* [Internet]. 2020;604(April):125097. Available from: https://doi.org/10.1016/j.colsurfa.2020.125097

181. Zotiadis C, Patrikalos I, Loukaidou V, Korres DM, Karantonis A, Vouyiouka S. Self-healing coatings based on poly(urea-formaldehyde) microcapsules: in situ polymerization, capsule properties and application. *Prog Org Coatings*. 2021;161(May):106475.
182. Kosarli M, Bekas DG, Tsirka K, Baltzis D, Vaimakis-Tsogkas D, Orfanidis S, et al. Microcapsule-based self-healing materials: healing efficiency and toughness reduction vs. capsule size. *Compos Part B Eng* [Internet]. 2019;171(April):78–86. Available from: https://doi.org/10.1016/j.compositesb.2019.04.030
183. Gaudin F, Sintes-Zydowicz N. Correlation between the polymerization kinetics and the chemical structure of poly(urethane-urea) nanocapsule membrane obtained by interfacial step polymerization in miniemulsion. *Colloids Surfaces A Physicochem Eng Asp* [Internet]. 2012;415:328–42. Available from: http://dx.doi.org/10.1016/j.colsurfa.2012.09.040
184. Yuan L, Gu A, Liang G. Preparation and properties of poly(urea-formaldehyde) microcapsules filled with epoxy resins. *Mater Chem Phys*. 2008;110(2–3):417–25.
185. Costa C, Santos AF, Fortuny M, Araújo PHH, Sayer C. Kinetic advantages of using microwaves in the emulsion polymerization of MMA. *Mater Sci Eng C* [Internet]. 2009;29(2):415–9. Available from: http://dx.doi.org/10.1016/j.msec.2008.08.013
186. Herrmann C, Crespy D, Landfester K. Synthesis of hydrophilic polyurethane particles in non-aqueous inverse miniemulsions. *Colloid Polym Sci*. 2011;289(10):1111–7.
187. Tzavidi S, Zotiadis C, Porfyris A, Korres DM, Vouyiouka S. Epoxy loaded poly(urea-formaldehyde) microcapsules via in situ polymerization designated for self-healing coatings. *J Appl Polym Sci*. 2020;137(43):1–11.
188. Udoh II, Shi H, Liu F, Han EH. Microcontainer-based waterborne epoxy coatings for AA2024-T3: effect of nature and number of polyelectrolyte multilayers on active protection performance. *Mater Chem Phys* [Internet]. 2020;241(July 2019):122404. Available from: https://doi.org/10.1016/j.matchemphys.2019.122404
189. Grundmeier G, Schmidt W, Stratmann M. Corrosion protection by organic coatings: electrochemical mechanism and novel methods of investigation. *Electrochim Acta*. 2000;45(15–16):2515–33.
190. Sørensen PA, Kiil S, Dam-Johansen K, Weinell CE. Anticorrosive coatings: a review. *J Coatings Technol Res*. 2009;6(2):135–76.
191. Ye Z, Zhang P, Zhang J, Deng L, Zhang J, Lin C, et al. Novel dual-functional coating with underwater self-healing and anti-protein-fouling properties by combining two kinds of microcapsules and a zwitterionic copolymer. *Prog Org Coatings*. 2019;127(January 2018):211–21.
192. Mirabedini SM, Farnood RR, Esfandeh M, Zareanshahraki F, Rajabi P. Nanocomposite coatings comprising APS-treated linseed oil-embedded polyurea-formaldehyde microcapsules and nanoclay, part 2: self-healing and corrosion resistance properties. *Prog Org Coatings* [Internet]. 2020;142(December 2019):105592. Available from: https://doi.org/10.1016/j.porgcoat.2020.105592
193. Moore JS, Kessler MR, Sottos NR, Brown EN, Geubelle PH, Sriram SR, et al. correction: autonomic healing of polymer composites. *Nature*. 2011;415(6873):794–7.
194. Parsaee S, Mirabedini SM, Farnood R, Alizadegan F. Development of self-healing coatings based on urea-formaldehyde/polyurethane microcapsules containing epoxy resin. *J Appl Polym Sci*. 2020;137(41):1–15.
195. Cho SH, White SR, Braun P V. Self-healing polymer coatings. *Adv Mater*. 2009;21(6):645–9.
196. Zhang C, Wang H, Zhou Q. Preparation and characterization of microcapsules based self-healing coatings containing epoxy ester as healing agent. *Prog Org Coatings* [Internet]. 2018;125(September):403–10. Available from: https://doi.org/10.1016/j.porgcoat.2018.09.028
197. da Cunha ABM, Leal DA, Santos LRL, Riegel-Vidotti IC, Marino CEB. pH-sensitive microcapsules based on biopolymers for active corrosion protection of carbon steel at different pH. *Surf Coatings Technol* [Internet]. 2020;402(May):126338. Available from: https://doi.org/10.1016/j.surfcoat.2020.126338
198. Wang L, Li SN, Fu JJ. Self-healing anti-corrosion coatings based on micron-nano containers with different structural morphologies. *Prog Org Coatings* [Internet]. 2023;175(December 2022):107381. Available from: https://doi.org/10.1016/j.porgcoat.2022.107381

199. Huang Y, Zhao C, Li Y, Wang C, Shen T, Cheng D, et al. Development of self-healing sol-gel anticorrosion coating with pH-responsive 1H-benzotriazole-inbuilt zeolitic imidazolate framework decorated with silica shell. *Surf Coatings Technol* [Internet]. 2023;466(February):129622. Available from: https://doi.org/10.1016/j.surfcoat.2023.129622
200. Rahimi A, Amiri S. Anticorrosion hybrid nanocomposite coatings with encapsulated organic corrosion inhibitors. *J Coatings Technol Res*. 2015;12(3):587–93.
201. Chen Z, Scharnagl N, Zheludkevich ML, Ying H, Yang W. Micro/nanocontainer-based intelligent coatings: synthesis, performance and applications — a review. *Chem Eng J*. 2023 Jan;451:138582.
202. McCafferty E. Relationship between the isoelectric point (pHpzc) and the potential of zero charge (Epzc) for passive metals. *Electrochim Acta*, 55(5):1630–7.
203. Cai H, Wang P, Zhang D. Smart anticorrosion coating based on stimuli-responsive micro/nanocontainer: a review. *J Oceanol Limnol*. 2020;38(4):1045–63.
204. Xiao H, Wang Y, Gu L, Feng Z, Lei B, Zhu L, et al. Smart sensing coatings for early warning of degradations: a review. *Prog Org Coatings* [Internet]. 2023;177(January):107418. Available from: https://doi.org/10.1016/j.porgcoat.2023.107418
205. Abu-Thabit NY, Hamdy AS. Stimuli-responsive polyelectrolyte multilayers for fabrication of self-healing coatings — a review. *Surf Coatings Technol* [Internet]. 2016;303(Part B):406–24. Available from: http://dx.doi.org/10.1016/j.surfcoat.2015.11.020
206. Chen T, Fu J. An intelligent anticorrosion coating based on pH-responsive supramolecular nanocontainers. *Nanotechnology*. 2012;23(50):505705.
207. Jiang S, Lv L, Li Q, Wang J, Landfester K, Crespy D. Tailoring nanoarchitectonics to control the release profile of payloads. *Nanoscale*. 2016;8(22):11511–7.
208. Syed JA, Tang S, Meng X. Intelligent saline enabled self-healing of multilayer coatings and its optimization to achieve redox catalytically provoked anti-corrosion ability. *Appl Surf Sci* [Internet]. 2016;383:177–90. Available from: http://dx.doi.org/10.1016/j.apsusc.2016.04.178
209. Pirhady Tavandashti N, Ghorbani M, Shojaei A, Gonzalez-Garcia Y, Terryn H, Mol JMC. PH responsive Ce(III) loaded polyaniline nanofibers for self-healing corrosion protection of AA2024-T3. *Prog Org Coatings* [Internet]. 2016;99:197–209. Available from: http://dx.doi.org/10.1016/j.porgcoat.2016.04.046
210. Lv LP, Zhao Y, Vilbrandt N, Gallei M, Vimalanandan A, Rohwerder M, et al. Redox responsive release of hydrophobic self-healing agents from polyaniline capsules. *J Am Chem Soc*. 2013;135(38):14198–205.
211. Chen S, Huang Z, Yuan M, Huang G, Guo H, Meng G, et al. Trigger and response mechanisms for controlled release of corrosion inhibitors from micro/nanocontainers interpreted using endogenous and exogenous stimuli: a review. *J Mater Sci Technol*. 2022;125:67–80.

3 Encapsulation of Metal/ Metal Oxide Nanoparticles as Corrosion Inhibitors

Shobhana Sharma, Ankit Sharma, Sushil Kumar Sharma, and Abhinay Thakur

3.1 INTRODUCTION

Corrosion is an electrochemical process in which metal converts into a chemically stable oxide [1]. The corrosion process brings the gradual deterioration of metal through the chemical reaction of a metallic surface with the environment [2]. Prevention methods to control corrosion are dedicated to corrosion engineering. Corrosion engineering applies technical and scientific skills to prevent metallic materials, structures, devices, and systems from destructive corrosion processes [3–4]. Thus, this technology prefers wear resistance which possesses products of high performance. The protective, relatively economical, and environmentally compatible anticorrosion coating is the simplest and easiest way to prevent corrosion [5–6]. Coating protects metallic components of various industries from degradation against moisture, oxidation, salt spray, chemicals, etc., in a drastic change of environmental conditions [7–8]. Coating acts as a barrier between a metallic surface and a corrosive environment, inhibiting contact of chemicals with the surface and protecting it from corrosion [9]. Thus, the anticorrosive nature of coating enhances the lifespan of metallic components and provides financial benefits to industries [10–11]. This anticorrosive coating used for corrosion protection is of three types: metallic, organic, and inorganic.

To provide dual protective modes of action and enhance their anticorrosion property, this coating can be used alone or in combination with other inhibitors. Recent developments in the thrust area of research incorporate hybrid and smart coating in the field of corrosion [12]. Nanotechnology provides a driving force for the development of smart hybrid coating material. Smart coating senses their environment and responds appropriately to the stimulus [13–14]. The coating composition design alters at the molecular level, changing the morphology to a nanometer scale [15]. These changes in the assembly of coating material at the sequential macromolecular level provide exceptional properties to coating materials [16]. These unusual properties of coating material influenced scientific research and applications in the field of industries [17]. The progress in innovative particle systems like nano-sol-gel material, liquid crystals, conductive polymers, and polyelectrolytes leads to the development of smart coating material [18]. Thus, altering the formulation of smart coating materials improved their properties, longevity, and durability. The various functional properties and enhanced performance possess smart coatings due to the presence of metal/metal oxide/ polymer composite nanoparticles [19]. The hybrid nanoparticles design multifunctional coatings, which include the remarkable properties of nanoparticles. Smart coating finds application in medicine, transport, defense, textile, electronics, construction, and other industries to protect against corrosion [20]. Factors like preparatory methods, the presence of functional groups and reactive components, and their applications can classify smart coating materials. Microcapsule healing, self-healing, self-cleaning, and anticorrosion coatings are typical examples of certain smart coatings

DOI: 10.1201/9781032677255-3

[21]. Smart coatings perform multiple tasks to protect metallic components from corrosive environments like healing, sensing, and protection [22]. This research article helps to learn about the advancement in smart coatings with their versatile uses and unique properties. The present chapter highlights the use of metal, metal oxides, and composites in developing smart coating material and improving its efficiency toward anticorrosion.

3.2 METAL/METAL OXIDE NANOPARTICLE AS CORROSION INHIBITOR

In the present scenario, nanotechnology is a fast-growing field with its application in research. Nanoparticles having a size range of 1–100 nm have found various approaches in different areas like environment [23], industries [24–25], agriculture [26], biomedical [27], and food [28–30] due to the presence of unique and superior physical and chemical properties [31–32]. The most crucial application of nanoparticles in the industrial field is having a tremendous and specific characteristic to protect the metal from corrosion in different environments [33–34]. Various techniques are used to reduce the corrosion rate in environmental modifications, such as coating on the metal surface, corrosion inhibitors, and changes in pH and potential by cathodic or anodic reaction [35]. The prominent option is to use inhibitors to protect the metal from corrosion. These inhibitors make a protective film on the metal surface and help to control the corrosion rate. Many organic and inorganic inhibitors are employed to reduce corrosion in metal industries. These inhibitors are very expensive and toxic to the environment and human beings, so their use is limited. Nanomaterials and their additives are suitable corrosion inhibitors due to their excellent specific-to-volume proportion concerning conventional macroscopic materials [36–37]. Nano compounds stop the reaction of the surface and control the corrosion rate by blocking active sites of metal surfaces and also provide hardness, durability, optical properties, and straight and thermal stability [38]. Nanomaterials are eco-friendly and degradable to the environment.

Metal and its oxide nanoparticles are excellent and effective for corrosion control. Corrosion inhibitors can easily be applied on metallic surfaces through the adsorption of metal nanoparticles and their oxides on metal surfaces. Many types of research in the literature offer corrosion inhibition phenomenon. Zhang et al. used Au nanoparticle and n-hexyl thiol mixed films to protect iron against corrosion in 0.5M H_2SO_4 solution [39]. Electrochemical impedance spectroscopy (EIS) and polarization curve methods indicate that self-assembled mixed films form on the iron surface and prevent it from corrosion. Shao et al. incorporated titania (TiO_2) nanoparticles in a nickel matrix and studied the anticorrosion properties of nanocomposite coating in a NaCl medium [40]. Their result showed that nanocomposite coating improves corrosion resistance and mechanical properties such as hardness and wear resistance performance.

Atta et al. used silver nanoparticles as a corrosion inhibitor for carbon steel in 1M HCl [41]. EIS study indicated that silver nanoparticles drastically reduced the corrosion rate of carbon steel and exhibited good performance as well as barrier properties against corrosion in 1M HCl solution. The maximum inhibition efficiency of Ag nanoparticles is observed at 400 ppm. Atta et al. investigated the poly (ethylene glycol) thiol-stabilized silver nanoparticles as corrosion inhibitors for carbon steel alloy in 1M HCl [42]. A polarization curve study reveals that Ag-poly (ethylene glycol) thiol behaves as a mixed-type inhibitor. The polarization measurements agree with EIS for increasing the efficiency of the inhibitor with inhibitor concentration. Palanisamy et al. reported the corrosion inhibition tendency of olive oil stabilized iron oxide nanoparticles (IONPs) on mild steel in 0.5M HCl solution [43]. The weight loss method is employed to study corrosion rate and inhibition efficiency. The olive oil stabilized IONPs show higher inhibition efficiency than paints-coated mild steel plates. The corrosion protection of mild steel by superhydrophobic silicon dioxide (SiO_2)/DTMS hybrid film was demonstrated by Wu et al. [44]. EIS, static contact angle, and iron dissolution measurement indicated that the superhydrophobic film significantly enhances the corrosion resistance of mild steel in NaCl solution. The corrosion behavior of the polyaniline (PANI), PANI/Zn, PANI/Ag, and PANI/TiO_2 nanocomposite films on Al1050 electrode was studied by Ates et al. in a 3.5% NaCl solution [45]. It was found that PANI/Ag nanocomposites film showed a higher inhibition efficiency of 97.4% than PANI (91.4%), PANI/TiO_2 (91.91%), and PANI/Zn (92.52%). The inhibition effect of magnetic nanofluid (MNF) on carbon steel in an acidic medium was investigated by Parekh et al. using scanning electron microscopy (SEM), gravimetric, and potentiodynamic measurements [46]. The inhibition efficiency increases as the concentration of MNF increases until it reaches an optimum concentration (51.7 mM). The observed inhibitory action is caused by the adsorption of MNF on the steel surface and the forming of a protection layer between the metal and corrosive medium. Hosseini et al. reported the corrosion resistance property of polypyrrole containing TiO_2, ZnO, and Mn_2O_3 nanoparticles on aluminum metal in an acidic medium [47]. TiO_2 nanoparticles on aluminum samples show a greater corrosion inhibition performance than ZnO and Mn_2O_3 nanoparticles. The inhibition effect of polydopamine (PDA) on mild steel in 1M HCl was investigated by Yang et al. using various methods such as weight loss, potentiodynamic polarization, EIS, SEM, and atomic force microscopy (AFM) [48]. Potentiodynamic polarization studies showed that corrosion inhibitors used a mixed-type inhibitor. The anticorrosion property of poly (aniline copolymer with N-ethylamine) containing ZnO nanoparticles on mild steel was studied by Mobin et al. in 0.1M HCl and 5% NaCl solution [49]. The nanocomposite coating showed excellent protection against mild steel corrosion, showing 89–96% inhibition efficiency after 30 days of immersion. The presence of scribed marks on the polymer coatings does not significantly affect the performance of the coating, indicating the self-passivating nature of copolymer and its nanocomposite. Hasnidawani et al. investigated the anticorrosion properties of ZnO nanoparticles in epoxy resin on mild steel in different media (e.g., freshwater, HCl solution, NaCl solution, and NaOH solution) [50]. The corrosion rate is calculated through the weight loss method. The result indicates the order of corrosion rate degradation in the following media: HCl>NaCl>NaOH>Fresh water trends. Mahvidi et al. evaluated the corrosion inhibition performance of zinc ferrite nanoparticles with different particle sizes on mild steel in a 3.5% NaCl solution [51]. EIS, polarization curve, and SEM were employed to evaluate the effect of zinc ferrite nanoparticles on steel corrosion. Kouhi et al. reported an atomic absorption spectroscopy technique to study the anticorrosion property of TiO_2 nanoparticles coating on blades of centrifugal pumps [52]. The study found that by increasing the coating thickness, the corrosion rate decreases and inhibition efficiency increases. The corrosion-preventing efficiency of titania nanocoating was 39.1%, 67.8%, and 73.6% concerning different thicknesses of the inhibitor coating in pump blades. Mahmood et al. developed copper corrosion resistance using an anodization technique [53]. Copper possesses a coating of nano copper oxide, which increased its corrosion inhibition efficiency to 86.2% and 74.5% in a test solution containing 3.5% NaCl and 2 mg/L NH_3, respectively. Rajendran et al. analyzed the performance of anticorrosion property of the combination of poly(methyl methacrylate), resin, and ZnO nanoparticles on mild steel in 3.5% NaCl solution at different exposure times [54]. The potentiodynamic study reveals that polymer combination with nanoparticles acts as a mixed-type corrosion inhibition with high inhibition efficiency due to inhibitor adsorption on metallic surfaces. The corrosion behavior of carbon steel in 0.5 M HCl, 0.5 M H_3PO_4, and 0.5 M H_2SO_4 was examined in the presence of a 20 ppm Mn_2O_3 nanoparticle by Najjar et al. [55]. In this research, the corrosion inhibition study was carried out by potentiodynamic polarization curve and electrochemical impedance spectroscopic method, and Mn_2O_3 nanoparticles enhanced the corrosion resistance of carbon steel in acidic solutions. The efficiency increased with decreasing pH at which synthesis of Mn_2O_3 occurs. Khamis et al. studied the corrosion inhibition effects of magnetite nanoparticles in polyvinyl pyrrolidone on carbon steel in 1M HCl solution [56]. The polarization curve and electrochemical impedance study showed that the inhibition efficiency increases with the concentration of magnetite nanoparticles. Armentia et al. observed that stainless steel plates coated with a silane-based coating containing zirconia (ZrO_2) nanoparticles show low wet ability, high wear resistance, and aesthetical properties [57]. The friction coefficient of the coating decreases with the addition of nanoparticles. The anticorrosion property of acyl coating containing silica nanoparticles on mild steel was studied by Rajkumar et al. in 1% HCl [58]. The potentiodynamic polarization curve and EIS examined the anti-corrosive behavior. The inhibition efficiency of SiO_2/acryl coating was 45.38% and 77.75% in 1% HCl and 3.5% NaCl solution, respectively.

Khodair et al. investigated the corrosion protection of mild steel via epoxy resin and epoxy resin containing magnesium oxide (MgO) nanoparticles in acidic and saline solutions [59]. In an acidic medium, the epoxy coating was very powerful and showed 97% inhibition efficiency. In saline solution, the coating performance was poor (29.8%). Adding MgO nanoparticles in epoxy coating increased the inhibition efficiency from 29.8% to 93.7% in saline solution. Moreover, green synthesis of Cr_2O_3 nanoparticles was reported by Sharma et al. using *Cannabis sativa* leaves extract [60]. The corrosion inhibition study of Cr_2O_3 nanoparticles on mild steel at HCl, HNO_3, and NH_3 solutions revealed that Cr_2O_3 shows the maximum inhibition performance (88.89 at 303K) in NH_3 solution. Al-Senani et al. synthesized CoO/Co_3O_4 nanoparticles from egg white [61]. CoO/Co_3O_4 nanoparticles' anticorrosion effectiveness was examined on low-carbon steel in 1M HCl. Electrochemical measurements showed that CoO/Co_3O_4 nanoparticles act as mixed inhibitors for the anodic and cathodic reactions. The efficacy to inhibit the corrosion was 93% at 80 ppm and 298K. Jabri et al. developed PANI and TiO_2 nanocomposites to assess their anticorrosion behavior on mild steel [62]. Potentiodynamic, wet/dry, and atmospheric tests investigated the corrosion behavior of coated and uncoated mild steel specimens. Polyaniline-TiO_2 composite thin film coating with minimum thickness controls the corrosion of mild steel with good film stability, high durability, cost-effectiveness, and environmentally friendly approach. Table 3.1 presents various metal and metal oxide nanomaterials as corrosion inhibitors with their efficiency.

3.3 NANOCOMPOSITES AS CORROSION INHIBITORS

Nanomaterials have dimensions between 1 nm and 100 nm. These are versatile materials having enormous applications in various fields, such as health and medicine [63–65], agriculture [66], engineering [67–70], food industry [71], etc. Their versatility in different applications is attributed to their high mechanical strength and less weight, fine grain size, and increased grain boundary volume fraction. Their unique property guarding metal surfaces against corrosion in multiple atmospheric conditions makes them the best candidate for anti-corrosion coating material over metal surfaces. About 4% of the gross national product is consumed in corrosion, making corrosion the biggest concerning issue around the globe. Conditions responsible for the corrosion include temperature, air, humidity, pH variation, and electrolyte concentration on the metal surface [72–73]. The reactions are carried out in controlled conditions through many processes, of which the nanocomposite material coating on the metal surface is an efficient way to inhibit the corrosion process [74]. The unique properties compared to macroscopic materials and nanomaterials are environmentally friendly.

3.3.1 Nanocomposite Coating

Nowadays, the use of nanocomposite materials as a coating on metal surfaces for the corrosion inhibition process is gaining momentum. They mainly comprise heterogeneous films having one phase either isolated or embedded in the second phase or maybe having diverse phases. The phase's configuration can be done in such a way that either both phases are crystalline or another possibility is that one phase is amorphous and the other is crystalline. The dimensions of phases may be in nanoscale, or one of the phases will be of nano dimensions. Nanocoating has vast measurements of uses, from optical devices, magnetic storage tools, biomedical areas, high-speed machining tools, and many more. The presence of specific characteristics of nanomaterial,

TABLE 3.1
Showing different metal/metal oxide nanomaterials used as corrosion inhibitors with their efficacy

S.No.	Metal/Metal Oxide Nanoparticle	Preparatory Methods	Techniques Used for Characterization	Techniques Used for Corrosion Measurements	Maximum Inhibition Efficiency (%)	References
1.	Au nanoparticle with n-hexyl thiol mixed film	Thermal process	TEM	PDP EDX EIS SEM	97.1%	39
2.	TiO_2 in Ni matrix	—	SEM	SEM	----	40
3.	Ag nanoparticle	Reduction of Ag_2O by $NaBH_4$ in the presence of modified NRN-10	H¹NMR TEM UV-Vis spectroscopy	EIS	----	41
4.	Poly (ethylene glycol) thiol stabilized Ag nanoparticle	Sol-gel	H¹NMR FTIR TEM Uv-Vis spectroscopy DLS	PDP EIS	90.95% (1MHCl)	42
5.	Olive oil stabilized iron oxide nanoparticle	Co-precipitation	SEM	WL	80.88%	43
6.	SiO_2/DTMS	Electrochemical Deposition	UV-Vis spectroscopy SEM	CA EIS	----	44
7.	Zn, Ag, TiO_2 nanoparticles in polyaniline	Electrochemical deposition	CV UV-Vis spectroscopy FTIR ATR EDX	Tafel Extrapolation EIS	PANI/Ag 97.54%	45
8.	Magnetic nanofluid (MNF)	Co-precipitation	SEM EDX	WL PDP	90% (1500 ppm) (1M HCl), 74.4% (1500 ppm) (1M H_2SO_4)	46
9.	TiO_2, ZnO, and Mn_2O_3 nanoparticle in polypyrrole	Electropolymerization	XRD SEM TEM	PDP EIS	98.09 pPy/TiO_2 (acidic solution)	47
10.	Polydopamine (PDA) nanoparticle	Polymerization	FTIR SEM AFM	WL PDP EIS	86.42 (40 mol/L) (1M HCl)	48

11.	ZnO nanoparticles with poly (aniline-co-N-ethylaniline)	Chemical oxidative polymerization	FTIR XRD SEM	WL PDP EIS	94.00 (0.1 M HCl), 82.46% (5% NaCl) 96% (distilled water)	49
12.	Zno nanoparticle in epoxy resin	----	----	WL	----	50
13.	$ZnFe_2O_4$ nanoparticle	Sol-gel	TEM	PDP ICP-OES EIS SEM	----	51
14.	TiO_2	Sol-gel	SEM	SEM AAS	----	52
15.	CuO nanoparticle	Anodization	FESEM EFX XRD		86.2% in 3.5% NaCl and 74.5% in 2mg/L NH_3	53
16.	ZnO nanoparticles in PMMA and resin	Wet chemical	FTIR XRD SEM EDX Zeta potential	CA WL PDP CV	99% in 3.5%NaCl at 25^{0}C (At 10% PRZn con.)	54
17.	Mn_2O_3 nanoparticles	Co-precipitation	FTIR XRD TEM	PDP EIS	23.06–55.60% (20 ppm) (0.5M HCl)	55
18.	Magnetite nanoparticles and polyvinyl pyrrolidine	Co-precipitation	FTIR XRD TEM	PDP EIS	98.39% (1M HCl) (0.1% Con.) at 30^{0}C	56
19.	Zirconia (ZrO_2) nanoparticles in silane	Sol-gel	FTIR DLS XRF SEM	CA Wear resistance	----	57
20.	SiO_2 nanoparticle in acrylic resin	Using rice husk in the eco-friendly method	FTIR XRD SEM EDX	PDP EIS	77.75% in 3.5% NaCl and 45.38% HCl	58
21.	MgO nanoparticle in epoxy	Sol-gel	XRD SEM EDX	WL	97% (acidic solution) 93.7 (saline solution)	59

(continued)

TABLE 3.1 (Continued)
Showing different metal/metal oxide nanomaterials used as corrosion inhibitors with their efficacy

S.No.	Metal/Metal Oxide Nanoparticle	Preparatory Methods	Techniques Used for Characterization	Techniques Used for Corrosion Measurements	Maximum Inhibition Efficiency (%)	References
22.	Cr_2O_3	Green synthesis using *Cannabis sativa* leaves extract	UV-Vis spectroscopy FTIR DLS Zeta potential	WL	88.88 at 1000 ppm (at 303K)	60
23.	CoO/Co_3O_4 nanoparticle	Egg white	FTIR XRD EDX SEM	PDP EIS	93% at 80 ppm (at 298K)	61
24.	TiO_2 nanoparticle In PANI	Sol-gel	FTIR XRD EDX DLS SEM	PDP	----	62

[1] HNMR = proton nuclear magnetic resonance, TEM= transmission electron microscopy, UV-Vis spectroscopy = ultraviolet-visible spectroscopy, SEM = scanning electron microscopy, EIS = electrochemical impedance study, AAS=atomic absorption spectroscopy, CV = cyclic voltammetry, FTIR = Fourier transform infrared, ATR-IR = attenuated total reflectance-infrared, EDX = energy dispersive X-ray analysis, XRD = X-ray diffraction, CA = contact angle, WL = weight loss, PDP = potentiodynamic polarization, AFM = atomic force microscopy, DLS = dynamic light scattering, XRF = X-ray fluorescence, ICP-OES = inductively coupled plasma-optical emission spectrometry.

such as its electronic, mechanical, and optical properties, is the basis of its broader application. Although nanocoating has immense utilization in various fields, its utility as the nano dimension corrosion inhibitor is of great significance in metal corrosion inhibition [75–77]. The nanocoating has heterogeneous phases: one phase is the filler component as inorganic material, and the second is embedded in the host matrix. The reason for taking inorganic nanomaterials as fillers for nanocomposite coating is that these fillers show strong inhibition efficiency as there is sufficient interaction between filler particles and the host matrix, imparting the least porosity to the composite structure.

3.3.2 Polymer Matrix Nanocomposite

In polymer matrix nanocomposite, out of two phases, the matrix phase is polymer, while the filler phase has nanofiller particles. Chemical uniformity is missing in polymers comprised of multiple elemental systems. The filler of nanocomposite material has replaced the conventional fillers owing to their nano size and dispersion in the composite material matrix. This unique property of nanofillers augments the properties of the polymer. It makes them specific to be used for multiple applications, for which this nanocomposite material is one of the best candidates for inhibiting metal corrosion [78–79]. The present chapter mainly pertains to the various types of research conducted on metal corrosion inhibition by nanocomposite material. It holds vital information catering to the knowledge of industries, researchers, and students regarding corrosion inhibition by nanocomposite material.

Fekry et al. studied the nanocomposite film coating over 316L stainless steel implants in imitated body fluids [80]. The film made implants resistant to corrosion, imitated body fluid treated with 10^{-3} M ibandronate sodium, a medicine that is corrosion resistant at pH 7.4. The designed nanocomposite polymer comprises multi-walled carbon nanotubes, hydroxyapatite nanoparticles, titanium oxide nanoparticles, and chitosan. SEM techniques revealed a homogeneous scattering of components in the chitosan matrix. EDX analysis proved the elemental occurrence over the metal surface in their adequate amount, viz., carbon (47.4%), oxygen (26.02%), calcium (7.8%), nitrogen (2.53%), phosphorus (0.96%), and titanium (0.86%). The high percentage of carbon nanotube provided rupture strength to the composite material. At the same time, hydroxyapatite nanoparticles imparted corrosion resistance behavior to the material, and the titanium oxide nanoparticles imparted toughness to the hydroxyapatite nanoparticles. Chitosan enhanced the dispersion of the film over the metal alloy. Potentiodynamic polarization, cyclic voltammetry, and EIS techniques were employed for studying the corrosion behavior of the nanocomposite layer. Tafel extrapolation evaluated I_{corr} and E_{corr} of the TONP/HANP/CNT-CS. Nanofilm coating on the alloy, after emersion of the alloy in the imitated body fluid for seven days with and without 10^{-3} ibandronate sodium drug. Corrosion current density fells to 1.7nA/cm^2, showing less corrosion rate. The corrosion resistance efficiency of the nanofilm came to 99% with the presence of 10^{-3} ibandronate sodium drug; without the drug, it was about 91.6%. Further, Zhang et al. opted for chemical oxidative polymerization techniques for developing PANI nanofibers [81]. They also used the electroplating technique for synthesizing styrene block copolymer polystyrene-b-(ethylene-co-butene)-b-styrene (SEBS) and SEBI/PANI fibers. The fiber structure was deformed when subjected to a binary solvent system of tetrahydrofuran (THF)/toluene. The application of an electrostatic field at high voltage charged PANI molecules. The shifting of molecules toward the Taylor cone was noticed as THF is volatile and evaporates fast, so SEBI/PANI unique nanofibers were developed. Later, PANI nanofibers got piled up and accumulated at the SEBI's interface with dimensions of 40–50 nm diameter and 700 +/– 12 nm. The heat sustainability of PANI is more as compared to the SEBI's coating. Electrospun technique used for coating SEBI/PANI on stainless steel provides excellent corrosion inhibition efficiency, and the coating's tensile bearing strength is more than 300% and without compromising with the corrosion inhibition efficiency. The corrosion outcome was examined through Tafel plots for coating (SEBI/PANI) over 304 stainless steel I_{corr} of 1.943×10^{-6} A/cm^2, and an E_{corr} value of 0.490 V was noticed for the bare 304 stainless steel with the rate of corrosion V_{corr}, 2.7×10^{-2} mm/year; while when basic 304 stainless steel was spin-coated with SEBI nanocoating, the value of V_{corr} noticed was $V_{corr} = 2.1\times10^{-3}$ mm/year; and while it was spin-coated with SEBI/PANI layer, V_{corr} was 4×10^{-4} mm/year with high value of E_{corr} and lowest magnitude of I_{corr} and V_{corr}, which signifies that SEBI/PANI spin-coated film is an excellent candidate for corrosion inhibition.

Mohamed et al. employed bis-benzamidine-based polybenzoxazine composite as a coating on mild steel using salt spray and electrochemical techniques [82]. These composites possess epoxidized soya bean oil and bentonite, a nano clay with 3%– 5% by weight; the protection efficiency raised to 98.16%, which makes the composite film a good candidate for the corrosion inhibition process. Compared with coating filling epoxidized soya bean oil, the corroded area was more prominent when only polybenzoxazine was employed as a composite film component. The best results found PBZ/E-SBO (polybenzoxazine/epoxidized soya bean oil) weight ratio 1:2 by wt%. The reason for efficient corrosion inhibition efficiency is intensive cross-linking between the polybenzoxazine and epoxidized soya bean oil while curing, decreasing the permeability. The SEM/TEM techniques revealed excellent dispersion of nano clay. So the appropriate composition for the best corrosion inhibition efficiency was seen with PBZ/E-SBO matrix with 3% by weight of nano clay. The anticorrosion behavior of poly(4-methyl-5-vinyl thiazole) PVTZ coating on mild steel and the corrosion inhibition efficiency were studied by Pugazhenthi et al. [83]. Potentiodynamic polarization and EIS techniques were used to evaluate its anti-corrosion property on the mild steel in 3.5(w/v) NaCl as the immersion medium. The composite layer was synthesized from the PVTZ matrix by incorporating TiO_2 nanoparticles of size 25 nm. Field-emission scanning electron microscopy (FESEM) technique revealed that TiO_2 nanoparticles dispersed in the matrix. Kumar et al. developed Cu, Ni, and zinc metal oxide nanoparticles (CNZMO) of size around 50–80 nm using the sol-gel method and used these nanofillers to prepare the composite material with a PANI matrix [84]. Thermogravimetric analysis technique implied the thermal stability of composite material is at 670 °C. X-ray diffraction (XRD) and FESEM results revealed the crystalline nature of the nanofiller and their homogeneous distribution in the matrix. The anti-corrosive study of CNZMO-PANI nanocomposite coating was done on mild steel using 0.1N HCl as the immersion medium, with excellent results even at a concentration of 80 ppm. The poly ortho-anisidine (PoA) nanoparticles via oxidative chemical polymerization, using nano SiO_2 with wt% of 5%, 10%, and 15% w/w, were synthesized by Sharma et al. [85]. The specific interaction between silicon oxide nanoparticles (SNP) and PoAchain was revealed through Fourier transform infrared (FTIR) results. The homogeneous distribution of SNP in the PoA matrix was shown through SEM results. The peak at 2θ, 22.45°, in XRD results revealed the crystalline structure of SNPs; however, gibbous pattern was noticed at 2θ, 17° and 24°, corresponding to the PoA matrix. The approximate size range of PoA and SiO_2 was around 38 and 20 nm, respectively. A hybrid composite material was synthesized using PoA/SiO_2 and acrylic system as a common matrix to be coated over the steel surface, to study the anticorrosive properties using 3.5% NaCl solution as the immersion medium for mild steel. Nanocomposites with 5–15% SiO_2 particles showed a strong rise in E_{corr} values. Another anticorrosive polypyrrole (PPy)/Fe_2O_3, a nanocomposite conducting polymeric material over cold spin substrate, was synthesized by Jadhav et al. [86]. The open-circuit potential and EIS technique were used to evaluate the anti-corrosive property of the nanomaterial. ASTM B117 conditions were implied using a salt fog experiment with exposure of 40 days. A comparative study was conducted using metal oxide/PPy film over the metal surface by exposing the samples into the salt spray experiments and drawing a graph between R_c (corrosion resistance) and the exposure time (in days). The graph revealed the exceptionally high value of corrosion resistance for metal oxide/PPy film over the metal oxide film on the metal surface, which is mainly attributed to the interface bonding interaction of the ions in PPy and PPy's passive film. Saleh et al. employed the electrochemical polymerization technique to prepare N-terminal tetrahydropthalimic acid over the stainless steel [87]. The in situ polymer synthesis was characterized by AFM, SEM, and FTIR techniques. They incorporated nanographene and nano ZnO in the polymeric coating to augment its anticorrosive properties. The corrosive medium taken was 0.2 M HCl and stainless steel coated with nanofiller-reinforced composite material showed the increased value of E_{corr} (in accordance with Tafel plots), which reveals the anodic protective property of the polymeric membrane over the stainless steel. Vinothkumar et al. coated copper metal with composite material using electrochemical polymerization of p-3-amino-5-mercapto-1,2,4-trazole(p-AMT) and added graphene oxide (GO) to it [88]. The resultant composite was characterized by FTIR, X-ray photoelectron spectroscopy (XPS), energy-dispersive X-ray spectroscopy , and Raman spectroscopy. The bonding interaction between azole's nitrogen atom and graphene's electrophilic carbon atom was revealed by broadening of N-O frequency, the formation of metal sulfur connection was provided by infrared spectra, and homogeneous distribution of GO and p-AMT were revealed by SEM images. Corrosion studies were carried out in 3.3% NaCl solution with an immersion period of 3 days using EIS and potentiodynamic polarization techniques, which revealed excellent anticorrosive efficiency of 78.48% with AMT/GO composite coating over copper metal in 3.5% NaCl solution. The anticorrosive behavior of chitosan nanocomposite coating over the copper metal was studied by Bahari et al. [89]. The incorporated fillers in the chitosan matrix were silica nanoparticles and 2-mercapto-benzothiazole, which augmented the barrier protection property of chitosan film and enhanced the anticorrosive property of the coating, as revealed by EIS studies depicting 85% efficiency of coating owing to mass transition and dispersion phenomenon in the corrosion mechanism. Chitosan was self-interlinked by adding glutaraldehyde and tripoly phosphate and studies were conducted on the swelling ratios of interconnected chitosan nanocoating over the metal surface evaluated by the water quantity uptake. The results revealed that by incorporating a higher concentration of silica nanoparticles in the chitosan matrix the interconnection of chitosan is highly resisted due to the interaction of silica nanoparticles with the amino chains, making the interconnected chitosan to decrease its coating inhibition efficiency.

Harb et al. improved the corrosion inhibition efficiency of Ti6Al4V metal alloy whose corrosion durability was very low [90]. They developed poly(methyl methacrylate) SiO_2 composite film i.e., PMMA-SiO_2, by sol-gel method, wherein radical polymerization leads to PMMA interconnections with silica phase through 3-(trimethyloxysilyl) propyl methacrylate. The coating of 5 μm thickness being nonporous and crack was a good candidate for corrosion inhibition of metal surface with a magnitude of 26 GΩcm^2 as revealed through EIS results. The immersion period was 100 days, and the medium used was simulated body fluids (SBF). PMMA-silica coating's anticorrosive behavior was also revealed by the 90° value of phase angle, which is an eminent quantity needed for the quasi-ideal capacitor. In contrast, the attack of electrolyte over the metal surface can easily be ruled out, with 21 days immersion period of alloy in SBF which attributes the bioactivity of the composite to the formation of apatite layer, which could improvise in the phenomenon of osseointegration, whose main cause was the increase in the interface free energy. The PMMA-silica-coated Ti6Al4V alloy was a good candidate for its bioactivity and corrosion resistance. Gallegos-Melgar et al. demonstrated the electrodeposition of Al_2O_3-reinforced chitosan–sodium alginate coating on the metal surface [91]. SEM technique was employed to characterize the coating. SEM results were also useful in analyzing the anodic aluminum oxide coating for the presence of defects, cavities, pores, and any deformation in the composition under imitated sea atmosphere. Potentiodynamic polarization curves studied AA6063 bare and coated alloy, which was compared with a similar specimen immersed in the electrolyte with an immersion period of 30 days. The high polarization efficiency was noticed for the chitosan alginate films, compared with the chitosan films alone, as there was ionic cross-linking due to the formation of polyion complex) owing to intermixing of two polymers. The research work envisages the protection of aluminum alloy from corrosive chloride ions by imparting an anticorrosion-resistant film on the alloy. Chen et al. synthesized sandwich-like structural composition for corrosion inhibition by deposition of 8-hydroxyquinoline (8-HQ) at the interface of GO and then incorporation of PDA to improvise the resistivity for avoiding the entrance of corrosion species at the interface due to self-polymerization of dopamine in GO/8-HQ matrix increasing the complexity of the matrix [92]. FTIR, XPS, and SEM techniques were employed to study the internal structural aspect of the sandwich composition of GO/8-HQ/PDA. The corrosion study of GO/8-HQ/PDA over AZ31b alloy followed by immersion in NaCl solution for 140 days. The corrosion potential E_{corr} raised to –0.580V, with I_{corr} declining to 9.207×10^{-8} A/cm^2. However, on the 80th day of the experiment, there happened the discharge of 8HQ from GO/8-HQ/PDA which led to passivization on AZ31b alloy's interface, making the deficient points more efficient, as revealed by the falling value of E_{corr} after the 80th day. Better results were observed on reinforcing GO/8-HQ/PDA in an epoxy matrix coated over AZ31b alloy. The alloy immersed in NaCl solution of 3.5% wt strength, and the corrosion efficiency was noticed when the epoxy film was coated without GO/8-HQ/PDA component reinforcement, revealing that GO/8-HQ/PDA components obstruct the passage of corrosion-causing species. GO and GO/8-HQ are not good candidates for corrosion inhibition when used alone. GO has a rigid structure toward the electrolyte as compared to GO/8-HQ. PDA species attributes excellent hydrophobicity and imparts physical hindrances avoiding direct contact with NaCl when reinforced in an epoxy matrix of GO/8-HQ/PDA. GO/8-HQ/PDA imparts excellent anti-corrosive properties, not allowing cracks or pores in the coating surface, resisting the corrosion species migration. With its structural uniqueness and hydrophobic properties, GO/8-HQ/PDA sandwich structure imparts excellent anti-corrosive properties, making it a good candidate for corrosion inhibition at the commercial level. Aung et al. demonstrated the synthesis of novel hybrid nanocomposite coating material made of acrylated epoxidized jatropha oil (AEJO) resin as the matrix components with nano zinc oxide as filler for coating on mild steel surface, wherein NaCl solution was used as the immersion medium [93]. The nanocomposite coating design showed variation in the wt% of zinc oxide reinforcement in the AEJO matrix. FTIR and FESEM results revealed that the zinc oxide particles were homogeneously distributed in the AEJO matrix. The physical properties of the hybrid composite film improved due to the large surface area of inorganic filler, which minimized the prominent circular pores in the new coating. With a higher concentration of filler, uniformity in distribution is not maintained, which enhances corrosion efficiency. A contact angle meter was employed to evaluate the wetting ability of ZnO. The pure specimen contact angle was 85.2° +/– 2.01°, which enhanced to 95° to 104.4° +/– 2.01° with the incorporation of ZnO, as the contact angle was greater than 90° of the hybrid nanocomposite, which significantly proves that water permeates into the coating's surface efficiently. The research examines corrosion in nano ZnO with varying concentrations. The highest barrier protection property was with 1–3 wt% of ZnO, as revealed by the results of EIS and Tafel plots. The leading cause of corrosion efficiency was the reduction of electrolyte penetration in the coating material owing to the efficient reinforcement of the nanoparticles with the matrix components.

3.4 CONCLUSION

This study concludes the advancement of smart coating material with the encapsulation of nanostructured metal, metal oxides, and composite moiety in the formation of coating materials. Incorporating nanomaterials in coating material provides remarkable properties with an enhanced anticorrosive nature of the material in multiple folds. The continuous demand for high-performance smart coating material, which improves the product's lifespan and reduces its maintenance cost, acts as a driving force for progress in the field of smart coating technology. The most exciting feature is that smart coating materials cope with substrate and environmental alterations. The main challenge is the economy required for developing and adopting additional smart coating technologies in the markets of developing countries. Smart coating technologies are cost-prohibitive to apply in various fields, which decreases their market adoption rate. The advancement in smart coating material is continuous in laboratories, but its practical conversion in industries is rare due to a slower adoption rate.

3.5 FUTURE PERSPECTIVES

The multifunctional smart coating creates a new category of research that responds dynamically to any environmental change. In the present scenario, smart coatings are critical to all unmet requirements and gaps in various sectors. The fabrication of advanced smart coating materials is essential for hi-tech applications and the development and establishment of modern industries. The setup of modern industries requires the protection of metallic components from corrosion in different environments. Thus, efforts have been made to design ecofriendly, economical, efficient, and durable smart coating material that enhances the longevity of machines in market industries. Attempts have been made to improve the market adoption rate of smart coatings technology. If the adoption rate of smart coatings technology increases in the market, it will benefit industrialists and improve developing countries' economies.

REFERENCES

1. Zheng L, Luo S. Fabrication of a durable superhydrophobic surface with corrosion resistance on copper. *Int. J. Electrochem. Sci.* 2023, 18(4), 100093. https://doi.org/10.1016/j.ijoes.2023.100093
2. Hossen A, Mahmud R, Islam A. Minimization of corrosion in aquatic environment–a review. *Int J Hydro*. 2023, 7(1), 9–16.
3. Monfared V, Ramakrishna S, Nasajpour-Esfahani N, Toghraie D, Hekmatifar M, Rahmati S. Science and technology of additive manufacturing progress: processes, materials, and applications. *Met. Mater. Int.* 2023, 1–29. https://doi.org/10.1007/s12540-023-01467-x
4. Javaherdashti R, Ghanbarzadeh A. A short review of some important aspects of the science of corrosion. Corrosion policy decision making: science. *Eng. Manag. Econ.* 2022, 7–45. https://doi.org/10.1002/9781119764342.ch2
5. Low EJ, Yusoff HM, Batar N, Nor Azmi IN, Chia PW, Lam SS, Kan SY, Liew RK, Lee GE, Venkateswarlu K, Ridwan Zulkifli MF. The use of food additives as green and environmental-friendly anti-corrosion inhibitors for protection of metals and alloys: a review. *Environ. Sci. Pollut. Res.* 2023, 1–11. https://doi.org/10.1007/s11356-023-27823-3
6. Song W, Zhao X, Jin Z, Fan L, Ji X, Deng J, Duan J. Poly (vinyl alcohol) for multi-functionalized corrosion protection of metals: a review. *J. Clean. Prod.* 2023,136390. https://doi.org/10.1016/j.jclepro.2023.136390
7. Ulaeto SB, Ravi RP, Udoh II, Mathew GM, Rajan TP. Polymer-based coating for steel protection, highlighting metal–organic framework as functional actives: a review. *Corros. Mater. Degrad*e. 2023, 4(2), 284–316. https://doi.org/10.3390/cmd4020015
8. Sathish M, Radhika N, Saleh B. Current status, challenges, and future prospects of thin film coating techniques and coating structures. *J. Bio- Tribo-Corros.* 2023, 9(2), 35. https://doi.org/10.1007/s40735-023-00754-9
9. Farh HM, Seghier ME, Zayed T. A comprehensive review of corrosion protection and control techniques for metallic pipelines. *Eng. Fail. Anal.* 2022, 106885. https://doi.org/10.1016/j.engfailanal.2022.106885
10. Das S, Bezbarua P, Das S. *Sustainable Nanomaterial Coatings for Anticorrosion. Nanomaterials for Sustainable Tribology*. CRC Press, 2023, 203.
11. Asim N, Badiei M, Samsudin NA, Mohammad M, Razali H, Soltani S, Amin N. Application of graphene-based materials in developing sustainable infrastructure: an overview. *Compos. B. Eng.* 2022, 110188. https://doi.org/10.1016/j.compositesb.2022.110188

12. Kathavate VS, Deshpande PP. *Smart Coatings: Fundamentals, Developments, and Applications*. CRC Press; 2022.
13. Dumontel B, Conejo-Rodríguez V, Vallet-Regí M, Manzano M. Natural biopolymers as smart coating materials of mesoporous silica nanoparticles for drug delivery. *Pharmaceutics*. 2023, 15(2), 447. https://doi.org/10.3390/pharmaceutics15020447
14. Yadav K, Kumar A. Introduction of smart coatings in various directions. In *Antiviral and Antimicrobial Smart Coatings*. Elsevier; 2023, 219–238. https://doi.org/10.1016/B978-0-323-99291-6.00017-7
15. Wang L, Li S, Fu J. Self-healing anti-corrosion coatings based on micron-nano containers with different structural morphologies. *Prog. Org. Coat.* 2023, 175, 107381. https://doi.org/10.1016/j.porgcoat.2022.107381
16. Maan AM, Graafsma CN, Hofman AH, Pelras T, de Vos WM, Kamperman M. Scalable fabrication of reversible antifouling block copolymer coatings via adsorption strategies. *ACS Appl. Mater. Interfaces*. 2023, 15(15), 19682–19694. https://doi.org/10.1021/acsami.3c01060
17. Sanjay SL, Annaso BG, Chavan SM, Rajiv SV. Recent progress in preparation of superhydrophobic surfaces: a review. *J. Surf. Eng. Mater. Adv. Technol.* 2012, 2012. https://doi.org/10.4236/jsemat.2012.22014
18. Ayyaril SS, Shanableh A, Bhattacharjee S, Rawas-Qalaji M, Cagliani R, Shabib AG. Recent progress in micro and nano-encapsulation techniques for environmental applications: a review. *Results Eng.* 2023, 101094. https://doi.org/10.1016/j.rineng.2023.101094
19. Arooj N, Khan TM, Mumtaz M, Rehman A, Ahmad I, Shah A, Hassan MU, Raffi M. Polymeric Fe3O4 nanoparticle/carbon nanofiber hybrid nanocomposite coatings for improved terahertz shielding. *ACS Appl. Nano Mater.* 2023, 6(7), 5264–5273. https://doi.org/10.1021/acsanm.2c05313
20. Wang Y, Ge-Zhang S, Mu P, Wang X, Li S, Qiao L, Mu H. Advances in sol-gel-based superhydrophobic coatings for wood: a review. *Int. J. Mol. Sci.* 2023, 24(11), 9675. https://doi.org/10.3390/ijms24119675
21. Kothari J, Iroh JO. Self-healing poly (urea formaldehyde) microcapsules: synthesis and characterization. *Polymer* 2023, 15(7), 1668. https://doi.org/10.3390/polym15071668
22. Chen Z, Scharnagl N, Zheludkevich ML, Ying H, Yang W. Micro/nanocontainer-based intelligent coatings: synthesis, performance and applications–a review. *J. Chem. Eng.* 2023, 451, 138582. https://doi.org/10.1016/j.cej.2022.138582
23. Dong F, Koodali RT, Wang H, Ho WK. Nanomaterials for environmental applications. *J. Nanomater.* 2014, 2014, 1–7. https://doi.org/10.1155/2014/276467
24. Stark WJ, Stoessel PR, Wohlleben W, Hafner AJ. Industrial applications of nanoparticles. *Chem. Soc. Rev.* 2015, 44(16), 5793–5805. https://doi.org/10.1039/C4CS00362D
25. Kim BH, Hackett MJ, Park J, Hyeon T. Synthesis, characterization, and application of ultrasmall nanoparticles. *Chem. Mater.* 2014, 26(1), 59–71. https://doi.org/10.1021/cm402225z
26. Sabir S, Arshad M, Chaudhari SK. Zinc oxide nanoparticles for revolutionizing agriculture: synthesis and applications. *Sci. World J.* 2014, 2014, 1–8. https://doi.org/10.1155/2014/925494
27. Salata OV. Applications of nanoparticles in biology and medicine. *J. Nanobiotechnology*. 2004, 2(1), 1–6. https://doi.org/10.1186/1477-3155-2-3
28. Singh T, Shukla S, Kumar P, Wahla V, Bajpai VK, Rather IA. Application of nanotechnology in food science: perception and overview. *Front. Microbiol.* 2017, 8, 1501. https://doi.org/10.3389/fmicb.2017.01501
29. Rodrigues SM, Demokritou P, Dokoozlian N, Hendren CO, Karn B, Mauter MS, Sadik OA, Safarpour M, Unrine JM, Viers J, Welle P. Nanotechnology for sustainable food production: promising opportunities and scientific challenges. *Environ. Sci. Nano.* 2017, 4(4), 767–781. https://doi.org/10.1039/C6EN00573J
30. King T, Osmond-McLeod MJ, Duffy LL. Nanotechnology in the food sector and potential applications for the poultry industry. *Trends Food Sci. Technol.* 2018, 72, 62–73. https://doi.org/10.1016/j.tifs.2017.11.015
31. Guo D, Xie G, Luo J. Mechanical properties of nanoparticles: basics and applications. *J. Phys. D Appl. Phys.* 2013, 47(1), 013001. https://doi.org/10.1088/0022-3727/47/1/013001
32. Kumbhakar P, Ray SS, Stepanov AL. Optical properties of nanoparticles and nanocomposites. *J. Nanomater.* 2014, 2014:1–1. https://doi.org/10.1155/2014/181365

33. Hameed RA, Abu-Nawwas AA, Shehata HA. Nano-composite as corrosion inhibitors for steel alloys in different corrosive media. *Adv. Appl. Sci. Res.* 2013, 4(3), 126–129.
34. Khodair ZT, Khadom AA, Jasim HA. Corrosion protection of mild steel in different aqueous media via epoxy/nanomaterial coating: preparation, characterization and mathematical views. *J. Mater. Res. Technol.* 2019, 8(1), 424–435. https://doi.org/10.1016/j.jmrt.2018.03.003
35. Fotovvati B, Namdari N, Dehghanghadikolaei A. On coating techniques for surface protection: a review. *J. Manuf. Mater. Process.* 2019, 3(1), 28. https://doi.org/10.3390/jmmp3010028
36. Roduner E. Size matters: why nanomaterials are different. *Chem. Soc. Rev.* 2006, 35(7), 583–592. https://doi.org/10.1039/B502142C
37. Abdeen DH, El Hachach M, Koc M, Atieh MA. A review on the corrosion behaviour of nanocoatings on metallic substrates. *Materials* 2019, 12(2), 210. https://doi.org/10.3390/ma12020210
38. Rathish RJ, Joany RD, Pandiarajan M, Rajendran S. Corrosion resistance of nanoparticle-incorporated nano coatings. *Eur. Chem. Bull.* 2013, 2(12), 965–970.
39. Zhang Z, Chen S, Ren H, Zhou J. Protection of iron against corrosion by coverage with Au nanoparticles and n-hexylthiol mixed films. *Appl. Surf. Sci.* 2009, 255(9), 4950–4594. https://doi.org/10.1016/j.apsusc.2008.12.043
40. Shao W, Nabb D, Renevier N, Sherrington I, Luo JK. Mechanical and corrosion resistance properties of TiO2 nanoparticles reinforced Ni coating by electrodeposition. *IOP Conf. Ser.: Mater. Sci. Eng.* 2012, 40(1), 012043. https://doi.org/10.1088/1757-899X/40/1/012043
41. Atta AM, El-Mahdy GA, Al-Lohedan HA. Corrosion inhibition efficiency of modified silver nanoparticles for carbon steel in 1 M HCl. *Int. J. Electrochem. Sci.* 2013, 8(4), 4873–4885.
42. Atta AM, Allohedan HA, El-Mahdy GA, Ezzat AR. Application of stabilized silver nanoparticles as thin films as corrosion inhibitors for carbon steel alloy in 1 M hydrochloric acid. *J. Nanomater.* 2013, 2013, 132. http://dx.doi.org/10.1155/2013/580607
43. Palanisamy K, Devabharathi V, Sundaram NM. Corrosion inhibition studies of mild steel with carrier oil stabilized of iron oxide nanoparticles incorporated into paint. *Environments* 2014, 1, 2.
44. Wu LK, Zhang XF, Hu JM. Corrosion protection of mild steel by one-step electrodeposition of superhydrophobic silica film. *Corros. Sci.* 2014, 85, 482–487. https://doi.org/10.1016/j.corsci.2014.04.026
45. Ates M, Topkaya E. Nanocomposite film formations of polyaniline via TiO2, Ag, and Zn, and their corrosion protection properties. *Prog. Org. Coat.* 2015, 82, 33–40. https://doi.org/10.1016/j.porgcoat.2015.01.014
46. Parekh K, Jauhari S, Upadhyay RV. Mechanism of acid corrosion inhibition using magnetic nanofluid. *Adv. Nat. Sci.: Nanosci. Nanotechnol.* 2016, 7(4), 045007. https://doi.org/10.1088/2043-6262/7/4/045007
47. Hosseini M, Fotouhi L, Ehsani A, Naseri M. Enhancement of corrosion resistance of polypyrrole using metal oxide nanoparticles: potentiodynamic and electrochemical impedance spectroscopy study. *J. Colloid Interface Sci.* 2017, 505, 213–219. http://dx.doi.org/10.1016/j.jcis.2017.05.097
48. Yang F, Li X, Dai Z, Liu T, Zheng W, Zhao H, Wang L. Corrosion inhibition of polydopamine nanoparticles on mild steel in hydrochloric acid solution. *Int. J. Electrochem. Sci.* 2017,12(8), 7469–7480. https://doi.org/10.20964/2017.08.52
49. Mobin M, Aslam J, Alam R. Corrosion protection of poly (aniline-co-N-ethylaniline)/ZnO nanocomposite coating on mild steel. *Arab. J. Sci. Eng.* 2017, 42, 209–224. https://doi.org/10.1007/s13369-016-2234-z
50. Hasnidawani JN, Hassan NA, Norita H, Samat N, Bonnia NN, Surip SN. ZnO nanoparticles for anticorrosion nanocoating of carbon steel. *Mater. Sci. Forum.* 2017, 894, 76–80. https://doi.org/10.4028/www.scientific.net/MSF.894.76
51. Mahvidi S, Gharagozlou M, Mahdavian M, Naghibi S. Potency of ZnFe 2 O 4 nanoparticles as corrosion inhibitor for stainless steel; the pigment extract study. *Mater. Res.* 2017, 20, 1492–1502. http://dx.doi.org/10.1590/1980-5373-MR-2016-0772
52. Kouhi S, Ghamari B, Yeganeh R. The effect of nanoparticle coating on anticorrosion performance of centrifugal pump blades. *Jordan J. Mech. Ind. Eng.* 2018, 12(2), 117–122.

53. Mahmood MH, Suryanto S, Al Hazza MH, Haidera FI. Developing of corrosion resistance nano copper oxide coating on copper using anodization in oxalate solution. *Int. J. Eng.* 2018, 31(3), 450–455. https://doi.org/10.5829/ije.2018.31.03c.07
54. Rajendran A, Sivalingam T, Narayanan P, Murugavel SC. Nano composite polymer resin coating to control mild steel corrosion in marine environment. *Am. J. Appl. Chem.* 2018, 6(3), 102–125. https://doi.org/10.11648/j.ajac.20180603.14
55. Najjar R, Abdel-Gaber AM, Awad R. Electrochemical corrosion behaviour of carbon steel in acidic media in presence of Mn_2O_3 nanoparticles synthesized at different pH. *Int. J. Electrochem. Sci.* 2018, 13, 8723–8735. https://doi.org/10.20964/2018.09.50
56. Khamis EA, Hamdy A, Morsi RE. Magnetite nanoparticles/polyvinyl pyrrolidone stabilized system for corrosion inhibition of carbon steel. *Egypt. J. Pet.* 2018, 27(4), 919–926. https://doi.org/10.1016/j.ejpe.2018.02.001
57. Lopez de Armentia S, Pantoja M, Abenojar J, Martinez MA. Development of silane-based coatings with zirconia nanoparticles combining wetting, tribological, and aesthetical properties. *Coatings* 2018, 8(10), 368. https://doi.org/10.3390/coatings8100368
58. Rajkumar R, Vedhi C. A study of corrosion protection efficiency of silica nanoparticles acrylic coated on mild steel electrode. *Vacuum* 2019, 161, 1–4. https://doi.org/10.1016/j.vacuum.2018.12.005
59. Khodair ZT, Khadom AA, Jasim HA. Corrosion protection of mild steel in different aqueous media via epoxy/nanomaterial coating: preparation, characterization and mathematical views. *J. Mater. Res. Technol.* 2019, 8(1), 424–435. https://doi.org/10.1016/j.jmrt.2018.03.003
60. Sharma UR, Sharma N. Green synthesis, anti-cancer and corrosion inhibition activity of Cr_2O_3 nanoparticles. *Biointerf. Res. Appl. Chem.* 2021, 11, 8402–8412. https://doi.org/10.33263/BRIAC111.84028412
61. Al-Senani GM, Al-Saeedi SI. The use of synthesized CoO/Co_3O_4 nanoparticles as a corrosion inhibitor of low-carbon steel in 1 M HCl. *Materials* 2022, 15(9), 3129. https://doi.org/10.3390/ma15093129
62. Al Jabri H, Devi MG, Al-Shukaili MA. Development of polyaniline–TiO_2 nano composite films and its application in corrosion inhibition of oil pipelines. *J. Indian Chem. Soc.* 2023, 100(1), 100826. https://doi.org/10.1016/j.jics.2022.100826
63. Ge L, Li Q, Wang M, Ouyang J, Li X, Xing MM. Nanosilver particles in medical applications: synthesis, performance, and toxicity. *Int. J. Nanomed.* 2014, 9, 2399–2407. https://doi.org/10.2147/ijn.s55015
64. Hebeish A, Hashem M, Abd El-Hady MM, Sharaf S. Development of CMC hydrogels loaded with silver nano-particles for medical applications. *Carbohydr. Polym.* 2013, 92(1), 407–413. https://doi.org/10.1016/j.carbpol.2012.08.094
65. Javidi M, Zarei M, Naghavi N, Mortazavi M, Nejat AH. Zinc oxide nano-particles as sealer in endodontics and its sealing ability. *Contemp. Clin. Dent.* 2014, 5(1), 20. doi: 10.4103/0976-237X.128656
66. Ali MA, Rehman I, Iqbal A, Din S, Rao AQ, Latif A, Samiullah TR, Azam S, Husnain T. Nanotechnology, a new frontier in agriculture. *Adv Life Sci.* 2014, 1(3), 129–138.
67. Gensler R, Gröppel P, Muhrer V, Müller N. Application of nanoparticles in polymers for electronics and electrical engineering. *Part Part Syst. Charact.* 2002, 19(5), 293–299. https://doi.org/10.1002/1521-4117(200211)19:5%3C293::AID-PPSC293%3E3.0.CO;2-N
68. Bi SS, Shi L, Zhang LL. Application of nanoparticles in domestic refrigerators. *Appl. Therm. Eng.* 2008, 28(14–15), 1834–1843. https://doi.org/10.1016/j.applthermaleng.2007.11.018
69. Anselmann R. Nanoparticles and nanolayers in commercial applications. *J. Nanopart. Res.* 2001, 3(4), 329–336. https://doi.org/10.1023/A:1017529712314
70. Umoren SA, Solomon MM. Protective polymeric films for industrial substrates: a critical review on past and recent applications with conducting polymers and polymer composites/nanocomposites. *Prog. Mater. Sci.* 2019, 104, 380–450. https://doi.org/10.1016/j.pmatsci.2019.04.002
71. Souza VG, Fernando AL. Nanoparticles in food packaging: biodegradability and potential migration to food—a review. *Food Packag. Shelf Life.* 2016, 8, 63–70. https://doi.org/10.1016/j.fpsl.2016.04.001
72. Sereda PJ. *Weather Factors Affecting Corrosion of Metals.* ASTM International; 1974.

73. Suzumura K, Nakamura SI. Environmental factors affecting corrosion of galvanized steel wires. *J. Mater. Civ. Eng*. 2004, 16(1), 1–7. https://doi.org/10.1061/(ASCE)0899-1561(2004)16:1(1)
74. Fotovvati B, Namdari N, Dehghanghadikolaei A. On coating techniques for surface protection: a review. *J. Manuf. Mater. Process*. 2019, 3(1), 28. https://doi.org/10.3390/jmmp3010028
75. Birjandi FC, Sargolzaei J. Super-non-wettable surfaces: a review. *Colloids Surf. A Physicochem. Eng. Asp*. 2014, 448, 93–106. https://doi.org/10.1016/j.colsurfa.2014.02.016
76. Jeon IY, Baek JB. Nanocomposites derived from polymers and inorganic nanoparticles. *Materials* 2010, 3(6), 3654–3674. https://doi.org/10.3390/ma3063654
77. Quarta A, Di Corato R, Manna L, Ragusa A, Pellegrino T. Fluorescent-magnetic hybrid nanostructures: preparation, properties, and applications in biology. *IEEE Trans. Nanobiosci*. 2007, 6(4), 298–308. https://doi.org/10.1109/TNB.2007.908989
78. Fathima Sabirneeza AA, Geethanjali R, Subhashini S. Polymeric corrosion inhibitors for iron and its alloys: a review. *Chem. Eng. Commun*. 2015, 202(2), 232–244. https://doi.org/10.1080/00986445.2014.934448
79. Winey KI, Vaia RA. Polymer nanocomposites. *MRS Bull*. 2007, 32(4), 314–322. https://doi.org/10.1557/mrs2007.229
80. Fekry AM, Azab SM. The development of an innovative nano-coating on the surgical 316 L SS implant and studying the enhancement of corrosion resistance by electrochemical methods using Ibandronate drug. *Nano-Struct. Nano-Objects*. 2020, 21, 100411. https://doi.org/10.1016/j.nanoso.2019.100411
81. Zhang R, Huang K, Zhu M, Chen G, Tang Z, Li Y, Yu H, Qiu B, Li X. Corrosion resistance of stretchable electrospun SEBS/PANi micro-nano fiber membrane. *Eur. Polym. J*. 2020, 123, 109394. https://doi.org/10.1016/j.eurpolymj.2019.109394
82. Mohamed MG, Kuo SW, Mahdy A, Ghayd IM, Aly KI. Bisbenzylidene cyclopentanone and cyclohexanone-functionalized polybenzoxazine nanocomposites: Synthesis, characterization, and use for corrosion protection on mild steel. *Mater. Today Commun*. 2020, 25, 101418. https://doi.org/10.1016/j.mtcomm.2020.101418
83. Pugazhenthi I, Ghouse SM. Corrosion protection performance of titania nanoparticles filled poly (4-methyl-5-vinylthiazole) applied on mild steel in 3.5% sodium chloride solution. *J. Plast. Film Sheeting*. 2021, 37(1), 17–32. http://dx.doi.org/10.1177/8756087920939301
84. Kumar H, Boora A, Yadav A. Polyaniline-metal oxide-nano-composite as a nano-electronics, opto-electronics, heat resistance and anticorrosive material. *Results Chem*. 2020, 2, 100046. https://doi.org/10.1016/j.rechem.2020.100046
85. Sharma K, Goyat MS, Vishwakarma P. Synthesis of polymer nano-composite coatings as corrosion inhibitors: a quick review. *IOP Conf. Ser.: Mater. Sci. Eng*. 2020, 983(1), 012016. https:/ doi.org/10.1088/1757-899X/983/1/012016
86. Jadhav N, Kasisomayajula S, Gelling VJ. Polypyrrole/metal oxides-based composites/nanocomposites for corrosion protection. *Front. Mater. Sci*. 2020, 7, 95. https://doi.org/10.3389/fmats.2020.00095
87. Saleh KA, Ali MI. Electro polymerization for (N-terminal tetrahydrophthalamic acid) for anti-corrosion and biological activity applications. *Iraqi J. Sci*. 2020, 1–12. https://doi.org/10.24996/ijs.2020.61.1.1
88. Vinothkumar K, Sethuraman MG. A robust method of enhancement of corrosion inhibitive ability of electrodeposited poly-3-amino-5-mercapto-1, 2, 4-triazole films over copper surface using graphene oxide. *J. Adhes. Sci. Technol*. 2020, 34(6), 651–669. https://doi.org/10.1080/01694243.2019.1674599
89. Bahari HS, Ye F, Carrillo EA, Leliopoulos C, Savaloni H, Dutta J. Chitosan nanocomposite coatings with enhanced corrosion inhibition effects for copper. *Int. J. Biol. Macromol*. 2020, 162, 1566–1577. https://doi.org/10.1016/j.ijbiomac.2020.08.035
90. Harb SV, Uvida MC, Trentin A, Lobo AO, Webster TJ, Pulcinelli SH, Santilli CV, Hammer P. PMMA-silica nanocomposite coating: effective corrosion protection and biocompatibility for a Ti6Al4V alloy. *Mater. Sci. Eng. C*. 2020, 110, 110713. https://doi.org/10.1016/j.msec.2020.110713
91. Gallegos-Melgar A, Serna SA, Lázaro I, Gutiérrez-Castañeda EJ, Mercado-Lemus VH, Arcos-Gutierrez H, Hernández-Hernández M, Porcayo-Calderón J, Mayen J, Monroy MD. Potentiodynamic polarization performance of a novel composite coating system of Al_2O_3/chitosan-sodium alginate,

applied on an aluminum AA6063 alloy for protection in a chloride Ions environment. *Coatings*. 2020, 10(1), 45. https://doi.org/10.3390/coatings10010045

92. Chen Y, Ren B, Gao S, Cao R. The sandwich-like structures of polydopamine and 8-hydroxyquinoline coated graphene oxide for excellent corrosion resistance of epoxy coatings. *J. Colloid Interface Sci.* 2020, 565, 436–448. https://doi.org/10.1016/j.jcis.2020.01.051
93. Aung MM, Li WJ, Lim HN. Improvement of anticorrosion coating properties in bio-based polymer epoxy acrylate incorporated with nano zinc oxide particles. *Ind. Eng. Chem. Res*. 2020, 59(5), 1753–1763. https://doi.org/10.1021/acs.iecr.9b05639

4 Release Behavior of Corrosion Inhibitor Depending on External Stimuli

Walid Daoudi, Kaoutar Zaidi, Salma Lamghafri, Abdellah Elyoussfi, Omar Dagdag, Adyl Oussaid, and Abdelmalik El Aatiaoui

4.1 INTRODUCTION

Control of corrosion is of paramount importance in various industrial sectors such as the petroleum industry, aerospace, shipbuilding, and chemical industry [1, 2]. The use of corrosion inhibitors is a commonly employed strategy to protect metallic materials from deterioration caused by corrosion [3]. Corrosion inhibitors function by interfering with the electrochemical reactions responsible for corrosion and forming a protective layer on the metal surface [3, 4]. Traditionally, corrosion inhibitors are added to protective systems through formulations such as coatings, paints, or liquid solutions [5]. However, this approach has certain limitations, including uncontrolled release of the corrosion inhibitor and the need for periodic reapplication of the coating or solution to maintain effective protection [6]. In recent years, significant research has been conducted to develop corrosion inhibition systems with controlled release. These systems enable the targeted release of the corrosion inhibitor in response to specific external stimuli, such as variations in pH, temperature, ionic strength, or electrochemical potential [7–9].

The release behavior of a corrosion inhibitor dependent on external stimuli offers numerous potential advantages [10]. Firstly, it allows for controlled and gradual release of the inhibitor, ensuring continuous and prolonged protection of the metal surface. Secondly, this approach reduces the amount of corrosion inhibitors required, which can lower costs and minimize undesirable environmental effects. Moreover, the selective release of the inhibitor in response to specific conditions in the corrosive environment enables targeted protection of the most corrosion-prone areas. Various approaches have been explored to achieve controlled release of corrosion inhibitors. This includes incorporating corrosion inhibitors into polymer matrices, microscopic capsules, or stimuli-responsive polymer systems [11]. These systems can be designed to release the corrosion inhibitor continuously, progressively, or triggered by specific stimuli. This chapter focuses on the release behavior of a corrosion inhibitor dependent on external stimuli. It examines different strategies and techniques used to achieve controlled release of corrosion inhibitors in response to specific stimuli. Furthermore, the advantages, limitations, and potential applications of these controlled release systems will be discussed. By gaining a better understanding of the release behavior of corrosion inhibitors dependent on external stimuli, it will be possible to design more efficient and durable protection systems for metallic materials exposed to corrosive environments.

DOI: 10.1201/9781032677255-4

4.2 BASICS OF CONTROLLED RELEASE

In general, controlled release is a method of regulating the gradual diffusion of an active substance from a carrier material. The main objective is to maintain optimal therapeutic or protective levels of the substance in the surrounding environment over a prolonged period of time [12]. The basic principles of controlled release involve the use of specific mechanisms to control the rate and duration of release in order to optimize the efficacy and safety of the active substance. On the other hand, there are various mechanisms of controlled release used in corrosion inhibitors, which include passive diffusion, controlled degradation, response to external stimuli, controlled release matrices, and electric or magnetic field-based controlled release systems [13]. Passive diffusion is based on the natural diffusion of the corrosion inhibitor through a support matrix. The release occurs as the inhibitor diffuses from the material through the pores or channels. The rate of diffusion depends on the properties of the support material, such as its permeability and porosity. In the controlled degradation mechanism, the release of the corrosion inhibitor is controlled by the degradation of the support material. The material is designed to degrade gradually, thereby releasing the inhibitor. The degradation rate can be adjusted according to specific needs by selecting the appropriate material and modifying its structure.

The response to external stimuli mechanism utilizes external stimuli such as pH variations, temperature changes, or ionic concentration to trigger the release of the corrosion inhibitor. The support material is designed to respond to these stimuli and release the inhibitor in a controlled manner. For example, a pH-sensitive polymer can deform or dissolve under specific conditions, enabling the release of the inhibitor [14]. Corrosion inhibitors can be encapsulated in controlled-release matrices or micro/nanocontainers. These matrices can be made from polymers, lipids, or composite materials. The release of the inhibitor occurs through degradation, diffusion, or matrix rupture. In electric or magnetic field-based controlled release systems, electric or magnetic fields are applied to control the release of the corrosion inhibitor. The support materials are sensitive to these fields and release the inhibitor when the field is applied. Each mechanism of controlled release has specific advantages and limitations, and the choice of mechanism depends on the specific requirements of the application and the properties of the corrosion inhibitor. Understanding these mechanisms allows for the design of effective controlled release systems tailored to corrosion protection needs. In this chapter, we specifically focus on the mechanisms as response to external stimuli using external stimuli.

4.3 EXTERNAL STIMULI FOR CONTROLLED RELEASE

External stimuli are external variables such as temperature, pH, radiation, etc., which can be used to trigger the controlled release of the corrosion inhibitor. Each external stimulus offers specific advantages and limitations, enabling precise controlled release tailored to corrosion protection needs [15].

4.3.1 pH

pH plays a crucial role in the release of corrosion inhibitors. The effect of pH on inhibitor release depends on the type of inhibitor used and its chemical properties. In some cases, an increase in pH can promote the release of corrosion inhibitors. For example, for certain organic inhibitors, a higher pH can cause dissociation or degradation of the inhibitor compounds, leading to their release into the corrosive environment. This can be particularly beneficial in areas where pH naturally increases due to the formation of a protective layer, as it ensures the continuous release of inhibitors to maintain corrosion protection.

On the other hand, in certain systems, a decrease in pH may be associated with an increased release of corrosion inhibitors. For example, for certain inhibitors based on inorganic salts, an acidic environment can facilitate the dissolution of inhibitor compounds, allowing their release into the corrosive medium. This increased release of inhibitors at low pH can be useful in acidic environments where corrosion is more active. For example, a study by Zea et al. [10] explored the use of mesoporous silica nanoparticles loaded with an environmentally friendly corrosion inhibitor, sodium phosphomolybdate, to offer controlled pH-dependent release, even in the absence of encapsulation. Figure 4.1 shows that significant amounts of the inhibitor are released throughout the pH range (4–5), with a particularly marked increase above pH 11. Under acidic loading conditions, the presence and incorporation of the $[PMo_{12}O_{40}]^{3-}$ species is observed, but above a certain pH (4–5), this species decomposes and the formation of molybdates and phosphates must be taken into account. The release curves for molybdenum (Mo) and phosphorus (P), in the presence or absence of the capsule, show almost parallel behavior. This observation strongly suggests that the pH of the solution has a significant influence on the Mo and P species present.

In a similar study, Li et al. [16] developed pH-sensitive poly(styrene sulfonate)-block-thiophene-acetic acid (PSS-BTA)/polyethyleneimine (PEI) shell-and-shell BTA-loaded smart

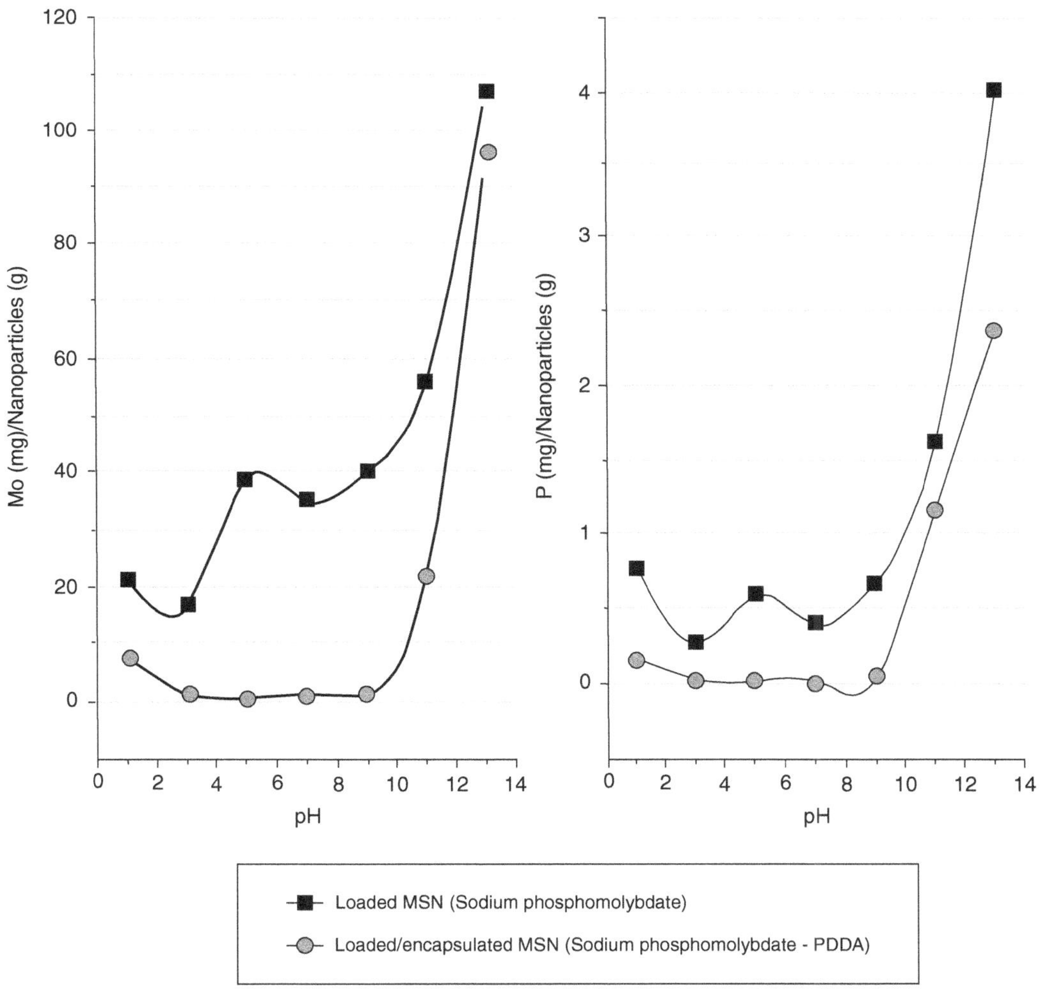

FIGURE 4.1 Inhibitor release as a function of pH for the cases of sodium phosphomolybdate and for Mo and P, respectively.

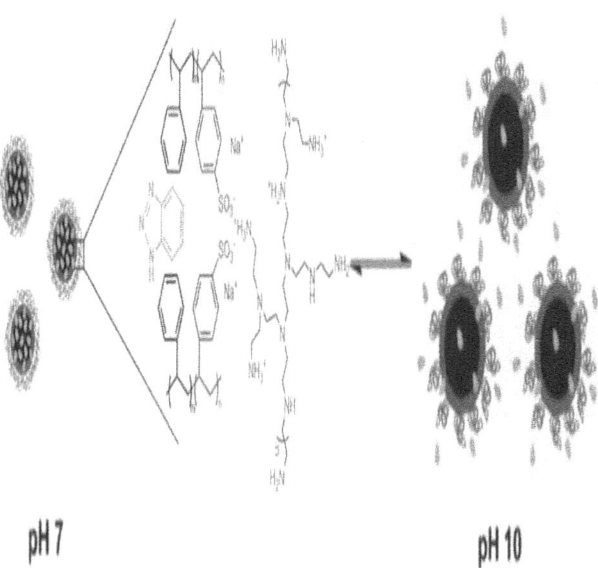

FIGURE 4.2 A schematic diagram depicting the composition of CISCs and the release of BTA by pH-responsive CISCs.

corrosion inhibitor micro/nanocontainers (CISCs) for self-healing organic epoxy coatings (Figure 4.2). Through electrostatic interaction between the PSS on the particle surfaces and the PEI chains, the highly branched PEI chains were adsorbed onto the PSS surfaces. When the pH value reached 10, the PEI chains shrank due to the decrease in PEI charge density, leading to subsequent desorption from the CISC surfaces. This desorption resulted in a rapid release of BTA from inside the CISCs.

Snihirova et al. [17] and Snihirova et al. [18] used calcium carbonate ($CaCO_3$) microbeads (Figure 4.3) and hydroxyapatite microparticles modified with different corrosion inhibitors, respectively, as actively reactive, pH-sensitive corrosion inhibiting agents to improve the corrosion resistance of an aluminum alloy. Microparticle dissolution was induced by local acidification resulting from the anodic half-reaction. The increase in acidity, caused by the anodic reaction, released corrosion inhibitors to suppress the corrosive activity of the unprotected metal. This approach demonstrates a novel strategy for the controlled release of corrosion inhibitors in response to specific corrosive conditions, opening up new perspectives in the design of more effective and adaptive anti-corrosion coatings.

In similar studies, Zhu et al. [9] investigated a pH-sensitive release system using mesoporous silica nanocontainers (MSNs) with an insoluble zinc-benzotriazole complex (Zn-BTA) as the closure material. The preparation and pH response of BTA-loaded nanocontainers are shown in Figure 4.4. Following the loading of BTA into the MSNs, Zn^{2+} species were introduced onto the MSN surface and reacted with BTA molecules to form insoluble Zn-BTA complexes, preventing leakage of BTA stored in the MSN pores. As the pH decreases, the Zn-BTA complexes that seal the nanocontainers break down, leading to the opening of the pores and the release of BTA.

However, it is important to note that the effect of pH on inhibitor release can vary depending on the specific characteristics of the inhibitor used, the nature of the corrosive environment, and the experimental conditions. Therefore, thorough experimental studies are required to assess the effect of pH on the release of corrosion inhibitors in specific systems. An increase or decrease in pH can result in increased inhibitor release, thus providing continuous corrosion protection under specific conditions.

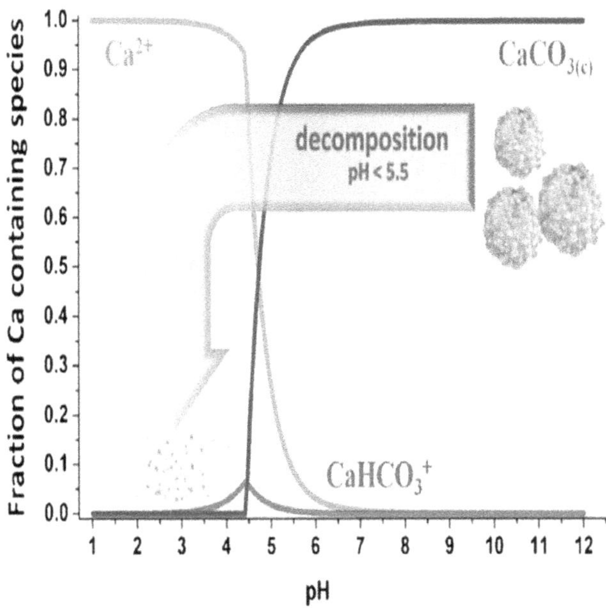

FIGURE 4.3 Schematic illustration of the corrosion inhibitor release mechanism by $CaCO_3$ microbeads based on different fractions of Ca^{2+} ions and pH value in the solution.

4.3.2 Temperature

The effect of temperature on the release of corrosion inhibitors is also an important factor to consider. Temperature can influence inhibitor release kinetics and play a crucial role in their corrosion protection efficacy. In many cases, an increase in temperature can accelerate the release of corrosion inhibitors. As temperature rises, inhibitor molecules can acquire greater kinetic energy, making it easier for them to diffuse out of micro/nanocontainers or dissociate from encapsulating structures. This can lead to a faster and greater release of inhibitors into the corrosive environment, thereby enhancing corrosion protection. However, it should be noted that the effect of temperature may vary according to the specific characteristics of the inhibitor and the release system used. Some inhibitors may exhibit particular sensitivity to temperature, with greater release at specific temperatures or in particular temperature ranges. In some cases, a lower temperature may also be beneficial, as it can slow the release of inhibitors, thus ensuring long-term protection.

Li et al. [19] used a double-walled SiO_2/polymer hybrid nanotube to protect against metallic corrosion (Figure 4.5). This nanotube consisted of a hollow cavity, a porous SiO_2 inner wall, and a stimuli-responsive polymeric outer wall. To prepare these hybrid nanotubes, poly(N-isopropylacrylamide) (PNIPAM) was loaded as temperature-reactive active molecules. At low temperatures (25°C), the PNIPAM outer wall swelled, while at high temperatures (50°C) it contracted. The transition from temperature-reactive to pH- and redox-reactive polymers also proved effective. When the pH of the solution reached 12, the carboxylic acid groups of the methacrylic acid outer wall were deprotonated, causing strong swelling of the anionic polymer chains. Using a reducing agent called DL-dithiothreitol, the disulfide bonds of the poly(ethylene glycol) methacrylate outer wall were broken, transforming the cross-linked outer wall into linear chains.

4.3.3 Redox Potential

Corrosion inhibition aims to prevent or slow down undesirable chemical reactions between a metallic material and its corrosive environment. Inhibiting agents are substances that are added to a

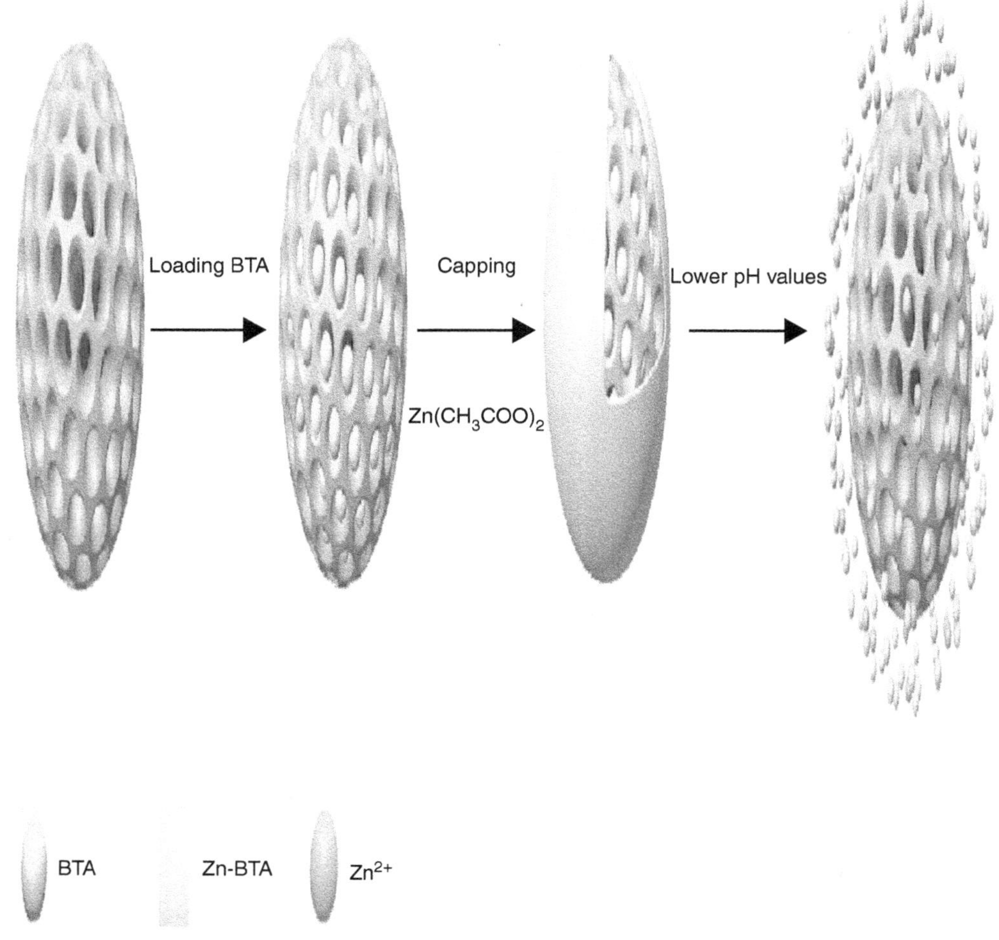

FIGURE 4.4 Schematic illustration of the preparation of pH-sensitive nanocontainers and the stimulus-response process.

system to form a protective layer on the metal surface, thereby reducing corrosion reactions. Redox stimuli are used to control the inhibitory activity of these agents on the metal surface. Specific reducing or oxidizing agents are introduced into the corrosive medium to regulate the formation or removal of the protective layer. When reducing agents are present, they can reduce the inhibiting species adsorbed on the metal surface, thereby reducing their ability to protect against corrosion. This makes it possible to control the intensity of inhibition by adjusting the concentration of reducing agents. Similarly, the addition of oxidizing agents can oxidize the inhibiting species, thereby increasing their protective power on the metal surface. This strengthens or regenerates the inhibitor layer and improves the effectiveness of corrosion inhibition. As an example, Jiang et al. [20] demonstrated that the core-shell structure of SiO_2 nanocapsules exhibited redox reactivity, accelerating the release of the corrosion inhibitor 2-mercaptobenzothiazole (MBT) upon chemical reduction while delaying its release upon oxidation (Figure 4.6). More specifically, the bis[3-(triethoxysilyl)propyl] tetrasulfide present in the SiO_2 shell exhibited redox-reactive permeability behavior. When the tetrasulfide bonds were reduced by tris(2-carboxyethyl)phosphine hydrochloride, acting as a reducing agent, the permeability of the SiO_2 shell increased. These CISC

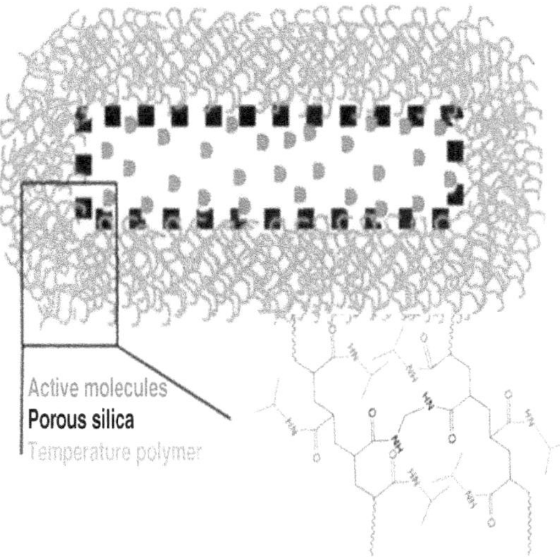

FIGURE 4.5 Representative diagram of temperature-responsive SiO_2/PNIPAM hybrid nanotubes.

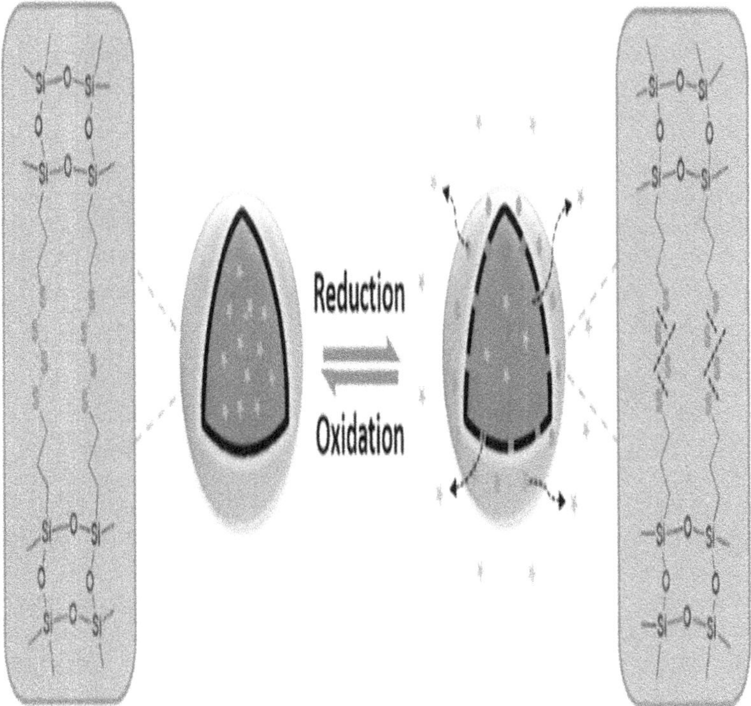

FIGURE 4.6 The diagram schematically represents the mechanism of MBT release from CISCs when they are subjected to reduction and oxidation processes.

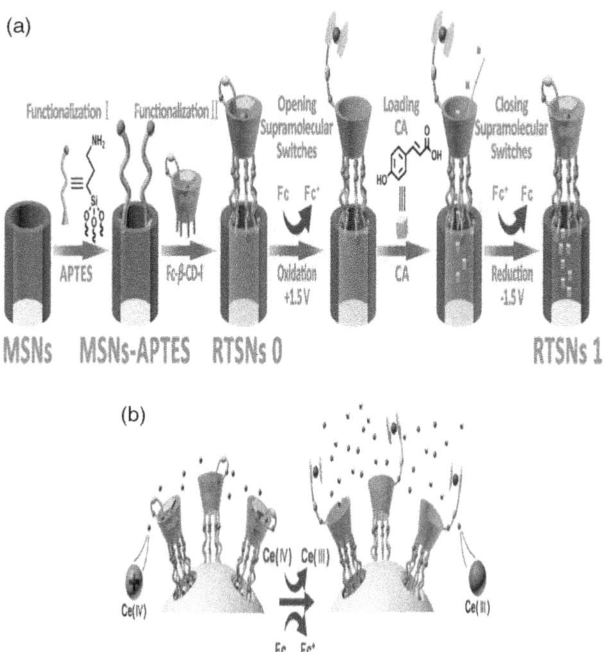

FIGURE 4.7 (a) Schematic preparation of RTSNs 1. (b) Schematic depiction of the release mechanism of redox-responsive CISCs.

systems are designed with redox-reactive supramolecular switches to regulate trapped corrosion inhibitors and prevent corrosion. Wang et al. [21] prepared redox-reactive smart nanocontainers 1 (RTSNs 1), which were redox-reactive CISCs, and developed a two-layer nanocomposite coating (a ZrO_2-SiO_2 coating doped with cerium(IV) ions (Ce^{4+}) combined with a ZrO_2-SiO_2 coating incorporating RTSNs 1), which was applied to the surface of an aluminum alloy (Figure 4.7).

4.3.4 Light

The release behavior of the corrosion inhibitor in response to light stimuli can be studied using a variety of approaches. One commonly used method is the use of photochemical systems where the corrosion inhibitor is designed to react specifically to light. When exposed to a specific wavelength of light, the inhibitor undergoes a chemical reaction which results in its release into the corrosive environment. This approach allows the controlled release of the inhibitor, which can be particularly useful in corrosion protection coatings. Another method involves the use of light-sensitive microcapsules containing the corrosion inhibitor. These microcapsules contain light-sensitive compounds which, when exposed to light, cause the capsules to rupture or open, releasing the inhibitor into the corrosive environment. This approach also allows controlled release of the inhibitor, as it depends on the specific light irradiation to which the microcapsules are subjected. In addition, there are materials with controlled light release that can be used to encapsulate corrosion inhibitors. These materials have the ability to store light energy and release it gradually over time. When exposed to light, these materials absorb the light energy and release it in a controlled manner, enabling the corrosion inhibitor to be gradually released into the corrosive medium. As an example, TiO_2 is commonly used as a container core in the ultraviolet (UV), visible, and NIR regions due to its powerful photocatalytic properties [22]. In addition, the layer-by-layer assembly method of polyelectrolytes has been employed to create smart shells for corrosion inhibitors,

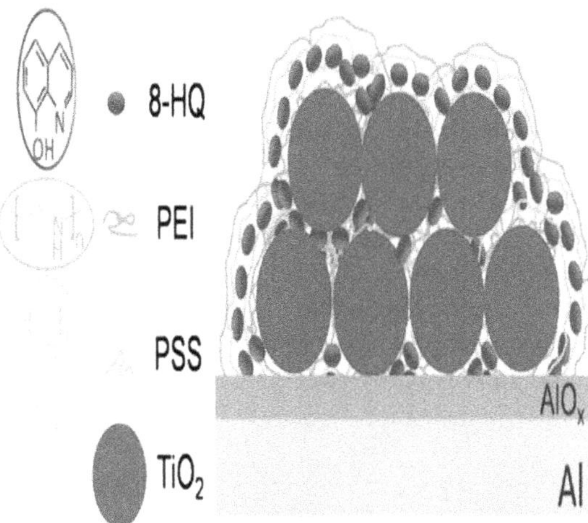

FIGURE 4.8 The cross-sectional diagram illustrates the smart coating loaded with successive layers of TiO_2/PEI/PSS/8-HQ/PSS/PEI.

enabling their release to be regulated and thus promoting advances in retroactive anticorrosion coatings [23]. For example, He et al. [24] described photocatalytic TiO_2 nanoparticles obtained by coating the corrosion inhibitor, 8-hydroxyquinoline (8-HQ), with polyelectrolyte nanostructures. These nanoparticles can be activated by UV light, resulting in a self-repair effect on an aluminum alloy (Figure 4.8). The functional groups of PSS and PEI undergo modifications in response to the surrounding electron (or ion) density, reducing the Van der Waals force with 8-HQ and resulting in the release of the inhibitor from the TiO_2-based coating. Skorb et al. [25] used visible and infrared laser irradiation to stimulate polyelectrolyte multilayer shells (PEI/PSS) or container cores (TiO_2) modified with silver nanoparticles (AgNP). The metal particles acted as light absorption centers, generating local heat that disrupted the local polymer shell and released the charged corrosion inhibitor. In addition, the deposition of two PEI/PSS bilayers on mesoporous TiO_2 particles under UV irradiation proved effective in protecting aluminum alloys against corrosion.

4.4 TRIGGER AND RESPONSE MECHANISMS

The triggering and response mechanisms used to control corrosion inhibitor release in response to external stimuli are varied. They include thermal, chemical, electrical, and light stimuli. Thermal stimuli are based on heat-sensitive materials that soften or melt when the temperature reaches a predefined threshold, enabling the inhibitor to be released. Chemical stimuli use specific reactions triggered by changes in pH, the presence of chemicals or redox reactions, which modify the properties of the encapsulating materials and release the inhibitor. Electrical stimuli involve changes in electrical potential or the application of electrical currents to induce changes in the encapsulation materials and enable the inhibitor to be released. Finally, light stimuli, such as UV or visible light, activate photosensitive materials which undergo chemical changes in response to the light, resulting in the release of the inhibitor. These triggering and response mechanisms enable controlled and tailored release of the corrosion inhibitor in response to external stimuli.

4.5 DISCUSSION OF FUTURE RESEARCH PERSPECTIVES

Future research into corrosion inhibitor release behavior as a function of external stimuli offers many promising prospects. Technological advances will enable the development of even more precise and controllable release systems to meet the specific needs of different corrosive environments. Studies could focus on the design of intelligent materials and innovative coatings capable of detecting corrosion conditions and releasing the inhibitor adaptively. In addition, the integration of multiple stimuli, such as a combination of light and chemical or electrical stimuli, could enable even more targeted and effective release. At the same time, research into the fundamental mechanisms of corrosion inhibitor release and retention will be needed to improve our understanding of the phenomena involved. By exploring new materials, developing advanced manufacturing techniques, and exploiting the possibilities offered by nanotechnology, it is possible to open up new avenues for the design of more effective and sustainable corrosion inhibitor release systems. This future research will help enhance corrosion protection in a variety of industrial applications, and develop more effective and environmentally friendly strategies for extending the life of metal structures.

4.6 CONCLUSION

In conclusion, corrosion inhibitor release behavior as a function of external stimuli represents a promising avenue for developing more effective and sustainable corrosion protection strategies. Triggering and response mechanisms, such as thermal, chemical, electrical, and light stimuli, offer possibilities for precise control of inhibitor release. These technological advances make it possible to design intelligent coatings capable of detecting corrosion conditions and releasing the inhibitor adaptively, thus ensuring optimum protection of metal structures. Future research prospects lie in improving the precision and durability of release systems, integrating multiple stimuli, and exploring new materials and manufacturing techniques. By combining these efforts, we will be able to develop innovative solutions to effectively prevent corrosion, extend the life of materials, and reduce maintenance costs, while preserving the environment.

REFERENCES

[1] W. Daoudi, A. El Aatiaoui, N. Falil, M. Azzouzi, A. Berisha, L.O. Olasunkanmi, O. Dagdag, E.E. Ebenso, M. Koudad, A. Aouinti, M. Loutou, A. Oussaid, Essential oil of dysphania ambrosioides as a green corrosion inhibitor for mild steel in HCl solution, *J. Mol. Liq.* 363 (2022) 119839. https://doi.org/10.1016/j.molliq.2022.119839

[2] A.E. Aatiaoui, M. Koudad, T. Chelfi, S. Erkan, M. Azzouzi, A. Aouniti, K. Savaş, M. Kaddouri, N. Benchat, A. Oussaid, Experimental and theoretical study of new Schiff bases based on imidazo(1,2-a) pyridine as corrosion inhibitor of mild steel in 1M HCl, *J. Mol. Struct.* 1226 (2021) 129372. https://doi.org/10.1016/j.molstruc.2020.129372

[3] S.M. Lashgari, H. Yari, M. Mahdavian, B. Ramezanzadeh, G. Bahlakeh, M. Ramezanzadeh, Application of nanoporous cobalt-based ZIF-67 metal-organic framework (MOF) for construction of an epoxy-composite coating with superior anti-corrosion properties, *Corros. Sci.* 178 (2021) 109099. https://doi.org/10.1016/j.corsci.2020.109099

[4] W. Daoudi, M. Azzouzi, O. Dagdag, A. El Boutaybi, A. Berisha, E.E. Ebenso, A. Oussaid, A. El Aatiaoui, Synthesis, characterization, and corrosion inhibition activity of new imidazo[1.2-a]pyridine chalcones, *Mater. Sci. Eng. B.* 290 (2023) 116287. https://doi.org/10.1016/j.mseb.2023.116287

[5] J. Aljourani, K. Raeissi, M.A. Golozar, Benzimidazole and its derivatives as corrosion inhibitors for mild steel in 1M HCl solution, *Corros. Sci.* 51 (2009) 1836–1843. https://doi.org/10.1016/j.corsci.2009.05.011

[6] X. Xing, J. Wang, W. Hu, Inhibition behavior of Cu-benzoltriazole-calcium alginate gel beads by piercing and solidification, *Mater. Des.* 126 (2017) 322–330. https://doi.org/10.1016/j.matdes.2017.04.024

[7] E. Lee, B. Kim, Preparation and characterization of pH-sensitive hydrogel microparticles as a biological on–off switch, *Polym. Bull.* 67 (2011) 67–76. https://doi.org/10.1007/s00289-010-0403-x

[8] K.A. Zahidah, S. Kakooei, M.C. Ismail, P. Bothi Raja, Halloysite nanotubes as nanocontainer for smart coating application: a review, *Prog. Org. Coat.* 111 (2017) 175–185. https://doi.org/10.1016/j.porgcoat.2017.05.018

[9] Q. Zhu, X. Xu, Y. Huang, S. Liu, A. Zuo, Y. Tang, pH-responsive mesoporous silica nanocontainers based on Zn-BTA complexes as stoppers for controllable release of corrosion inhibitors and application in epoxy coatings, *Prog. Org. Coat.* 181 (2023) 107581. https://doi.org/10.1016/j.porgcoat.2023.107581

[10] C. Zea, R. Barranco-García, J. Alcántara, J. Simancas, M. Morcillo, D. de la Fuente, pH-dependent release of environmentally friendly corrosion inhibitor from mesoporous silica nanoreservoirs, *Microporous Mesoporous Mater.* 255 (2018) 166–173. https://doi.org/10.1016/j.micromeso.2017.07.035

[11] R. Ghamsarizade, A.A. Sarabi, Sh. Roshan, H. Eivaz Mohammadloo, Study on release and inhibition action of mixed ZAPP and 8-HQ corrosion inhibitors loaded in pH-sensitive microcapsules for Mg AZ31, *Colloids Surf. Physicochem. Eng. Asp.* 644 (2022) 128883. https://doi.org/10.1016/j.colsurfa.2022.128883

[12] T. Biswick, D.-H. Park, Y.-G. Shul, J.-H. Choy, P-coumaric acid–zinc basic salt nanohybrid for controlled release and sustained antioxidant activity, *J. Phys. Chem. Solids.* 71 (2010) 647–649. https://doi.org/10.1016/j.jpcs.2009.12.058

[13] S. Chen, Z. Huang, M. Yuan, G. Huang, H. Guo, G. Meng, Z. Feng, P. Zhang, Trigger and response mechanisms for controlled release of corrosion inhibitors from micro/nanocontainers interpreted using endogenous and exogenous stimuli: a review, *J. Mater. Sci. Technol.* 125 (2022) 67–80. https://doi.org/10.1016/j.jmst.2022.02.037.

[14] D.V. Andreeva, D.G. Shchukin, Smart self-repairing protective coatings, *Mater. Today.* 11 (2008) 24–30. https://doi.org/10.1016/S1369-7021(08)70204-9

[15] C. Zhou, X. Xie, H. Yang, S. Zhang, Y. Li, C. Kuang, S. Fu, L. Cui, M. Liang, C. Gao, Y. Yang, C. Gao, C. Yang, Novel class of ultrasound-triggerable drug delivery systems for the improved treatment of tumors, *Mol. Pharm.* 16 (2019) 2956–2965. https://doi.org/10.1021/acs.molpharmaceut.9b00194

[16] G.L. Li, M. Schenderlein, Y. Men, H. Möhwald, D.G. Shchukin, Monodisperse polymeric core–shell nanocontainers for organic self-healing anticorrosion coatings, *Adv. Mater. Interfaces.* 1 (2014) 1300019. https://doi.org/10.1002/admi.201300019

[17] D. Snihirova, S.V. Lamaka, M.F. Montemor, "SMART" protective ability of water based epoxy coatings loaded with CaCO3 microbeads impregnated with corrosion inhibitors applied on AA2024 substrates, *Electrochimica Acta.* 83 (2012) 439–447. https://doi.org/10.1016/j.electacta.2012.07.102

[18] D. Snihirova, S.V. Lamaka, M. Taryba, A.N. Salak, S. Kallip, M.L. Zheludkevich, M.G.S. Ferreira, M.F. Montemor, Hydroxyapatite microparticles as feedback-active reservoirs of corrosion inhibitors, *ACS Appl. Mater. Interfaces.* 2 (2010) 3011–3022. https://doi.org/10.1021/am1005942

[19] G.L. Li, Z. Zheng, H. Möhwald, D.G. Shchukin, Silica/polymer double-walled hybrid nanotubes: synthesis and application as stimuli-responsive nanocontainers in self-healing coatings, *ACS Nano.* 7 (2013) 2470–2478. https://doi.org/10.1021/nn305814q

[20] S. Jiang, L. Lv, Q. Li, J. Wang, K. Landfester, D. Crespy, Tailoring nanoarchitectonics to control the release profile of payloads, *Nanoscale.* 8 (2016) 11511–11517.

[21] T. Wang, L. Tan, C. Ding, M. Wang, J. Xu, J. Fu, Redox-triggered controlled release systems-based bi-layered nanocomposite coating with synergistic self-healing property, *J. Mater. Chem. A.* 5 (2017) 1756–1768. https://doi.org/10.1039/C6TA08547D

[22] N. Joseph, P. Ahmadiannamini, R. Hoogenboom, I.F.J. Vankelecom, Layer-by-layer preparation of polyelectrolyte multilayer membranes for separation, *Polym. Chem.* 5 (2014) 1817–1831. https://doi.org/10.1039/C3PY01262J

[23] N.N. Taheri, B. Ramezanzadeh, M. Mahdavian, Application of layer-by-layer assembled graphene oxide nanosheets/polyaniline/zinc cations for construction of an effective epoxy coating anti-corrosion system, *J. Alloys Compd.* 800 (2019) 532–549. https://doi.org/10.1016/j.jallcom.2019.06.103

[24] X. He, C. Chiu, M.J. Esmacher, H. Liang, Nanostructured photocatalytic coatings for corrosion protection and surface repair, *Surf. Coat. Technol.* 237 (2013) 320–327. https://doi.org/10.1016/j.surfcoat.2013.06.038

[25] E.V. Skorb, A.G. Skirtach, D.V. Sviridov, D.G. Shchukin, H. Möhwald, Laser-controllable coatings for corrosion protection, *ACS Nano.* 3 (2009) 1753–1760. https://doi.org/10.1021/nn900347x

5 Encapsulation of Nanoparticles as Corrosion Inhibitors for Smart Coatings

Luana B. Furtado, Carolina Garín Correa, and Rafaela C. Nascimento

5.1 INTRODUCTION

Organic coatings are highly effective in protecting metallic structures against corrosion [1]. Coatings act as a physical barrier that isolates the material from the specific media. Depending on the specific protection mechanism, there are three types of organic coating: barrier, sacrificial, and inhibitive [2]. Problems associated with coatings are surface defects such as microcracks, bubbles, and micropores, which usually occur during application. As such, the protective effect of the coating gradually degrades, leading to performance deficiencies and localized corrosion on the substrate, especially in marine environments.

In order to effectively extend the protective service life of coatings, intelligent self-healing coating systems have attracted considerable interest in the scientific world. These coatings have been classified into two main types: damage repair coatings and intelligent corrosion inhibitor release coatings. The coatings commonly referred to as repair coatings respond intelligently to damage by delaying substrate corrosion via active agents that autonomously cure the aged matrix with no external trigger [3].

Likewise, in so-called intelligent or smart coatings, natural substances such as pigments, functionalized capsules, or particles are generally used. The most innovative format is the use of nanoparticles such as mesoporous silica, layered double hydroxide (LDH) nanocontainers, polymeric particles, porous TiO_2, graphene, carbon nanotubes, halloysite nanotubes (HNT), multilayer systems, zinc phosphate ion-exchanged particles, and LDH pigments. The corrosion inhibitor is generally released into the polymeric organic coatings in a controlled manner [4]. Their protective mechanism can be attributed to chemical reactions or physisorption, producing a protective film on the substrate.

Inserting corrosion inhibitors into coatings can reduce their mechanical and barrier properties due to inhibitor interference with the functional groups, curing reactions of the coating, and/or uncontrollable inhibitor leaching into the polymer matrix [2, 5, 6]. Storing the corrosion inhibitor in nanocontainers prevents the problems that arise when corrosion inhibitors are added directly to the coating layer, including uncontrollable release, rapid depletion of the inhibiting substance, and coating damage such as osmotic blistering [7].

In this respect, this chapter provides an overview of smart coatings based on nanocontainers that encapsulate a corrosion inhibitor. The different types of nanocontainers were divided into three main categories: inorganic, organic, and hybrid. The release of corrosion inhibitors is

 DOI: 10.1201/9781032677255-5

pH-dependent. Throughout the chapter, the anticorrosion performance and characterization of these nanostructures are addressed. Finally, some gaps are pointed out for future studies.

5.2 DESIGNING THE ENCAPSULATION RESERVOIR (ACCORDING TO MATRIX TYPES)

The development of efficient protective systems that can self-heal is important for industrial applications [7]. One promising approach is to combine a passive coating (containing a matrix with nanometric dimensions) with an active corrosion inhibitor (mostly organic or inorganic compounds). The search for controlled methods for active substance release arose from barrier problems in organic coatings [8]. With an undesirable polymer matrix, inhibitor interactions can be prevented by trapping these molecules inside nanocontainers [9]. The active compounds can rapidly leach from the coating, weakening the protective barrier [6]. Encapsulating the corrosion inhibitor is fundamental to overcoming these problems while providing a more sustainable approach to substitute the efficient but unsafe chromates [7].

The design of the encapsulation system usually falls into one of the following categories: inorganic nanocontainers, polymeric capsules, and hybrid inorganic nanocontainers coated with responsive polymeric shells [9] (see Figure 5.1). The controlled and sustained release of the active compound will strongly depend on the structure of the nano/micro-container matrix, its chemical nature, and the media pH. Other vital aspects are the chemical characteristics of the corrosion inhibitor, its solubility and performance in the media, and its ability to repair the corroded surface areas.

5.2.1 Inorganic Nanoreservoirs

A nanocontainer, also called a nanoreservoir, is a nanosized matrix filled with an active compound. This active substance can be confined inside a porous material, intercalated or entrapped between layers, or a combination of one of these materials with a polymeric shell. The purpose is to incorporate the nanoreservoir into the polymeric matrix. The nanoreservoir contains active substances, which, once released, will interact with the surface damage [10].

Among the inorganic reservoirs is the group of nanocrystalline LDHs, which, due to their lamellar structure, can store and release anionic inhibitors such as vanadates, molybdates, etc. [10]. LDHs are inorganic solids with a structure similar to that of hydrotalcite, where the partial substitution of trivalent cations for divalent cations results in a positive sheet charge. The

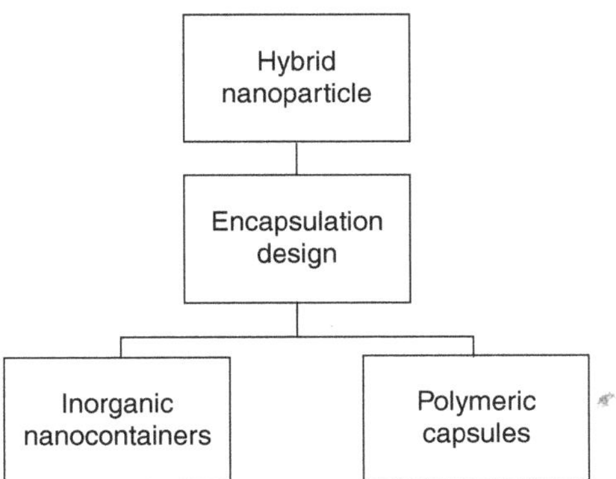

FIGURE 5.1 Diagram representing the encapsulating design according to the matrix.

accumulated charge is offset by anions in the interlayer space [11]. Molybdate ions were successfully intercalated within ZnAlCe LDH to inhibit the corrosion of Q235 steel in 3.50% NaCl solution. Although the dimensions were in micrometers (particle size of 0.1–2.0 μm), the anion exchanged for molybdate reduced chloride concentration. Corrosion resistance increased due to a passive film composed of ferrous or iron molybdate. This means that molybdate ions can act as ion exchangers, thereby lowering chloride concentration and forming stable oxides with Fe^{2+} ($FeMoO_4$) and Fe^{3+} ($Fe_2Mo_3O_{12}$). The presence of Zn and Ce was responsible for the deposition of zinc and cerium hydroxide film, which act as an additional protective layer [12]. A similar strategy could be adopted to synthesize new nanocontainers via co-precipitation to produce environmentally friendly smart coatings.

The use of natural clays, such as bentonites and halloysites, as micro- and/or nanoreservoirs for corrosion inhibitors to formulate anticorrosion coatings has also been reported. Similarly to LDHs, they retain aggressive ions from the corrosive media and release corrosion inhibitors only when the corrosive species permeate the coating. The presence of a permanent or intrinsic charge is a common characteristic that differs from clays. While the latter are negative, hydrotalcites (LDHs) are positive. Coating systems based on calcium-exchanged silica (Si/Ca), hydrotalcite/vanadate (HT/V), and calcium bentonite (bentonite/Ca) were applied as ion exchangers to assess their anticorrosive behavior. The results for the metal/coating adhesion of the primer paint showed intermediate performance for films formulated with bentonite/Ca, between that of Si/Ca and HT/V coatings. In addition, bentonite/Ca films to protect cold rolled low-carbon steel exhibited low permeabilities for oxygen and chloride ions. The drawback is the significant delamination observed after analyzing the evolution of coating resistance with immersion time in 0.10 M NaCl [13]. Although the study was not primarily based on the use of clays as nanoreservoir, it shows that the potential for ion exchange could be useful to future investigations. The ability to retain/release ions and the enhanced barrier properties should be considered.

Inorganic nanoreservoirs include halloysites, which are naturally available tubular nanocontainers that have been combined with organic corrosion inhibitors. HNTs are two-layered hollow aluminosilicate cylinders with a lumen diameter of approximately 10–150 nm and 1–15 μm long [14]. Corrosion inhibitors can be incorporated into the hollow lumens of halloysites (generally in acetone medium). When filled with active anticorrosive substances, these nanocontainers can be dispersed in a primer coating. Their potential as prospective nanocontainers is attributed to their availability, low cost, durability, and high mechanical strength [15]. HNTs) must be prepared before loading the corrosion inhibitor in order to increase their volume for encapsulation, which can be achieved by exposing them to sulfuric acid solution (2 mol L^{-1}). Falcón, Sawczen, and Aoki [15] applied this procedure to load dodecylamine into HNTs to prevent corrosion of AISI 1020 carbon steel in 0.01 mol L^{-1} NaCl solution. Next, dodecylamine was embedded into the halloysite inner gallery by adding an ethanolic solution containing dodecylamine to the previously prepared HNTs. The ethanolic solution was then evaporated and the organic compound was loaded into halloysite. The nanocontainer was then washed, centrifuged, and dried. The loaded HNTs (shown in Figure 5.2) were dispersed in the paint diluent before being added to the epoxy coating. The corrosion resistance of carbon steel samples prepared with nanotubes containing dodecylamine was evaluated by electrochemical impedance spectroscopy (EIS). Bode diagrams for the two-layered coated samples containing 0 and 10 wt.% of HNTs in the primer showed capacitive behavior in the absence of defects. A slight decrease in impedance modulus at low frequencies demonstrated that adding HNTs did not change the barrier properties of the coating. This indicates that good dispersion was achieved, preventing agglomeration. For the samples with defective coatings, 10 wt.% of HNTs resulted in corrosion inhibitor release into the affected area after 4 h of immersion. After 8 h of immersion, self-healing was observed due to the increased impedance modulus at low frequencies, caused by the release, migration, and adsorption of dodecylamine at the defect site [15].

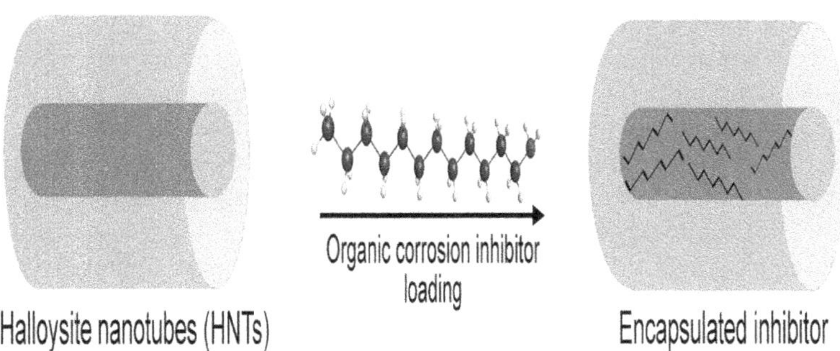

FIGURE 5.2 Halloysite nanotubes (HNTs) preparation with dodecylamine as a corrosion inhibitor.

With regard to nanotubes, Wang, Wang, and Hu [3] used 8-hydroxyquinoline (8-HQ) as a corrosion inhibitor, which was incorporated into HNTs under negative pressure. Encapsulation was then performed using Cu-8-HQ complex stopper. Two coatings were obtained by adding 10 wt.% 8-HQ@HNTs or 10 wt.% Cu-8-HQ@HNTs to the epoxy. 8-HQ@HNTs and Cu-8-HQ@HNTs were, respectively, mixed with the epoxy, and then epoxy hardener B solution was added to the mixture by stirring vigorously. Next, the as-prepared blank epoxy, 8-HQ@ HNT and Cu-8-HQ@ HNT epoxies were used to coat a steel substrate, resulting in a final coating thickness of about 50 ± 5 μm. The water contact angle for the epoxy was 65 ± 1°. After adding 8-HQ to the epoxy coating, the contact angle increased to 81 ± 1°, and the contact angle of the Cu-8-HQ@HNTs/epoxy coating increased to 88 ± 1°. It can be concluded that the addition of 8-HQ@HNTs or Cu-8-HQ@HNTs reduces the hydrophilicity of the epoxy coating, thereby preventing penetration of the electrolyte solution. EIS tests were performed in 3.50% NaCl solution for 90 days. Two capacitive areas were observed in the EIS spectra of the non-modified epoxy coating, but the capacitive area ratio decreased continuously. These results demonstrate that the corrosion resistance of both the blank and 8-HQ@HNT/epoxy coatings declined as immersion time increased, indicating corrosion of the steel substrate. By contrast, EIS of the Cu-8-HQ@HNT/epoxy coating for the same immersion time showed only a capacitive arc, demonstrating that this coating provided better protection, and the electrolyte solution did not diffuse into the steel. The Cu-8-HQ@HNT/epoxy coating exhibited better results than its 8-HQ@HNTs counterpart due to complex formation in the former, smoother coating, and higher hydrophobicity.

Since silica nanoparticles are inorganic nanostructures that produce smart coatings with anticorrosion properties, pH-responsive nanocontainers can release the entrapped corrosion inhibitor from hollow silica nanoparticles. Alternatively, silica nanoparticles may be covered layer-by-layer (L-b-L) with corrosion inhibitor and polyelectrolyte layers (this example will be discussed in the hybrid matrix section) [10]. Yeganeh et al. [9] used silica as nanocontainer for organic corrosion inhibitor (sulfamethazine) to prevent corrosion of mild steel in 3.50 wt.% NaCl. First, previously prepared silica was loaded with the corrosion inhibitor by mixing silica powder with sulfamethazine using acetone as solvent. The loaded nanocontainers subsequently release the corrosion inhibitor in the epoxy coating, enhancing the barrier effect against corrosive species and healing the scratched areas. Based on the charge transfer resistance of epoxy and epoxy/mesoporous silica coatings, it is clear that nanocontainers provided better protection than the epoxy coating even without the inhibitor. This means that after adding silica, the barrier effect improved. The organic corrosion inhibitor improved charge transfer resistance even more. The results indicated that the slow release of this molecule, associated with its adsorption via sulfur atoms in the scratched corroded zone, was responsible for enhanced corrosion protection [9].

Other studies have investigated silica nanoparticles, such as that by Yang et al. [16]. These nanostructures were functionalized with organic substances for mild steel in 3.50% NaCl. The silica nanoparticles were prepared by mixing ammonium hydroxide, anhydrous ethanol, and distilled water with orthosilicate. The sol-gel reaction lasted 4 h at room temperature, reaching 200 nm in diameter. Next, the hydrolysis reaction was conducted between the active substances and silica nanostructures, resulting in surface functionalization. Finally, the functionalized nanoparticles were centrifugated, washed with water to remove free components, and vacuum-dried overnight. The epoxy resin coating was prepared by mechanical stirring, using pure silica nanoparticles, corrosion inhibitor and functionalized corrosion inhibitor. The structure of the corrosion inhibitor was confirmed by Fourier-transform infrared spectroscopy (FTIR). The nanoparticle showed Si-OH bands near 3384 and 956 cm^{-1} and, for the inhibitor, bands at 3292 cm^{-1} are related to N-H stretching vibration and C=O stretching at 1666 cm^{-1}. The nanoparticles functionalized with the inhibiting structure showed bands for N-H, C-H, and carbonyl groups. Thermogravimetric analysis (TGA) was carried out to quantify the inhibitor incorporated into silica nanoparticles, resulting in 8.40%. Field emission scanning electron microscopy revealed that the nanoparticles were more dispersed in the coating due to inhibitor functionalization onto the nanoparticle surface. This is related to a decrease in non-covalent bond interaction between nanoparticles. Mechanical properties were also investigated. Tensile strength decreased with the incorporation of modified nanoparticles, reaching elongation at break. The silica nanoparticles filled the macro- and nanoscale gaps in the polymeric matrix, leading to external energy absorption during deformation. However, this improvement occurred until an aggregation effect was reached due to greater nanoparticle concentration.

Zhao et al. [17] chemically modified silica nanoparticles with 1 H, 1 H, 2 H, 2 H-perfluorodecyltrimethylsilane (FAS-17), producing fluorosilane-modified SiO_2 nanoparticles (FS). The inorganic nanoparticles were dispersed in an ethanol-water mixture at room temperature for 30 min. Next, FAS-17 was added to the system and reacted at 363 K for 6 h. The product was then washed, centrifuged, and dried at 353 K. For coating, the FS nanoparticles were spray-coated at a concentration of 0.10 mg L^{-1}. Next, epoxy resin (EP) and curing agent (mass ratio of 4:1) were mixed, diluted in ethyl acetate, and dispersed using ultrasound in order to obtain an epoxy paint of 0.25 g mL^{-1}. FS was added to the epoxy paint to obtain FSEP paint. Nanoparticle functionalization was confirmed by FTIR, which indicated bands at 1245 and 566 cm^{-1} related to C-F and C-F2 stretching, respectively, confirming the successful incorporation of –CF2 and –CF3 low-surface energy groups responsible for hydrophobicity on the SiO_2 surface. The ratio between the functionalization agent and inorganic nanoparticles is important since a 3:10 ratio led to a higher hydrophobicity (contact angle of 162.0º). An additional ratio increase did not increase the hydrophobicity, since no further reaction takes place between hydroxyl groups and the silica surface due to hindrance effects. Moreover, nanoparticle size affects hydrophobicity, which is higher for smaller particles, probably due to better surface coverage. Another important aspect is the ratio between the modified nanoparticles and EP, which influences coating stability and durability. Analysis of different ratios indicated that the best result was achieved at 7:10, which produced the best adhesion between EP and the modified nanoparticles. Thus, a further modified nanoparticle increase will not lead to higher stability due to adhesion loss.

As such, different inorganic nanostructures can be used to incorporate and release corrosion inhibitors. Nanocrystalline structures and natural clays provided layers that enable ion exchange. Halloysite is composed of tubular structures that require previous preparation to increase the inner space into which the corrosion inhibitor is loaded. The efficiency of this process can be confirmed by techniques such as TGA and FTIR. Silica nanoparticles can be employed layer by layer or by surface modification, which can also be confirmed by FTIR. The amount of inhibitor incorporated into these structures can be determined by TGA. The modifying agent: silica nanoreservoir ratio should also be considered, since after reaching a limit, no further incorporation is achieved.

5.2.2 Organic/Polymeric Shells

Nanocontainers can be designed and synthesized to encapsulate the corrosion inhibitor inside a polymeric shell. The organic shell efficiently isolates the active compound and is highly compatible with the coating. In this respect, Farzi et al. [8] encapsulated cerium nitrate within poly(urea-formaldehyde) (PUF) microcapsules. These microcapsules have a number of advantages, including easy preparation and chemical and thermal stability. They are also compatible with epoxy-based coating. The PUF spheres were prepared in two synthesis steps. The first was a polycondensation reaction to form the urea-formaldehyde prepolymer. Since urea-formaldehyde is a water-soluble gas (at room temperature), it was employed in an aqueous solution (37%) called formalin. The prepolymer was obtained by mixing and stirring urea and formalin at 343 K for 1 h. Decreasing the temperature to 298 K turned the homogeneous solution into suspended solids. The second step resembled direct emulsion polymerization. The prepolymer was added to a solution containing polyethylene maleic anhydride, an emulsifying agent, to stabilize the spheres. Stirring produces fine bubbles for size-controlled sphere production. The PUF particles form around the bubbles, acquiring a spherical shape. These polymeric shells are initially hollow. After coming into contact with the inhibiting solution ($Ce(NO_3)_3$ in water), the polymeric material is gradually filled due to its thin shells. TGA and FTIR confirmed the high $Ce(NO_3)_3$ content. The self-healing ability of the modified coating was studied in 0.60 M NaCl using EIS. The electrochemical results confirmed inhibitor release in the damaged area, resulting in efficient self-healing of the epoxy-based coating. Chlorine and cerium ions precipitated in the form of cerium hydroxides and oxides in the damaged area (as shown in Figure 5.3), resulting in a passive layer. In addition, the EIS results demonstrated that healing performance is a function of the microcapsules and self-healing agent concentration. The best self-healing result was obtained at the highest microcapsule concentration (10 wt%). However, an increase in microcapsule concentration decreases adhesion to the substrate.

Another study using emulsion polymerization was performed by Choi et al. [18], who investigated polymeric nanocapsules loaded with six types of amine corrosion inhibitors, which were synthesized by multistage emulsion polymerization. Amines with different molecular weights,

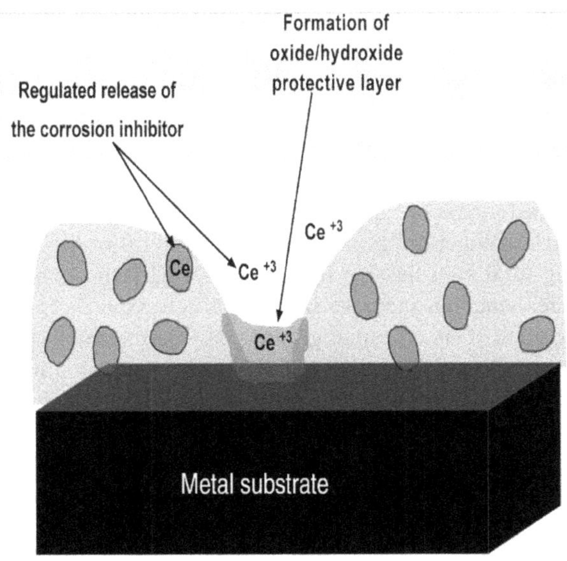

FIGURE 5.3 Schematic representation of poly(urea-formaldehyde) (PUF) nanospheres as encapsulating agent for cerium.

functional groups, and basicity were used in encapsulation: diethanolamine (DEA), ethanolamine (ETA), propylamine (PPA), triethanolamine (TEA), dipropylamine (DPA), and 5-amino-1-pentanol (5AP). A polyurethane resin was used as the main binder, and the synthesized nanocapsules in aqueous dispersion were directly added to the coating resin with a curing agent and a small concentration of a wetting agent. The concentration of the incorporated capsules was 33.30 wt% of the total solid in the coating system. The first step of the synthesis procedure consisted of the formation of a poly(MMA-co-MA-co-BA) (MMA – methyl methacrylate; MA – methacrylic acid, BA – butyl acrylate) shell on the carboxylated seed material. The hydrophilic characteristics of seed materials make it difficult to fabricate the hydrophobic polystyrene shell with concentric core-shell particle morphology. Thus, the intermediate shell, which is amphiphilic, was forced to increase the affinity between the hydrophilic core and hydrophobic polystyrene shell. The next step involved the formation of the outermost shell, composed of polystyrene. The prepared nanoparticles only have sufficient physical stability after the complete formation of the styrene shell and can be used as additives in the coating resin. The final step was swelling, achieved by neutralizing the core with amine species.

Yang and Ooij [19] developed a method to encapsulate an organic triazole inhibitor using plasma polymerization. The resulting plasma polymerized perfluorohexane, and polypyrrole layers were applied to prevent corrosion of the AA2024-T3 alloy, which is suitable for aerospace applications. Plasma-treated triazole inhibitors were mixed with a water-based epoxy coating, providing long-term corrosion protection. The slow release of three types of organic inhibitors was provided by the formation of plasma-treated layers. The multilayer structure was selected because of the super hydrophobic plasma-polymerized perfluorohexane layer. When a hydrophilic layer is applied to the coating, it helps disperse the particles while maintaining coating hydrophobicity. The inner layer of the double-layered samples was composed of perfluorohexane and the outer layer of pyrrole. The organic triazole derivative tolyl triazole (TTA), benzotriazole (BTA), and 1,2,4-triazole (TA) are effective corrosion inhibitors in solution. After incorporating them into the paint, it was observed that the encapsulated and water-soluble corrosion inhibitors could be controllably released in 3.50 wt.% NaCl, thereby protecting AA2024-T3 panels. They are as effective as chromate-based inhibitors for preventing corrosion but are more environmentally friendly.

In terms of an organic inhibitor incorporated into an organic nanostructure, Choi et al. [20] produced nano-sized capsules with a shell containing methyl methacrylate, butyl acrylate, and methacrylic acid. The initiator was potassium persulfate and the anionic emulsifier was ammonium nonylphenol ether sulfate. Synthesis was performed in deionized water at 353 K. The second shell had the same constituents, but in different proportions, with the addition of styrene. Since corrosion inhibitor (triethanolamine) penetration into the shell is not easy, triethanolamine was inserted into the system with styrene in the pre-emulsion mixture in order to create the shell at the same time as the active substance is encapsulated. The second shell was hydrophobic and provided stability to the nanoparticles. Water diffusion with the inhibitor through the outer shell caused particle swelling, increasing total capsule size by 100–300 nm. This phenomenon should be better evaluated in future studies since an increase in nanoparticle volume is correlated with inhibitor concentration inside the shell. Gas chromatography-mass spectrometryrevealed that the volume occupied by the inhibitor increased from 2.19 to 5.04% with a rise in inhibitor concentration from 0.08 to 0.15 M. This characterization was important because a further increase to 0.20 M did not result in greater inhibitor incorporation, demonstrating a limit for inhibitor incorporation into the nanoparticle structure. Thus, self-release activity should be more thoroughly studied in order to investigate nanocapsule concentration in the coating considering the amount of inhibitor released into the media.

In this respect, it is important to note that polymeric nanocapsules are obtained through emulsion and plasma polymerization. More than one shell is produced, which may be associated with inhibitor release and nanocapsule stability. Given that inhibitor incorporation into the shell is sometimes difficult, the best strategy may be shell formation at the same time as inhibitor incorporation. In addition, the limit for inhibitor incorporation into the polymeric shell should be determined in order to evaluate the number of nanocapsules required for a specific medium and application, as a function of the amount of inhibitor released in the medium.

5.2.3 Hybrid Nanoreservoirs

Another type of nanocontainer is the hybrid variety, that is, based on organic and inorganic origins. Ashrafi-Shahri, Ravari, and Seifzadeh [21] chemically modified mesoporous silica nanoparticles by introducing a nitrogenous functional group. The nanoparticles were then functionalized using 3-aminopropyl triethoxysilane with toluene, under a vacuum. Next, the particles were added to 1-propanol containing $FeCl_3$ in order to produce an $N{=}Fe^+$ bond. This bond was used to incorporate the corrosion inhibitor (Eriochrom Black T), which contains aromatic rings, nitrogen, oxygen, and sulfur heteroatoms. These sites are responsible for corrosion inhibition. Additionally, the porosity of silica particles enables them to incorporate the inhibiting molecules. BET tests were conducted to measure the surface area and porosity of the nanoparticles before and after corrosion inhibitor incorporation. The authors reported a decrease in specific surface area due to inhibitor incorporation into the mesoporous spaces of the nanoparticles. FTIR was used to confirm the different modification steps, with bands at around 3400 and 1620 cm^{-1} related to Si-OH stretching and bending vibrations, respectively. The decreased intensity of the band at 933 cm^{-1} confirmed functionalization. This band is related to Si-OH vibration, whose intensity declined due to the chemical reaction between amine and silanol functional groups. With regard to the incorporation of the inhibiting molecules, the bands at 1336 and 1053 cm^{-1} correspond to nitro and HSO_3 groups in the inhibitor structure.

In relation to the hybrid inhibitor incorporated into a hybrid nanocontainer, Manasa et al. [22] investigated the loading of $Ce(NO_3)_3.6H_2O$ and zirconium n-propoxide into nanotubes for aluminum alloy A356.0 protection in NaCl. The nanotubes were etched in H_2SO_4 solution for 3 days at 333 K, followed by washing with deionized water and drying to achieve a larger pore volume. 3-glycidoxypropyltrimethoxysilane was hydrolyzed with tetraethoxysilane in the presence of HCl, using a molar ratio of 3:5:1 to prepare the sol-gel matrix. The inhibitors were then incorporated into the nanotubes at a Ce^{3+}/Zr^{4+} ratio of 1:23 in a stirred system. The surface area was determined by BET analysis, which indicated a pore volume of 0.145 cm^3/g for the nanotubes before treatment, 0.320 cm^3/g for the treated nanotubes and 0.029 cm^3/g for the nanostructure after inhibitor incorporation. The 80% decrease in the pore volume confirmed the loading of the active substances into the nano reservoir.

In terms of inorganic particles covered with natural organic compounds, Izadi, Shahrabi and Ramezanzadeh [23] prepared a bilayer on Fe_3O_4 nanoparticles. The L-b-L process is a multistage procedure to simultaneously precipitate the three layers on the nanoparticle. The nettle extract (NE) was previously prepared with nettle leaves in an aqueous solution at 343 K for 4 h, followed by filtration and concentration. Next, the nanoparticles were transferred to the first layer for polyaniline deposition in an ultrasound reactor for 1 h at 277 K, followed by centrifugation (10,000 rpm, 15 min) and drying (323 K, 5 h). The green extract was then incorporated into the nanoparticles under ultrasound irradiation for 1 h. The last step consisted of a polyacrylic acid layer in a similar procedure. FTIR revealed bands at 1294 and 1315 cm^{-1} related to C-N of aromatic amine in the leaf extract and at 1415 and 1622 cm^{-1} corresponding to amide groups. A second layer was confirmed by a band at 1728 cm^{-1}. This approach is interesting because it adds value to the leaf extract, which is incorporated as an intermediate shell of the inorganic nanoparticles.

With respect to the aforementioned L-b-L film formation technique, Zheludkevich et al. [7] used this method to produce a hybrid inorganic nanocontainer (silica), covered by polymeric layers and benzotriazole (corrosion inhibitor). Since SiO_2 nanoparticles are negatively charged,

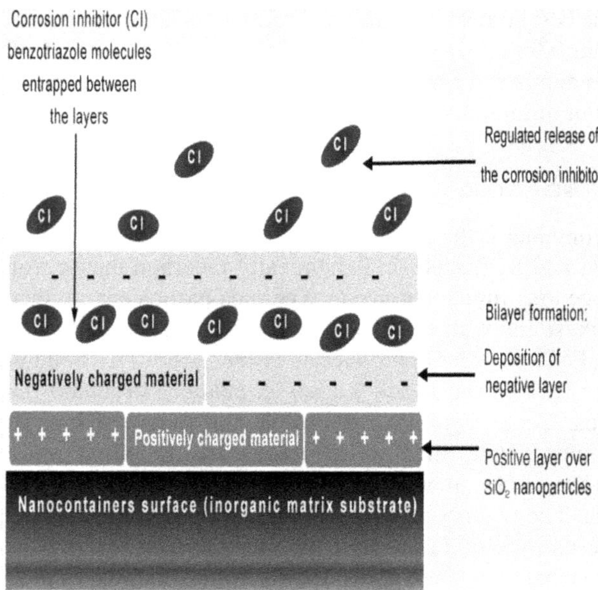

FIGURE 5.4 Layer-by-layer film structure for silica nanocontainers. Layer structure intercalating the corrosion inhibitor (SiO_2/PEI/PSS/benzotriazole/PSS/benzotriazole). PEI: poly(ethylene imine), PSS: poly(styrene sulfonate).

the first stage is layer deposition of the positive poly(ethylene imine) (PEI), followed by a negative poly(styrene sulfonate) (PSS) layer and a third inhibitor layer of benzotriazole. The last two steps (PSS and benzotriazole) were repeated to achieve maximum inhibitor incorporation in the final L-b-L nanostructure. The nanocontainers mitigate the negative consequence of directly adding the inhibitor to the matrix and provide corrosion inhibitor delivery on demand. When corrosion begins, the pH of the affected area changes, opening the nanocontainer shell to deliver benzotriazole. The resulting nanocontainers are presented in Figure 5.4, with a layer structure of SiO_2/PEI/PSS/benzotriazole/PSS/benzotriazole.

In terms of silica nanoparticles combined with organic layers, Qian et al. [24] studied the dry nanocontainer powers of SiO_2/chitosan/alginate/polyaspartic acid/alginate/polyaspartic acid, which were added (5 wt.%) to the epoxy matrix. A curing agent was then applied (polyamide). The metallic samples (mild steel) were coated using an OSP applicator to achieve 50 ± 5 μm thicknesses. Nanoreservoir potential after each layer was monitored, confirming their deposition due to opposite electrical charges. The potential of the initial SiO_2 nanoparticles was –32 mV. The chitosan layer increased the surface charge to –5 mV. Once the alginate layer was deposited, the potential decreased to –33 mV. Polyaspartic acid (PASP) deposition in an acidic solution yields a more positive surface potential (1 mV). A strategy to incorporate a higher inhibitor concentration in the nano reservoir was to deposit alginate/PASP bilayers onto SiO_2 nanoparticles twice, leading to a potential of 9 mV.

In another study, Rahsepar Mohebbi and Hayatdavoudi [5] used the 2-mercaptobenzothiazole (MBT)-loaded mesoporous silica nanocontainers at concentrations of 1, 3, and 5 wt% in the coating. The coatings were air sprayed, achieving a 55 μm thickness. In addition, the nanostructures exhibited a narrow size distribution (75–120 nm), which increased slightly after the inhibitor incorporation. In the ultraviolet-visible (UV-Vis) spectrum, the peak associated with the inhibitor was observed at 320 nm, confirming successful encapsulation.

Mahmoudi et al. [25] investigated the anti-corrosive performance of silane coating with praseodymium (Pr) encapsulated in HNTs. A similar approach was used in another study [26] to protect magnesium alloy. Praseodymium is a lanthanide and, like cerium, can form protective oxide/hydroxide films. HNT preparation to increase the inner lumen and the loading step of the inhibitor was similar to that of other reports [15]. The loaded HNTs were then encapsulated with urea-formaldehyde capsules to controllably release the Pr ions onto the scratched areas exposed to 3.50% NaCl solution. Surface characterization confirmed the presence of a protective oxide/hydroxide layer provided by the release of Pr^{3+} ions.

In this respect, studies have focused on modifying inorganic structures such as silica, nanotubes, and lanthanides with organic compounds, providing corrosion inhibitor encapsulation. Another reported strategy is the L-b-L technique, where inorganic structures such as silica receive polymeric or simple organic layers, with the middle layer as the organic inhibitor to be released. It is also important to note that the inhibitor inside the hybrid container can be organic, inorganic, or hybrid. The inhibition mechanism of the hybrid option likely involves complex formation between inorganic ions such as cerium and the organic inhibiting molecules.

5.3 ENCAPSULATION AND CHARACTERIZATION TECHNIQUES

Corrosion inhibitors can be incorporated into the coating system as pretreatment, primer or top-coat. It should be noted that, in order to be efficient, inhibiting agents should be soluble in the studied media and applied at a specific concentration. If solubility is low, there are not enough active substances on the metallic substrate to protect it. However, if it is too high, protection will last less due to rapid leaching from the coating. Another issue related to high solubility is delamination blisters of the coating caused by water permeating through the coating [7].

Sol-gel is one of the most widely used methods to obtain hybrid organic–inorganic materials, typically with metal oxide nanoparticles. It involves a molecular precursor, such as metal alkoxide, dissolution in a solvent (water or alcohol), followed by gel transformation caused by hydrolysis/alcoholysis under heating and stirring [27]. This process manipulates the synthesis parameters to produce silica particles with controlled particle size, size distribution, and morphology [28].

Inhibitor-loaded deposits using the L-b-L strategy have several advantages over other methods, such as on-demand inhibitor release, the absence of harmful interactions between inhibitor molecules and coating components, and long-term inhibitor feeding at corrosion sites. In this case, the core of the nanostructure is usually solid, consisting of metal oxides and/or inorganic/organic compounds. Solid cores have several advantages, such as maintaining the barrier properties of the coating after electrolyte diffusion, reducing deposit damage to the organic coating during the dispersion stage and facilitating the synthesis of deposits in the nanoscale range. Different layers can be deposited on the core via L-b-L and corrosion inhibitors can be loaded between the layers. Finally, a polymeric shell covers the nanoreservoir layers before application. The nanoreservoir coating can be obtained by in situ emulsion polymerization. Nanocontainers or nanoreservoirs, such as sandwiches, are prepared by alternating deposition of oppositely charged polyelectrolytes or inhibitors on the surface of nanoparticles such as silica nanoparticles. Adding inhibitors to basic polyelectrolyte multilayers may hinder chemical interactions between the active substances and the matrix. This technique releases the inhibitor only when the local pH changes during corrosion.

Qian et al. [24] used SiO_2 nanoparticles coated with chitosan/alginate layers and shell-trapping PASP inhibitor between layers, as prospective nanocontainers embedded in epoxy coatings on steel substrates. The results demonstrated that the self-healing coating protects the steel against corrosion by releasing PASP, which is adsorbed onto the electrochemically corroded steel. Gradual corrosion can also be prevented by multilayers of swollen polyelectrolytes that cover the defect caused by the scratching test.

In another study by Izadi, Shahrabi and Ramezanzadeh [2], an active nanoreservoir of Fe_3O_4-NS was studied and obtained based on NE. A polyelectrolyte layer of polyaniline encapsulated the corrosion inhibitor on the Fe_3O_4 nanoparticles (Fe_3O_4-NP) and prevented inhibitor release from Fe_3O_4-NS before application. The authors demonstrated that the good inhibitory performance of Fe_3O_4-NS is related to • electrons and oxygen functional groups within the green NE, such as quercetin, quinic acid, histamine, and serotonin.

5.4 INHIBITOR RELEASE FROM THE NANOPARTICLES

Applying unmodified sol-gel coatings to substrates usually provides little protection due to porosity and low thickness. In this respect, inhibiting molecule release from nanoparticles, also called nanocarriers, is a suitable strategy to overcome this problem. This release is typically associated with pH changes. Choi et al. [20] produced organic nano-sized capsules containing triethanolamine as a corrosion inhibitor, which is released in the medium depending on pH. This release, as a function of pH, is linked to local corrosion simulation. If the coating is damaged, the affected region acts as an anode. Cathodic regions are associated with oxygen reduction in undamaged areas, raising the pH to 8. These pH changes are responsible for triethanolamine release through microchannels on the polymeric shell due to osmotic swelling. Additionally, as a function of pH and time, both acidic and alkaline environments lead to inhibitor release. This release is independent of the residence time in solution, a desirable characteristic for corrosion protection.

Ashrafi-Shahri, Ravari, and Seifzadeh [21] prepared silica nanoparticles based on functionalization with 3-aminopropyl triethoxysilane, followed by the incorporation of a corrosion inhibiting molecule. The inhibiting molecules are released as a function of alkaline regions. In order to confirm the pH sensitivity of the functionalized nanoparticles, UV-Vis spectra were recorded between 6.5 and 13 pH. Maximum absorption was observed at more alkaline values, confirming inhibitor release sensitivity to pH changes.

Haddadi et al. [4] studied the release of 2-mercaptobenzimidazole (MBI) from carbon hollow spheres (CHSs) using the silica method. The inhibitor was evaluated for mild steel protection in saline media. MBI does not release well in acidic solution (pH ~3), since it is concentrated in the capsules, as opposed to basic pH (pH ~11), where release is the highest. Thus, MBI release is pH-dependent, following the same strategy of the aforementioned studies.

5.5 COATING EVALUATION TECHNIQUES

5.5.1 Anticorrosion Properties

Choi et al. [20] produced nano-sized capsules containing triethanolamine as a corrosion inhibitor, which is released in the medium depending on the pH. Metallic samples were investigated in 3.50% NaCl solution and observed under an optical microscope. The specimens immersed without inhibitor release were corroded over the entire surface area, while in the presence of the inhibitor, only the X-cut region was affected. After the coating layers were removed, the inhibitor protected the surface against rust. After 24 h of immersion, EIS tests revealed resistance below $10^5\ \Omega\ cm^2$, which continued to decrease over time. For the tests performed with an inhibitor, total impedance increased from 8×10^5 to $5 \times 10^6\ \Omega\ cm^2$ after two days of immersion. This indicates the self-healing action of the encapsulated inhibitor. After two days of immersion, the Z_{mod} value was still high but started to decline, which may correspond to inhibitor desorption and/or molecule rearrangement. In this case, the study of nanocapsule concentration would address protection over time.

Ashrafi-Shahri, Ravari, and Seifzadeh [21] chemically modified mesoporous silica nanoparticles by introducing a nitrogenous functional group. An organic inhibitor was then incorporated into this nanostructure. The different coatings produced were analyzed by EIS. The high-frequency loop was related to the outermost part of the sol-gel film, while the inner and intermediate layers corresponded to the low-frequency loop. It is important to note that the impedances are significantly higher than those of the uncoated surfaces. In order to better understand the different coating behavior, absolute impedance was recorded at 0.001 Hz. For unmodified sol-gel coating, the impedance modulus decreased as a function of immersion time due to corrosive electrolyte diffusion through pores and defects. This behavior was also observed for coating containing 0.20% wt. of non-functionalized nanoparticles because of defective sites caused by stress inside the coating. Impedance modulus increased from 8 to 24 h in the coating containing 0.20% wt. of the nanoparticles with incorporated corrosion inhibitor. These values are 2.70 and 6.40 times higher than those observed for uncoated surfaces and coated with unmodified nanoparticles, respectively. The corrosion inhibitor is released after corrosion onset, leading to local alkaline sites due to reduced water and oxygen. This pH change releases the inhibiting molecules owing to the unstable chemical bond between N and Fe atoms in alkaline media. Thus, sol-gel-containing nanoparticles with incorporated inhibiting molecules are pH sensitive.

Manasa et al. [22] investigated the loading of $Ce(NO_3)_3.6H_2O$ and zirconium n-prop-oxide into nanotubes to prevent aluminum alloy A356.0 corrosion in NaCl. The authors observed that at initial immersion periods (1 h), the sol-coated and nanotube-coated matrix without inhibitor exhibited higher impedance than that of the self-healing coated matrix. For longer immersion times, the self-healing coated substrates exhibited higher values, with the highest obtained for exposure periods of 168 and 216 h. The lower barrier characteristics of the pure and nanotube-coated matrices without inhibitor incorporation are related to random nanotube arrangement, increasing preferential water permeability in some regions of the coating. The nanotube-coated surface containing inhibiting molecules leads to a change in pH due to corrosive reactions, resulting in inhibitor release. Inhibitors form a passive layer in the damaged region, thereby mitigating corrosion. Potentiodynamic polarization (PDP) tests revealed that the higher corrosion current at longer exposure times is linked to the nanotube-coated matrix without inhibitor incorporation and the lowest current density is achieved with coatings containing inhibitor. This confirms that the application of non-modified nanostructures does not inhibit corrosion, since the effects of agglomeration or non-uniform dispersion may lead to preferential water permeability through the coating, thus increasing corrosion. As such, the nanostructure is a way for the inhibitor to be released when an undesired corrosion reaction takes place.

Izadi, Shahrabi, and Ramezanzadeh [23] prepared a polymeric bilayer on Fe_3O_4 nanoparticles, containing phenolic compounds extracted from nettle leaves between the polymeric layers. The anticorrosive performance of nettle-NC extract in the presence of zinc acetate in the coating was investigated in 3.50% NaCl. The results indicated that Zn^{2+} cations interact with negative sites, forming $Zn(OH)_2$. Organic compounds from the extract, including serotonin, histamine, and caffeic acid, which have functional groups such as NH_2, OH, and C=O, are electron-rich and share a lone pair of electrons with empty orbitals of iron. The higher impedance at low frequencies after 24 h when compared to shorter immersion times is related to inhibitor release. These compounds were evaluated in the silane bilayer coating. The silane coating was prepared using a mixture of tetraethylorthosilicate (TEOS) and teriethoxymethylsilane (TEMS). A second time constant appeared in the Bode-phase plot of neat silane-coated samples, which is related to corrosion product deposition. This layer is porous and weakly adhered, resulting in corrosion inhibition 4 h before other times. The coating prepared with zinc acetate and silane bilayer showed two relaxation times, which are associated with Zn^{2+} cation interaction with hydroxyl anions and a decrease in cathodic reaction. After 1 h, impedance decreased due to Zn^{2+} cation consumption, reducing inhibition. Thus, only the presence of zinc cations is not sufficient to inhibit corrosion. The nettle-NCs/silane bilayer exhibited the best results due to the synergism between extracted compounds and Zn^{2+} cations, forming chelates that can adsorb onto metallic samples.

Yang et al. [16] investigated silica nanoparticles functionalized with a hybrid organic inhibitor for mild steel in 3.50% NaCl. PDP tests were performed after 48 h immersion for bare mild steel and different concentrations of functionalized silica nanoparticles with corrosion inhibitors on their surface. E_{corr} changed to –0.612 V, –0.508 V, and –0.373 V for 1.25, 2.50, and 5.00% wt. of functionalized nanoparticles, respectively. In addition, I_{corr} changed from 1.158×10^{-6} A/cm^2 for mild steel substrate to 3.999×10^{-8} A/cm^2 and 1.570×10^{-10} A/cm^2 in the presence of 0.50 and 2.50% wt of functionalized nanoparticles, respectively, confirming the high protection provided by the modified nanoparticles as a function of their concentration. However, a further concentration increase may lead to agglomeration, resulting in decreased efficiency. In addition, the nanoparticles exhibited better mechanical and corrosion properties, thereby providing a synergistic contribution. For the samples with nanoparticles or inhibitors, salt spray tests revealed small corrosion sites due to ion penetration. The modified nanoparticle was able to fill microcracks, forming an efficient barrier, leading to optimized micromorphology, mechanical behavior, and a barrier effect, the last even after 100 h exposure.

Tang et al. [29] chemically modified silica nanoparticles using the L-b-L method. The organic layers consisted of phytic acid and poly(allylamine hydrochloride). The EIS technique was performed to compare undoped SiO_x/ZrO_x-based film, SiO_x/ZrO_x-based film doped with phytic acid nanoparticles, and SiO_x/ZrO_x-based film doped with benzotriazole nanoparticles, after 5 h immersion in 3.50% NaCl. The phytic acid-doped film exhibited better results. Electrolyte penetration through the coating produces a change in pH, resulting in inhibitor release. The better results with phytic acid when compared to benzotriazole are attributed to the larger number of adsorption sites in the phytic acid structure. Moreover, phosphorous is an adsorption site of phytic acid, which results in better inhibition efficiency. In this respect, the structure of the corrosion inhibitor being released also interferes with overall mitigation.

In this respect, it is important to compare the pure matrix coating with that exhibiting non-modified nanocontainers and the coating filled with corrosion inhibitors. This effect should be evaluated over time to better understand corrosion protection in the presence of different coatings. In general terms, coatings with nanocontainers without corrosion inhibitor release cause a decrease in capacitive arcs by EIS due to the random arrangement, agglomeration, and defects on the film, resulting in preferential pathways for water permeability. Another noteworthy aspect is that an increase in functionalized nanocontainer concentration may lead to agglomeration, which causes preferential water permeability pathways and local corrosion. Finally, the structure of the released inhibitor will affect inhibition potential because heteroatoms and high-density structures provide better inhibition.

5.5.2 Wettability, Self-Cleaning Ability, Mechanical Durability, and Adhesion Strength

Dong et al. [30] studied inorganic aluminum phosphate (AP) adhesive containing organic polydimethylsiloxane (PDMS) and silica/halloysite composite nanotubes (HNTs) modified with 1H,1H,2H,2H-perfluorodecyltriethoxysilane. The F-SiO2/HNTs-AP@PDMS coating showed good hydrophobicity with a contact angle of 158° and a sliding angle of 2°, preventing water droplet adhesion, possibly due to a stable air layer. In addition, water droplets roll down the surface, removing contaminant powder, and leaving it clean. In this respect, the coating maintained a high-water contact angle, a low sliding angle, and its surface morphology after 200,000 water drops, confirming good resistance to water droplet impact. Wear resistance is also a consequence of the bonding between inorganic nanoparticles and organic polymers. The AP adhesive also binds the modified nanoparticles tightly together. The nanoparticles may then interact with PDMS, enhancing coating adhesion. This interaction with PDMS reduces shrinkage stress during PDMS curing, preventing cracking.

5.6 MECHANISM

Some explanations can be put forward for the inhibiting mechanism. In general terms, pH modification releases the corrosion inhibitor from the nanocontainer. In this sense, Qian et al. [24] investigated the reduction processes of electrochemical corrosion that increase localized pH. The homogeneous distribution of nanocontainers (SiO_2/chitosan/alginate/PASP/alginate/PASP) in the coating matrix allows their presence in the micrometer-scale scratch. pH modifications cause functional group dissociation in chitosan, alginate, and PASP, resulting in release stimulation. Chitosan is a weak polybase and alginate is a weak polyacid, serving as a pH-buffering system for the scratch area. Consequently, pH is maintained at around 5.2–7.2 on metallic substrates, favoring the release mechanism.

Another explanation was provided by Izadi, Shahrabi, and Ramezanzadeh [2], who observed that the corrosion products created on the substrate tested in the solution containing the mixture of Fe_3O_4-NS (nanoreservoir) and zinc acetate (96 h) are negligible. The better performance of the Fe_3O_4-NS in combination with zinc acetate is related to chelation between the zinc cations and nettle structure. Polar functions (S, O, or N) and π electrons in the heterocyclic compounds are two main factors that affect the corrosion inhibition behavior of a green corrosion inhibitor. The high chelating capability of NE as a green corrosion inhibitor provides a high affinity to form bonds with transition metal ions and zinc cations, which contain empty orbitals. The N-H and O-H bonds in these compounds are effective chelating agents with zinc and iron cations.

Roselli et al. [31] associated the corrosion inhibition of lanthanides (Ce and La) with cation concentration in the solution. Cations react with the hydroxyl ions of the cathodic oxygen reduction reaction, forming highly insoluble hydroxides and oxides, according to equations 5.1 and 5.2:

$$Ln^{3+} + 3OH \rightarrow Ln(OH)_3 \tag{5.1}$$

$$2Ln(OH)_3 \rightarrow Ln_2O_3 + 3H_2O \tag{5.2}$$

These compounds precipitate on the metal surface and form a passive layer that inhibits the cathodic current flow, resulting in prolonged corrosion resistance.

For more concise information, a literature summary is presented in Table 5.1.

5.7 CONCLUSION

The nanocontainer, also called a nanoreservoir, is a nanosized matrix filled with an active compound that can be confined inside a porous material, intercalated or entrapped between layers or by combining one of these materials with a polymeric shell. In terms of inorganic nanoreservoirs, nanocrystalline structures and natural clays provide layers that make ion exchange possible. Halloysite is composed of tubular structures previously prepared to increase the inner space into which the corrosion inhibitor is loaded. Silica nanoparticles can be applied L-b-L or by surface modification. In relation to organic nanostructures, polymeric shells must be compatible with the active substance being released. Another important aspect is that sometimes more than one shell is designed to provide stability or incorporate more corrosion inhibitors. Finally, the nanostructures may be hybrid. In this case, one strategy is to organically modify the inorganic nanoparticle.

In terms of corrosion evaluation techniques, EIS has been widely employed to compare the pure matrix, and the matrices containing the non-modified and modified nanostructures. Tests performed as a function of time focused on evaluating the capacitive arcs and their evolution over immersion time.

TABLE 5.1
Summary of the nanostructures containing corrosion inhibitor, preparation methods, and corrosion tests

Encapsulating System	Matrix Type	Encapsulation Material (Matrix Name)	Inhibitor Type	Active Compound (Inhibitor)	Inhibitor Distribution through Matrix	Media	Corrosion Test	Ref.
Inorganic	Nanoreservoir	Fe_3O_4 nanoparticles	Organic	Nettle extract (NE)	Layer-by-layer (LbL)	3.5% NaCl	PDP	2
Inorganic	Nanotubes	Halloysite nanotubes (HNTs)	Hybrid	8-Hydroxyquinoline (8-HQ) and Cu	Loaded in the nanotubes	3.5% NaCl	EIS	3
Inorganic	Microcapsules	Carbon hollow spheres (CHSs)	Organic	2- mercaptobenzimidazole (MBI)	Loaded in the microcapsules	3.5% NaCl	EIS	4
Inorganic	Nanocontainer	Mesoporous silica and nanocontainer	Organic	2-mercaptobenzothiazole (MBT)	Loaded in the nanocontainer	3.5% NaCl	PDP and EIS	5
Hybrid	Inorganic	Halloysite nanotubes	Organic	Benzotriazole	Loaded in the nanotubes	3.5% NaCl	PDP, EIS, Scanning vibrating electrode technique (SVET)	6
Hybrid	Inorganic	Silica nanoparticles (silica-zirconia)	Organic	Benzotriazole	Layer-by-layer (LbL)	0.05 M NaCl	SVET	7
Organic	Microcapsules	Poly (urea-formaldehyde) (PUF)	Rare-earth metals, lanthanide	Cerium nitrate	Inside PUF spheres	0.6 M NaCl	EIS	8
Hybrid	Inorganic	Mesoporous silica	Organic	Sulfamethazine	Loaded into the pores of mesoporous silica	3.5% NaCl	EIS	9
Inorganic	Microcontainer	Zeolites	Rare-earth metals, lanthanide	Cerium ions (Ce(III))	Adsorbed inside the pores	NaCl (0.025-0.5 M)	LPR, EIS	10
Inorganic	Functionalized nanoparticles	SiO_2 nanoparticles	Organic	1-(3′-aminopropyl) imidazole, CH_3CN, 3-(triethoxysilyl) propyl isocyanate in a nitrogen atmosphere, refluxed for 4 h at 354 K, followed by n-bromohexadecane.	Surface functionalization of silica nanoparticles via hydrolysis condensation	3.5% NaCl	PDP	16

Inorganic	Functionalized nanoparticles	SiO_2 nanoparticles	Organic	1 H, 1 H, 2 H, 2 H-perfluorodecyltrimethylsilane	Chemical reaction between SiO_2 and the corrosion inhibitor	3.5% NaCl	EIS and PDP	17
Organic	Nanocapsules	Polystyrene	Organic	Diethanolamine (DEA), ethanolamine (ETA), propylamine (PPA), triethanolamine (TEA), dipropylamine (DPA) and 5-amino-1-pentanol (5AP)	Loaded in the nanocapsules	0.05 M NaCl	SVET and EIS	18
Organic	Organic (polymer)	Plasma-polymerized perfluorohexane, pyrrole and acetylene	Organic	Tolyltriazole (TTA), benzotriazole (BTA), 1,2,4-triazole (TA)	Deposit of an ultrathin polymer film over the inhibitor particles surface	3.5% NaCl	DC (direct current) polarization, Contact angle	19
Organic	Polymeric capsules	First shell: butyl acrylate, methyl methacrylate, methacrylic acid; Second shell: styrene	Organic	Triethanolamine	Loaded inside the capsule	3.5% NaCl	SVET, EIS	20
Hybrid	Nanocontainer	Mesoporous silica nanoparticles functionalized with 3-aminopropyl triethoxysilane	Hybrid	Eriochrom black T	Loaded in silica nanoparticles	3.5% NaCl	EIS	21
Hybrid	Microcapsules	Halloysite nanotubes	Organic and inorganic	$Ce(NO_3)_3.6H_2O$, Zirconium n-propoxide	Loaded inside the nanotubes	SVE	EIS, PDP, SVET	22
Hybrid	Nanocontainer	Fe_3O_4 nanoparticles (NPs) + Polyaniline layer + Nettle extract layer (corrosion inhibitor) + Polyacrylic acid layer	Organic	Phenolic and amide groups (Plant extract)	Layer-by-layer (LbL)	3.5% NaCl	EIS	23
Hybrid	Nanotube within polymer spheres	Halloysite + poly(urea-formaldehyde)	Rare-earth metals, lanthanide	Praseodymium	Pr ions inside halloysite lumen, encapsulated in polymer spheres	3.5% NaCl	EIS, PDP, Contact angle	25

(continued)

TABLE 5.1 (Continued)
Summary of the nanostructures containing corrosion inhibitor, preparation methods, and corrosion tests

Encapsulating System	Matrix Type	Encapsulation Material (Matrix Name)	Inhibitor Type	Active Compound (Inhibitor)	Inhibitor Distribution through Matrix	Media	Corrosion Test	Ref.
Hybrid	Nanocontainer	Multilayers as follows: SiO_2/ Polyallylamine hydrochloride/ corrosion inhibitor/ Polyallylamine hydrochloride/ corrosion inhibitor/ Polyallylamine hydrochloride	Organic	Phytic acid and benzotriazole	Layer-by-layer (LbL)	3.5% NaCl	EIS	29
Inorganic	Composite nanotubes	Halloysite composite nanotubes (HNTs) with SiO_2	Organic	Polydimethylsiloxane (PDMS)	Nanoparticle surface functionalization	3.5% NaCl	PDP and EIS	30

In this respect, some important parameters in this application are the concentration of the inhibitor incorporated or functionalized on the nanoparticle, inhibitor solubility in the medium, inhibitor release from the nanostructure as a function of pH, nanostructure dispersion on the matrix, effect of nanoparticle surface and volume area to incorporate the corrosion inhibitor, nanostructure arrangement, and chemical characteristics such as the functional groups of the corrosion inhibitor.

Some gaps have been identified in the literature, such as the most qualitative analysis that has been performed in order to evaluate coating efficiency. Thus, future studies should focus on quantitative investigations. Additionally, studies do not mention the costs involved in applying these self-healing coatings. Another point to address is the response of the corrosion inhibitor release as a function of medium salinity and its correlation with medium pH. The use of real conditions to evaluate coating performance for at least one year of exposure should also be investigated. The corrosion inhibitor should be stable and soluble in the medium, eco-friendly and used at low concentrations. The nanocontainer must also be eco-friendly.

REFERENCES

[1] G.E. Rezvani, K.S. Nouri, K.M. Sadegh, M. Dinari, A. Shahla, E. M. Hossein, O. Das, N. R. Esmaeely, Synthesis of TiO_2 nanogel composite for highly efficient self-healing epoxy coating, *Journal of Advanced Research*. 43 (2023) 137–146. https://doi.org/10.1016/j.jare.2022.02.008

[2] M. Izadi, T. Shahrabi, B. Ramezanzadeh, Synthesis and characterization of an advanced layer-by-layer assembled Fe3O4 /polyaniline nanoreservoir filled with Nettle extract as a green corrosion protective system, *Journal of Industrial and Engineering Chemistry*. 57 (2018) 263–274. https://doi.org/10.1016/j.jiec.2017.08.032

[3] M. Wang, J. Wang, W. Hu, Preparation and corrosion behavior of Cu-8-HQ@HNTs /epoxy coating, *Progress in Organic Coatings*. 139 (2020) 105434. https://doi.org/10.1016/j.porgcoat.2019.105434

[4] S.A. Haddadi, S.A.A. Ramazania, M. Mahdavian, P. Taheri, J.M.C. Mol, Fabrication and characterization of graphene-based carbon hollow spheres for encapsulation of organic corrosion inhibitors, *Chemical Engineering Journal*. 352 (2018) 909–922. https://doi.org/10.1016/j.cej.2018.06.063

[5] M. Rahsepar, F. Mohebbi, H. Hayatdavoudi, Synthesis and characterization of inhibitor-loaded silica nanospheres for active corrosion protection of carbon steel substrate, *Journal of Alloys and Compounds*. 709 (2017) 519–530. https://doi.org/10.1016/j.jallcom.2017.03.104

[6] A. Gautam, T. Silva, S. Sathiyanarayanan, K.V. Gobi, R. Subasri, Capped inhibitor-loaded halloysite nanoclay-based self-healing silica coatings for corrosion protection of mild steel, *Ceramics International*. 48 (2022) 30151–30163. https://doi.org/10.1016/j.ceramint.2022.06.288

[7] M.L. Zheludkevich, D.G. Shchukin, K.A. Yasakau, H. Mohwald, M.G.S. Ferreira, Anticorrosion coatings with self-healing effect based on nanocontainers impregnated with corrosion inhibitor, *Chemistry of Materials*. 19 (2007) 402–411.

[8] G. Farzi, A. Davoodi, A. Ahmadi, R.E. Neisiany, M.K. Anwer, M.A. Aboudzadeh, Encapsulation of cerium nitrate within poly(urea-formaldehyde) microcapsules for the development of self-healing epoxy-based coating, *ACS Omega*. 6 (2021) 31147–31153. https://doi.org/10.1021/acsomega.1c04597

[9] M. Yeganeh, N. Asadi, M. Omidi, M. Mahdvian, An investigation on the corrosion behavior of the epoxy coating embedded with mesoporous silica nano container loaded by sulfamethazine inhibitor, *Progress in Organic Coatings*. 128 (2019) 75–81. https://doi.org/10.1016/j.porgcoat.2018.12.022

[10] S. Roselli, C. Deyá, M. Revuelta, A.R. Di Sarli, R. Romagnoli, Zeolites as reservoirs for Ce(III) as passivating ions in anticorrosion paints, *Corrosion Reviews*. 36 (2018) 305–322. https://doi.org/10.1515/corrrev-2017-0090

[11] J.I. Velasco, M. Ardanuy, M. Antunes, 4-Layered double hydroxides (LDHs) as functional fillers in polymer nanocomposites, *Advances in Polymer Nanocomposites*. (2012) 91–130. https://doi.org/10.1533/9780857096241.1.91

[12] H. Yan, J. Wang, Y. Zhang, W. Hu, Preparation and inhibition properties of molybdate intercalated ZnAlCe layered double hydroxide, *Journal of Alloys and Compounds.* 678 (2016) 171–178. https://doi.org/10.1016/j.jallcom.2016.03.281

[13] N. Granizo, J.M. Vega, I. Díaz, B. Chico, D. de la Fuente, M. Morcillo, Paint systems formulated with ion-exchange pigments applied on carbon steel: Effect of surface preparation, *Progress in Organic Coatings.* 70 (2011) 394–400. https://doi.org/10.1016/j.porgcoat.2010.09.035

[14] D.G. Shchukin, S.V. Lamaka, K.A. Yasakau, M.L. Zheludkevich, M.G.S. Ferreira, H. Mohwald, Active anticorrosion coatings with halloysite nanocontainers, *Journal Physical Chemistry* C. 112 (2008) 958–964. https://doi.org/10.1021/jp076188r

[15] J.M. Falcón, T. Sawczen, I.V. Aoki, Dodecylamine-loaded halloysite nanocontainers for active anticorrosion coatings, *Frontiers in Materials.* 2 (2015) 1–13. https://doi.org/10.3389/fmats.2015.00069

[16] W. Yang, W. Feng, Z. Liao, Y. Yang, G. Miao, B. Yu, X. Pei, Protection of mild steel with molecular engineered epoxy nanocomposite coatings containing corrosion inhibitor functionalized nanoparticles, *Surface & Coatings Technology.* 406 (2021) 126639. https://doi.org/10.1016/j.surfcoat.2020.126639

[17] Y. Zhao, M. Huo, J. Huo, P. Zhang, X. Shao, X. Zhang, Preparation of silica-epoxy superhydrophobic coating with mechanical stability and multifunctional performance via one-step approach, *Colloids and Surfaces A: Physicochemical and Engineering Aspects.* 653 (2022) 129957. https://doi.org/10.1016/j.colsurfa.2022.129957

[18] H. Choi, K.K. Young, P. J. Myung, Encapsulation of aliphatic amines into nanoparticles for self-healing corrosion protection of steel sheets, *Progress in Organic Coatings.* 76 (2013) 1316–1324. https://doi.org/10.1016/j.porgcoat.2013.04.005

[19] H. Yang, W.J. van Ooij, Plasma deposition of polymeric thin films on organic corrosion-inhibiting paint pigments: a novel method to achieve slow release, *Plasmas and Polymers.* 8 (2003) 4. https://link.springer.com/article/10.1023/A:1026389311431

[20] H. Choi, Y.K. Song, K.Y. Kim, J.M. Park, Encapsulation of triethanolamine as organic corrosion inhibitor into nanoparticles and its active corrosion protection for steel sheets, *Surface & Coatings Technology.* 206 (2012) 2354–2362. https://doi.org/10.1016/j.surfcoat.2011.10.030

[21] S.M. Ashrafi-Shahri, F. Ravari, D. Seifzadeh, Smart organic/inorganic sol-gel nanocomposite containing functionalized mesoporous silica for corrosion protection, *Progress in Organic Coatings.* 133 (2019) 44–54. https://doi.org/10.1016/j.porgcoat.2019.04.038

[22] S. Manasa, A. Jyothirmayi, T. Siva, S. Sathiyanarayanan, K.V. Gobi, R. Subasri, Effect of inhibitor loading into nanocontainer aditives of self-healing corrosion protection coatings on aluminum alloy A356.0, *Journal of Alloys and Compounds.* 726 (2017) 969–977. https://doi.org/10.1016/j.jallcom.2017.08.037

[23] M. Izadi, T. Shahrabi, B. Ramezanzadeh, Electrochemical investigations of the corrosion resistance of a hybrid sol-gel film containing green corrosion inhibitor-encapsulated nanocontainers, *Journal of the Taiwan Institute of Chemical Engineers.* 81 (2017) 356–372. https://doi.org/10.1016/j.jtice.2017.10.039

[24] B. Qian, Z. Song, L. Hao, W. Wang, D. Kong, Self-healing epoxy coatings based on nanocontainers for corrosion protection of mild steel, *Journal of the Electrochemical Society.* 164 (2017) C54–C60. https://iopscience.iop.org/article/10.1149/2.1251702jes

[25] R. Mahmoudi, P. Kardar, A.M. Arabi, R. Amini, P. Pasbakhsh, The active corrosion performance of silane coating treated by praseodymium encapsulated with halloysite nanotubes, *Progress in Organic Coatings.* 138 (2020) 105404. https://doi.org/10.1016/j.porgcoat.2019.105404

[26] R. Mahmoudi, P. Kardar, A.M. Arabi, R. Amini, P. Pasbakhsh, Acid-modification and praseodymium loading of halloysite nanotubes as a corrosion inhibitor, *Applied Clay Science.* 184 (2020) 105355. https://doi.org/10.1016/j.clay.2019.105355

[27] D. Bokov, A.T. Jalil, S. Chupradit, W. Suksatan, M.J. Ansari, I.H. Shewael, G.H. Valiev, E. Kianfar, Nanomaterial by sol-gel method: synthesis and application, *Advances in Materials Science and Engineering.* 21 (2021) 5102014. https://doi.org/10.1155/2021/5102014

[28] S.H. Soytaş, O. Oğuz, Y.Z. Menceloğlu, Polymer nanocomposites with decorated metal oxides, in micro and nano technologies, *Polymer Composites with Functionalized Nanoparticles.* (2019) 287–323. https://doi.org/10.1016/B978-0-12-814064-2.00009-3

[29] F. Tang, X. Wang, X. Xu, L. Li, Phytic acid doped nanoparticles for green anticorrosion coatings, *Colloids and Surfaces A: Physicochemical and Engineering Aspects*. 369 (2010) 101–105. https://doi.org/10.1016/j.colsurfa.2010.08.013
[30] K. Dong, L. Bian, Y. Liu, Z. Guan, Superhydrophobic coating based on organic/inorganic double component adhesive and functionalized nanoparticles with good durability and anti-corrosion for protection of galvanized steel, *Colloids and Surfaces A: Physicochemical and Engineering Aspects*. 640 (2022) 128360. https://doi.org/10.1016/j.colsurfa.2022.128360
[31] S. Roselli, M. Revuelta, C. Deya, R. Romagnol, Halloysites as carriers for Ce(III) and La(III) ions in anti-corrosive paint for steel, *Progress in Organic Coatings*. 174 (2023) 107287. https://doi.org/10.1016/j.porgcoat.2022.107287

6 Encapsulation of Natural Products or Eco-Friendly Materials as Corrosion Inhibitors for Smart Coatings

Mahboobeh Azadi

6.1 INTRODUCTION

In nature, metals and alloys are found in the ore state or a combination of oxides, carbonate, halides, sulfides, etc. By extracting metals to purified conditions, and forming those into the structural shape for various applications, their free energy increases (from the point of view of thermodynamics). Thus, structural materials are not in the lowest state of free energy and they are unstable. In this situation, the corrosion phenomenon can occur based on metals' tendency to recombine with environmental components to reach a lower energy state. Therefore, gradual deterioration or destruction of metals will happen [1, 2]. Through corrosion reactions, the metal atoms change to ions, and metals are exposed to weight loss. Thus, the failure or deterioration of materials is performed through physical, chemical, and electrochemical reactions [3]. It is notable that most pure, refined, and structural metals, except gold, silver, and platinum, show a natural tendency to exhibit a low energy state through corrosion phenomena. In this situation, corrosion reactions are spontaneous processes, and oxide and hydrate compounds are formed on the metal's surface. Notably, when the metal is iron, the corrosion event is commonly called "rusting" or "oxidizing," which is a well-known example of electrochemical corrosion. This type of iron deterioration forms oxides or salts and results in orange-colored precipitations on the iron surface. In this manner, the strength and other mechanical properties are extremely weakened [1]. Corrosion would be defined as the destruction or irreversible natural deterioration of metals through chemical or electrochemical reactions with the exposure media (like moisture, air, seawater, acids, and other aggressive solutions). The exposure environments include both aqueous and gas media. However, most of the corrosion reactions happen in aqueous environments [1, 4]. The term corrosion is deduced from the Latin word *corrōdere,* which means "to gnaw" [5].

6.2 CORROSION CONTROL AND PREVENTION

Corrosion is one of the reasons for large industrial damages or failure and is a costly degradation process. The National Association of Corrosion Engineers and other related companies published a report in 2001 about the costs of corrosion (higher than $300 billion annually in the United States of America). Thus, it is necessary to find various ways to protect metals when exposed to different environments [1, 5, 6]. The main method to control corrosion is preventing the metal surface from direct interaction with an aggressive medium [4]. These ways include utilizing organic and inorganic protective coatings as physical barriers, corrosion inhibitors, cathodic and anodic protections, and proper material selection such as utilizing various types of stainless steels in

DOI: 10.1201/9781032677255-6

certain environments. However, selecting one or two of these methods as a final way of corrosion protection should be affordable and accessible [1, 2]. In all cases of corrosion control methods, a change in either the metal surface or the environment is usually required. Details of applying coatings and consuming inhibitors are explained in section 6.2.1 and 6.2.2, respectively.

6.2.1 Coatings

One of the suitable, easiest, well-known, cheapest, and most common ways to control metal degradation is creating a barrier to the direct exposure of metals with their environments. This barrier can be created through the application of various coatings on the surface of metals. Coatings can protect metals by physically isolating of metals against the environment or by preventing the penetration of aggressive species toward the surface of metals [7, 8].

However, it is notable that coatings are exposed to deterioration by weathering and mechanical agents. Thus, such chemical and mechanical agents result in the formation of microcracks, micro-scratches, pores, and gradual failure of coatings. In this situation, metal substrates will be exposed to aggressive mediums and start to corrode. In various industries, maintenance, and repairs of failed coatings are time-consuming and expensive works, especially, for large structures such as ships and offshore oil rigs. Thus, developing a new type of intelligent coating that is well known as "smart or self-healing coatings" limited such problems. These coatings have a longer service life that can partially or fully repair or recover their small defects, without any external source or help [1]. More explanation about the performance will be discussed in section 6.2.3. It is found that such coatings can be classified based on their chemical compositions, and they include (1) nonmetallic coatings like ceramic or organic coatings such as paints and plastics and (2) metallic coatings.

6.2.2 Inhibitor

"Inhibitors" or "anti-corrosive" materials are chemical substances that are added in small concentrations in corrosive media to prevent, decrease, or control the corrosion rate of metals to an acceptable level. Inhibitors can be added in both liquid and gas media. They can create such reduction through three mechanisms:

(1) Forming a thin passive layer on the metal surface to prevent access of corrosive species to the metal surface. As shown in Figure 6.1(a), inhibitor species and metal surfaces create continuous corrosion products in a protective layer. Such inhibitors are usually compounds of chromates, nitrates, and dichromate of metals.

(2) Forming a physical barrier at the interface of metal/corrosive media to retard the diffusion of the corrosive species into the surface of metals, as indicated in Figure 6.1(b). The chemical composition of such inhibitors is usually large organic molecules.

(3) Trapping, consuming, or changing of corrosive species of the aggressive mediums, as shown in Figure 1(c). However, an insignificant change should have happened in the concentration of the corrosive medium [1, 9]. Examples of such inhibitors are hydrazine, silica gel, zeolite, etc.

Based on inhibitor mechanisms, they can be classified as anodic, cathodic, or mixed-type [1]. Moreover, both organic and inorganic materials can be used as inhibitors. Organic corrosion inhibitors have a high molecular weight and usually contain heteroatoms with high electron density such as oxygen (C=O, O–H), phosphorus, nitrogen (–NH, $-NH_2$), thiol (C=S), and sulfur or polar groups and π bonds. These are candidate agents for adsorbing at the surface of metals since the free orbitals of metals have a high tendency to adsorb electrons [10, 11]. On the other hand, it is notable that inorganic corrosion inhibitors are usually various compounds of chromate, dichromate,

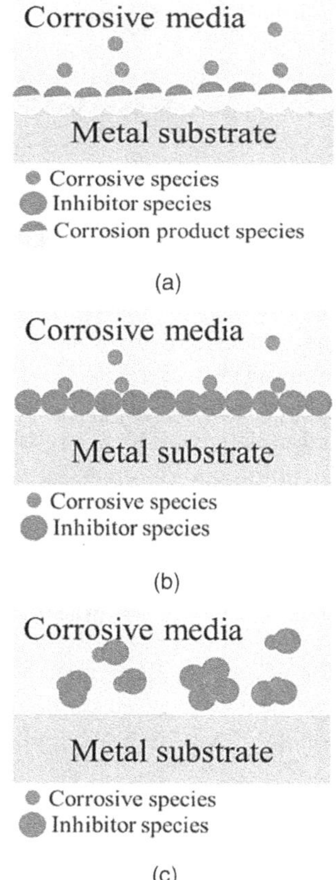

FIGURE 6.1 Schematic views of inhibitor mechanisms, included (a) forming a passive layer, (b) forming a physical barrier, and (c) trapping or consuming corrosive species in the aggressive medium.

tungstate, nitrite, molybdate, phosphate, borate, and also basic salt of bromine, arsenic, and iodine. However, some of these compounds are toxic and are not suitable due to the hazardous environmental implications. Thus, researchers suggest using green or eco-friendly corrosion inhibitors as biodegradable materials. They do not include toxic compounds or heavy metals [12, 13]. Such green materials are introduced in section 6.2.2.3.

It was found that the first corrosion inhibition that was reported as a patent belonged to Baldwin (British Patent 2327). This inhibitor was extracted from molasses with vegetable oils and was used to protect sheet steels in acids [5].

6.2.2.1 Natural Products as a Green Corrosion Inhibitor

Natural products contain a large family of chemical materials produced naturally by any organism such as marine animals, insects, plants, bacteria, and fungi that are found in nature. These materials have been utilized in various applications over the years as pure compounds, essential oils, and plant extracts. Natural sources have led to basic studies on potential bioactive components for commercial and industrial developments [14, 15].

Recently, natural products such as extracts from various plants, seeds, leaves, roots, fruits, and flowers have been utilized for corrosion control of metals against corrosive media. This is based on the cost-effective, economical, and acceptable anticorrosive performance of these materials in various industrial applications such as in anti-freezing solutions, pickling or cooling liquids, and

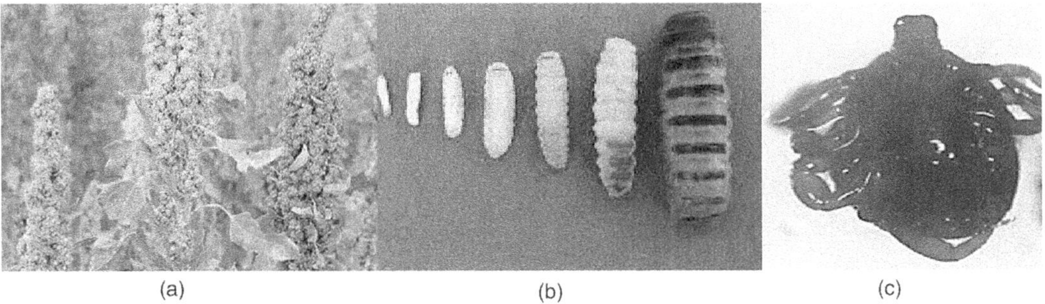

FIGURE 6.2 Examples of utilizing natural products and biomaterials as eco-friendly corrosion inhibitors for steel protection, included, (a) Quinoa seed, (b) *Oestrus ovis* larvae, and (c) *Hyalomma* tick (Photos were taken by the author).

concretes to retard corrosion reactions on the metal substrate [9]. Most of these products contain complex constituents, such as flavonoids, polyphenols, terpenes, phthalocyanines, tannins, anthraquinones, organic acids, aromatic rings, alkaloids, and nitrogen-base compounds. These compounds can be adsorbed at the metallic surface to create a physical barrier and retard corrosion reactions [5, 10–13]. Examples of natural products as green inhibitors are extracts of plants such as *Nicotiana tabacum*, Propolis, *Medicago sativa*, horse-chestnut, chitosan oligosaccharide, natural glucomannan, *Rhizophora apiculate*, Octanoate, *Eugenia jambolana*, Sarang rice straw, bitter leaf, *Ficus carica*, Ginkgo, *Salvia officinalis*, *Foeniculum vulgare*, *Mentha rotundifolia*, *Sida acuta* [16], and Quinoa seed [17], ionic liquids, and extracts of insects such as *Oestrus ovis* larvae, leeches, and *Hyalomma* tick [10–13, 16], as shown in Figure 6.2.

In most of the studies based on the electrochemical measurements, the inhibition efficiency was about 80–97% in various aggressive environments when the inhibitor concentrations range was about 500–3000 ppm. These green corrosion inhibitors were added directly to the corrosive medium without any carrier [10–13].

6.2.3 Self-Healing or Smart Coatings

Based on the restrictions of most corrosion inhibitors to use directly in the environment, smart coatings are widely utilized for the corrosion protection of various metals. These coatings can be used in both wet and dry conditions and act as an insulated layer on metal surfaces which blocks corrosive ions, water, and oxygen diffusion. On the contrary, when conventional coatings are applied to metal surfaces, they serve as a protective barrier against environmental factors. However, over time, these coatings may experience localized damage caused by mechanical abrasion or chemical exposure during service. In this situation, corrosive species would diffuse fast to the metal surface through these defects. In addition, the corrosion of the metal surface under the coating will be accelerated based on the reduction of the anodic-to-cathodic area ratio. Finally, the failure of coatings will happen and result in the dis-bonding of coatings and the metal surface. Therefore, there are many efforts to develop "self-healing" coatings. In this situation, these coatings can recover their defects on demand [1, 9, 18].

Thus, the word "Self-healing" coating will be defined as the healing ability of destructed coating autonomously without any external agent [19]. Moreover, "Self-healing" coatings can be defined as intelligent materials that are structurally capable of repairing ruptured parts on their ability [20]. For the past two decades, smart coatings have been developed for metal protection. Various types of polymeric coatings such as elastomers, thermoset/thermoplastic polymers, biopolymer matrices, and composites can be developed for self-healing applications [2].

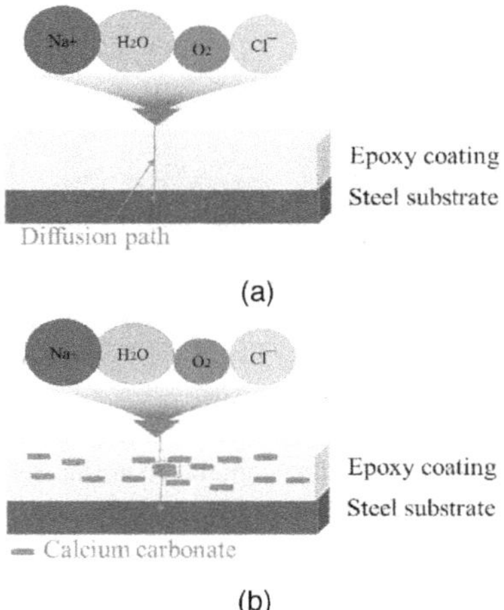

FIGURE 6.3 Schematic views for corrosion protection of steel substrates (a) in the absence of corrosion inhibitor in epoxy coatings and (b) in the presence of corrosion inhibitor (a natural product) embedded in epoxy coating.

One simple approach is adding corrosion inhibitors directly into the smart coating chemical composition to provide additional corrosion resistance. In this situation, the inhibitor molecules will be released and will act as healing species and will fill defects that are anodic sites [9, 21, 22].

In cases of studies, Azadi et al. [8] used rice husk ash as a natural product and corrosion inhibitor in epoxy paints to protect carbon steel substrates in acidic solutions. The obtained results showed that the corrosion resistance of filled coatings would be increased three times compared to unfilled coatings. Then, they fabricated epoxy coatings that contained modified calcium carbonate as a corrosion inhibitor for corrosion protection of steel in a 3.5% wt NaCl solution. Electrochemical impedance spectroscopy (EIS) results showed that this natural product could act as physical particles to increase the diffusion path of corrosive species in coatings from the aggressive media toward the steel substrate [23]. Figure 6.3 compares schematically the diffusion path for filled and unfilled epoxy coatings. In this situation, the corrosion resistance increased up to 99%. Moreover, Giuliani et al. [17] utilized chitosan-based coatings as an eco-friendly and natural polymeric layer embedding various corrosion inhibitors, such as benzotriazole and mercaptobenzothiazole , to protect copper-based alloy in the indoor environment. Wu et al. [24] used lignin nanoparticles (produced from black liquor [a waste stream from the wheat straw pulping process]) as a green inhibitor in polyurethane coatings. Banerjee et al. [25] used three eco-friendly corrosion inhibitors including hydroxyquinoline, polystyrene sulfonate, and polyethyleneimine in polymeric coatings for the protection of aluminum alloy 2024 to achieve a long lifetime.

It is notable that in all the above cases when mechanical damage to the coating occurs, the green inhibitor is released to passivate or protect the surface. However, such a direct addition of inhibitor in coatings has its limitation, which is explained in section 6.2.4.

6.2.4 Encapsulation of Corrosion Inhibitors in Smart Coatings

To increase the performance of corrosion inhibitors as healing agents in coatings, it should be better to use them in the proper carrier. Since the low solubility of inhibitors can cause a reduction in the wettability of molecules at the metal interface, as a result, the healing effect will be decreased. If the solubility is high, the effective time of healing action is reduced. Moreover, the high solubility of inhibitor molecules causes blistering and delamination of coatings based on the water adsorption of coatings. Therefore, the release of inhibitor molecules should be in a controllable way. Thus, the direct addition of inhibitor molecules to the coating chemical composition is not suitable [9]. To reduce such limitations various ways such as injection of inhibitors in nanoporous metal oxides like titanium oxide and silica, layered clays, double-layer hydroxide, polymer and glass fiber containers, core–shell, inorganic halloysite nanotubes, nano-valves, polyelectrolyte shells, nano-containers, micro-carrier, tubular reservoirs, and ion-exchanged particles are used to create effective self-healing mechanisms [9, 21, 22, 26, 27]. However, among all types of carriers for inhibitors, the encapsulation of corrosion inhibitors in coatings has proved to be an effective way to enhance the lifetime of coatings, increase the compatibility of the inhibitor with the matrix of coatings, and control the inhibitor release rate [1].

Encapsulation is a new method in which healing agents are imprisoned in a "wall" material or a biodegradable matrix, forming nano/micro-systems. Notably, the encapsulation of natural compounds has been utilized in various industries such as food, agriculture, and pharmaceutical applications when a controllable release of the encapsulated compound is needed [1, 14]. Nanoencapsulation can be produced by two main methods: top-down and bottom-up. The top-down process involves emulsification–solvent evaporation and emulsification. However, bottom-up methods include, nano-precipitation, conservation, supercritical fluid techniques, and inclusion complexation. Microencapsulation can be created through physicomechanical, chemical, physicochemical methods such as spray drying, coacervation, emulsification, emulsion solvent, ionic gelation, evaporation, centrifugal extrusion, melt extrusion, spray congealing, fluid bed coating, and pan coating method. However, it is found that among all these encapsulation methods (in both scales), coacervation, ionic gelation, emulsification, and spray-drying are widely used [14].

Based on microcapsule content types, three release mechanisms are suggested, as follows: (1) mechanical agents such as scratch, friction, and wear force are factors that damage the wall of microcapsule materials; (2) physical agents such as ultraviolet (UV) radiation, light, and temperature changes are factors that rupture the microcapsule wall; and (3) in the chemical state, diffusion of corrosive ions toward the microcapsule wall or pH changes or water uptake are parameters that cause a failure in microcapsule walls.

In mechanical mode, a sudden scratch, or other damage to coatings, results in the release of the encapsulated self-healing agent toward damaged regions, and the recovery of the barrier property of coatings will be done. Thus, the durability and the coating's lifetime increase. In physical mode, the adsorption of light like UV and heat causes the wall rupture of the microcapsule, and the release of a healing agent like corrosion inhibitors will happen as shown in Figure 6.4. In chemical mode, localized pH changes or diffusion of corrosive ions would result in the release of the encapsulated self-healing agent [2, 28, 29].

However, for the release of corrosion inhibitors as a healing agent, changes in pH are the most used releasing mechanism since pH changes are associated with corrosion phenomena [26]. It is notable that in some cases, pH-sensitive coatings are limited since pH sensitivity needs special issues such as the identification of pH range, which is complex when the pH values of an environment change continuously [6].

In 1993, for the first time, Dry et al. [30] fabricated a self-healing material, utilizing fibers as healing carriers in a polymer matrix composite. The releasing of the healing agent was done through a thermo-mechanical loading. Then, White et al. [3] encapsulated a healing agent in an epoxy matrix composite. They also added a catalyst in the coating system for the polymerization of the healing agent during the releasing process to form a protective layer in created cracks. After that, Brown et al. [31] found that microencapsulation of dicyclopentadiene as a corrosion inhibitor caused long shelf-life, proper strength, and excellent bonding to epoxy coatings.

Choi et al. [9] encapsulated organic inhibitors into nanoparticles in polymer coating to protect steel sheets. They showed that the release rate of the inhibitor depended on the pH of the corrosive media. Razaghi Kashani et al. [20] used 2-mercapto benzimidazole as an anticorrosion healing

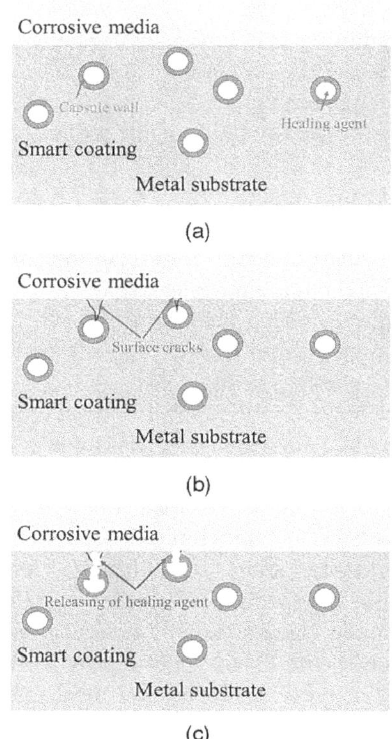

FIGURE 6.4 Schemes of releasing corrosion inhibitor as a healing agent in smart coating, steps containing (a) before creating damage in the smart coating, (b) after cracks formed at the surface of the coating and scratch the capsule walls, and (c) releasing the corrosion inhibitor as a healing agent.

agent and trapped it into polyurethane nano-capsules in an epoxy coating for corrosion protection of AA2024 substrate in a 5% NaCl solution. The incorporation of polyurethane nano-capsules into epoxy-based coatings not only enhances their resistance to damage but also contributes to their overall longevity and performance. As a result, these smart coatings have the potential to offer extended protection to various substrates in challenging operating environments, thereby reducing maintenance requirements and prolonging the service life of coated surfaces. EIS measurements showed a higher resistance of fabricated smart coatings of epoxy systems. Veedu et al. [32] loaded green corrosion inhibitor (Ixora leaf extract) into the epoxy coating for steel protection in marine solution. EIS test results revealed that in the presence of a 2000 ppm inhibitor, the inhibition efficiency was about 80%. Smart coatings can achieve better performance than traditional coatings [33].

6.2.5 Case Studies of Encapsulation of Natural Products or Eco-Friendly Materials as Corrosion Inhibitors in Smart Coatings

Pearman et al. [21] encapsulated three organic eco-friendly corrosion inhibitors (including 8-hydroxyquinoline, phenyl phosphonic acid, and 2-mercaptobenzothiazole) in organic microparticles in desired coatings. Then, the release mechanism of inhibitors in a basic solution (0.01 M KOH) for carbon steel substrate was investigated. It was found that 2-mercaptobenzothiazole has a low release time (1 hour) which was followed by a continued release.

Li et al. [34] also encapsulated benzotriazole as a corrosion inhibitor in mesoporous silica as a natural microcapsule for aluminum protection. It was found that the triggering agent for releasing the healing agent was pH change. Moreover, Zea et al. [35] also used sodium phosphomolybdate as an eco-friendly and natural inhibitor in mesoporous silica to protect carbon steel in 1 M NaCl solution. In this situation, the anticorrosive behavior of smart coatings in the presence of artificial defects was studied. It was found that the releasing action happened at pH with a value of 4–9. It was notable that Liu et al. [36] encapsulated saline compounds as a green corrosion inhibitor in hollow mesoporous silica (HMS). These epoxy coatings contained 0.5 wt% HMS to protect the magnesium alloy. Such coatings showed a high corrosion resistance after a long immersion time (about 2,688 h). Moreover, Yeganeh et al. [37] investigated the corrosion behavior of epoxy coatings that contained mesoporous silica nanocontainer filled with sulfamethazine (an eco-friendly inhibitor). The results indicated that the utilized inhibitor adsorbed in mesoporous silica pores and released in the corroded regions. In this situation, links between sp^2 electron pairs presented on O and N atoms of sulfamethazine and iron atoms would be formed.

In addition, Habib et al. [6] fabricated epoxy smart coatings containing halloysite nanotubes (HNTs) as carriers for hydroxyquinoline. The release mechanism of this eco-friendly inhibitor was by pH variations and the inhibition efficiency of smart coatings was about 98% compared to epoxy coatings without smart agents. Moreover, Subhalakshmi et al. [38] used HNT as a nano-container for vanillin and thyme oil (extracted natural products) as inhibitors. They showed that the inhibition efficiency was kept up to 900 °C. The increase in pH value was also the release mechanism of both inhibitors. It was notable that HNT was a natural clay, consisting of silica-water-alumina layers.

Mohammadi et al. [39] produced a smart epoxy coating with an eco-friendly corrosion inhibitor (sodium diethyldithiocarbamate in silane sol-gel layer as a carrier) for protection of the AA 2024-T3 substrates. The releasing mechanism was ion exchange. The obtained results showed that the incubation of active protection performance for such coatings increased significantly. Moreover, Javidparvar et al. [40] investigated the corrosion mechanism of epoxy coatings in the presence of Ce nanoparticles (eco-friendly corrosion inhibitor) and graphene oxide (GO) and found that GO acted as a smart pH-sensitive reservoir for releasing green corrosion inhibitors to the active sites of the metal.

Gao et al. [41] prepared porous carbon nano-sheets (PCN) derived from fallen leaves as natural carriers. Two nontoxic and green chelating agents extracted from a plant called phytic acid and aminopropyltriethoxysilane (APTES) were used as eco-friendly inhibitors for water-based epoxy coatings. Salt spray and EIS test results showed that APTES could easily adsorb at the surface of the steel, create a protective layer, and repair the cracks. In addition, pull-out measurements displayed that the adhesion strength increased by about 100% when PCN was added to the epoxy coatings.

Leal et al. [42] prepared various microcapsules embedded with linseed oil as a natural product that was synthesized through an in situ polymerization process. The percentage of loaded microcapsules was about 4.8 wt% in epoxy resins to protect carbon steel substrates. It was found that in a modified NaCl solution with a pH value of 5.7, the rate of inhibitor release was higher than in a pH value of 9.0. Notably, Liu et al. [43] synthesized mesoporous chitosan microspheres that were loaded with an eco-friendly inhibitor (sodium phytate). The utilized percentage of chitosan microspheres was about 5 wt% in polyacrylate coatings as smart water-based coatings.

Nawaz et al. [44] fabricated smart epoxy coatings for steel in the marine medium. They used gum Arabic as a natural product (eco-friendly inhibitor) embedded into cerium oxide nanoparticles (CONP). Brunauer-Emmett-Teller and thermogravimetric measurements indicated about ~30.0 wt% loading of gum Arabic into the CONP. EIS results also showed that the impedance would be increased up to 99.7% compared to epoxy coatings without healing agents. However, the polarization measurement showed an increase of about 76%. Asaad et al. [45] used 5% of *Elaesis guineensis*/silver nanoparticles (EG/AgNP) as an eco-friendly inhibitor in multi-walled carbon nanotubes (MWCNTs) as nano-container in epoxy coating. This fabricated smart coating was designed to protect mild steel in seawater. The obtained results indicated that a high inhibition efficiency of about 97.87% was achieved for smart coatings.

Nawaz et al. [46] prepared cellulose microfibers as a natural product carrier for polyethyleneimine (PEI) and dodecylamine (DOC) as autonomous inhibitor. These agents were used in epoxy coatings. UV-Vis results showed the pH sensitivity and time-dependent release of utilized inhibitors.

Additionally, some other studies are summarized in Table 6.1. The utilized corrosion inhibitors were eco-friendly and extracted from natural materials.

6.3 FUTURE DEMAND AND CHALLENGES

This contribution reports the most recent developments attributed to smart or self-healing coatings encapsulated natural products or eco-friendly materials as corrosion inhibitors, applied for corrosion protection. Self-healing coatings based on the utilization of capsulation of corrosion inhibitors as healing agents are attractive for various industries. However, there are still some limitations related to smart coating technologies.

For instance, the encapsulation of healing agents in smart coatings in a cost-effective process is a challenge. Often, incorporating microcapsule coatings will affect the homogeneity of coatings. For example, these materials act as sites for nucleation of pores and cracks that result in delamination of coating or form a diffusion way for higher water uptake, etc. Since the inhibition efficiency of natural products in small concentrations was not too high, more concentrations of them are needed. In this situation, the mentioned limitations expose more challenges.

Moreover, in some cases, the extraction process of plants to produce eco-friendly inhibitors is a time- and cost-consuming step that cannot be ignorable. On the other hand, until now, the low loading percentage of inhibitors (10–20%) was encapsulated successfully. This percentage was not enough for proper protection in some cases and an increase in adding more inhibitor carriers results in more limitations, as mentioned.

6.4 CONCLUSION

In this chapter, the basic concept of corrosion, coatings, and inhibitors was introduced briefly. Then, the details of smart coatings as innovative and protective layers were studied. These coatings can increase the lifetime of the coating compared to custom layers as they can heal or rebuild themselves. Since they contain corrosion inhibitors as healing agents that will be released when exposed to degradation agents. Recently, green or eco-friendly processes have been used based on environmental limitations. These materials can be natural products such as extracts of various plants, flowers, and insects. Thus, papers with topics of encapsulation of natural products or eco-friendly materials as corrosion inhibitors for smart coatings are reviewed in this chapter. It was found that despite the benefits of such coating there are still some limitations and challenges about them.

TABLE 6.1
Summarizations of case studies for encapsulation of releasing mechanism in smart coatings

Corrosive Medium	Substrate	Coating	Corrosion Inhibitor	Container	Release Mechanism	Reference
3.5%wt NaCl	Mild steel	Epoxy	Chitosan	Silica	-	[4]
35g/l NaCl	Steel	Epoxy	Hydroxyquinoline	Halloysite nanotubes	pH change	[6]
0.01 M KOH	Carbon steel	Polymer	Hydroxyquinoline	Organic microparticles	-	[21]
1.5 M NaCl	Carbon steel	Zinc	Molybdate	Mesoporous silica	pH change	[29]
-	Al	Polymer	Benzotriazole	Mesoporous silica	pH change	[34]
1 M NaCl	Carbon steel	PDDA	Sodium phosphomolybdate	Mesoporous silica	pH change	[35]
-	Mg alloy	Epoxy	Silane	HMS	Scratch	[36]
NaCl	Mild steel	Epoxy	Sulfamethazine	Mesoporous silica	-	[37]
H_2SO_4	Steel	-	Vanillin and thyme oil	Halloysite nanotubes	pH change	[38]
3.5%wt NaCl	Mild steel	Epoxy	Cerium oxide	Graphene oxide	pH change	[40]
3.5%wt NaCl	Steel	Epoxy	PA and APTES	PCN	Micro-scratch	[41]
0.05 M NaCl With pH= 5.7 and 9	Carbon Steel	Epoxy	Linseed oil	Microcapsules	pH change and scratch	[42]
-	Steel	Polyacrylate	Sodium phytate	Chitosan microspheres	pH change	[43]
Marine solution	Steel	Epoxy	Gum Arabic	CONP	-	[44]
Sea water	Mild steel	Epoxy	EG/AgNP	MWCNT	-	[45]
-	-	Epoxy	PEI and DOC	CMF	pH change	[46]
3.5%wt NaCl	Steel/Al (7075)	PMMA	Cerium oxide	Inorganic colloid	-	[47]
-	Stainless steel	Polyurethane	Nicotinic acid	Lignin	Micro-scratch	[48]
Water	AA2024	Zirconia	Benzotriazole	Mesoporous silica	pH change	[49]
3.5%wt NaCl	Magnesium	Epoxy	Mercaptobenzothiazole	Mesoporous silica	-	[50]
0.5 M NaCl	Mild steel	Alkyd	Benzotriazole	ZnO nanocontainers	pH change	[51]
-	-	EPBG	Polydopamine	Graphene	-	[52]
NaCl	AA2024-T3	Polyvinyl butyral	Chitosan	Cerium nitride	pH change	[53]
-	Mild steel	Urethane	Polyaniline	Coconut oil-based	-	[54]
3% NaCl	Mild steel	Epoxy	Polythiophene	Silicone	-	[55]
3.5% NaCl, 0.5 M HCl	Steel	Polyurethane	Quinoline and acetylation of neem oil	Polyurea	Scratch	[56]
3.5% NaCl	Mild steel	Epoxy- silane	Zinc acetate	Nano container	Scratch	[57]
3.5% NaCl	Steel	Polyester-epoxy	Toluidines and phenylenediamines	$BaSO_4$	Scratch	[58]
3.5%wt NaCl	Mild steel	Epoxy	Quinoline and cerium	Urea-formaldehyde microcapsule	-	[59]
NaCl	Steel	Epoxy	Linalyl acetate	Urea-formaldehyde microcapsule	Scratch	[60]

REFERENCES

1 Palumbo G (2015) Smart coatings for corrosion protection by adopting microcapsules. *Phys Sci Rev* 2015:20150006.
2 Bothi Raja P, Assad MA, Ismail M (2020) Inhibitor-encapsulated smart nanocontainers for the controlled release of corrosion inhibitor. *Corrosion Protection at the Nanoscale*, Elsevier: 91–106.
3 White SR, Sottos NR, Geubelle PH (2001) Autonomic healing of polymer composites. *Nature* 409:794–797.
4 Wonnie Ma IA, Ammar S, Shahid B, Sachin SA K, Ramesh K, Ramesh S (2021) Development of active barrier effect of hybrid chitosan/silica composite epoxy-based coating on mild steel surface. *Surf Interfaces* 25:101250.
5 Payal R, Jain A (2021) Green corrosion inhibitors for coatings. *Mater Res Foundations* 107:147–174.
6 S Habib S, Khan A, Ismail SM (2023) Polymeric smart coatings containing modified capped halloysite nanotubes for corrosion protection of carbon steel. *J Mater Sci* 58: 6803–6822.
7 Azadi M, Bahrololoom ME, Heidari F (2011) Enhancing the mechanical properties of an epoxy coating with rice husk ash, a green product. *J Coat Technol Res* 8:117–123.
8 Azadi M, Olya MJ, Bahrololoom ME (2016) EIS study of epoxy paints in two different corrosive environments with a new filler: rice husk ash. *Prog Color, Colorants Coat* 9:53–60.
9 Choi H, Kyun Song Y, Young Kim K, Myung Park J (2012) Encapsulation of triethanolamine as organic corrosion inhibitor into nanoparticles, and its active corrosion protection for steel sheets. *Surf Coat Technol* 206:2354–2362.
10 Bidi MA, Azadi M, Rassouli M (2021) An enhancement on corrosion resistance of low carbon steel by a novel bio-inhibitor (leech extract) in the H_2SO_4 solution. *Surf Interfaces* 24:101159.
11 Mobtaker H, Azadi M, Hassani N, Neek-Amal M, Rassouli M, Bidi MA (2021) The inhibition performance of quinoa seed on corrosion behavior of carbon steel in the HCl solution; theoretical and experimental evaluations. *J Mol Liq* 335:116183.
12 Bidi MA, Azadi M, Rassouli M (2021) Comparing the inhibition efficiency of two bio-inhibitors to control the corrosion rate of carbon steel in acidic solutions. *Anal Bioanal Electrochem* 13:52–66.
13 Mobtaker H, Azadi M, Rassouli M (2022) The corrosion inhibition of carbon steel in 1 M HCl solution by Oestrus ovis larvae extract as a new bio–inhibitor. *Heliyon* 8:12297.
14 Detsi A, Kavetsou E, Kostopoulou E (2020) Nanosystems for the encapsulation of natural products: the case of chitosan biopolymer as a matrix. *Pharmaceutics*. 12:669.
15 Zakeri A, Bahmani E, Sabour Rouh Aghdam A (2022) Plant extracts as sustainable and green corrosion inhibitors for protection of ferrous metals in corrmini-review: a mini review. *Corros Commun* 5:25–38.
16 Bidi MA, Azadi M, Rassouli M (2020) A new green inhibitor for lowering the corrosion rate of carbon steel in 1 M HCl solution: Hyalomma tick extract. *Mater Today Commun* 24:100996.
17 Giuliani C, Pascucci M, Riccucci C (2018) Chitosan-based coatings for corrosion protection of copper-based alloys: a promising more sustainable approach for cultural heritage applications. *Prog Org Coat* 122:138–146.
18 Yin Y, Schulz M, Rohwerder M (2021) Optimizing smart self-healing coatings: investigating the transport of active agents from the coating towards the defect. *Corr Sci* 190:10966.
19 X Zhao X, Jiang D, Ma L, Zeng X, Li Z, Huang G (2022) Corrosion effects and smart coatings of corrosion protection. *Coatings* 12:1378.
20 Razaghi Kashani O, Amiri S, Hosseini Zori M (2022) Self-healing and anti-corrosion nanocomposite coatings based on polyurethane nanocapsules containing mercapto benzimidazole. *J Nanostruct* 12: 726–737.
21 Pearman BP, Calle LM, Zhang X (2014) Characterization of encapsulated corrosion inhibitors for environmentally friendly smart coatings. *NASA Technical Reports Server 20140017326.*
22 Olivieri F, Castaldo R, Cocca M, Gentile G, Lavorgna M (2021) Mesoporous silica nanoparticles as carriers of active agents for smart anticorrosive organic coatings: a critical review. *Nanoscale* 13:9091–9111.
23 Azadi M, Ferdosi Heragh M, Bidi MA (2020) Electrochemical characterizations of epoxy coatings embedded by modified calcium carbonate particles. *Prog Color, Colorants Coat* 13:213–222.

24 Wu L, Liu S, Wang Q (2022) High strength, and multifunctional polyurethane film incorporated with lignin nanoparticles. *Ind Crops Prod* 177: 114526.

25 Banerjee D, Guo X, Benavides J, Rameau B, Cloutier S (2020) Designing green self-healing anticorrosion conductive smart coating for metal protection. *Smart Mater Struct* 29:105027.

26 Montemor MF (2014) Functional and smart coatings for corrosion protection: a review of recent advances. *Surf Coat Technol* 258:17–37.

27 Zea C, Barranco-Garcia R, Alcantara J, Simancas J, Morcillo M, Fuente D (2018) pH-dependent release of environmentally friendly corrosion inhibitor from mesoporous silica nanoreservoirs. *Microporous Mesoporous Mater* 255:166–173.

28 Udoh II, Shi H, Daniel EF (2022)Active anticorrosion and self-healing coatings: a review with focus on multi-action smart coating strategies. *J Mater Sci Technol* 116: 224–237.

29 Alipour K, Nasirpouri F (2017) Smart anti-corrosion self-healing zinc metal-based molybdate functionalized-mesoporous-silica (MCM-41) nanocomposite coatings. *RSC Adv* 7(82):51879–51887.

30 Dry CM, Sottos NR (1993) Passive smart self-repair in polymer matrix composite materials. *Smart Struct Mater* 1916: 438–444.

31 Brown EN, Kessler MR, Sottos NR, White SR (2003) In situ poly (urea-formaldehyde) microencapsulation of dicyclopentadiene. *J Microencapsul* 20:719–730.

32 Veedu KK, Mohan S, Somappa SB, Gopalan NK (2022) Eco-friendly anticorrosive epoxy coating from Ixora leaf extract: a promising solution for steel protection in marine environment. *J Clean Prod* 340:130750.

33 Khan MAA, Irfan OM, Djavanroodi F, Asad M (2022) Development of sustainable inhibitors for corrosion control. *Sustainability* 14:9502.

34 Li GL, Zheng Z, Mohwald H, Shchukin DG (2013) Silica/polymer double-walled hybrid nanotubes: synthesis and application ass-responsive nanocontainers in self-healing coatings. *ACS Nano* 7:2470–2478.

35 Zea C, Alcantara J, Barranco-Garcia R, Simancas J, Morcillo M, Fuente D (2017) Anticorrosive behavior study by localized electrochemical techniques of sol–gel coatings loaded with smart nanocontainers. *J Coat Technol Res* 14(4):841–850.

36 Liu J, Yan D, Zhang Z (2022) Eco-friendly silane as corrosion inhibitor for dual self-healing anticorrosion coatings. *J Coat Technol Res* 19:1381–1391.

37 Yeganeh M, Asadi N, Omidi M, Mahdavian M (2019) An investigation on the corrosion behavior of the epoxy coating embedded with mesoporous silica nanocontainer loaded by sulfamethazine inhibitor. *Prog Org Coat* 128:75–81.

38 Subhalakshmi Suresh Kumar SS, Kakooei S, Ismail MC, Harris M (2020) Synthesis, and characterization of metal ion end capped nanocontainer loaded with duo green corrosion inhibitors. *J Mater Res Technol* 9(4):8350–8354.

39 Mohammadi I, Shahrabi T, Mahdavian M, Izadi M (2022) Construction of an epoxy coating with excellent protection performance on the AA 2024-T3 using ion-exchange materials loaded with eco-friendly corrosion inhibitors. *Prog Org Coat* 166:106786.

40 Javidparvar AA, Naderi R, Ramezanzadeh B, Bahlakeh G (2019) Graphene oxide as a pH-sensitive carrier for targeted delivery of eco-friendly corrosion inhibitors in chloride solution: experimental and theoretical investigations. *J Ind Eng Chem* 72:196–213.

41 Gao X, Yan R, Lv Y, Ma H (2020) In situ pretreatment and self-healing smart anti-corrosion coating prepared through eco-friendly water-base epoxy resin combined with non-toxic chelating agents decorated biomass porous carbon. *J Clean Prod* 266:121920.

42 Leal DA, Riegel-Vidotti I, Ferreira M, Marino C (2018) Smart coating based on double stimuli-responsive microcapsules containing linseed oil and benzotriazole for active corrosion protection. *Corros Sci* 130:56–63.

43 X Liu X, Li W, Wang W (2018) Synthesis, and characterization of pH-responsive mesopo-rous chitosan microspheres loaded with sodium phytate for smart water-based coatings. *Mater Corros* 60:736–748.

44 Nawaz M, Shakoor RA, Kahraman R, Montemor MF (2021) Cerium oxide loaded with Gum Arabic as environmentally friendly anti-corrosion additive for protection of coated steel. *Mater Des* 198:109361.

45 Asaad MA, Raja PB, Huseien GF, Fediuk R, Ismail M, Alyousef R (2021) Self-healing epoxy coating doped with Elaesis guineensis/silver nanoparticles: a robust corrosion inhibitor. *Constr Build Mater* 312:125396.

46 Nawaz M, Habib S, Khan A, Shakoor RA, Kahraman R (2020) Cellulose microfibers (CMFs) as a smart carrier for autonomous self-healing in epoxy coatings. *New J Chim* 44(15):5702–5710.

47 Harb V S,Rodrigues SM, de Souza ACT (2021) Smart PMMA-cerium oxide anticorrosive coatings: effect of ceria content on structure and electrochemical properties. *Prog Organic Coat* 161:106548.

48 Wang J, Seidi F, Huang Y, Xiao H (2022) Smart lignin-based polyurethane conjugated with corrosion inhibitor as bio-based anticorrosive sublayer coating. *Ind Crops Prod* 188:115719.

49 Chen T, Fu J (2012) An intelligent anticorrosion coating based on pH-responsive supramolecular nanocontainers. *Nanotechnology* 23(50):505705.

50 Qiao Y, Li W, Wang G, Zhang X, Cao N (2015) Application of ordered mesoporous silica nanocontainers in an anticorrosive epoxy coating on a magnesium alloy surface. *RSC Adv* 5:47778–47787.

51 Sonawane SH, Bhanvase BA, Jamali AA (2012) Improved active anticorrosion coatings using layer-by-layer assembled ZnO nanocontainers with benzotriazole, *Chem Eng J* 189:464–472.

52 Chen G, Jin B, Li Y, He Y, Luo J (2022) A smart healable anticorrosion coating with enhanced loading of benzotriazole enabled by ultra-highly exfoliated graphene and mussel-inspired chemistry. *Carbon* 187:439–450.

53 Li C, Guo X, Franke GS (2021) Smart coating with dual-pH sensitive, inhibitor-loaded nanofibers for corrosion protection. *Mater Degrad* 5:54.

54 Riaz U, Nwaoha C, Ashraf SM (2014) Recent advances in corrosion protective composite coatings based on conducting polymers and natural resource derived polymers. *Prog Org Coat* 77:743–756.

55 Palraj S, Selvaraj M, Vidhya M, Rajagopal G (2012) Synthesis, and characterization of epoxy–silicone–polythiophene interpenetrating polymer network for corrosion protection of steel. *Prog Org Coat* 75:356–363.

56 Marathe R, Tatya P, Chaudhari A (2015) Neem acetylated polyester polyol-renewable source based smart PU coatings containing quinoline (corrosion inhibitor) encapsulated polyurea microcapsules for enhance anticorrosive property. *Ind Crops Prod* 77:239–250.

57 Izadi M, Shahrabi T, Ramezanzadeh B (2018) Active corrosion protection performance of an epoxy coating applied on the mild steel modified with an eco-friendly sol-gel film impregnated with green corrosion inhibitor loaded nanocontainers. *Appl Surf Sci* 440:491–505.

58 Abd El-Ghaffar MA, Abdelwahab NA, Fekry AM, Sanad MA, Sabaa MW, Soliman SMA (2020) Polyester-epoxy resin/conducting polymer/barium sulfate hybrid composite as a smart eco-friendly anti-corrosive powder coating. *Prog Org Coat* 144:105664.

59 Eivaz Mohammadloo H, Mirabedini SM, Pezeshk-Fallah H (2019) Microencapsulation of quinoline and cerium based inhibitors for smart coating application: anti-corrosion, morphology and adhesion study. *Prog Org Coat* 137:105339.

60 Khan A, Sliem MH, Arif A (2019) Designing and performance evaluation of polyelectrolyte multi-layered composite smart coatings. *Prog Org Coat* 137:105319.

7 Encapsulation of Organic Compounds as Corrosion Inhibitors for Smart Coatings

Anjali Peter, Amita Somya, Abhinay Thakur, and Ashish Kumar

7.1 CORROSION: A UNIVERSAL DEGRADATION PHENOMENON

Corrosion, commonly known as rust, has been present since the earliest days of Earth's existence. It is an undesirable process that tarnishes metals, diminishes their lifespan, and robs them of their beauty. Corrosion [1] occurs when metals react with their surrounding environment, resulting in their destruction and deterioration, as depicted in Figure 7.1. This pervasive issue poses significant challenges in various fields, including chemical, mechanical, metallurgical, biochemical, and medical engineering. It particularly affects the design and longevity of mechanical parts that are variable in size, function, and, employable life [2].

7.2 HISTORICAL PERSPECTIVE

In 1834, Becquerel proposed the idea that differences in metal ion concentrations were accountable for corrosion procedure [3]. De la Rive [4] suggested that impurities played a central role in deterioration. Researchers like Tweney [5] established a interrelation betwex chemical action and electric currents. In 1847, Ball [6] discovered that variations in the concentration of oxygen (O_2) in a flowing stream can generate a current between different portions of Fe or Zn. Corrosion research has undergone rapid evolution, initially focusing on polarization in the mid-20th century [7], and subsequently broadening to encompass dissolution, localized corrosion, and high-temperature corrosion, with developments continuing beyond the 1970s [8–11].

7.3 CORROSION CHARACTERISTICS

The corrosion process is influenced by various significant factors, as outlined below.

7.3.1 Metal Characteristics

Different metals exhibit distinct corrosion behavior. Highly reactive metals like Mg and K are prone to corrosion, whereas noble metals such as Ag, Au, Pt, and Pd remain unreactive in their environments. However, few metals like Ti, Al, Cr, Zn, and Ta demonstrate passivity characteristics.

7.3.2 Surrounding Environment

The surrounding conditions, including pH, humidity, temperature, and environmental impurities, play a crucial role in corrosion. The exposure of metals to these factors influences the corrosion process.

DOI: 10.1201/9781032677255-7

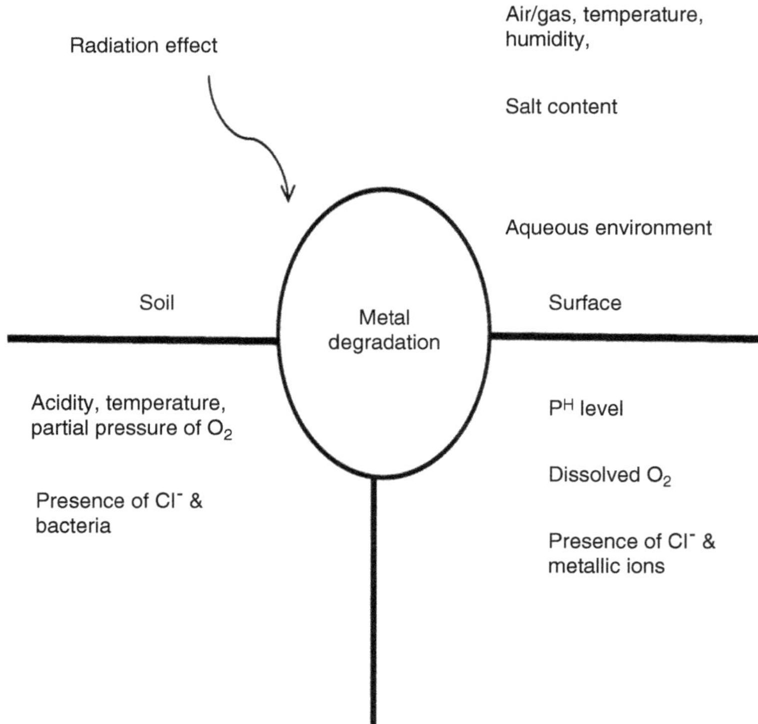

FIGURE 7.1 Environmental degradation of material.

7.3.3 Electrolyte Concentration

Corrosion is known as an electrochemical phenomenon involving the processes of the donation and acceptance of electrons. Electrolytes act as mediators, facilitating the movement of electrons from one location to another. The concentration of electrolytes directly affects the rate and extent of corrosion.

7.3.4 Temperature

Corrosion reactions can occur at both low and high temperatures. Reactive metals may undergo corrosion at low temperatures, while both reactive and passive metals can experience corrosion at high temperatures.

7.3.5 Electrode Potential

Electrode potential arises from polarization, which occurs during the flow of current. Polarization can manifest as anodic polarization at the anode or cathodic polarization at the cathode. The direction of current flow determines the behavior of electrode potential. Anodic polarization (positive direction) reduces the formation of oxide layers (oxidation process) and slows down the corrosion rate. Conversely, cathodic polarization (negative direction) accelerates the corrosion process.

7.3.6 Hydrogen Overvoltage

Hydrogen overvoltage is associated with the oxidation-reduction process and the anode-cathode reaction. The vulnerability of corrosion is influenced by hydrogen overvoltage. Lower values of hydrogen overvoltage enhance the corrosion reaction, whereas higher values decrease the rate of corrosion [9].).

7.3.7 Impurities and Conductive Species

Certain impurities, such as NO_2, HCl, and CO_2, act as catalysts, creating favorable conditions for corrosion. Conductive elements like a few chemical salts can increase the corrosion reaction. The presence of conducting species facilitates ion exchange processes, thereby promoting corrosion.

7.4 TYPES OF CORROSION

Corrosion reveals itself in distinct sorts and it is crucial to understand their appearances for proper identification and characterization. The different forms of corrosion are outlined in Figure 7.2.

7.4.1 General Corrosion

This type of corrosion occurs uniformly on the metal's surface, where the corrosive environment affects all areas of the material [10]. It is commonly encountered and leads to a general reduction in material properties—for example, steel in acid solutions.

7.4.2 Intergranular Corrosion

Intergranular corrosion primarily occurs in alloyed materials and is typically triggered by the occurrence of galvanic components. In the above, specific areas of the alloy, such as grain boundaries, act as anodes, while other regions act as cathodes. The large surface area ratio between the anode and cathode promotes a higher corrosion rate. Intergranular corrosion can penetrate deeply into the metal, causing damage and, in some cases, even failure.

7.4.3 Pitting Corrosion

Pitting corrosion results in the localized formation of holes or pits on the metal surface, giving it a rough and irregular appearance. It is considered a form of localized corrosion. In some

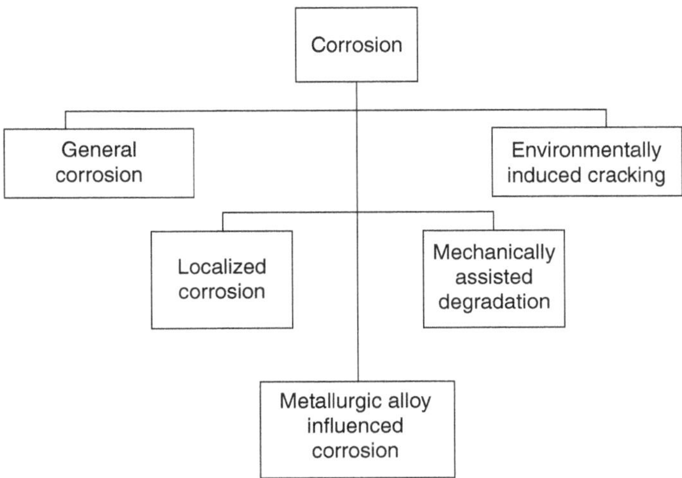

FIGURE 7.2 Types of corrosion.

instances, the pits may be confined to certain areas of the metal surface, while other regions remain unaffected. Pitting corrosion can also be associated with crevice corrosion, water-line attack, and erosion corrosion attack [11].

7.4.4 Exfoliation

Exfoliation refers to the separation of the layers of metal away from a solid metal or an alloy. This phenomenon occurs in distorted objects and is influenced by factors such as corrosion supporting "fiber orientation." It is commonly observed in wrought products with elongated structures.

7.4.5 Dealloying/Selective Corrosion

Dealloying or selective corrosion occurs in alloyed materials in which one element is relatively more active (less noble) as compared to other elements. The less noble metal undergoes preferential dissolution, leaving behind porous structures, which may subsequently be filled in by the corrosion products.

7.4.6 Galvanic Corrosion

Galvanic corrosion arises in electrically conductive environments due to potential differences between dissimilar metals. When a more noble metal acts as the cathode with a larger surface area, and a less noble metal acts as the anode with a smaller surface area, significant cathodic and anodic reactions take place, leading to an accelerated corrosion rate. The galvanic series is applied to assess the likelihood of galvanic corrosion—for instance, aluminum and steel alloys.

7.4.7 Waterline Corrosion

Waterline corrosion is commonly noticed at the interfaces between water and air. The formation of O_2 -concentration cells play a significant role in this type of corrosion. For instance, steel pipes driven into the soil are particularly susceptible to attack just below the groundwater level, while steel pipes immersed in seawater experience the strongest attack beneath the waterline.

7.4.8 Filiform Corrosion

Filiform corrosion appears as worm-like or filamentous structures under organic coatings on metal surfaces, particularly in environments with a humidity level of 78–80%. It is relatively easier to treat when detected at an early stage. An example is corrosion observed in oil and petroleum fields.

7.4.9 Stress Corrosion Cracking

Stress corrosion cracking is a combination of corrosion and mechanical stress, occurring in various types of cracks when the applied stress is below the metal's tensile strength.

7.5 THERMODYNAMICS OF CORROSION

Thermodynamics is the field of study that explores energy transfer at the atomic level and the biochemical processes involved. Energy plays a crucial role in maintaining the stability of metals, and thermal energy is a natural source of this energy, as revealed in Figure 7.3. As the temperature

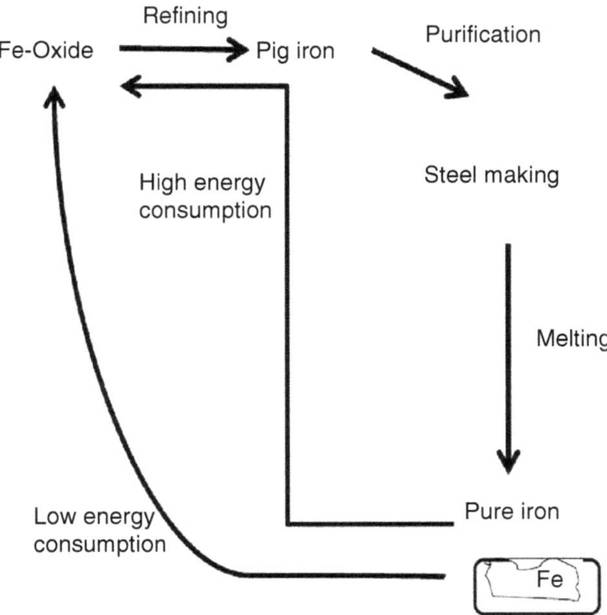

FIGURE 7.3 Schematic diagram of the thermodynamic process.

increases, it triggers vibrations in atoms with larger amplitudes. Adequate temperature levels result in the translational movement of atoms, connecting energy with their fluctuations. The rotational translation also facilitates the transition of a matter from a metastable state to a stable state by utilizing the surplus energy available. In practical situations, when a metal block, such as iron, is exposed to the environment, it will eventually undergo degradation or "corrosion."

7.6 CORROSION INHIBITION

In the realm of chemistry, an inhibitor is a fundamental component that impedes or reduces the speed of a chemical reaction. Specifically, a corrosion inhibitor is a sort of chemical material [12–14] that, when introduced to an environment, mitigates the rate of attack by the surroundings. The utilization of corrosion inhibitors has become a prominent strategy in addressing corrosion challenges. To optimize their effectiveness, three crucial factors need to be considered[15].

- Identification of decomposition complexities: It is crucial to identify and understand the potential challenges associated with the breakdown of the inhibitor.
- The commercial viability of the preservation process: This involves assessing whether the cost of implementing the inhibitor and maintaining the protective measures outweighs the potential damages caused by corrosion and the expenses of manual intervention.
- Compatibility of the inhibitor with the specific process at hand: It is imperative to ensure that the chosen inhibitor is compatible with the particular system or process to avoid any adverse consequences.

By carefully considering these factors, corrosion inhibitors can be effectively employed to mitigate the damaging effects of corrosion.

7.7 CORROSION PREVENTION METHODS

Several corrosion prevention methods were being explored and utilized. Here are some corrosion prevention methods that were relevant at the time:

1. **Protective Coatings:** Various types of protective coatings, such as paints, epoxy coatings, and galvanization, are commonly used to provide a barrier between the metal surface and the surrounding environment, preventing direct contact and corrosion.
2. **Cathodic Protection:** This method involves the use of sacrificial anodes or impressed current to protect the metal structure. Sacrificial anodes, typically made of zinc or aluminum, corrode instead of the protected metal, effectively preventing corrosion.
3. **Corrosion Inhibitors:** These are substances that can be added to the environment or applied directly to the metal surface to reduce corrosion rates. Inhibitors work by forming a protective layer on the metal surface or altering the electrochemical processes involved in corrosion.
4. **Alloying:** By alloying metals with other elements, it is possible to enhance their resistance to corrosion. For example, stainless steel, an alloy of iron, chromium, and other elements, gives it excellent corrosion resistance properties.
5. **Environmental Modifications:** Modifying the environment around the metal structure can help prevent corrosion. For instance, controlling humidity levels, maintaining proper pH levels, or implementing desiccants to remove moisture can reduce the likelihood of corrosion.
6. **Design Considerations:** Incorporating corrosion-resistant design features into structures can minimize the corrosion risk. This may include proper drainage systems, coatings on critical areas, and avoiding crevices where moisture can accumulate.

7.8 ENCAPSULATION

Encapsulation is a procedure used to entrap or encapsulate organic corrosion inhibitors within micro- or nano-sized containers, often made of polymers or other protective materials. These containers, called microcapsules or nanoparticles, act as reservoirs for the corrosion inhibitors. The encapsulation process provides several advantages, such as controlled release of the inhibitors, protection against degradation or leaching, and targeted delivery to the corrosion-prone areas. Smart coatings incorporate these encapsulated corrosion inhibitors to provide enhanced corrosion protection and functionality. Encapsulation allows for the controlled release of the inhibitors when triggered by specific environmental conditions, such as changes in pH, temperature, humidity, or exposure to corrosive agents. This controlled release mechanism ensures that the inhibitors are released when and where they are needed, optimizing their effectiveness and prolonging the coating's protective performance. Additionally, smart coatings can be designed to have self-healing capabilities, where the encapsulated corrosion inhibitors are released when damage or microcracks occur on the coating surface. The released inhibitors can fill the cracks and prevent further corrosion, thus enhancing the coating's durability. In addition to material selection, design, cathodic protection, and coatings, organic corrosion inhibitors serve as a vital method for safeguarding materials against corrosion. Corrosion is a widespread phenomenon that negatively impacts various materials, including metals, plastics, glass, and concrete. The economic costs associated with corrosion are substantial, accounting for 4.0% of the gross domestic product (GDP) in the United States, and similar losses are observed globally. While it is currently impossible to completely halt the corrosion process, certain approaches can effectively slow it down. Organic corrosion inhibitors are extensively utilized in industries due to their effectiveness over a wide temperature range, compatibility with protected materials, water solubility, cost-effectiveness, and relatively low toxicity. These inhibitors adsorb onto surfaces, forming protective films that displace water and shield against degradation. Effective organic corrosion inhibitors often contain nitrogen, oxygen, sulfur, and phosphorus, which possess lone electron pairs, as well as structural components with π-electrons that interact favorably with metals during the adsorption process [16–20].

7.9 CORROSION INHIBITORS

A corrosion inhibitor is a material when supplied in minimum quantities to a situation, decreases or prevents corrosion [21]. Corrosion is a natural process that can deteriorate the integrity of metals and other materials, leading to structural damage and reduced lifespan. Corrosion inhibitors are substances that can be added to coatings or applied directly to the surface of a material to prevent or slow down the corrosion process. They work by forming a protective barrier or by modifying the electrochemical reactions that cause corrosion. Many efforts have been made throughout history to find effective corrosion inhibitors of a crucial starting location in various destructive conditions [22–23].

Aldehydes, thioaldehydes, acetylenic composites, nitrogen-based components and their secondary, sulfur-holding complexes, strychnine, quinine, and nicotine inhibitors are some of the alkaloids that may be found in corrosive conditions. In neutral environments, some alkaloids (such as palavering and nicotine, quinine, and strychnine inhibitors) as well as some ionic salts (like arsenates, chromates, benzoates, chromates, and phosphates) act as composites and in addition to their ion exchange characteristics [24–25], they all exhibit positive inhibitory behavior in neutral conditions. Inhibitors limit or stifle the metal's reaction to the media [26–27].

There are several types of corrosion inhibitors available, and the choice of the best inhibitor depends on various factors such as the type of corrosion, the material being protected, the environmental conditions, and the specific application. Below are some common types of corrosion inhibitors.

7.9.1 Inorganic Inhibitors

Inorganic inhibitors include compounds such as chromates, phosphates, and silicates. They form protective films on the metal surface and provide effective corrosion inhibition. However, due to their environmental concerns and toxicity, their usage has decreased in recent years.

7.9.2 Organic Inhibitors

Organic inhibitors, as discussed earlier, are compounds derived from organic substances. They form a protective film or adsorb onto the metal surface, inhibiting corrosion. Organic inhibitors are versatile, environmentally friendly, and compatible with various coating systems.

7.9.3 Volatile Corrosion Inhibitors (VCIs)

VCIs are organic or inorganic compounds that release vapor-phase inhibitors to protect metal surfaces from corrosion. VCIs are particularly useful for protecting enclosed or hard-to-reach spaces, such as inside machinery or packaging materials.

7.9.4 Passivating Inhibitors

Passivating inhibitors promote the formation of a passivation layer on the metal surface, which acts as a barrier against corrosion. Examples include compounds like sodium nitrite, sodium molybdate, and cerium salts.

7.9.5 Cathodic Inhibitors

Cathodic inhibitors enhance the cathodic reaction of the electrochemical corrosion process. They facilitate the reduction of corrosive agents and help maintain a protective cathodic potential on the metal surface. Sodium benzoate and amines are examples of cathodic inhibitors.

7.9.6 Mixed Inhibitors

Mixed inhibitors are formulations that combine different types of corrosion inhibitors to achieve synergistic effects. By combining multiple inhibitors, mixed inhibitors offer improved corrosion protection and can target multiple corrosion mechanisms simultaneously.

7.9.7 Organic Synthetic Inhibitors

There are various organic substances that can be used as corrosion inhibitors and encapsulated within smart coatings. The selection of the organic corrosion inhibitor depends on the specific application, the type of corrosion being addressed, and compatibility with the coating system. Here are some examples of commonly used organic substances:

- **Amines:** Amines, such as aliphatic amines or aromatic amines, are widely used as corrosion inhibitors. They form a protective layer on the metal surface, which inhibits the corrosion process. Examples include cyclohexylamine, benzylamine, and dodecylamine.
- **Organic salts:** Organic salts, particularly quaternary ammonium salts, are effective corrosion inhibitors. They form a protective film on the metal surface, preventing the access of corrosive species. Examples include benzalkonium chloride, dodecylbenzenesulfonic acid, and alkylpyridinium salts.
- **Organic acids and their derivatives:** Organic acids and their derivatives, such as carboxylic acids and their esters, are commonly used as corrosion inhibitors. They work by creating a protective shield or reacting with metal ions to prevent corrosion. Examples include acetic acid, benzoic acid, and their esters.
- **Imidazolines:** Imidazolines are organic compounds derived from imidazole. They possess excellent corrosion inhibition properties and are commonly used in oil and gas applications. Examples include 2-aminoethylaminomethyl-4-methylimidazoline and dodecylbenzylamine.
- **Polyamines:** Polyamines are organic compounds with multiple amine groups. They are effective corrosion inhibitors due to their ability to form a protective layer on the metal surface. Examples include ethylenediamine, diethylenetriamine, and triethylenetetramine.
- **Benzotriazole (BTA):** BTA is a versatile organic compound widely used as a corrosion inhibitor for various metals. It creates a protective shield on the surface of the metal, inhibiting corrosion. BTA and its derivatives can be encapsulated for controlled release within smart coatings.

The research facility has been equipped with artificial inhibitors to function as a suitable replacement for the actual entities. Different synthetic corrosion inhibitors shown in Table 7.1 [27] have been used by all kinds of various companies.

The selection of corrosion inhibitors in a scientific context is contingent upon the specific application and the type of corrosion under consideration. Typical examples of organic corrosion inhibitors encompass organic salts, amines, organic acids, and their derivatives. The process of encapsulation facilitates the inclusion of a broad spectrum of organic compounds, thereby enabling customization and tailoring of the corrosion protection characteristics exhibited by the coating.

Organic coatings [28–33] serve as an effective physical barrier, affording protection to metal structures against corrosive environments. These coatings combine organic resin and barrier pigments, thereby impeding the migration of corrosive ions toward the metal substrate and

TABLE 7.1
Types of organic synthetic inhibitors

Synthetic Corrosion Inhibitor	Chemical Structure	Commercial Name
Benzyl chloride quaternary compounds	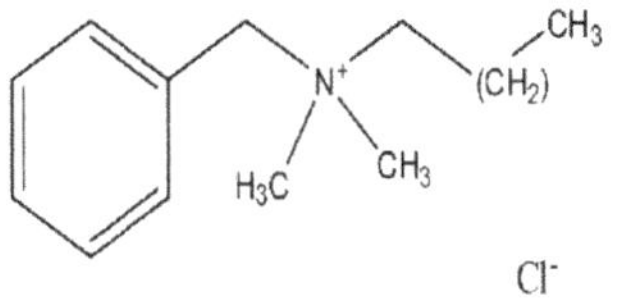 CH_3, N^+, (CH_2), H_3C, CH_3, Cl^-	Alpha 1018, 1038, 1458, 1505, 3013, 3444
Diethylsulfate quaternary compounds	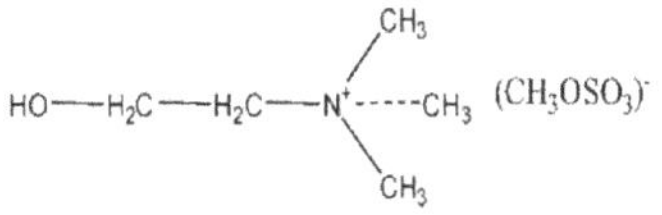 $HO—H_2C—H_2C—N^+$----CH_3 $(CH_3OSO_3)^-$; CH_3, CH_3	Alpha 1080
Amine ethoxylates compounds	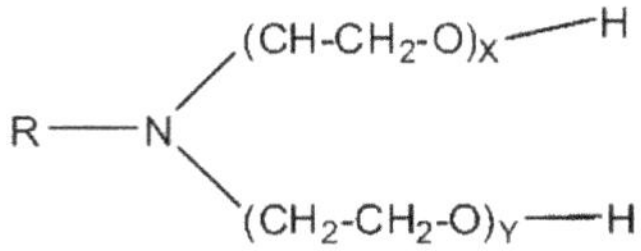 $R—N$ $(CH\text{-}CH_2\text{-}O)_X—H$ $(CH_2\text{-}CH_2\text{-}O)_Y—H$	C1815
Imidazoline compounds	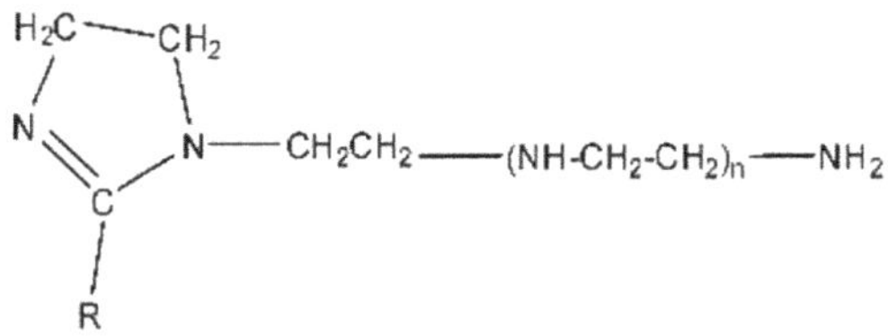 $H_2C—CH_2$, N, C, R, $N—CH_2CH_2—(NH\text{-}CH_2\text{-}CH_2)_n—NH_2$	Alpha 1153,1158
Organic boron compounds	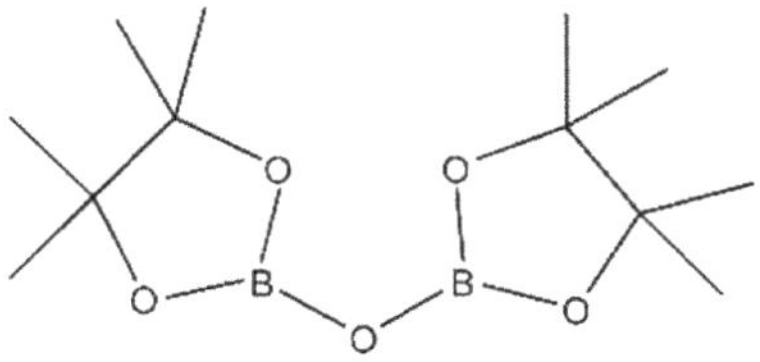 O, B, O, B, O, O, O	Alpha 3220

enhancing the resistance between local anodes and cathodes present on the surface of the substrate [34–35]. Nevertheless, attaining flawless coating films is a rarity. Moreover, extended exposure to the environment and various interactions can exacerbate preexisting defects or introduce new ones, thereby compromising the protective properties of the coating [36–40].

An alternative approach to ensure an adequate supply of inhibitive species and minimize undesired losses involves the implementation of controlled release mechanisms through the utilization of intelligent carriers. Intelligent carriers such as polymeric capsules, halloysites, or mesoporous nanoparticles serve as reservoirs for the inhibitive species. Upon the occurrence of corrosion reactions, the localized rise in pH triggers the liberation of inhibitive species from the carrier. Coatings endowed with the ability to adapt their properties in response to environmental changes, including pH, temperature, pressure, surface tension, and mechanical forces, are commonly referred to as smart coatings. The controlled release of pigments can extend the longevity of the coating by impeding the spontaneous leaching of inhibitive species from the film and enhancing its barrier properties [41–47].

It's crucial to mention that the selection of an organic corrosion inhibitor should consider factors such as compatibility with the coating system 72-73, target metal or substrate, environmental conditions, and the desired release mechanism within the encapsulated coating.

TABLE 7.2
List of some potent encapsulated corrosion inhibitors used for metals along with their findings

Encapsulation	Corrosion Protection Metal	Findings	References
Organoclays in epoxy latex particles	Clay, a plate-like layered silicate, is a well-known effective nanofiller that improves the gas barrier performance of various polymers, including epoxy coatings	The best oxygen barrier efficiency was approximately 14% at 2 wt% clay loading.	[48]
Encapsulated corrosion inhibitors[2-mercaptobenzothiazole (MBT) and 2-mercaptobenzimidazole (MBI)] in the presence of β-cyclodextrin (β-CD)	corrosive electrolytes	Long-term corrosion protection	[49]
Halloysite nanotubes loaded with the corrosion inhibitor benzotriazol	admixed into the paint to improve its anticorrosion performance as well as the coating tensile strength	Improved	[50]
Benzotriazole (BTA) corrosion inhibitor in a Zn-BTC metal-organic framework (MOF)	Aluminum	The durability of this smart coating is significantly prolonged in a 3.5 wt % NaCl solution.	[51]
Carboxymethylcellulose (CMC-Na) as encapsulating material with benzotriazole (BTA)	envisaged as responsive materials,	Verified (smart coatings)	[52]
Superhydrophobic imidazole encapsulated hollow Zn phosphate nanoparticles	mild steel	Grafted with octanol to get superhydrophobicity.	[53]
Zeolite imidazole framework-8 (ZIF-8) with (RGO@ZIF)	solid corrosion inhibitor	Explored to upgrade the rate of loading of inhibitors.	[54]
Levamisole (LMS) and 4-phenylimidazole (PIZ)	copper in sulfuric acid solution	LMS had better corrosion inhibition performance than PIZ.	[55]
Biocarbon supported Ni catalysts	hydrogenation of nitrobenzene to cyclohexylamine, 10%Ni/CSC-II	Greatly active.	[56]
Bonding of five imidazole-type molecules—imidazole, 1-methyl-imidazole, benzimidazole, 2-mercapto-1-methyl-imidazole, and 2-mercaptobenzimidazole—with copper surfaces	copper	Mercapto-substituted imidazoles are found resistant to deprotonation upon adsorption.	[57]
Encapsulated cerium nitrate with thin boehmite nanoparticles	aluminum alloys	The corrosion inhibitor can be discharged within the coating where defects appear retarding the corrosion reactions occurring on those sites.	[58]
Coating formulations based on hybrid organic-inorganic sol-gel coatings	Zr^{4+} (Z) and various rare-earth elements (RE: Ce^{3+}, La^{3+}, Gd^{3+}, Er^{3+} on mild steel substrates.	Self self-healing property of cerium and zirconium-based coatings was demonstrated.	[59]

TABLE 7.2 (Continued)
List of some potent encapsulated corrosion inhibitors used for metals along with their findings

Encapsulation	Corrosion Protection Metal	Findings	References
Nettle-NCs and zinc acetate (ZA)	mild steel	Nettle-NCs extract confirmed the successful release of Nettle molecules from Nettle-NCs.	[60]
Cyclodextrin-assisted molecular encapsulation.	aluminum alloy 2024-T3	The activity of encapsulated inhibitors in the presence of corrosive media has been explored.	[58]
Polyaniline (PANI) capsules + encapsulation of corrosion inhibitor 2-Mercaptobenzothiazole (MBT).	AA2024-T3	Significantly improved its corrosion protection performance due to the interesting smart characteristics of PANI capsules.	[61]
Bentonite/acrylamide capsule loaded with the corrosion inhibitor (HCTS)	downhole corrosion environment of oil wells	Prevented the CO_2-NaCl corrosion of downhole pipelines under high temperatures and pressures.	[62]
Water-based epoxy coatings modified with nanocapsules loaded with 2-Mercaptobenzothiazole (MBT)	AA5083 and galvanneal substrates	Exploration of self-healing characteristics of water-based epoxy coatings.	[63]
Benzotriazole and cetyltrimethyl ammonium bromide (CTAB) co-loaded silica nanocontainers	chromium	One-step synthetic method was efficient, and energy-saving.	[64]
Self-assembled Nanophase Particle (SNAP)+ β-cyclodextrin	aluminum alloys	Long-term corrosion protection properties.	[58]
Inhibitors 2-mercaptobenzimidazole and 2-mercapatobenzothiozole encapsulated in polymerized urea and formaldehyde.	chromium	2-MBT is superior to 2-MBI.	[65]
Calcium alginate gel (Alg-Ca^{2+}) capsules	P110 steel	Inhibitor-loaded capsules facilitate their sinking processes and corrosion protection performance.	[66]
Halloysite nanotubes (HNT) were incorporated into polybenzoxazine coating	carbon steel	The performance of polybenzoxazine coatings was effectively improved.	[67]
Halloysite clay nanotubes (HNTs) using (poly (diallyldimethylammonium chloride)) (PDDA) and poly (styrene sulfonate) (PSS) polyelectrolytes to form core (inhibitor-loaded HNTs)-shell (polyelectrolytes layers (PDDA/PSS)) structured smart microcapsules or micro-containers.	Al alloy 2024	Compatible	[68]

(continued)

TABLE 7.2 (Continued)
List of some potent encapsulated corrosion inhibitors used for metals along with their findings

Encapsulation	Corrosion Protection Metal	Findings	References
Urea-formaldehyde (UF) microcapsules loaded with linseed oil (LO) and benzotriazole (BTA)	mild steel specimens in 3.5% sodium chloride solution	The addition of microcapsules enhances the corrosion resistance.	[69]
Zeolitic imidazole framework (ZIF) and silica	Aluminum	Improved inhibition	[70]
Imidazole from CZM nanocontainer	Mild Steel	Significant improvement in the anti-corrosion performance.	[71]

7.10 CONCLUSION

The encapsulation of smart coatings as corrosion inhibitors is a cutting-edge strategy in materials science, promising enhanced corrosion resistance and prolonged lifespan for materials like metals, alloys, and composites. Smart coatings, designed with unique functionalities, encapsulate corrosion inhibitors to ensure a controlled and sustained release over time. This controlled release mechanism provides efficient and targeted protection against corrosion, particularly in industries such as automotive, aerospace, infrastructure, and marine applications. Moreover, smart coatings often exhibit self-healing capabilities, autonomously repairing damage or microcracks to maintain their protective properties. This self-healing ability significantly extends the lifespan of coated materials, reducing the need for frequent maintenance and repairs. The applications of these coatings are diverse, ranging from protecting vehicle components in the automotive industry to preserving critical aerospace components and infrastructure subjected to harsh environmental conditions.

In summary, the encapsulation of corrosion inhibitors within smart coatings offers a forward-looking solution to corrosion challenges. The controlled release and self-healing features not only enhance the performance but also contribute to cost savings by reducing maintenance needs across various industries. As technology advances, smart coatings are poised to play a pivotal role in safeguarding critical infrastructure and extending the life cycle of materials.

REFERENCES

[1] Alavi A, Cottis RA. The determination of pH, potential and chloride concentration in corroding crevices on 304 stainless steel and 7475 aluminium alloy. Corrosion Science. 1987 Jan 1;27(5):443–51.

[2] Bhaskaran R, Palaniswamy N, Rengaswamy NS, Jayachandran M. A review of differing approaches used to estimate the cost of corrosion (and their relevance in the development of modern corrosion prevention and control strategies). Anti-Corrosion Methods and Materials. 2005 Feb 1;52(1):29–41

[3] Amin HM, Galal A, editors. Corrosion protection of metals and alloys using graphene and biopolymer-based nanocomposites. CRC Press; 2021 Feb 14.

[4] de La Rive AA. Note relative à l'action qu'exerce sur le zinc l'acide sulfurique étendu d'eau. Verlag nicht ermittelbar; 1830.

[5] Tweney RD. Inventing the field: Michael Faraday and the creative "engineering" of electromagnetic field theory. Inventive minds: Creativity in technology. 1992:31–47.

[6] Ball P. The elements: a very short introduction. OUP Oxford: 2004 Apr 8 (published in United State by Oxford University Press Inc., New York).

[7] Frankel GS, Isaacs HS, Scully JR, Sinclair JD. Wagner-Traud to Stern-Geary; development of corrosion kinetics. Corrosion Retrospective. 2002. [8] Peter A, Sharma SK. Corrosion: Introduction.

In Corrosion Protection of Metals and Alloys Using Graphene and Biopolymer Based Nanocomposites 2021 Feb 14 (pp. 3–15). CRC Press.
[9] Kucera V, Mattsson E. Atmospheric corrosion. In Corrosion mechanisms 2020 Nov 25 (pp. 211–284). CRC Press.
[10] Konishi H, Yamashita M, Uchida H, Jun IM. Characterization of rust layer formed on Fe, Fe-Ni and Fe-Cr alloys exposed to Cl-rich environment by Cl and Fe K-Edge XANES measurements. Materials transactions. 2005;46(2):329–336.
[11] Leyerzapf H. Meaning and history of the term "Corrosion". Werkst. Korros. 1985 Feb;36(2):88–96.
[12] Sowmyashree AS, Somya A, Kumar S, Rao S, Jayaprakash GK. Discotic anthraquinones as novel corrosion inhibitor for mild steel surface. Journal of Molecular Liquids. 2022 Feb 1;347:118194.
[13] Sowmyashree AS, Somya A, Kumar CP, Rao S. Novel nano corrosion inhibitor, integrated zinc titanate nano particles: Synthesis, characterization, thermodynamic and electrochemical studies. Surfaces and Interfaces. 2021 Feb 1;22:10081.
[14] Sowmyashree AS, Somya A, Rao S, Kumar CP, Al-Romaizan AN, Hussein MA, Khan A, Marwani HM, Asiri AM. Potential sustainable electrochemical corrosion inhibition study of Citrus limetta on mild steel surface in aggressive acidic media. Journal of Materials Research and Technology. 2023 May 1;24:984–994
[15] Askari M, Aliofkhazraei M, Jafari R, Hamghalam P, Hajizadeh AJ. Downhole corrosion inhibitors for oil and gas production–a review. Applied Surface Science Advances. 2021 Dec 1;6:100128
[16] Veronika BK. Knowledge about metals in the first century. korrozios figyelo. 2008 Jan 1;48(5):133.
[17] ISO E. 8044. Corrosion of Metals and Alloys: Basic Terms and Definitions. European Committee for Standardization (CEN): Brussels, Belgium. 1999.
[18] Torsner E. Solving corrosion problems in biofuels industry. Corrosion Engineering, Science and Technology. 2010 Feb 1;45(1):42–48.
[19] Davis JR, editor. Corrosion: Understanding the basics. Asm International; 2000.
[20] Fontanana MG. A Critical analysis of pitting corrosion. Corrosion Engineering Digest. 1959 Jul 25;8(7):298–307.
[21] Al-Otaibi MS, Al-Mayouf AM, Khan M, Mousa AA, Al-Mazroa SA, Alkhathlan HZ. Corrosion inhibitory action of some plant extracts on the corrosion of mild steel in acidic media. Arabian Journal of Chemistry. 2014 Jul 1;7(3):340–346.
[22] Verma C, Ebenso EE, Quraishi MA. Ionic liquids as green and sustainable corrosion inhibitors for metals and alloys: an overview. Journal of Molecular Liquids. 2017 May 1;233:403–414.
[23] Hossain N, Asaduzzaman Chowdhury M, Kchaou M. An overview of green corrosion inhibitors for sustainable and environment friendly industrial development. Journal of Adhesion Science and Technology. 2021 Apr 3;35(7):673–690.
[24] Somya A. Hybrid ion exchangers. Hybrid Nanomaterials-Flexible Electronics Materials. 2020 Apr 25. DOI: http://dx.doi.org/10.5772/intechopen.92116
[25] Somya A. Metal Phosphates: Their Role as Ion Exchangers in Water Purification. InMetal Phosphates and Phosphonates: Fundamental to Advanced Emerging Applications 2023 Mar 30 (pp. 341–356). Cham: Springer International Publishing. DOI: https://doi.org/10.1007/978-3-031-27062-8_19
[26] Umoren SA, Obot IB, Ebenso EE, Obi-Egbedi NO. Synergistic inhibition between naturally occurring exudate gum and halide ions on the corrosion of mild steel in acidic medium. International Journal of Electrochemical Science. 2008 Sep 1;3(9):1029–1043.
[27] Peter A, Obot IB, Sharma SK. Use of natural gums as green corrosion inhibitors: an overview. International Journal of Industrial Chemistry. 2015 Sep;6:153–164.
[28] Lyon SB, Bingham R, Mills DJ. Advances in corrosion protection by organic coatings: What we know and what we would like to know. Progress in Organic Coatings. 2017 Jan 1;102:2–7.
[29] Sørensen PA, Kiil S, Dam-Johansen K, Weinell CE. Anticorrosive coatings: a review. Journal of coatings technology and research. 2009 Jun;6:135–176.
[30] Ma IW, Ammar S, Kumar SS, Ramesh K, Ramesh S. A concise review on corrosion inhibitors: types, mechanisms and electrochemical evaluation studies. Journal of Coatings Technology and Research. 2022 Jan 1:1–28.
[31] Rammelt U, Reinhard G. Characterization of active pigments in damage of organic coatings on steel by means of electrochemical impedance spectroscopy. Progress in Organic Coatings. 1994 Jun 1;24(1-4):309–322.

[32] Hao Y, Liu F, Han EH, Anjum S, Xu G. The mechanism of inhibition by zinc phosphate in an epoxy coating. Corrosion Science. 2013 Apr 1;69:77–86.

[33] Furman SA, Scholes FH, Hughes AE, Lau D. Chromate leaching from inhibited primers: Part II: Modelling of leaching. Progress in Organic Coatings. 2006 May 1;56(1):33–38.

[34] Grundmeier G, Schmidt W, Stratmann MJ. Corrosion protection by organic coatings: electrochemical mechanism and novel methods of investigation. Electrochimica Acta. 2000 May 3;45(15–16):2515–2533.

[35] Ashrafi-Shahri SM, Ravari F, Seifzadeh D. Smart organic/inorganic sol-gel nanocomposite containing functionalized mesoporous silica for corrosion protection. Progress in Organic Coatings. 2019 Aug 1;133:44–54.

[36] Ding C, Fu J. Smart anticorrosion coatings based on nanocontainers. InSmart Nanocontainers 2020 Jan 1 (pp. 413–429). Elsevier.

[37] Wei H, Wang Y, Guo J, Shen NZ, Jiang D, Zhang X, Yan X, Zhu J, Wang Q, Shao L, Lin H. Advanced micro/nanocapsules for self-healing smart anticorrosion coatings. Journal of Materials Chemistry A. 2015;3(2):469–480.

[38] Zahidah KA, Kakooei S, Ismail MC, Raja PB. Halloysite nanotubes as nanocontainer for smart coating application: A review. Progress in Organic Coatings. 2017 Oct 1;111:175–185.

[39] Hollamby MJ, Fix D, Dönch I, Borisova D, Möhwald H, Shchukin D. Hybrid polyester coating incorporating functionalized mesoporous carriers for the holistic protection of steel surfaces. Advanced materials. 2011 Mar 14;11(23):1361–1365.

[40] Yeganeh M, Asadi N, Omidi M, Mahdavian M. An investigation on the corrosion behavior of the epoxy coating embedded with mesoporous silica nanocontainer loaded by sulfamethazine inhibitor. Progress in Organic Coatings. 2019 Mar 1;128:75–81.

[41] Shchukina E, Grigoriev D, Sviridova T, Shchukin D. Comparative study of the effect of halloysite nanocontainers on autonomic corrosion protection of polyepoxy coatings on steel by salt-spray tests. Progress in Organic Coatings. 2017 Jul 1;108:84–89.

[42] Ma IW, Ammar S, Bashir S, Selvaraj M, Assiri MA, Ramesh K, Ramesh S. Preparation of hybrid chitosan/silica composites via ionotropic gelation and its electrochemical impedance studies. Progress in Organic Coatings. 2020 Aug 1;145:105679.

[43] Zheludkevich ML, Tedim J, Ferreira MG. "Smart" coatings for active corrosion protection based on multi-functional micro and nanocontainers. Electrochimica Acta. 2012 Nov 1;82:314–323.

[44] Yeganeh M, Marashi SM, Mohammadi N. Smart corrosion inhibition of mild steel using mesoporous silica nanocontainers loaded with molybdate. International Journal of Nanoscience and Nanotechnology. 2018 May 1;14(2):143–151 .

[45] Nazeer AA, Madkour M. Potential use of smart coatings for corrosion protection of metals and alloys: A review. Journal of Molecular Liquids. 2018 Mar 1;253:11–22.

[46] Falcón JM, Otubo LM, Aoki IV. Highly ordered mesoporous silica loaded with dodecylamine for smart anticorrosion coatings. Surface and Coatings Technology. 2016 Oct 15;303:319–329.

[47] Sinko J. Challenges of chromate inhibitor pigments replacement in organic coatings. Progress in organic coatings. 2001 Sep 1;42(3–4):267–282.

[48] de Lima-Neto P, de Araujo AP, Araujo WS, Correia AN. Study of the anticorrosive behaviour of epoxy binders containing non-toxic inorganic corrosion inhibitor pigments. Progress in Organic Coatings. 2008 May 1;62(3):344–350.

[49] Pham VH, Ha YW, Kim SH, Jeong HT, Jung MY, Ko BS, Hwang YJ, Chung JS. Synthesis of epoxy encapsulated organoclay nanocomposite latex via phase inversion emulsification and its gas barrier property. Journal of Industrial and Engineering Chemistry. 2014 Jan 25;20(1):108–112.

[50] Abdullayev E, Lvov Y. Clay nanotubes for corrosion inhibitor encapsulation: release control with end stoppers. Journal of Materials Chemistry. 2010;20(32):6681–6687.

[51] Mohammadpour Z, Zare HR. Fabrication of a pH-sensitive epoxy nanocomposite coating based on a Zn-BTC metal–organic framework containing benzotriazole as a smart corrosion inhibitor. Crystal Growth & Design. 2021 Jun 15;21(7):3954–3966.

[52] Calegari F, da Silva BC, Tedim J, Ferreira MG, Berton MA, Marino CE. Benzotriazole encapsulation in spray-dried carboxymethylcellulose microspheres for active corrosion protection of carbon steel. Progress in Organic Coatings. 2020 Jan 1;138:105329.

[53] Jadhav AJ, Holkar CR, Pinjari DV. Anticorrosive performance of super-hydrophobic imidazole encapsulated hollow zinc phosphate nanoparticles on mild steel. Progress in Organic Coatings. 2018 Jan 1;114:33–39.
[54] Zhong F, He Y, Wang P, Chen C, Wu Y. Novel pH-responsive self-healing anti-corrosion coating with high barrier and corrosion inhibitor loading based on reduced graphene oxide loaded zeolite imidazole framework. Colloids and Surfaces A: Physicochemical and Engineering Aspects. 2022 Jun 5;642:128641.
[55] Yan T, Zhang S, Feng L, Qiang Y, Lu L, Fu D, Wen Y, Chen J, Li W, Tan B. Investigation of imidazole derivatives as corrosion inhibitors of copper in sulfuric acid: combination of experimental and theoretical researches. Journal of the Taiwan Institute of Chemical Engineers. 2020 Jan 1;106:118–129.
[56] Lu X, He J, Jing R, Tao P, Nie R, Zhou D, Xia Q. Microwave-activated Ni/carbon catalysts for highly selective hydrogenation of nitrobenzene to cyclohexylamine. Scientific Reports. 2017 Jun 1;7(1):2676.
[57] Jadhav AJ, Holkar CR, Pinjari DV. Anticorrosive performance of super-hydrophobic imidazole encapsulated hollow zinc phosphate nanoparticles on mild steel. Progress in Organic Coatings. 2018 Jan 1;114:33–39.
[58] Khramov AN, Voevodin NN, Balbyshev VN, Mantz RA. Sol–gel-derived corrosion-protective coatings with controllable release of incorporated organic corrosion inhibitors. Thin Solid Films. 2005 Jul 1;483(1-2):191–196.
[59] Gautam A, Raju KS, Gobi KV, Subasri R. Effect of Transition Metal and Different Rare-Earth Inhibitors-Based Sol–gel Coatings on Corrosion Protection of Mild Steel. Metals and Materials International. 2023 Oct;29(10):2909–2925.
[60] Wang H, Akid R. Encapsulated cerium nitrate inhibitors to provide high-performance anti-corrosion sol–gel coatings on mild steel. Corrosion Science. 2008 Apr 1;50(4):1142–1148.
[61] Tavandashti NP, Ghorbani M, Shojaei A, Mol JM, Terryn H, Baert K, Gonzalez-Garcia Y. Inhibitor-loaded conducting polymer capsules for active corrosion protection of coating defects. Corrosion Science. 2016 Nov 1;112:138–149.
[62] Sun A, Cui G, Liu Q. Capsule corrosion inhibitor loaded with hyperbranched chitosan: Carbon dioxide corrosion protection for downhole pipelines in oil fields. Colloids and Surfaces A: Physicochemical and Engineering Aspects. 2023 May 5;664:131106.
[63] Plawecka M, Snihirova D, Martins B, Szczepanowicz K, Warszynski P, Montemor MF. Self healing ability of inhibitor-containing nanocapsules loaded in epoxy coatings applied on aluminium 5083 and galvanneal substrates. Electrochimica Acta. 2014 Sep 10;140:282–293.
[64] Xu JB, Cao YQ, Fang L, Hu JM. A one-step preparation of inhibitor-loaded silica nanocontainers for self-healing coatings. Corrosion Science. 2018 Aug 1;140:349–362.
[65] Marathe RJ, Chaudhari AB, Hedaoo RK, Sohn D, Chaudhari VR, Gite VV. Urea formaldehyde (UF) microcapsules loaded with corrosion inhibitor for enhancing the anti-corrosive properties of acrylic-based multi-functional PU coatings. Rsc Advances. 2015;5(20):15539–15546.
[66] Wang L, Zhang C, Xie H, Sun W, Chen X, Wang X, Yang Z, Liu G. Calcium alginate gel capsules loaded with inhibitor for corrosion protection of downhole tube in oilfields. Corrosion Science. 2015 Jan 1;90:296–304.
[67] Xu D, Lou C, Huang J, Lu X, Xin Z, Zhou C. Effect of inhibitor-loaded halloysite nanotubes on active corrosion protection of polybenzoxazine coatings on mild steel. Progress in Organic Coatings. 2019 Sep 1;134:126–133.
[68] Cui M, Njoku DI, Li B, Yang L, Wang Z, Hou B, Li Y. Corrosion protection of Aluminium Alloy 2024 through an epoxy coating embedded with smart microcapsules: The responses of smart microcapsules to corrosive entities. Corrosion Communications. 2021 Mar 1;1:1–9.
[69] Fadil M, Chauhan DS, Quraishi MA. Smart coating based on urea-formaldehyde microcapsules loaded with benzotriazole for corrosion protection of mild steel in 3.5% NaCl. Russian Journal of Applied Chemistry. 2018 Oct;91:1721–1728.
[70] Chen Y, Li X, Luan Z, Zhu L, Hu J, Xia X, Ma K, Geng B, Yan M. Fabrication of intelligent anticorrosion waterborne aluminum pigments based on encapsulation of composited zeolitic imidazole framework layers. Journal of Materials Science. 2023 May;58(19):8143–8156.

[71] Bhanvase BA, Patel MA, Sonawane SH. Kinetic properties of layer-by-layer assembled cerium zinc molybdate nanocontainers during corrosion inhibition. Corrosion Science. 2014 Nov 1;88:170–177.

[72] Somya, A. and Varshney, A. P., 2023. Functionalized thin film coatings for automotive coatings. In: Kumar, A., Verma, C. & Thakur, A. (eds.) *Corrosion Mitigation Coatings: Functionalized Thin Film Fundamentals and Applications*. Walter de Gruyter GmbH & Co KG, Germany. DOI: https://doi.org/10.1515/9783111016160-016

[73] Anjali P, Amita S. Anticorrosion Nano coatings and Potential Role of Simulation in Nano Coatings. Journal of Applied Optics. 2024 Mar 13:113–126. DOI: https://appliedopticsjournal.net/index.php/JAO/article/view/91

8 Measurements of Self-Releasing of the Encapsulated Corrosion Inhibitor in Response to Solution pH, Temperature, and Concentration

Ruby Aslam, Qihui Wang, and Zhitao Yan

8.1 INTRODUCTION

Corrosion inhibitors are chemicals used to protect metals from corrosion [1–3]. In the world of protective coatings, corrosion inhibitors play a pivotal role in safeguarding materials from the detrimental effects of corrosion [4, 5]. Corrosion inhibitors are added to coatings to extend the lifespan of the material underneath. They work by slowing down the corrosion process, protecting the material from environmental factors like moisture, oxygen, and salts, etc. However, the addition of corrosion inhibitors directly to the coating matrix might react prematurely with the environments or the coating material itself, which could decrease their effectiveness [6, 7]. They might also be released too quickly, providing strong initial protection but failing to protect the material over the long term. Encapsulation helps to control these issues, ensuring a steady, long-lasting release of the inhibitors [8, 9]. These encapsulated inhibitors added to the coating's matrix form a critical line of defense against environmental factors that can accelerate corrosion. The release behavior of these inhibitors from the coatings, however, is not a constant phenomenon [10]. It is influenced by a variety of external stimuli, including temperature fluctuations, humidity, pH, pressure, light, and exposure to corrosive elements, etc. This chapter aims to delve into the intricate dynamics of how these external stimuli impact the release behavior of corrosion inhibitors from coatings, providing a comprehensive understanding of this critical aspect of corrosion protection. Understanding the release behavior of corrosion inhibitors is crucial in optimizing the performance of protective coatings. The release mechanism can be either passive, where the inhibitor is released upon contact with a corrosive agent, or active, where the release is triggered by specific environmental changes. The effectiveness of these inhibitors is largely dependent on their release rate, which in turn is heavily influenced by external stimuli. Investigating these stimuli will help us gain deeper insights into the complex interplay between the inhibitors, the coatings, and the environment. This will not only enhance our knowledge of corrosion prevention strategies but also pave the way for the development of more efficient and responsive protective coatings. In this chapter, we will explore these external stimuli and their impact on the release behavior of corrosion inhibitors in detail.

DOI: 10.1201/9781032677255-8

Before delving into the main theme of this chapter, it's beneficial to first explore the history of corrosion inhibitors. Understanding their evolution gives context to their current applications. In the following section, we will discuss how these inhibitors have played a pivotal role in the development of smart coatings.

8.2 THE EVOLUTION AND FUNCTIONALITY OF CORROSION INHIBITORS: FROM ANCIENT CIVILIZATIONS TO NANOTECHNOLOGY

When corrosion damage fails to be detected and addressed, it can cause structural failures that can be catastrophic and are an expensive issue for a variety of industries and aspects of our lives [11]. When applied to a liquid or the environment, a corrosion inhibitor effectively slows down the rate of corrosion on metal surfaces. A protective coating is produced on the metal surface by corrosion inhibitors, which then stops or delays the electrochemical reactions that lead to corrosion. By decreasing system failures and shutdowns, eliminating contamination of products, and restricting heat loss, corrosion inhibitors help a structure appear intriguing [3, 12]. Corrosion inhibitors are extensively used in oil and gas exploration and production, petroleum refining, chemical production, heavy manufacturing, water treatment, and product additive industries.

Ancient civilizations recognized the need to preserve metals from corrosion, and this is how corrosion inhibitors gained their start [13]. Applying oils, greases, and waxes to metal surfaces was one of the approaches that the Egyptians, Greeks, and Romans employed to stop metal from corroding. They also discovered that copper alloys' corrosion resistance may be increased by adding lead or tin to them. Due to the extensive use of metals in machinery and infrastructure throughout the Industrial Revolution in the 18th and 19th centuries, corrosion protection became increasingly important [14]. Researchers started looking into how different compounds affect metal corrosion. The importance of oxygen in the corrosion process was identified by Antoine Lavoisier in 1781, which paved the path for the development of more potent corrosion inhibitors. Chromates were found to be efficient steel corrosion inhibitors in the middle of the 19th century. Early in the 20th century, researchers started producing more specialized corrosion inhibitors as their understanding of the mechanics underlying corrosion increased. Chromate inhibitors for steel in cooling water systems were found in the 1920s [15]. In the 1930s, the use of zinc salts as cathodic inhibitors became popular, and the first volatile corrosion inhibitors (VCIs) were introduced. In the mid-20th century, the development of organic corrosion inhibitors, such as amines, phosphonates, and benzotriazole (BTA), expanded the range of available corrosion protection options. These inhibitors offered improved performance and reduced environmental impact compared to earlier inorganic inhibitors [16]. In the 1960s, the use of silicates as corrosion inhibitors in water treatment became common. From the late 20th century to the present with growing concerns about the environmental impact of traditional corrosion inhibitors, researchers have focused on developing environmentally friendly alternatives, such as green corrosion inhibitors derived from natural sources like plant extracts [17]. Additionally, advancements in nanotechnology have led to the development of nano-sized corrosion inhibitors that can provide enhanced protection by forming ultra-thin protective layers on metal surfaces [18].

8.3 TYPES OF CORROSION INHIBITORS

There are several types of corrosion inhibitors, including the following.

8.3.1 Anodic Inhibitors

These inhibitors work by promoting the formation of a passive oxide film on the metal surface, which reduces the anodic reaction and corrosion rate. Examples include chromates, phosphates, and molybdates.

8.3.2 Cathodic Inhibitors

These inhibitors slow down the cathodic reaction by either adsorbing onto the metal surface or forming a barrier layer. Examples include zinc salts, calcium salts, and magnesium salts.

8.3.3 Mixed Inhibitors

These inhibitors act on both anodic and cathodic reactions and provide overall corrosion protection. Examples include BTA and some organic amines.

8.3.4 Volatile Corrosion Inhibitors

These inhibitors are released into the environment as vapor and adsorb onto metal surfaces, forming a protective layer. VCIs are commonly used for temporary protection during storage or transportation of metal components.

8.4 INNOVATIVE SMART COATINGS: THE ROLE OF CORROSION INHIBITORS

A good corrosion protective coating can delay corrosion of the metal substrate but cannot prevent it completely. Thus, routine coating inspections are necessary for the maintenance of critical structures. The development of smart corrosion protective and other advanced multifunctional coatings often requires the incorporation of various active components into a coating system [19]. Sometimes, these active ingredients can be incorporated directly into the coating system, but more often than not, the incorporation requires a delivery system to avoid undesired interactions between the active ingredient and the coating.

In the 1970s, the first stimuli-responsive carrier was proposed [20]. Since then, the development and use of stimuli-responsive carriers have received extensive research interest. Nanosized stimuli-responsive systems may be sensitive to particular external stimuli like changes in temperature or magnetic field as well as endogenous stimuli like pH variations [21, 22]. The stored substance can be released by stimuli-responsive nanoparticles in response to a particular external or endogenous stimulus [23]. The stored substance may be released in a particular location as a result of structural or chemical changes in the nanocarrier brought on by the release.

8.5 CONTROLLED RELEASE OF CORROSION INHIBITOR

There are different types of systems that are used to hold and release the corrosion inhibitors. Each type has its own unique properties and release mechanism, allowing for a wide range of possibilities in tailoring the coating's response to various conditions and stimuli. These systems are discussed in detail in the following section.

8.5.1 Reservoir Systems

A reservoir system for the controlled release of corrosion inhibitors helps to protect materials from corrosion over an extended period of time. These systems typically involve encapsulating the corrosion inhibitor within a protective matrix, which gradually releases the inhibitor. The first reservoir controlled-release device was unveiled decades ago. The reservoir of this system, which was contained in a silicon elastomer tube, was made to release tiny molecules gradually. It can give relatively consistent release rates and lengthy service life due to the inhibitor's huge reservoir capacity. In these systems, the confined material diffuses across planar or cylindrical membranes.

There are several common types of these reservoir systems, as discussed in the following.

8.5.1.1 Polymer Coatings

Most reservoir systems follow a three-step process for releasing the enclosed material. Initially, the material stored inside dissolves into the polymer. Then, it diffuses across the polymer membrane. Finally, the material dissolves into the external phase.

8.5.1.2 Microcapsules

Corrosion inhibitors can be encapsulated within microcapsules, which are then dispersed in a coating or embedded in a material. The microcapsules release the inhibitor when triggered by environmental factors such as a change in pH, temperature, or the presence of corrosion products. Because of the huge surface area-to-volume ratios that promote activities related to the interaction with the surrounding environment, the development of nanotechnology had a significant impact on the development of nanoscale reservoirs [24].

Zea et al. [25] loaded sodium phosphomolybdate into hollow mesoporous silica nanoparticles (HMSN). To achieve a pH-responsive release, loaded HMSN have been encapsulated by a simple deposition of one layer of poly(diallyldimethylammonium chloride) (PDDA) (Figure 8.1).

The loaded inhibitor did not discharge in the pH range of 3 to 9, according to the findings of inductively coupled plasma optical emission spectrometry. The amount released increased gradually at pH levels higher than 9 and was released completely at pH levels equal to 13. At a specific, basic pH, the corrosion inhibitor's smart release enhances the anticorrosive protection of carbon steel substrates. Nanoporous TiO_2 interlayer was used by Lamaka et al. [26] as a corrosion inhibitor

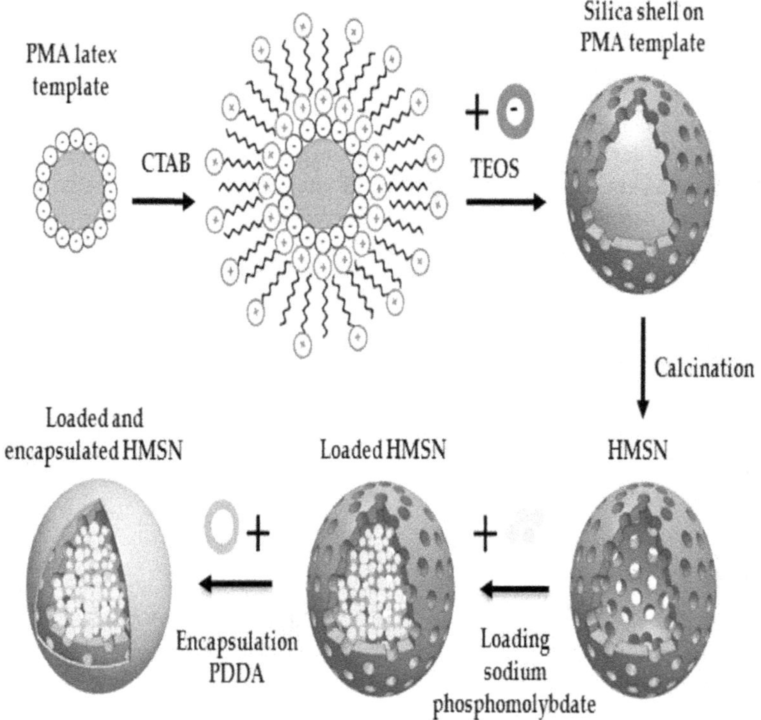

FIGURE 8.1 Synthesis of HMSN, loading with sodium phosphomolybdate and deposition of an external PDDA layer. (Reprinted with permission from reference [25], MDPI, 2018, Distributed under Creative Common Attributes based license (CCBY)).

reservoir at the metal-coating interface. It was discovered that the use of TiO_2 enhances corrosion prevention and results in self-healing properties. Zirconia nanoparticles have been proposed by Zheludkevich et al. as an inhibitor reservoir and it was discovered that zirconia particles could offer effective self-healing properties [27]. A cerium ion doped chitosan film served as an inhibitor reservoir for the coating utilized to protect aluminum alloy 2024 in another study conducted by Zheludkevich et al. It has been proven that this reservoir offers more effective corrosion protection and causes the coating to behave in a self-healing manner [28].

8.5.1.3 Mesoporous Materials

The best reservoirs are made out of hollow or porous particles because they are able to release the inhibitor when the electrolytes come in contact with the material's surface and begin the corrosion process [29]. Corrosion inhibitors can be loaded into mesoporous materials like silica or metal-organic frameworks, which have a high surface area and tuneable pore sizes. These materials can then be incorporated into coatings or used as additives in materials, providing a controlled release of the inhibitor. BTA, a corrosion inhibitor for aluminum substrates, was loaded into the shells of the HMSN that Zhao et al. synthesized with magnesium hydroxide precipitate. They verified that BTA releases under acidic circumstances and that aluminum had effective anticorrosive protection [30]. Additionally, Chen et al. [31] created HMSN that was loaded with BTA as a corrosion inhibitor and had nano-valves based on supramolecular complex stalks installed to give it a pH-sensitive function. A sol-gel coating that was doped with both kinds of pH reactive nanocapsules demonstrated some corrosion resistance on a substrate made of aluminum alloy AA2024.

On the outside of the HMSN, Fu et al. implanted bistable pseudorotaxanes as supramolecular nano-valves and filled them with caffeine molecules. The cutting-edge nanocontainers successfully encapsulate caffeine molecules at a neutral pH and release them either under acidic or under alkaline conditions. The damaged layer of aluminum oxide was successfully repaired while the rate of corrosive species' penetration was postponed [32]. The resistance of MSN nanocontainers at neutral, slightly acidic, or basic pH and the release of the corrosion inhibitor in more acidic and alkaline circumstances were shown by Shchukina et al.'s [33] and Shi et al.'s [34] pH-controlled release of 8-HQ from MSN. Based on tests made using ultraviolet-visible (UV-Vis) spectroscopic data, the loading capacity of 8-HQ in the SBA-8HQ is estimated to be around 80 mg/g. The pH dependence of 8-HQ release from SBA-8HQ was minimal at 6 and 7, according to Figure 8.2. However, after 12 hours of incubation, the release percentage of 8-HQ at pH 0 and 13.5 was 70% and 66%, respectively.

Kermannezhad et al. [35] found a similar behavior in that the release of 3-methoxybenzylamine (MBA) was very low at neutral pH (after 50 hours, only 40% of MBA was released), intermediate at slightly acidic or basic pH (approximately 60%), and nearly complete at extremely low and high pH (over 80%). Castaldo et al. [36] and Zheng et al. [37] investigated the kinetic release of BTA complexed with copper ions under acidic conditions and discovered that the release of BTA increased as the acidity rose. Wen et al. [38] developed a pH-sensitive carboxyl-functionalized mesoporous silica nanomaterial (MSNs-COOH) loaded with BTA by chemically altering mesoporous silica nanoparticles (MSNs). The authors discovered that MSN was discharged alkaline with BTA loaded in polyethyleneimine (PEI). The percentage of the loaded inhibitor BTA is around 10.09%, according to the thermal gravimetric analysis (TGA) analysis. The high pH sensitivity of BTA@MSNs-COOH-PEI and its Fickian diffusion-based self-release behavior allow it to release at a higher rate when the environment is more alkaline. BTA@MSNs-COOH-PEI's chemical makeup and valence information were determined using XPS. It is possible to identify the elements C, O, N, and Si from the survey spectrum in Figure 8.3. Additionally, the outcomes demonstrated that the sustained distribution of BTA from BTA@MSNs-COOH-PEI can continue for at least 25 hours, proving its effectiveness.

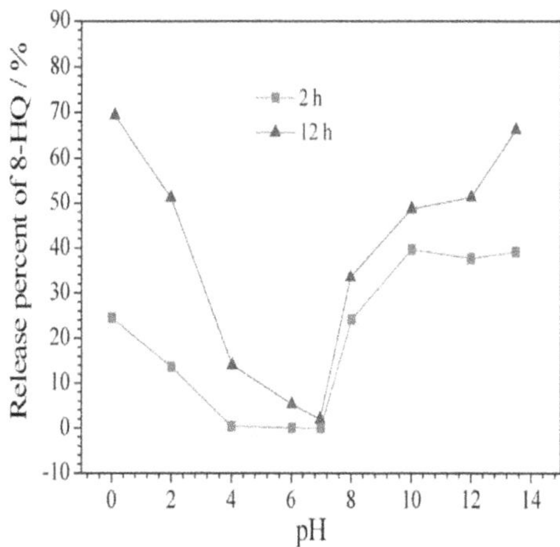

FIGURE 8.2 Release percentage of 8-HQ from SBA-8HQ in aqueous solutions with different pH values. (Adapted with permission from reference [34], Copyright•2018 Elsevier.)

Chen et al. [39] developed several systems designed for controlled BTA release in both acidic and alkaline environments. This was part of their research on BTA release from surface-modified MSN with CB [6] and cyclodextrin-based nanovalves. According to Wang et al. [40], they used CA as a corrosion inhibitor and -CD as a capping system to create redox-triggered nanoparticles activated by ferrocene moieties on an aluminum alloy substrate.

Ding et al. [41] developed a pillar arenes nanovalve that combines with magnesium ions to generate a complex that prevents the capping of the nanoparticles and results in the release of the corrosion inhibitor on a 2-hydroxy-4-methoxy-acetophenone (HMAP) loaded MSN. The coating showed exceptional anticorrosion efficiency for magnesium alloys after 15 days in 0.05 M NaCl.

By modifying the nanoparticles with octyltriethoxysilane, Hollamby et al. [42] enhanced the MSN dispersion in a polyester-based coating. They also stressed that the interactions between free dispersed BTA and the polymer network during the curing process make it ineffective at preventing corrosion phenomena. However, using corrosion inhibitor nanocarriers, like MSN, prevents BTA interactions with the polymer network and enhances corrosion protection. MSNs function as a smart reservoir of inhibitors, and they have little effect on the coating's adhesion and barrier qualities. Using MCM-41-type MSNs filled with corrosion inhibitor (i.e., NaF), Xie et al. [43] developed self-healing Ni-coating. Following prolonged exposure to a corrosive environment, it has been demonstrated that the F@MSNs improve the bare Mg alloy's ability to resist corrosion. This was attributed to the MgF_2 precipitation layer that was created when Mg ions were combined with fluoride ions that were produced from F@MSNs.

8.5.1.4 Layer-by-Layer (LbL) Assembly

This technique involves the sequential deposition of alternating layers of oppositely charged polymers and corrosion inhibitors onto a material's surface. The LbL assembly allows for precise control over the release rate of the inhibitor and can be tailored to specific applications.

For their reservoir, Lamaka et al. used stacked layers of polyelectrolytes having the opposite charge. On-demand release of inhibitors is provided by polyelectrolyte films that have been added to the hybrid sol-gel coating on aluminum alloy 2024 [44]. Divanadate anions were reportedly inserted into the interlayer space of nanoscaled double Zn/Al and Mg/Al hydroxides, which were used as the reservoir by Zheludkevich et al. [45]. The nanocrystalline layered double hydroxides (LDHs) can release vanadate ions in a regulated way. Using a porous oxide interlayer that has been

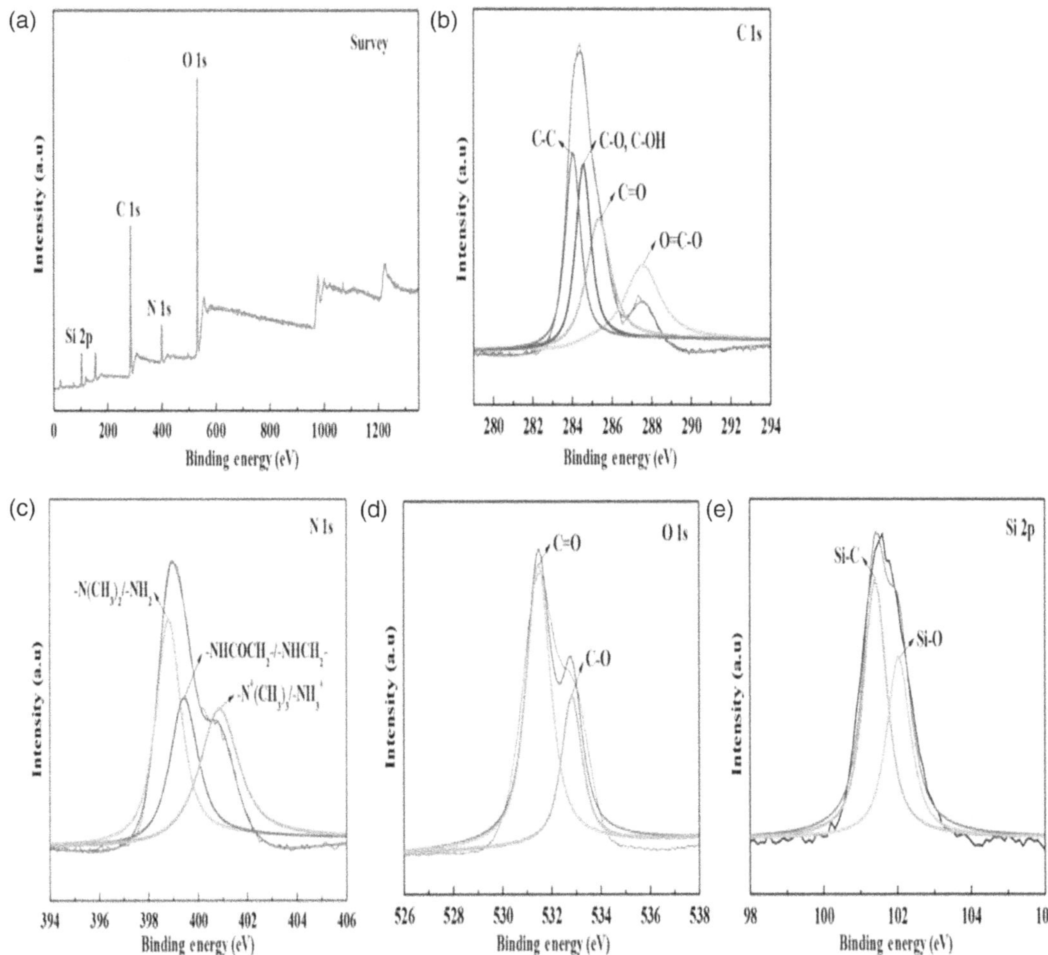

FIGURE 8.3 XPS (a) survey spectrum, and high-resolution spectra of (b) C 1s, (c) N 1s, (d) O 1s, and (e) Si 2p of BTA@MSNs-COOH-PEI. (Adapted with permission from reference [38], Copyright•2020 Elsevier.)

applied to the metal surface and doped with an organic corrosion inhibitor, thin hybrid sol-gel coatings on an aluminum alloy substrate are given further active corrosion protection [46, 47]. The oxide nanoparticles that were added to the hybrid coating to hold the inhibitor were also utilized as nanocarriers for the corrosion inhibitor [48]. Yang and van Ooij produced PP-perfluorohexane and PP-pyrolle layers utilizing radiofrequency plasma discharge to encapsulate triazole inhibitors via plasma polymerization [49]. Utilized as a pigment in a water-based epoxy coating, the plasma-treated triazole offered long-term corrosion protection by gradually releasing the inhibitor. In all situations, the inhibitor can only be released from the capsule when it has sustained mechanical damage. Decavanadate-intercalated Al-Zn LDH pigments were created by Buchheit et al. and added to the polymer coatings used to cover the AA2024 aluminum alloy [50, 51].

8.5.2 Hydrogel Systems

Hydrogel systems are another promising approach for the controlled release of corrosion inhibitors. Hydrogels are cross-linked, three-dimensional networks of polymers that can absorb a lot of water and expand to form a gel-like structure. They are frequently utilized for controlled medication delivery because of their biocompatibility and tunable characteristics. Hydrogels can also be used to release corrosion inhibitors gradually. When water-soluble polymers are cross-linked, hydrogels are created [52]. In response to certain environmental stimuli, such as variations in temperature, pH, or the presence of particular ions, hydrogels can be formed [53]. These responsive hydrogels can provide a more targeted release of corrosion inhibitors in response to specific environmental conditions that may trigger corrosion. Some hydrogels are designed to degrade over time [54], either through hydrolysis or through enzymatic degradation. These degradable hydrogels can provide a controlled release of corrosion inhibitors as the hydrogel matrix breaks down.

In hydrogels, the diffusion rate of entrapped materials is largely determined by the level of cross-linking. When polymer chains are interlinked or cross-linked, they form a macromolecular network. Although individual chains in hydrogels are water-soluble, crosslinks prevent them from dissolving in the aqueous phase. Instead, the material swells as water diffuses into the network's interstices [55]. The physical integrity of the network and osmotic forces limit this swelling.

The degree of cross-linking determines the swelling extent in hydrogels, which in turn defines the distance between chains in the network. This impacts how entrapped materials diffuse within the network, with the size and interchain separation of the material influencing the diffusion rate. Release from hydrogels can often be managed by copolymerizing compatible monomers. Hydrogels can be used in both dried and hydrated states, with dried hydrogels absorbing water from their surroundings. The release of agent molecules usually follows the water's diffusion within the polymer due to the slow diffusion rate of most agents in dried polymers. In certain circumstances, the release process can be controlled by designing the hydrogel materials to regulate the water entry rate. The swelling dynamics are influenced by the ionic environment, as ionic polymers often make up the swellable hydrogels. This unique property could enable the creation of controlled-release systems that can respond predictably to environmental changes [56].

Hydrogel systems work for controlled release of corrosion inhibitors in two ways.

8.5.2.1 Incorporation of Inhibitors

Corrosion inhibitors are either physically entrapped within the hydrogel network or chemically bound to the polymer chains. The choice of hydrogel material and the method of incorporation depend on the type of inhibitor and the desired release profile.

8.5.2.2 Swelling and Release

When the hydrogel comes into contact with water or moisture, it swells, allowing the corrosion inhibitor to diffuse out of the hydrogel matrix and into the surrounding environment. The release rate of the inhibitor can be controlled by adjusting the hydrogel's composition, cross-linking density, and degree of swelling. Yang et al. [53] fabricated a pH-controlled release solid inhibitor using polyacrylamide (PAM) hydrogel and oleate imidazoline (OIM). They used a scanning electron microscope, Fourier-transform infrared spectroscopy, TGA, and mechanical tests to examine the structure and characteristics of OIM@PAM. They also studied the OIM release behavior of OIM@PAM in different pH-value water environments using a UV-Vis spectrophotometer and had an in-depth discussion about the release mechanism.

The release of the material that has been trapped can be simply controlled in reservoir devices by adjusting the composition and thickness of the membrane that encloses the reservoir, according to the explanation above. As a result, reservoir devices can easily achieve alternate rates of release [56]. By minimizing contact between the coating matrix and the inhibitor, reservoir systems prevent undesired coating-inhibitor interaction. The effectiveness of the reservoirs depends on a number of elements, including sensing the start of the corrosion process, effective inhibitor storage, sufficient loading capacity, compatibility with the coating, and mechanical and chemical stability [29]. Using poly(2-dimethylaminoethyl methacrylate) hydrogels that were photo-cross-linked and pH-sensitive, Panpan et al. [57] investigated an intelligent corrosion inhibitor. These hydrogels exhibited greater swelling in an acidic environment (pH = 2), and as a result, the corrosion inhibitor (BTA) was released at a higher level (88%) for corrosion protection. In another study, Li et al [58] fabricated a smart hydrogel corrosion inhibitor, namely 1-H-benzotriazole, with pH-controlled release capability. The cross-linking reaction involving acrylic acid, 2-hydroxy-ethyl-methacrylate, and starch radical occurred to produce three-dimensional networks of pH-sensitive hydrogel. In terms of controlled release and corrosion-scale inhibition, polyvinyl alcohol hydrogel produced by Ting et al. [59] performed excellently. During the commercial procedure, particle polyvinyl alcohol (PVA) will make the synthesis of PVA hydrogel more practical and quicker (1 g to 3 mL distilled water).

8.5.3 Matrix Systems

For some chemicals, it is impossible to find a membrane with enough permeability to permit release from a reservoir. The complexity of reservoir device structures necessitates many manufacturing procedures, which raises the cost. Matrix systems are therefore offered as a substitute approach to ensure controlled release. The integrated element in a matrix system is disseminated or dissolved in the polymer. Most of the polymers employed in matrix systems are poly[ethylene-co-(vinyl acetate)] or elastomers. The loaded ingredient is dissolved and disseminated throughout the polymer in the matrix system. In this case, the substance diffuses through the polymer, causing the release of the integrated material into the external environment. The matrix's macroscopic geometry can have an impact on the pattern and rate of the incorporated material's release. By giving the particles greater direct access to the outside environment, increasing the matrix's surface area to volume ratio increases the rate at which the particles are released. In contrast, the release rate slows down over time because the particles closest to the surface are the first to leave the matrix. This is because the diffusion causes the particles close to the surface to go the shortest distance. In hydrophobic polymers, many particles, especially those that are wet soluble, do not properly dissolve. They cannot therefore be incorporated into the matrix system mentioned above. For these particles, it is possible to build matrix systems by scattering them across the matrix. In order to enable the controlled release of both large and small molecules, various strategies have been put forth. It has been held that the releasing mechanism is unaffected by the size of the distributed particles. The most common method of loading water-soluble chemicals into a matrix is to disperse them throughout the polymer matrix. The water-soluble chemicals in this matrix are released slowly when submerged in water. The higher initial rate of release is seen in the matrices with more loadings. It has been reported that the matrix structure is made up of connected pores. Channels or smaller pores connect the enormous pores in this structure, and connected pores that come into contact with the matrix boundary can release the water-soluble compounds into the environment. By making the particles of water-soluble compounds larger, the rate of release can be accelerated. The size of these particles affects the size of the water-filled channels that are created as a result of their breakdown. Larger particles take up more space in the matrix, which increases the connection between pores and aids in the diffusion of the particles. Incomplete release occurs in matrix systems that contain particles that won't dissolve in the polymer. Particles within the volume cannot reach the surface because some of these particles are enveloped or isolated by a polymer matrix, resulting in partial release. Only the pores that are close to the point where a water-soluble component and water converge can become wet in low-porosity materials; internal pores, which are encircled by the polymer material's backbone, are never able to become wet. By increasing the porosity of the polymer material, one can increase the release of incorporated water-soluble compounds and the number of pores that can be wetted which results in the formation of large-sized clusters of pores and connections of a large number of interior pores to the aqueous surrounding environment [55].

8.6 EXTERNAL STIMULI AFFECTING RELEASE BEHAVIOR

External stimuli affecting the release behavior of corrosion inhibitors can include temperature, pH, and electrochemical potential. These factors can influence the effectiveness and efficiency of the inhibitors in protecting metal surfaces from corrosion.

8.6.1 Temperature

The release behavior of some corrosion inhibitors can be temperature-dependent. As the temperature increases, the rate of chemical reactions and diffusion processes may also increase, leading to a more rapid release of the inhibitor. This can be beneficial in situations where higher temperatures lead to more aggressive corrosion environments. The influence of temperature on the release of corrosion inhibitors is not widely studied and is often linked to light exposure [60]. However, Kermannezhad and team [35] showed that temperature can indeed affect the release of MBA from pH-dependent MSN. They found that as the surrounding environment's temperature changed, so did the release of the corrosion inhibitor, with an increase in temperature leading to an increased release.

8.6.2 pH

This is the most studied stimulus for corrosion inhibitor release. The acidity or alkalinity of the environment can also affect the release behavior of corrosion inhibitors. Some inhibitors are more effective in acidic conditions, while others perform better in alkaline environments. Changes in pH can cause a change in the chemical state of the inhibitor, which may alter its adsorption and desorption behavior on the metal surface. In their research, Yabukia et al. [61] developed a polymer covering with cellulose nanofibers and a corrosion inhibitor to prevent carbon steel from corroding. To regulate the corrosion inhibitor's adsorption and desorption on cellulose nanofibers, the pH of the polymer was adjusted. SiO_2 nanoparticles, polyelectrolyte capsules, and polyelectrolyte-modified halloysite nanotubes were the three nanocontainers that Shchukin and Mohwald [62] examined for their release and reloading properties. They discovered that in either acidic or alkaline water, all three nanocontainers displayed a greater release of the inhibitor BTA. The halloysite-based nanocontainers had the highest reloading efficiency (up to 80%), but this dropped to 20% after five cycles. In a separate study by Chen and team, they loaded BTA into hollow mesoporous silica spheres and found no leakage of BTA in a neutral solution, but in an alkaline solution, BTA was rapidly released and the release rate increased with higher pH values [39]. According to Kermannezhad et al. [35], a neutral medium had the lowest anticorrosion agent release rate, while acidic and basic media had the highest rates. The silica particles and the inhibitor molecules have the same charge at pH values other than neutral (positive at pH 6, negative at pH > 6), according to Borisova et al. [63]. This causes faster release and stronger electrostatic repulsion forces. In order to demonstrate the stability of these nanocontainers at neutral, slightly acidic, or basic pH levels as well as the release of the corrosion inhibitor in more acidic and alkaline settings, Shchukina et al. [33] and Shi et al. [64] established a pH-controlled release of 8-HQ from MSN. Both Qian et al. [65] and Xu et al. [66] accomplished a one-step synthesis and loading of mesoporous silica nanoparticles loaded with benzotriazole- cetyltrimethylammonium bromide (MSN-BTA-CTAB) nanoparticles (Figure 8.4), which showed an enhanced corrosion inhibitor release with ambient acidification. In this paper, the scientists found that the presence of CTAB in the MSN pores stimulates release under acidic conditions; however, the release rate was noticeably reduced as pH increased until basic values. The release largely began when the MSN was etched by exposure to OH^- ions, which the CTAB forbade in alkaline media.

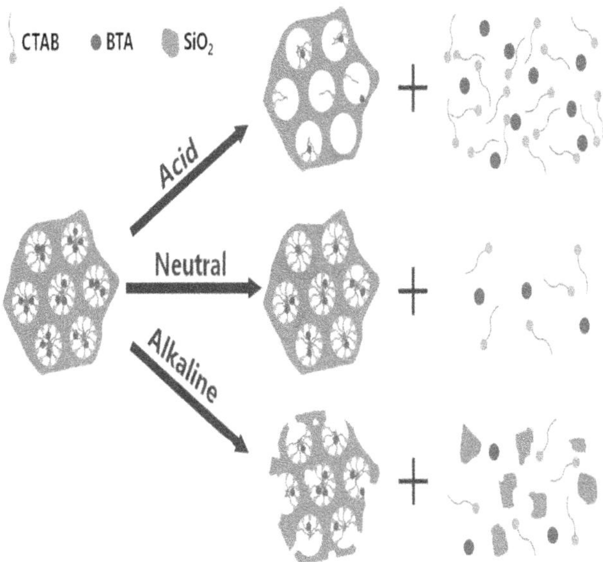

FIGURE 8.4 Release mechanism from MSN-BTA-CTAB: release rate decreases with the increase of pH. (Adapted with permission from reference [66], Copyright•2018 Elsevier.)

According to Wen et al. [38], BTA loaded in PEI-covered MSN showed an encouraged alkaline release. Encapsulated caffeine took advantage of the pH-responsive system using the carbon black (CB) [7], hexamethylenediamine (HDA), and fumarylacetoacetate hydrolase deficiency (FCDCA) pseudorotaxane nanovalve that Fu et al. [32] installed on the MSN. At neutral pH, CB [7] acted as gatekeepers, while lowering or raising the pH the interactions between CB [7], HDA, and FCDCA promoted the caffeine release. Particularly toward very low or high pH, the release of corrosion inhibitor is strongly encouraged.

8.6.3 Electrochemical Potential

The electrochemical potential of the metal surface can influence the release of corrosion inhibitors. Some inhibitors are designed to respond to changes in the electrochemical potential, releasing more of the inhibitor when the potential is more favorable for corrosion. This can help to provide better protection for the metal surface under varying conditions. Understanding how these external stimuli affect the release behavior of corrosion inhibitors is crucial for selecting the appropriate inhibitor for a specific application and ensuring that it provides effective protection against corrosion [67].

Ding and colleagues [67] created smart nanocontainers triggered by corrosion potential (CP-SNCs). In order to accomplish this, they used disulfide linkers to glue the supramolecular assemblies to the outer surface of magnetic nanovehicles (Fe3O4@mSiO2), and then they enclosed the organic corrosion inhibitor 8-hydroxyquinoline (8-HQ) within the mesopores of Fe3O4@mSiO2. The corrosion potential stimulus-feedback anticorrosion coating, which was used to coat the magnesium alloy AZ31B, was made using the CP-SNCs and a hybrid organic-inorganic sol-gel. 8-HQ is quickly released when the magnesium alloy's corrosion potential (-1.5 V vs. SHE) is applied because the disulfide linkers break and the supramolecular assemblies are eliminated. The authors additionally observed that as the reductive potential (-1 V vs. SHE) was gradually changed to an oxidative potential (+1 V vs. SHE), the rate of 8-HQ release was reduced.

8.6.4 Light Exposure

Some corrosion inhibitors are sensitive to light exposure, particularly UV radiation. UV radiation can cause photodegradation of the inhibitor, leading to a change in its release behavior and effectiveness. Ma and his team [68] created plasmonic titanium nitride nanoparticles (TiN)@mesoporous SiO_2 core-shell nanocontainers to store BTA, a corrosion inhibitor. These were then incorporated into a shape memory epoxy coating matrix. The coating, which contained 2 wt.% TiN-BTA@ SiO_2 NPs, showed optimal corrosion resistance and significant self-healing performance, activated within 30 seconds of exposure to near-infrared (NIR) light. Without light, BTA was released slowly, with only about 40% being released after 300 minutes. However, when intermittently exposed to a laser, the BTA release rate increased due to the photothermal effect of the plasmonic TiN NPs. Once the light was turned off, the release rate slowed again, confirming that encapsulating TiN NPs allows for light-responsive control of inhibitor release (Figure 8.5).

Chen et al. developed a self-healing, pH-responsive anticorrosion coating in another work [69]. When exposed to UV light (365 nm), the transform changes to the cis form, allowing corrosion inhibitors to escape into the solution. After 4 hours, 90% of the BTA released was seen in the neutral solution as the amount of BTA released grew noticeably.

8.6.5 Ion Concentration Triggers

Changes in ion concentration can trigger the release of corrosion inhibitors. Maia and team [70, 71] found that increasing sodium chloride concentration in water can boost the release of molecules from MSN due to the interaction between the corrosion inhibitor and NaCl, forming sodium salts that are more soluble than the corrosion inhibitor itself. According to research conducted by Zheng and colleagues [37], the BTA-Cu complex is sensitive to variations in the concentration of sulfide ions and dissolves as the Na_2S content in water rises, releasing BTA. In order to release the corrosion inhibitor, Ding and colleagues [41] created a pillararenes nanovalve on an HMAP-loaded MSN that combines with magnesium ions to form a compound that influences the capping of the nanoparticles.

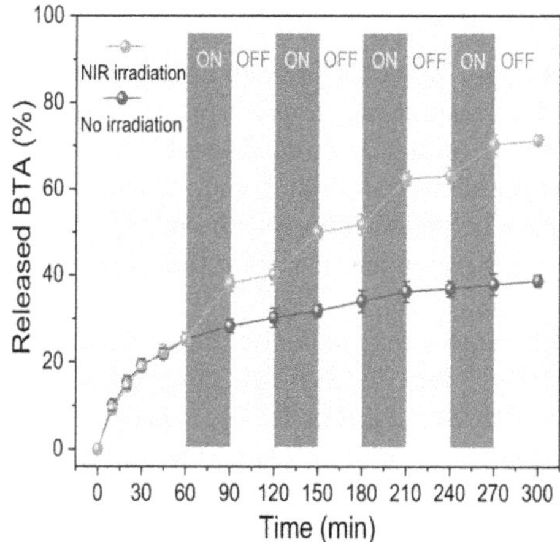

FIGURE 8.5 The release ratio of BTA from TiN-BTA@mSiO2 NPs in neutral NaCl solution with/without NIR irradiation. (Adapted with permission from reference [68], Copyright•2021 Elsevier.)

8.6.6 Redox Reactions Release Systems

Using ferrocene and β-CDs, Wang and team [72] developed a supramolecular switch that responds to redox stimuli. Cerium(IV) salts react with ferrocene to open the nanovalve and cause the release of CA, activating this switch. They discovered that CA release increased along with cerium concentration. Additionally, cerium salts were converted into cerium(III) salts following the redox reaction, which has a twofold anticorrosive effect. Cerium(III) salts are an inorganic corrosion inhibitor. In Sun's study [73], the interaction of dithiothreitol (DTT) with ZnO quantum dots, which served as the nanoparticles' external nanovalves, caused redox stimulation. They discovered that raising the DTT concentration sped up the release of MBT.

8.7 OTHER TRIGGERING MECHANISMS

8.7.1 Mechanical Stress

Mechanical stress, such as vibrations, pressure changes, or fluid flow, can influence the release of inhibitors. These factors can cause the inhibitor film to become disrupted, leading to a change in the adsorption and desorption rates.

8.7.2 Presence of Other Chemicals

The presence of other chemicals in the environment can also affect the release behavior of corrosion inhibitors. For example, the presence of certain ions or contaminants can compete with the inhibitor for adsorption sites on the metal surface, altering the effectiveness of the inhibitor.

8.7.3 Microbial Activity

In some cases, the presence of microorganisms can affect the release behavior of corrosion inhibitors. Microbes can either produce substances that enhance the inhibitor's performance or degrade the inhibitor, reducing its effectiveness.

8.7.4 Oxygen Concentration

The presence and concentration of oxygen in the environment can influence the release of inhibitors. Oxygen can react with the inhibitor, altering its chemical structure and effectiveness. Additionally, the presence of oxygen can also affect the corrosion process itself, which may indirectly impact the release behavior of the inhibitor.

8.7.5 Relative Humidity

The amount of moisture in the environment can also affect the release behavior of corrosion inhibitors. High humidity levels can promote the dissolution and transport of the inhibitor, while low humidity levels may slow down these processes, potentially affecting the inhibitor's performance.

8.7.6 Fluid Velocity

The velocity of the fluid (e.g., water or oil) in contact with the metal surface can influence the release of inhibitors. High fluid velocities can cause erosion or abrasion of the inhibitor film, leading to a faster release rate and potentially reducing the inhibitor's effectiveness.

8.7.7 Surface Roughness

The roughness of the metal surface can also impact the release behavior of corrosion inhibitors. Rough surfaces may have more adsorption sites for the inhibitor, leading to a stronger interaction between the inhibitor and the surface. However, rough surfaces can also cause uneven distribution of the inhibitor, which may affect its overall performance.

8.7.8 Time

The release behavior of some corrosion inhibitors can change over time due to factors such as aging, chemical degradation, or changes in the inhibitor's physical state.

8.8 MECHANISM OF RELEASE

An external stimulation, such as a change in temperature, radiation, pH, pressure, or mechanical activity, causes the coating to begin self-healing. Changes in electrochemical potential brought on by pH variations are the most typical catalyst for self-healing coating with capsules [43, 64, 73–77]. Because corrosion causes a local pH fall in anodic areas and an increase in cathodic ones, researchers have discovered that pH variations in the corrosion zone are particularly important. Therefore, it makes sense to detect and prevent corrosion by including pH-sensitive inhibitor capsules in the coating. The pH content has an impact on how MSNs interact with the substances or active agents they transport.

Changes in pH can affect the stability of the complex, triggering the release of loaded compounds in the affected area. The release of active agents from nanocontainers is influenced by their physicochemical properties and the way they interact with MSNs [30]. There have been varied observations regarding the release of corrosion inhibitors based on pH. With MSNs, more release was observed in alkaline conditions, potentially due to the surface potential of these powders in solution or the dissociation of the silica framework. One study [69] introduced photothermal-responsive inhibitor-loaded TiN@mSiO_2 nanocontainers into an epoxy coating matrix. Few studies showed that potential changes [68], changes in the ion concentration [37, 71–72], and lightning [69, 70] are also responsible for the triggering mechanism. The release of corrosion inhibitors by stimuli often involves a combination of physical and chemical processes. The following is a general overview of the mechanisms involved in inhibitor release in response to various stimuli.

8.8.1 Adsorption/Desorption

The inhibitor molecules initially adsorb onto the metal surface, forming a protective film. When a stimulus is applied, it can cause the desorption of the inhibitor molecules, disrupting the protective film and releasing the inhibitor into the surrounding environment. The adsorption and desorption rates are influenced by factors such as temperature, pH, and the presence of other chemicals.

8.8.2 Solubility Changes

Some stimuli can cause changes in the solubility of the inhibitor, affecting its release behavior. For instance, changes in temperature or pH can alter the solubility of the inhibitor, leading to its precipitation or dissolution. This, in turn, affects the concentration of the inhibitor in the environment and its ability to protect the metal surface.

8.8.3 Chemical Reactions

Certain stimuli can cause chemical reactions between the inhibitor and other components in the environment. These reactions can alter the chemical structure of the inhibitor, affecting its release behavior and overall effectiveness. For example, exposure to light or oxygen can cause the inhibitor to undergo oxidation or photodegradation, leading to a change in its release behavior.

8.8.4 Physical Processes

Some stimuli, such as mechanical stress or fluid velocity, can cause physical changes to the inhibitor film on the metal surface. These physical changes can disrupt the film, leading to the release of the inhibitor. For example, high fluid velocities can cause erosion or abrasion of the inhibitor film, while mechanical stress can cause the film to crack or peel.

8.9 METHODS FOR STUDYING RELEASE BEHAVIOR

There are several methods used to study the release behavior of corrosion inhibitors from coatings as follows.

8.9.1 Electrochemical Measurements

Electrochemical measurements can help to study the release behavior of corrosion inhibitors from coatings by monitoring changes in the electrochemical properties of the coating as the inhibitor is released. For example, electrochemical impedance spectroscopy can be used to measure changes in the coating's impedance, which provides information on the coating's corrosion resistance. As the inhibitor is released, the coating's impedance will change, allowing researchers to track the release behavior of the inhibitor over time.

8.9.2 UV-Vis Spectroscopy

UV-Vis spectroscopy can help to study the release behavior of corrosion inhibitors from coatings by measuring changes in the absorbance of light by the coating as the inhibitor is released. The inhibitor may absorb light at a specific wavelength, and as it is released from the coating, the absorbance at that wavelength will decrease. By monitoring changes in absorbance over time, researchers can track the release behavior of the inhibitor and determine the rate and extent of release.

8.9.3 X-Ray Photoelectron Spectroscopy

X-ray photoelectron spectroscopy (XPS) can help to study the release behavior of corrosion inhibitors from coatings by analyzing the chemical composition of the coating surface as the inhibitor is released. As the inhibitor is released, the chemical composition of the coating surface will change, and XPS can be used to identify and quantify these changes. For example, XPS can be used to measure changes in the concentration of elements such as nitrogen or sulfur, which may be present in the inhibitor. By monitoring changes in the chemical composition of the coating surface over time, researchers can track the release behavior of the inhibitor and determine the rate and extent of release.

8.9.4 Atomic Force Microscopy

Atomic force microscopy (AFM) can help to study the release behavior of corrosion inhibitors from coatings by imaging the coating surface and measuring changes in the coating's topography and roughness as the inhibitor is released. As the inhibitor is released, the coating's topography and roughness will change, and AFM can be used to visualize and quantify these changes with high resolution. For example, AFM can be used to measure changes in the height and distribution of surface features, such as pits or cracks, which may be indicative of corrosion. By monitoring changes in the coating's topography and roughness over time, researchers can track the release behavior of the inhibitor and determine its effectiveness in preventing corrosion.

8.10 CHALLENGES AND FUTURE DEVELOPMENTS

Despite improvements, there are still several challenges in the development of stimuli-responsive nanoparticles. The complicated architectures of these particles, particularly with regard to manufacturing and producibility, make it challenging to translate them from experimental models to practical systems. These systems need to be improved because so many stimuli-responsive systems have undergone in vivo testing. Because the energy and stimuli employed could harm the region around them, their use needs to be carefully monitored and applied. Although these technologies appear promising, much work remains before they can be thoroughly tested. MSNs could help develop novel coating systems based on on-demand trigger mechanisms for corrosion prevention. However, there is still a long way to go before this material is widely accepted and becomes inexpensive for use in damaged instruments.

8.11 CONCLUSIONS

Even with recent advancements in stimuli-responsive nanoparticles, many obstacles remain. The complexity of these particles makes their transition from experimental models to practical systems challenging, particularly in terms of manufacturing and producibility. Many of these systems, having been tested in vivo, require further optimization. The use of different stimuli and energies could potentially harm nearby areas, necessitating careful monitoring and application. Despite the potential of these systems, they require extensive testing. MSNs could contribute to the development of new corrosion protection coatings based on on-demand triggers. Yet, there's a long road ahead to make this material both standard and cost-effective for broad use in corrosion-prone tools. These nanoparticles can shield the active agent from degradation, ensuring prolonged effectiveness. Single or mixed nanocarriers utilized in coatings can exhibit enhanced mechanical, thermomechanical, and barrier properties. Additionally, they can react to numerous stimuli, protecting the substrate from deterioration both chemically and physically by releasing corrosion inhibitors as needed. The coating industry has started to consider nanoparticles as a way to tailor the characteristics of the matrices they are embedded in, as well as through their optical, chemical, and morphological qualities or their capacity for self-assembly. The application of MSNs in intelligent anticorrosive coatings has a lot of space for development. A new class of intelligent anticorrosive coatings may be developed by following established research trends and perspectives or by coming up with novel approaches to make use of MSNs' fascinating characteristics. These might possess better qualities than commercial products that are now used in a variety of fields, including transportation, civil and industrial infrastructure, and cultural heritage. This would help preserve our historical and cultural past and protect our future. The ability of certain nanostructured materials to carry and deliver active agents has sparked a growing interest in their application for coatings. This innovative concept of incorporating a nanocarrier into the coating, capable of delivering active agents in response to specific stimuli, has led to the emergence of a new generation of highly intelligent coating systems. These systems offer unparalleled levels of protection effectiveness and durability compared to traditional passive coatings. In recent years, there's been a surge in research focused on developing smart coatings containing various types of nanocarriers. These nanocarriers serve multiple purposes in coatings. Firstly, they are designed to release specific substances directly at the site where they are needed. The design of the nanocontainer is crucial for controlling the release of the active agent under different conditions or in response to specific external stimuli.

Additionally, they can shield the active agent from degradation, ensuring its long-lasting effectiveness. Furthermore, using a single type of nanocarrier or a combination of different ones can result in coatings with multifunctional properties. These coatings not only exhibit improved mechanical, thermomechanical, and barrier properties but can also respond to a variety of stimuli, providing both chemical and physical protection to the substrate through the on-demand release of corrosion inhibitors.

REFERENCES

1. Aslam R., Mobin M., Zehra S., Aslam J., (2022) A comprehensive review of corrosion inhibitors employed to mitigate stainless steel corrosion in different environments. *J. Mol. Liq.* 364:119992.
2. Aslam R., Mobin M., Aslam J., Aslam A., Zehra S., Masroor S., (2021) Application of surfactants as anticorrosive materials: A comprehensive review. *Adv. Colloid Interface Sci.* 295:102481.
3. Aslam R., Mobin M., Huda, Shoeb M., Murmu M., Banerjee P., (2021) Proline nitrate ionic liquid as high temperature acid corrosion inhibitor for mild steel: Experimental and molecular-level insights. *J. Indus. Eng. Chem.* 100:333.
4. Liu C., Qian B., Hou P., Song Z., (2021) Stimulus responsive zeolitic imidazolate framework to achieve corrosion sensing and active protecting in polymeric coatings. *ACS Appl. Mater. Interfaces* 13:4429.
5. White S. R., Sottos N. R., Geubelle P. H., Moore J. S., Kessler M. R., Sriram S. R., Brown E. N., Viswanathan S., (2021) Autonomic healing of polymer composites. *Nature* 409:794.
6. Zhong F., He Y., Wang P., Chen C., Wu Y., (2022) Novel pH-responsive self-healing anti-corrosion coating with high barrier and corrosion inhibitor loading based on reduced graphene oxide loaded zeolite imidazole framework. *Colloids Surf. A: Physicochem. Eng.* 642:128641.
7. Yap J. Y., Yaakob S. M., Rabat N. E., Shamsuddin M. R., Man Z., (2020) Release kinetics study and anti-corrosion behaviour of a pH-responsive ionic liquid-loaded halloysite nanotube-doped epoxy coating. *RSC Adv.* 10:13174.
8. Raja, P. B., Assad, M. A., Ismail, M., (2020). Inhibitor-encapsulated smart nanocontainers for the controlled release of corrosion inhibitors. In *Corrosion Protection at the Nanoscale* (pp. 91–105). Elsevier..
9. Sampedro-Guerrero J., Vives-Peris V., Gomez-Cadenas A., Clausell-Terol C, (2023) Efficient strategies for controlled release of nanoencapsulated phytohormones to improve plant stress tolerance. *Plant Methods* 19:47.
10. Taghavikish M., Dutta N. K., Roy N. C., (2017) Emerging corrosion inhibitors for interfacial coating. *Coatings* 7:217.
11. Bhandari J., Khan F., Abbassi R., Garaniya V., Ojeda R., (2015) Modelling of pitting corrosion in marine and offshore steel structures — A technical review. *J. Loss Prev. Process.* 37:39–62.
12. Mobin M., Parveen M., Huda, Aslam R., (2021) Effect of different additives, temperature and immersion time on the corrosion inhibition behaviour of L-valine for mild steel corrosion in 5% HCl solution. *J. Phys. Chem. Sol.* 161:110422.
13. Rani B. E., Basu B. B. J., (2012) Green inhibitors for corrosion protection of metals and alloys: An overview. *International Journal of corrosion*, Article ID 380217.
14. Singh, A., Ansari, K.R., Quraishi, M.A. and Lin, Y., 2019. *Investigation of Corrosion Inhibitors Adsorption on Metals using Density Functional Theory and Molecular Dynamics Simulation.* London: IntechOpen..
15. Vaghefinazari B., Wierzbicka E., Visser P., Posner R., Arrabal R., Matykina E., Mohedano M., Blawert C., Zheludkevich M. L., Lamaka S. V., (2022) Chromate-free corrosion protection strategies for magnesium alloys—A review: Part III—corrosion inhibitors and combining them with other protection strategies materials (basel). *Materials* 15: 8489.
16. Hackerman, N., Snaveley, E. S., (1984) Inhibitors. In: de Brasunas, A. S. (ed.), *Corrosion Basics.* Houston, TX., NACE International, 127.
17. Marzorati S., Verotta L., Trasatti S. P., (2018) Green corrosion inhibitors from natural sources and biomass wastes. *Molecules* 24:48.

18. Idumah C. I., Obele C. M., Emmanuel E. O., Hassan A., (2020) Recently emerging nanotechnological advancements in polymer nanocomposite coatings for anti-corrosion, anti-fouling and self-healing. *Surf. Interfaces* 21:100734.
19. Ulaeto S. B., Rajan R., Pancrecious J. K., Rajan T. P. D., Pai B. C., (2017) Developments in smart anticorrosive coatings with multifunctional characteristics. *Prog. Org. Coat.* 111:294.
20. Wei M., Gao Y., Li X., Serpe M. J., (2017) Stimuli-responsive polymers and their applications. *Polym. Chem.* 8:127.
21. Mura S., Nicolas J., Couvreur P., (2013) Stimuli-responsive nanocarriers for drug delivery. *Nat. Mater.* 12:991.
22. Gooneh-Farahani S., Naimi-Jamal M.R., Naghib S.M., (2019) Stimuli-responsive graphene-incorporated multifunctional chitosan for drug delivery applications: A review. *Expert Opin. Drug Deliv.* 16:79.
23. Kamaly N., Yameen B., Wu J., Farokhzad O. C., (2016) Degradable controlled-release polymers and polymeric nanoparticles: Mechanisms of controlling drug release. *Chem. Rev.* 116:2602.
24. Zhao Y., Tavares A. C., Gauthier M. A., (2016) Nano-engineered electro-responsive drug delivery systems. *J. Mater. Chem. B* 18:3019.
25. Zea C., Alcántara J., Barranco-García R., Morcillo M., Fuente D., (2018) Synthesis and characterization of hollow mesoporous silica nanoparticles for smart corrosion protection. *Nanomaterials.* 8:478.
26. Lamaka S. V., Zheludkevich M. L., Yasakau K. A., Serra R., Poznyak S. K., Ferreira M. G. S., (2007) Nanoporous titania interlayer as reservoir of corrosion inhibitors for coatings with self-healing ability. *Prog. Org. Coat.* 58:127.
27. Zheludkevich M. L., Serra R., Montemor M. F., Ferreira M. G. S., (2005) Oxide nanoparticle reservoirs for storage and prolonged release of the corrosion inhibitors. *Electrochem. Commun.* 7:836.
28. Zheludkevich M. L., Tedim J., Freire C. S. R., Fernandes S. C. M., Kallip S., Lisenkov A., Gandini A., Ferreira M. G. S., (2011) Self-healing protective coatings with "green" chitosan based pre-layer reservoir of corrosion inhibitor. *J. Mater. Chem.* 21:4805.
29. Snihirova D., Lamaka S. V., Taryba M., Salak A. N., Kallip S., Zheludkevich M. L., Ferreira M. G. S., Montemor M. F., (2010) Hydroxyapatite microparticles as feedback-active reservoirs of corrosion inhibitors. *ACS Appl. Mater. Interfaces* 2:3011.
30. Zhao D., Liu D. Hu Z., (2017) A smart anticorrosion coating based on hollow silica nanocapsules with inorganic salt in shells. *J. Coat. Technol. Res.* 14:85.
31. Chen T., Fu J., (2012) An intelligent anticorrosion coating based on pH-responsive supramolecular nanocontainers. *Nanotechnology* 23:505705.
32. Fu J., Chen T., Wang M., Yang N., Li S., Wang Y., Liu X., (2013) Acid and alkaline dual stimuli-responsive mechanized hollow mesoporous silica nanoparticles as smart nanocontainers for intelligent anticorrosion coatings. *ACS Nano.* 7:11397.
33. Shchukina E., Shchukin D., Grigoriev D., (2017) Effect of inhibitor-loaded halloysites and mesoporous silica nanocontainers on corrosion protection of powder coatings. *Prog. Org. Coat.* 102:60.
34. Shi H., Wu L., Wang J., Liu F., Han E. H., (2017) Submicrometer mesoporous silica containers for active protective coatings on AA 2024-T3. *Corros. Sci.* 127: 230.
35. Kermannezhad K., Najafi C. A., Momeni M. M., Rezaei B. (2016) Application of amine-functionalized MCM-41 as pH-sensitive nano container for controlled release of 2-mercaptobenzoxazole corrosion inhibitor. *Chem. Eng. J.* 306:849.
36. Castaldo R., Salzano de Luna M., Siviello C., Gentile G., Lavorgna M., Amendola E., Cocca M., (2020) On the acid responsive release of benzotriazole from engineered mesoporous silica nanoparticles for corrosion protection of metal surfaces. *J. Cult. Herit.* 44:317.
37. Zheng Z., Huang X., Schenderlein M., Borisova D., Cao R., Möhwald H., Shchukin D., (2013) Self–healing and antifouling multifunctional coatings based on pH and sulfide ion sensitive nanocontainers. *Adv. Funct. Mater.* 23:3307–3314
38. Wen J., Lei J., Chen J., Liu L., Zhang X., Li L., (2020) Polyethylenimine wrapped mesoporous silica loaded benzotriazole with high pH-sensitivity for assembling self healing anti-corrosive coatings. *Mater. Chem. Phys.* 253:123425.
39. Chen T., Fu J., (2012) pH-responsive nanovalves based on hollow mesoporous silica spheres for controlled release of corrosion inhibitor. *Nanotechnology* 23:235605.

40. Wang T., Tan L., Ding C., Wang M., Xu J., Fu J., (2017) Redoxtriggered controlled release systems-based bi-layered nanocomposite coating with synergistic self-healing property. *J. Mater. Chem. A* 5:1756–1768.
41. Ding C., Liu Y., Wang M., Wang T., Fu J., (2016) Self-healing, superhydrophobic coating based on mechanized silica nanoparticles for reliable protection of magnesium alloys. *J. Mater. Chem. A* 4:8041–8052.
42. Hollamby M. J., Fix D., Dönch I., Borisova D., Möhwald H., Shchukin D., (2011) Hybrid polyester coating incorporating functionalized mesoporous carriers for the holistic protection of steel surfaces. *Adv. Mater.* 23:1361.
43. Xie Z. H., Li D., Skeete Z., Sharma A., Zhong C. J., (2017) Nanocontainer-enhanced self-healing for corrosion-resistant Ni coating on Mg alloy. *ACS Appl. Mater. Interfaces* 9: 36247.
44. Lamaka S. V., Shchukin D. G., Andreeva D. V., Zheludkevich M. L., Möhwald H., Ferreira M. G. S., (2008) Sol-gel/polyelectrolyte active corrosion protection system. *Adv. Funct. Mater.* 18:3137.
45. Zheludkevich M. L., Poznyak S. K., Rodrigues L. M., Raps D., Hack T., Dick L. F., Nunes T, Ferreira M. G. S., (2010) Active protection coatings with layered double hydroxide nanocontainers of corrosion inhibitor. *Corros. Sci.* 52:602.
46. Lamaka S. V., Zheludkevich M. L., Yasakau K. A., Serra R., Poznyak S. K., Ferreira M. G. S., (2007) Nanoporous titania interlayer as reservoir of corrosion inhibitors for coatings with self-healing ability. *Prog. Org. Coat.* 58:127.
47. Lamaka S. V., Zheludkevich M. L., Yasakau K. A., Montemor M. F., Cecílio P., Ferreira M. G. S., (2006) TiOx self-assembled networks prepared by templating approach as nanostructured reservoirs for self-healing anticorrosion pre-treatments. *Electrochem. Commun.* 8:421.
48. Zheludkevich M. L., Serra R., Montemor M. F., Yasakau K. A., Salvado I. M. M., Ferreira M. G. S., (2005) Nanostructured sol–gel coatings doped with cerium nitrate as pre-treatments for AA2024-T3: corrosion protection performance. *Electrochim. Acta* 51:208.
49. Yang H., van Ooij W. J., (2004) Plasma-treated triazole as a novel organic slow-release paint pigment for corrosion control of AA2024-T3. *Prog. Org. Coat.* 50:149–161.
50. Buchheit R. G., Guan H., Mahajanam S., Wong F., (2003) Active corrosion protection and corrosion sensing in chromate-free organic coatings. *Prog. Org. Coat.* 47:174–182.
51. Mahajanam S. P. V., Buchheit R. G., (2008) Characterization of inhibitor release from Zn–Al-$[V_{10}O_{28}]^{6-}$ hydrotalcite pigments and corrosion protection from hydrotalcite-pigmented epoxy coatings. *Corrosion* 64:230.
52. Kalkhoran A. H. Z., Vahidi O., Naghib S. M., (2018) A new mathematical approach to predict the actual drug release from hydrogels. *Eur. J. Pharm. Sci.* 111:303.
53. Yang Q., Lin Bing, Tang J., Wang Y., Zheng H., Zhang H., Nie Z., Zhang Y., (2023) A pH-controlled solid inhibitor based on PAM hydrogel for steel corrosion protection in wide range pH NaCl medium. *Molecules* 28:1314.
54. Lei L., Bai Y., Qin X., Liu J., Huang W., Lv Q., (2022) Current understanding of hydrogel for drug release and tissue engineering. *Gels* 8:301.
55. Kalkhoran A. H. Z., Naghib S. M., Vahidi O., Rahmanian M., (2018) Synthesis and characterization of graphene-grafted gelatin nanocomposite hydrogels as emerging drug delivery systems. *Biomed. Phys. Eng. Exp.* 4:055017.
56. Saltzman W. M., (2001) *Drug Delivery: Engineering Principles for Drug Therapy*. Oxford University Press.
57. Panpan R., Dawei Z., Chaofang D., Xiaogang L., (2015) Preparation and evaluation of intelligent corrosion inhibitor based on photo-crosslinked pH-sensitive hydrogels. *Mater. Lett.* 160:480.
58. Li L., Chaofang D., Liqin L., Jiankuan L., Xo K., Dawei Z., Xiaogang L., (2014) Preparation and characterization of pH-controlled-release intelligent corrosion inhibitor. *Mater. Lett.* 116:318.
59. Ting G., Xiaoyan L., Wenbo C., Beibei L., Haoyuan S., (2014) A preliminary research on polyvinyl alcohol hydrogel: A slowly-released anti-corrosion and scale inhibitor. *J. Pet. Sci. Eng.* 122:453.
60. Skorb E. V., Skirtach A. G., Sviridov D. V., Shchukin D. G., Möhwald H., (2009) Laser-controllable coatings for corrosion protection. *ACS Nano* 3:1753.
61. Yabukia A., Shiraiwaa T., Fathonaba I. W., (2016) pH-controlled self-healing polymer coatings with cellulose nanofibers providing an effective release of corrosion inhibitor. *Corros. Sci.* 103:117.

62. Shchukin D. G., Möhwald H., (2007) Surface-engineered nanocontainers for entrapment of corrosion inhibitors. *Adv. Funct. Mater.* 17:1451.
63. Borisova D., Möhwald H., Shchukin D. G., (2011) Mesoporous silica nanoparticles for active corrosion protection. *ACS Nano* 5:1939.
64. Shi H., Wu L., Wang J., Liu F., Han E. H., (2017) Sub-micrometer mesoporous silica containers for active protective coatings on AA 2024-T3. *Corros. Sci.* 127:230.
65. Qian B., Michailidis M., Bilton M., Hobson T., Zheng Z., Shchukin D., (2019) Tannic complexes coated nanocontainers for controlled release of corrosion inhibitors in self-healing coatings. *Electrochim Acta* 297:1035.
66. Xu J.B., Cao Y.Q., Fang L., Hu J. M., (2018) A one-step preparation of inhibitor-loaded silica nanocontainers for self-healing coatings. *Corros. Sci.* 140:349.
67. Ding C. D., Xu J. H., Tong L., Gong G. C., Jiang W., Fu J. J., (2017) Design and fabrication of a novel stimulus-feedback anticorrosion coating featured by rapid self-healing functionality for protection of magnesium alloy. *ACS Appl. Mater. Interfaces* 9: 21034.
68. Ma L., Wang J., Zhang D., Huang Y., Huang L., Wang P., Qian H., Li X., Herman A., (2021) Dual-action self-healing protective coatings with photothermal responsive corrosion inhibitor nanocontainers. *Chem. Eng. J.* 404:127118.
69. Chen T., Chen R., Jin Z., Liu J., (2015) Engineering hollow mesoporous silica nanocontainers with molecular switches for continuous self-healing anticorrosion coating. *J. Mater. Chem. A* 3:9510.
70. Maia F., Tedim J., Lisenkov A. D., Salak A. N., Zheludkevich M. L., Ferreira M. G., (2012) Silica nanocontainers for active corrosion protection. *Nanoscale* 4:1287.
71. Maia F., Tedim J., Bastos A. C., Ferreira M. G. S., Zheludkevich M. L., (2013) Nanocontainer-based corrosion sensing coating. *Nanotechnology* 24:415502.
72. Wang T., Tan L., Ding C., Wang M., Xu J., Fu J., (2017) Redox-triggered controlled release systems-based bi-layered nanocomposite coating with synergistic self-healing property. *J. Mater. Chem. A* 5:1756.
73. Sun S., Zhao X., Cheng M., Wang Y., Li C., Hu S., (2019) Facile preparation of redox-responsive hollow mesoporous silica spheres for the encapsulation and controlled release of corrosion inhibitors. *Prog. Org. Coat.* 136:10530.
74. Zhang X. F., Chen R. J., Liu Y. H., Hu J. M., (2015) Electrochemically generated sol-gel films as inhibitor containers of superhydrophobic surfaces for the active corrosion protection of metals. *J. Mater. Chem. A* 4:649.
75. Stankiewicz A., Szczygieł I., Szczygieł B., (2013) Self-healing coatings in anti-corrosion applications. *J. Mater. Sci.* 48:8041.
76. Shchukin D. G., Möhwald H., (2011) Smart nanocontainers as depot media for feedback active coatings. *Chem. Commun.* 47:8730.
77. Grigoriev D. O., Köhler K., Skorb E., Shchukin D. G., Möhwald H., (2009) Polyelectrolyte complexes as a "smart" depot for self-healing anticorrosion coatings. *Soft Matter.* 5:1426.

9 Enhancing Corrosion Protection with Nano-Inhibitors

Amita Somya, Amit Prakash Varshney, Abhinay Thakur, and Ashish Kumar

9.1 INTRODUCTION

Corrosion is a natural process in which any unstable metal reacts with the surroundings (oxygen, moisture, or impurities) to form a stable product on the exterior. This stable byproduct is a consequence of the chemical process that results in metal loss and is used to estimate corrosion. Due to temperature and relative humidity fluctuations, the environment predominantly damages metals during transit and storage. Another important corrosion feature is that it does not occur evenly under all situations. Corrosion is a natural process that changes metals in the refining process to more unreactive states such as oxide, hydroxide, or sulfide.

The primary process through which almost all metals deteriorate is corrosion. Many aspects of our life are exposed to corrosion [1]. Most of us are personally aware of the significance of corrosion. However, corrosion often refers to the destructive result of a chemical interaction between metals or metal alloys and their environment. The most prevalent type of corrosion, also known as general corrosion or uniform corrosion, develops from electrochemical reactions in air or water on a totally exposed metal surface and spreads uniformly to completely degrade the metal. The appearance of holes due to rust in the body panels of recently manufactured cars has caused a lot of individuals to shudder [2]. Appliances made of steel, such as those used in the home and garden, frequently rust outside. We all have seen the stresses hot food can cause on cooking utensils or tasted the metallic flavor of acidic foods that have been left in open cans for too long. Corrosion is generally recognized to be the root cause of these consequences [3]. The Taj Mahal's fissures, which appeared in some areas and were caused by steel dowels that were completely deteriorated and rusted within and caused fractures in the stone, is the most obvious evidence of corrosion. But other types of material, such ceramics, plastics, and rubber, also experience corrosion often. Since almost all surroundings are corrosive to some extent and are important contributors to material failure as well as a significant economic burden on society [4], they also pose a significant risk to human health.

Although when we think of corrosion, we typically only think of metals, nonmetallic materials like plastic, ceramics, rubber, and concrete are all prone to corrosion when they are subjected to a variety of corrosive environments. The variance between potential energies of a corrosion product and corroding metal itself is a main component which controls this corrosion process. A specific amount of energy must be given to metal ores in order to extract metals from their naturally occurring ores. It follows that when exposed to their surroundings, these metals have a propensity to revert to the state from which they were brought about. It is interesting to note that the energy

DOI: 10.1201/9781032677255-9

requirements, energy storage capabilities, and energy released during corrosion vary widely among metals. The more energy required to extract a metal, the more thermodynamically unsteady it is, and hence the shorter it's interim existence in metallic state. Corrosion is, thus, described as the opposite of the extraction of metals [5].

The most important process for corroding any metal is electrochemical disintegration, which serves as the basis for both localized and uniform corrosion in all its forms. On the other hand, many forms of corrosion, like oxidation, fret corrosion, melting salt corrosion, etc., may be explained omitting the need of electrochemistry. Electrochemistry will always play a role in defining how corrosion attacks take place in both air and aquatic environments, regardless of the environment. Because of the exceptional capacity of metals to electrochemically react with aqua molecules, oxygen (O_2) and, some specific chemicals in an aqueous environment, electrons (e^-s) would be travelling out of particular parts of the metallic surface to others via an electrolyte which can conduct ions. A cathode is a part which receives electrons produced by a corrosion reaction while an anode is the portion of a metal surface that one gets corroded owing to the loss of electrons in an electrochemical corrosion process. We are all conscientious of how corrosion damages metal structures like homes, ships, buildings, pillars, etc. Unfortunately, most of us are ignorant of the internal corrosion which is degrading the capabilities of the water, gas, and oil pipelines beneath properties of all of us as well as the water pipes within our houses, where corrosion typically begins. A ubiquitous and omnipotent occurrence, corrosion may be seen as a global phenomenon. Every environment we come into contact with has it, including the air, water, soil, and other elements. Corrosion, sometimes known as rust, is a harmful phenomenon that shortens the lifespan of materials and removes their luster and beauty. Successful organizations dedicate a significant amount of time and resources to corrosion control during the design and functioning phases of their operations in order to prevent significant collapses caused by corrosion like personal injuries, unintended shutdowns, fatalities, and environmental contamination in the current business environment. Nonetheless, a finest design approach cannot take into consideration all the potential causes of corrosion that might shorten a system's lifespan. Rebar, or steel-reinforced concrete, can corrode in concrete without being seen at all, leading to the demolition of buildings, parking structures, bridges, the collapse of transmission towers, and certain parts of roadway, among other things. This is extremely expensive to repair and puts public safety at risk. As a result, it is crucial to regularly repair any metallic parts that are susceptible to corrosion. For corrosion losses, the Indian government spends about 3.5% of its gross domestic product (GDP) annually [6]. According to recent studies, the need for corrosion inhibitors is increasing not only in India but also in other countries [7]. Costs associated with corrosion show up as early degradation or failure that calls for upkeep, repair, and replacement of broken parts. Corrosion has a significant negative influence on the environment and the economy on practically all metallic items used in daily life as well as on the surfaces of our nation's infrastructure, including roads, bridges, buildings, chemical processing facilities, and waste water treatment [8]. Corrosion adversely impacts not only the environment but also human safety and industrial processes in addition to material loss. The key to reducing corrosion failures is raising awareness of corrosion and implementing timely, suitable control measures [9].

9.2 HISTORIC GROUND

One of the most perilous and conferred about corrosion catastrophes is a collapse of the Silver Bridge [10]. This bridge, which connected Point Pleasant, West Virginia, with Kanauga, Ohio, abruptly collapsed into the Ohio River on December 15, 1967, killing 46 people. This catastrophe was shown to be caused by corrosion fatigue and stress corrosion cracking (SCC). In terms of the number of people killed and injured, the Bhopal tragedy, which occurred in nighttime on 2nd–3rd December 1984 in Bhopal, India, is one of the deadliest industrial catastrophes. The exposure of methyl isocyanate, and several noxious reaction yields into the nearby regions, which resulted in the unfortunate seepage of water (500 liters) from corroded valves, pipelines, and other safety tool at Union Carbide India Limited, resulted in the deaths of 3,000 people and injuries to an estimated five lakh people [11]. Another noteworthy corrosion catastrophe that occurred in Uster, Switzerland, in 1985 was the swimming pool roof collapse. Twelve persons died when the stainless steel rods supporting the ceiling of this swimming pool broke due to SCC [12].

9.3 REPERCUSSIONS OF CORROSION

More so than the straightforward loss of a quantity of metal, corrosion has a significant impact on the efficient, safe, and effective functioning of machinery and buildings. Even if the amount of metal damaged is fairly little, failures of all kinds of machinery and the need for costly replacements may nonetheless happen. Figure 9.1 lists out some of the corrosion's most detrimental effects.

As indicated in Figure 9.2, there are three main categories [5] that may be used to categorize damages from corrosion or the costs connected with it.

9.3.1 Materials and Energy

When it comes to growing an industry, the impacts of corrosion on the equipment and its surroundings must be meticulously taken into consideration. Corrosion is one among the toughest concerns for the majority of industrialized countries to address. Corrosion of pipelines, equipment components made of metal, bridges, marine ships, tanks, etc., can result in major material and monetary losses for any nation. Additionally, corrosion-related failure might put the safety of operational equipment like boilers, pressure containers, steel vessels for dangerous chemicals, turbine blades, bridges and rotors, as well as steering systems for cars and aeroplane parts at risk [13]. In addition to the metals themselves, corrosion has a disastrous impact on water, energy, and the manufacturing process for metal fixtures. Despite the fact that one ton of steel is claimed to rust

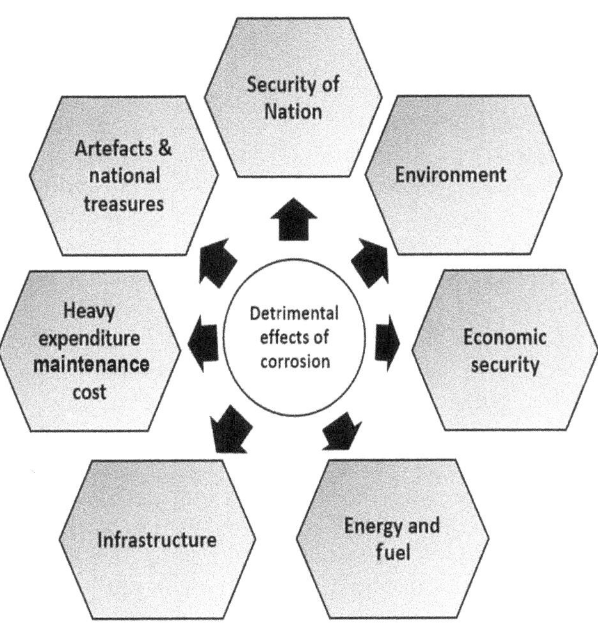

FIGURE 9.1 Detrimental effects of corrosion.

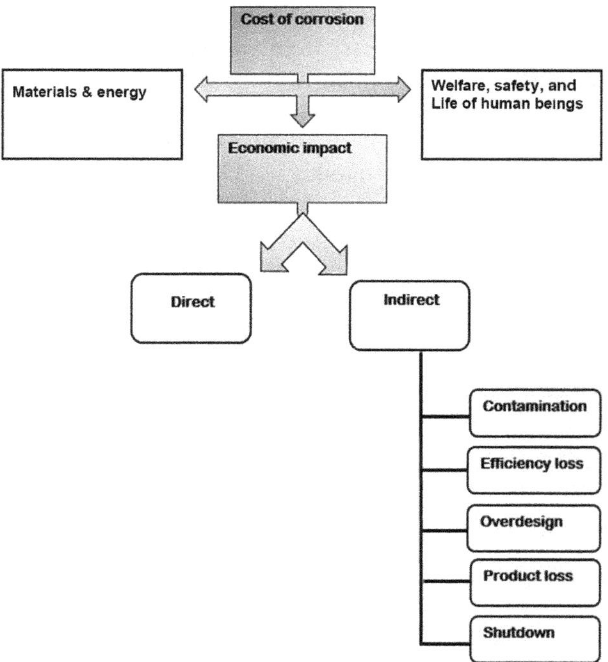

FIGURE 9.2 Impact of corrosion.

every 90 seconds, the energy required to make one ton of steel is about similar to the energy utilized by an average household over the course of three months. About half of the time, every ton of steel produced worldwide is replaced with new steel [14].

9.3.2 Economic Impact

Economic losses may be differentiated into two groups: direct and indirect losses. Direct loss incorporates the reinstatements of integrant of various machineries owing to rusting like pipelines, mufflers, metal roofing, condenser tubes, refurbishing the surfaces of metal articles to avoid corrosion, maintenance of underground oil pipelines, replacements of domiciliary hot water tanks, usage of corrosion-resistant metals and alloys, application of various processes to avoid or prevent corrosion like galvanizing, Ni-plating on steels, incorporation of corrosion inhibitors and dehumidification process in metal-equipped warehousing rooms which require extra labor and additional costs sometimes in millions. Despite the fact that it may be difficult to assess indirect losses, it is believed that direct losses have grown by many billions of dollars as a result. Unexpected plant shutdowns; the loss of oil, water, and gas due to corroded pipelines; the reduction in efficiency of energy conversion tools brought on by corrosion reactions; the defiling of aqua and foodstuffs in metal pipes and vessels, respectively; and, overdesign, which requires tools to be designed to be much heavier than necessary to withstand operating pressure or applied stress in order to increase equipment lifespan [15], are a few examples of indirect losses. The very first in-depth research on the expense of corrosion was done by Uhlig in 1949 [16]. According to Uhlig's research, the yearly toll of corrosion in the United States is projected to be \$5.5 billion or 2.1% of the gross national product (GNP) in 1949. In this study, the entire costs of corrosion-induced sustenance and reinstatement were calculated by adding the costs spent by holders and operators (direct costs) and those borne by consumers (indirect costs). Numerous developed nations, including Finland, Australia, China, India, Kuwait, Germany, Japan, the United Kingdom, and the United States [17] have carried out corrosion cost investigations using a variety of techniques, including the Uhlig procedure developed by Uhlig in 1950 [16], the Hoar procedure developed by Hoar in 1976[18], and the economic input/output model developed by the National Bureau of Standards in collaboration with the Battelle Memorial Institute in 1978 [19]. The cost of corrosion ranged between 1% and 5% of each nation's GNP, according to the studies that commonly made this point. The differences in corrosion cost relative to GNP were attributed to the different methodological approaches used in each study and the distinctive characteristics of each country. The expenses of corrosion were estimated to be $2.5 trillion in 2013, which is 3.4% of the global GDP, according to research carried out by the National Association of Corrosion Engineers as part of their Worldwide Measures of Prevention, Application, and Economics of Corrosion Technologies Study [17]. Utilizing data from the World Bank's economic sector and GDP, the costs of corrosion studies were contrasted with global expenses of corrosion in this study. According to the World Bank, the global economy was divided into economic zones with comparable economies in order to manage the numerous economic sectors throughout the world. United States, European Region, India, the Arab World, China, Japan, the Four Asian Tigers plus Macau, and the Rest of the World are how World Bank classified them. The predicted costs, however, usually fail to account for the impacts on the environment or individual welfare. Therefore, it is crucial that after obtaining added funds for corrosion research or, the most recent data on these studies, more comprehensive and exact estimations of the global costs may be established.

9.3.3 Welfare, Safety, and Lives of Human

Corrosion might have a terrible effect on people's safety and quality of life. This disastrous phenomenon has gone undiscovered as the root cause of almost fatal occurrences due to liability concerns or, simply just like the documents disappeared after these disastrous events. Corrosion has a number of negative impacts, such as a considerable decline in the quality of natural and historic monuments and a rise in the possibility of catastrophic equipment failures due to which a number of life-threatening accidents occur such as Bhopal chemical plant explosion [20], Failure of Aloha Airline Flight [21], Failure of Livestock Facilities [22], etc.

9.4 CORROSION MITIGATION

The operational mechanisms of corrosion are destructive and quiet. Both large and small enterprises are hampered by it. Since corrosion cannot be completely prevented, it can only be minimized by using certain anticorrosion techniques [23]. A variety of effective techniques are available to preserve metal from corrosion. They might broadly be based on

- Tune-up of metals
- Tempering of designs
- Change in corrosive environs
- Coating on metallic surfaces
- Applying inhibitors

These techniques can be applied alone or together. Using corrosion inhibitors in place of the numerous techniques to preclude the destruction or deterioration of metallic surfaces is one of the most well-known ways to guard against corrosion. Because of the low cost and practice approach, utilizing inhibitors is similar to standing up [24].

9.5 CORROSION INHIBITORS

A corrosion inhibitor is usually any chemical substance that, when given in very little amount to the metal environment, can reduce, prevent, and regulate corrosion. Inhibitors of corrosion are thought of as a first line of safeguarding against corrosion in the chemical and oil industries [25]. The forms of corrosion inhibitors might be in solids, liquids, or gases. Corrosion inhibitors are chosen based on the inhibitors' solubility or dispersibility in fluids. It has been discovered that corrosion inhibitors are a reliable and adaptable method of mitigating corrosion. Chemical inhibitors are employed in a variety of ways to slow the pace at which corrosion processes occur. Oil and gas exploration and production, petroleum refineries, chemical manufacturing, heavy manufacturing, water treatment, and product additive sectors all utilize corrosion inhibitors. Corrosion inhibitors prevent corrosion by either forming an adsorbed layer that serves as a barrier or by delaying the cathodic, anodic, or combined processes. Corrosion inhibitors are responsible for any corrosion-retarding processes or for any decrease in the rate of metal oxidation brought on by the addition of a chemical substance to the system. Inhibitors frequently don't require much training to use, and they have the benefit of in situ application without seriously interfering with the process. One of the best methods for preventing corrosion is the use of corrosion inhibitors.

The well-played role of an inhibitor might be due to:

- An increase in polarization of the anode (anodic inhibition)
- An increase in polarization of the cathode (cathodic inhibition)
- An increase in electrical resistance of circuit during the formation of a thin/thick deposit on surface of the metals (resistance inhibition)
- Restriction in the diffusion of depolarizers (diffusion restriction).

Depending on the chemical composition they have, inhibitors are divided into organic and inorganic categories [5]. Further, based on their modes of action, organic and inorganic inhibitors can also be further classified as neutralizing, scavenging, barrier or film generating, and other unidentified inhibitors. However, there are new categories like nano-inhibitors [26] and green corrosion inhibitors [27]. Figure 9.3 depicts the different classes of corrosion inhibitors.

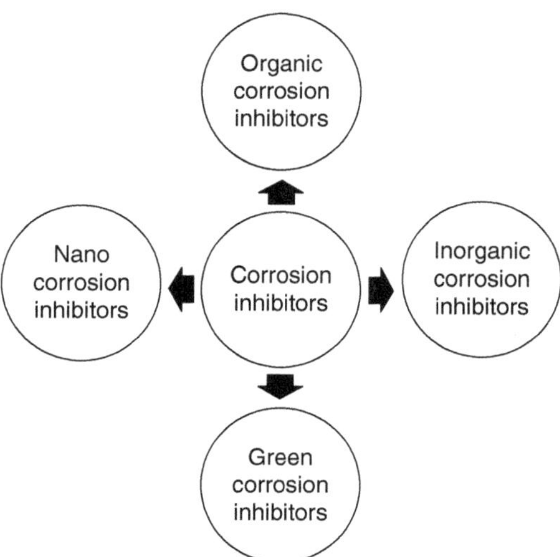

FIGURE 9.3 Different types of corrosion inhibitors.

9.5.1 Organic Inhibitors

Organic inhibitors, which might be anodic, cathodic, or mixed inhibitors, function by depositing a layer of material on the metals' surfaces. It is made possible by strong interactions like orbital adsorption, electrostatic adsorption, and chemisorption, which prevent the corrosive species from corroding the metallic surface [27]. The adsorption described above often only forms a single-molecule thick layer, and it seldom penetrates all the way to the core of the metal [28]. Physical and chemical criteria that regulate the adsorption process include aromaticity, functional groups, steric restrictions, electron density at donor atoms, orbital characteristics of donating electrons, and an electronic structure of the adsorbing molecules. The adsorption capacity of an organic inhibitor and the structural, mechanical, and chemical features of the adsorbed layers established under certain circumstances all affect how well it prevents corrosion [27]. An efficient organic inhibitor generally contains a hydrophobic section that repels the aqueous corroding species apart from the metallic surface and polar functional groups made up of sulfur, oxygen, and/or nitrogen atoms in the molecule. It is thought that the polar head is in charge of forming the adsorption layer. Organic inhibitors include chemical groups such as pyridines, 1,3-azoles, imidazolines, and fatty amides [29].

9.5.2 Inorganic Inhibitors

Inorganic inhibitors are the materials that use an inorganic substance as their active ingredient. Incorporating electropositive salts of a metal into a corrosive solution is one among the simplest ways to improve a metal's passivity. However, it must compulsorily have a redox potential that is greater than that of the metal that needs to be protected and maybe higher than the one needed for the protons to be discharged in order to prevent metal ions from being discharged on the metallic surface in need of protection. Through overvoltage reduction and the subsequent formation of an adherent deposit, the coating of the preventive metal on the surface of the corrosion-prone metal results in cathodic depolarization. The same purpose is achieved by using the metals mercury, palladium, rhodium, platinum, iridium, and rhenium. By integrating into the oxide layer, a variety of inorganic anions, such as phosphates, chromates, silicates, nitrates, and molybdates, provide passivation preservation to the metallic surfaces [29]. Toxicity, price, accessibility, and environmental friendliness are a few factors that need to be regarded seriously when choosing an inhibitor in a certain circumstance. Due to their toxicity, biodegradability, and bioaccumulation, conventional corrosion inhibitors pose substantial health hazards when discharged into the environment. It may not yet be clear how commercial corrosion inhibitors affect the environment, although it is known that certain of their chemical components are dangerous [30]. There are various salts such as arsenates, chromates, dichromates, and phosphates which are not only well known as efficient ion exchangers [31–32], but also have shown encouraging inhibitory efficiency. However, there are certain health hazards on long-term exposure and usages. Owing to the biological and health hazards associated with organic inhibitors, they have compelled us to seek out or use nontoxic or green corrosion inhibitors (GCIs), which will provide the greatest preservation for metallic buildings while having the least detrimental effects on people and the environment. The development of environmentally friendly inhibitors is, therefore, necessary.

9.5.3 Green Corrosion Inhibitors

Since it is standard practice to protect metals and equipment against corrosion, corrosion inhibitors must be long-lasting, non-toxic, and advantageous to the environment. Commercially accessible organic and inorganic inhibitors are costly and have unfavorable side effects, raising significant awareness regarding the subject of corrosion prevention. Therefore, economical, accessible, environmentally and ecologically acceptable, and renewable GCIs are more appealing to corrosion scientists and engineers for application. A substance that is naturally biocompatible is referred to as a "green inhibitor" or "eco-friendly inhibitor." The inhibitors, like plant extracts, are anticipated to be biocompatible due to their biological origins. Modern procedures make advantage of chemical compounds that are present in out-of-date medications, mushroom extracts, fruit peel extracts, and even plant extracts [33–34]. Numerous organic compounds that are benign to the environment have good qualities for protecting metal surfaces and function as corrosion inhibitors.

Although the term "green corrosion inhibitors" is fascinating and has many benefits to the environment, there are certain restrictions as well, which promote the potentiality of taking on obstacles when utilizing these inhibitors [35]. It is important to consider if a green corrosion inhibitor is still nontoxic after 28 days while evaluating them. In addition, it is important to monitor the inhibitor's bioaccumulation. Their lifespan is a significant disadvantage. It is challenging to keep them for an extended period of time since they are often extremely biodegradable. Another crucial factor is the type of extraction chosen. It is important to keep in mind that a less toxic solvent must be used while extracting plant fragments. Additionally, certain processes call for lengthy processing times and high temperatures, both of which work against green corrosion inhibition.

9.5.4 Nano-Inhibitors

One of the most exquisite methods for solving the technical issues of the 21st century is nanotechnology [36]. Significant progress has been made in the disciplines of materials design and production due to the achievement of a basic knowledge of the procedures and behaviors of matters at the nanoscale sizes. The term "nanomaterials" refers to substances with at the minimum one dimension smaller than 100 nm [37] in which structural flaws are greatly reduced and aspect ratio is markedly increased. The improvements in specific surface area and aspect ratio result in improved molecular level interlinkage at phase confluences in nanocomposite platings, which can significantly enhance their thermal, physical, electrical, mechanical, and anticorrosion attributes as well as cut the need for additives in coating formulations [37–38]. According to a literature review, nanoparticles are widely utilized as corrosion inhibitors; however, because of their extremely low or nonexistent solubility, their applicability in the aqueous phase is quite constrained. Because of this, current research in corrosion science and technology is focused on advancing the burgeoning of water solvable nanomaterials made of carbon allotropes and their derivatives. One of the most popular and often utilized techniques has been functionalization of covalent/noncovalent surface of the nanomaterials. According to the general maxim, adding nanomaterials to anticorrosion plating may significantly enhance their barrier conducts by a reduction in porosity and lengthening in meander path for corrosion creating species like oxygen, water, and chloride ions [39–40]. However, it is not an only effectual process but still this corrosion prevention technique using nanoparticles is the most significant one. In order to efficiently reject water and marine microorganisms, anticorrosion coatings can also be tuned to have superhydrophobic surface qualities using low surface energy nanomaterial, for example, silane modified nanoparticles [41–44]. The grafting of functional groups from modified nanomaterials on a polymeric resin can result in the production of an opaque, strongly cross-linked framework exhibiting a low permeability morphology [45]. Additionally, functionalized nanoparticles have the ability to strengthen the bond between the coating and the substrate and lessen the likelihood of coating delamination [46]. Conductive nanoparticles in zinc-rich coatings can boost the coating's electrical activity in addition to its inherent barrier function. As a result, zinc powder is consumed more successfully and its sacrificial protection is improved [47–48]. By using sacrificial or anodic passivation approaches [49–50] the metallic substrate can be protected by the incorporation of low-nobility nanometals into a polymeric resin like zinc and aluminum nanopowders. On comparing with the application of micro size or large particles as fillers [51], an efficacy of nano-sized particles permits the usage of a considerably smaller amount of nanomaterials for changing micro-level structure of the coatings in order to obtain the same

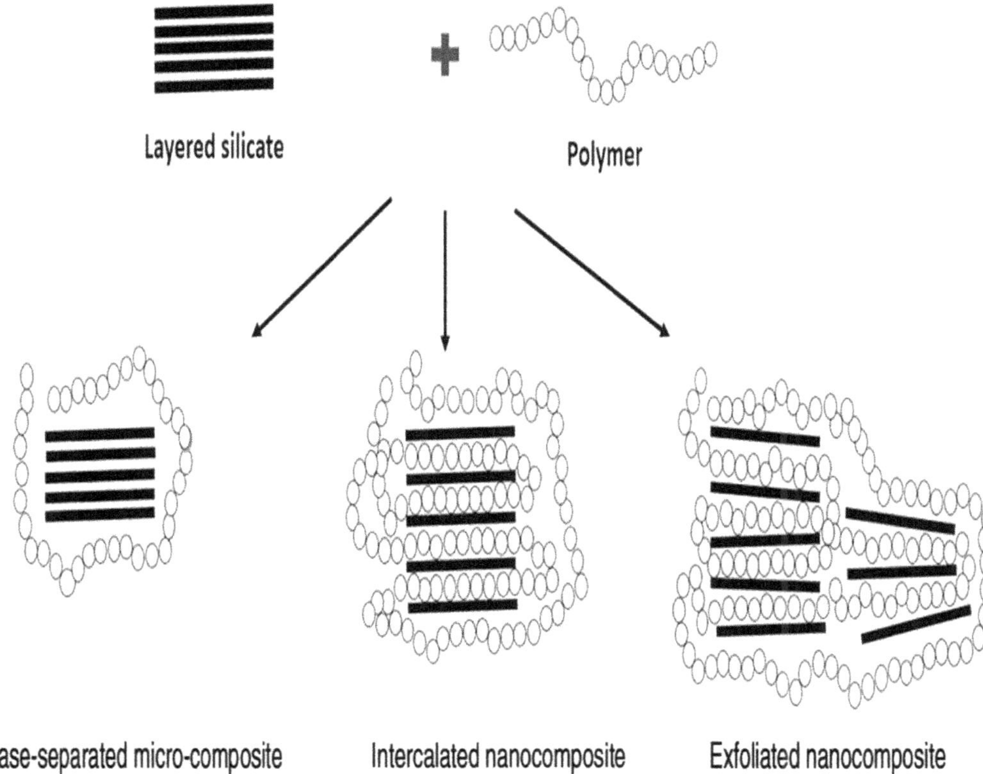

FIGURE 9.4 Different categories of nanocomposites based on nanosheets as per their dispersion states.

degree of protective qualities. The performance of the entire system's corrosion protection is significantly impacted by a dispersion state of nanoparticles in coating procedures. The anticorrosion property of a coating is improved by adding well-dispersed nanomaterials as opposed to coatings that use poorly dispersed nanofillers. There are three different ways that nanosheets are dispersed in polymer matrix, as shown in Figure 9.4. Among all three types, the most ideal morphological structure for greatly enhancing the anticorrosion characteristics of coatings is exfoliation [52]. Because the nanoparticles they include can plug holes [53–55] and, through the numerous toughening mechanisms [40], an application of nanomaterials in coatings offers environmentally friendly ways to improve the coherence and longevity of coatings. Additionally, nanoparticles can prevent disaggregation in the course of thermosetting polymer coatings' curing process, resulting in an improvement in the coating's homogeneity. Nanomaterials can pervade the porosity left by specific shrinkage during the polymer's curing procedure owing to their tiny size and large surface area. Finally, nanoparticles have the ability to form bridges between polymer molecules, increasing the crosslinking density of the polymer after curing [56–58].

9.6 DIFFERENT TYPES OF NANOMATERIALS USED AS CORROSION INHIBITORS

9.6.1 Smart Nanocomposites Used in Coatings

The coating shortcomings surface scuffs, microcracks, perforations, and desquamation brought on by heat strokes, mechanical strains, and weathering over the course of the coating's life can significantly reduce the barrier qualities of organic coatings. These inescapable damages need periodic assessment and physical mending that requires setup and supplies [40]. Designing and synthesizing nanostructured materials increases the potential for creating intelligent coatings which can be utilized to liberate mending or protective chemicals in a controlled way, primarily to the corrosion faults, or mangled parts of the coating once they are activated by external driving force. With appropriate design, the coated metal is given long-lasting, active corrosion protection [59]. Implementing intelligent anticorrosion coatings with self-healing capabilities can be done in one of the two ways. In contrast to the second strategy, which involves protecting an exposed metallic object [60] employing the corrosion inhibitor contained in intelligent nanocontainers (NCs), the first way involves localized restoration of a mangled coating by a restrained liberation of the healing agent. These two procedures fall under the category of exterior self-healing coatings. Another kind of intelligent protective coating is termed an intrinsic self-healing coating, and it has a polymeric matrix that already has a sensitive molecular structure that can repair itself when it sustains harm. The primary obstacles to the widespread use of intrinsic self-healing coatings, which may be utilized in several series of mending and have not been dependent on an external inhibitor, or healing agent, are the expensive cost and the scarcity of resin types with an appropriate responsive molecular structure [61]. It is quite difficult to guarantee the uniform distribution of the healing agent throughout the thickness of self-healing coatings based on external healing reagent because they require the large amount of healing reagent stored, often inside the microcapsules [62]. In contrast, self-healing coatings based on external inhibitor for anticorrosion applications have apparently been simple to use and inexpensive since the nanoscale reservoirs offer consistent and prolonged manumit of corrosive inhibitors [63]. The main justification for storing corrosion inhibitor in nanocomposites rather than simply inserting it within the coating of polymers is that the former procedure reduces the likelihood of early inhibitor leaching while still supplying the inhibitor wherever and whenever it is required. Additionally, the last-mentioned method increases the possibility of unfavorable interconnections between a corrosion inhibitor and the matrix of polymer, which might impair performance of the coating [63]. The nanoparticles that function like reservoirs to the corrosion inhibitor must possess a number of features including, unanimously along with the matrix of coating, strong mechanical, chemical stability, and impermeable through responsive shell wall which liberates the inhibitor mostly when necessary [63]. Corrosion inhibitors have typically been encapsulated inside NCs using one of three major methods. Encapsulation by polymer or shell layers utilizing a layer-by-layer (LbL) assembly, coacervation, or internal phase segregation approach is the first scenario [64]. The second case involves the use of a sol-gel technique for encapsulation by inorganic shells [65]. The third case necessitates the employment of a nanovalve which is created at the top of nanochannels. The loaded inhibitor reacts with transition-metal ions to produce the nanovalve (commonly known as metal-inhibitor complex) [66]. For instance, corrosion inhibitors have been housed in nanochanneled structures such as mesoporous silica nanoparticles and halloysite nanoclays in polymer-based anticorrosion coatings [67–68]. Using nanovalves or LbL assembled from polyelectrolyte multilayers, as given in the literature [40], controlled release of corrosion inhibitors can be achieved. The second way involves forming metal-inhibitor complexes and liberating transition or heavy metal ions at the channel apertures. These complexes serve as nanovalves that release the inhibitor [66, 69]. Stirring, vacuum, soaking, and LbL are only a few of the other methods employed to fill the nanocapsules [64]. To load inhibitors, few researchers had employed surface-functionalized nanoreservoirs such as nano-halloysites and mesoporous silica [70]. Mechanical injury, desorption, pH, ion exchange [32], and temperature are a few possible stimulation mechanisms for the release of the inhibitor [70]. Another way for reserving corrosion inhibitors is the layered double hydroxide (LDH) technique. In the early phases of corrosion, nanoreservoirs produced using the LDH approach work well because of the nature of fast ion exchange [60].

9.6.2 Carbon Nanotubes (CNTs) Used as Corrosion Inhibitors

The application of carbon-based allotropes in the form of corrosion inhibitors [71] is receiving significant attention these days, particularly CNTs, graphene, graphene oxides, and their derivatives. The widespread use of CNTs and their derivatives is ascribed to a number of intriguing properties, such as their high surface-to-volume ratio, extraordinary mechanical stability, high chemical and thermal stability, and excellent ability to interact with the metallic surface. In addition to their usage as anticorrosive materials, CNTs [71] and their derivatives are effectively applied as electrochemical sensors, catalysts for several chemical reactions, surfaces for the elimination of hazardous toxic metal ions, and composite materials made of polymers [72–73]. Since CNTs are effective heat- and electricity-conductors, it is believed that their pure form will speed up corrosion. CNTs may be produced using a number of chemical approaches, including arc discharge evaporation, thermal synthesis, and chemical vapor deposition [74]. CNT derivatives, particularly polymer composites, are being extensively researched as anticorrosive substances for ferrous and nonferrous alloys because of their wide range of intriguing features. Ionita and Pruna [75] used molecular mechanics and molecular dynamics approaches to predict the mechanical properties of polypropylene (PPy)/polyaminobenzene sulfonic acid-functionalized single-walled CNTs (CNTs-PABS) and PPy/carboxylic acid-functionalized single-walled CNTs (CNTs-CA). They also employed it as anticorrosive composite coatings for carbon steel (OL 48-50) alloys in 3.5% of sodium chloride were polypropylene film, polyaminobenzene sulfonic acid-functionalized single-walled CNTs, and carboxylic acid-functionalized single-walled CNTs. The anticorrosive impact of various formulations was measured using the potentiodynamic polarization (PDP), scanning electron microscope (SEM), and transmission electron microscope methods. As per the findings CNTs-PABS and CNTs-CA are distributed consistently across composite materials and are more effective in preventing corrosion than pure PPy. The results also indicated that the order in which the different formulations inhibited corrosion was PPy, CNTs-CA, and then CNTs-PABS. A different study has looked at the PPy/MW CN nanocomposite's anticorrosive effect for corrosion of 304 stainless steel in 3% of sodium chloride solution. A polyaniline (PANI)/f-MW-CNT nanocomposite [76] was created and used in another study [83] as an anticorrosive coating for mild steel in a neutral sodium chloride solution (3.5% of sodium chloride). In order to characterize the synthesized nanocomposite, Raman, attenuated total reflectance-infrared, and field emission scanning electron microscopy (FESEM) investigations were performed. MW CNTs showed good dissipation in polymer matrix of PANI, according to FESEM research. In the PANI matrix, the f-CNTs were securely retained in place and evenly dispersed. The perimeter of the f-CNTs dramatically increases in the appearance of PANI, according to subsequent FESEM research, which is explained by the PANI polymer coating of the f-CNT surface. In order to assess how hydrophobic the nanocomposite is, a contact angle study was conducted. According to the PDP investigation, the addition of PANI and P CNTs greatly lowers the corrosion current density, with the P CNT-induced reduction being more evident. According to the results of the electrochemical impedance spectroscopy investigation, PANI and P CNTs cause the charge transfer resistance (Rct) value to rise. It was discovered through this experiment that PANI and P CNTs both function as interface-type organic inhibitors, which means that they work by adhering to the surface of metals to become effective. When P CNTs 3 was present, the value of Rct increased most significantly. Rct values rose in the following order: PCNT3 > PCNT2 > PCNT1 > PANI. The value of corrosion current densities was only found to be in inverse order. It has also been looked into if the inclusion of alkyd coating based on PANI-MW CNT composite for mild steel corrosion in 1 M of hydrochloric acid solution increases the Rct values [77]. For alkyd, PANI/alkyd, and PANI-MW CNTs/alkyd based anticorrosive coatings, the Rct values of 151, 2172, and 10300 cm^2, respectively, were determined. Qiu and colleagues [78] produced carbon nanotube-polyaniline (c-PANI) through achieving an in situ oxidative polymerization process of aniline on exterior of MW CNTs. Research revealed that the perimeter of the nanobrushes is influenced by the mass ratio of aniline to MW CNTs.

The synthetic nanobrushes have strong anticorrosive action against epoxy resin with an amine cure. The nanobrush coating has good anticorrosive properties in neutral, acidic, and alkaline conditions. Other investigations have also been conducted to explore the inhibition potential of the nanocomposites based on PANI-CNTs [79]. Likewise, corrosion inhibition characteristics of multifarious CNT-based nanocomposites like [80] silver substituted hydroxyapatite/functionalized MW-CNT nanocomposite for 316 L stainless steel in simulated body fluid (SBF), graphitic filamentous nanocarbon-aligned carbon thin layer for copper in simulated sea water [81], conductive polyurethane-MW CNT composites for stainless steel in 3% of sodium chloride [82], nickel/multi-walled CNT composite for mild steel in 3.5% of sodium chloride [83], hydroxyapatite/multi-walled CNT composite for 316 stainless steel in SBF [84], poly(o-phenylenediamine)@ multi-walled CNT nanocomposites for steel in 3.5% wt.% of sodium chloride [85], CNT and zinc-rich epoxy primers for carbon steel in simulated concrete pore solutions in the presence of chloride ions [86], semiconductor CdS/multi-walled CNT quantum dots (CdS/MW-CNTs) for zinc plate in 3.5% of sodium chloride, 1 M of hydrochloric acid and 1 M of potassium hydroxide [87], TiO_2 coated multi-wall CNT (MW-carbon nanotubes)/bis-[triethoxysilylpropyl] tetrasulfide (BTESPT/TiO_2/MW-carbon nanotubes) [88], 4,5-diphenylimidazole-functionalized CNTs for nickel alloy in sulfuric acid [89] and poly(3-aminobenzoic acid) @ MW-CNT nanocomposite for copper in 3.5% of sodium chloride solution. The effectiveness of nanocomposites made from CNTs employed as anticorrosive coatings has been demonstrated in the literature [70], which reveals how effective they are. The passage of corrosive species including water, oxygen, and corrosive ions is prevented by the uniform distribution and filling of CNTs in the micropores in surface coatings. The toughness and longevity of anticorrosive coatings [90-91] made of polymers are also improved by the presence of CNTs. The hydrophobic properties of CNTs make them anticorrosive barriers for aqueous electrolytes. Most of the time, it has been shown that nanocomposites based on CNTs function as interface-type corrosion inhibitors because they are adsorbed at the metal-electrolyte interface, where they become effective.

9.7 CONCLUSION

Coatings are one of the most efficacious, adaptable, simple, and cost-effective methods for protecting metals and alloys against corrosion. The sectors still require coatings with excellent performance and negligible impact on the environment despite the fact that several anticorrosion coatings have been invented to safeguard metals and alloys. In this aspect, nanomaterials have played a significant role in terms of corrosion inhibition. The mechanism behind the usage of nanomaterials states when nanoparticles are applied in the organic coatings, they enhance barrier qualities by appropriate decrement in porosity and by lengthening the tortuous path for any corrosive species such as aqua, oxygen, and chloride anions. The following organic coating characteristics can be enhanced by nanomaterials: water adsorption/absorption, thermo-mechanical properties, substrate adhesion, chemical and ultraviolet stability, wear resistance, electrical conductivity, crosslinking density, uniformity, and biocompatibility. Although there are many nanomaterials invented and employed in the above aspects, however, there may be certain hazardous impact on human health and environments. Hence, these limitations must be explored in the future studies along with the applications.

REFERENCES

[1] Palanisamy G., Corrosion inhibitors, In: Ambrish Singh (ed.), *Corrosion Inhibitors*, Intechopen (2019). https://doi.org/10.5772/intechopen.80542

[2] Holcomb R., Corrosion in supercritical water: Ultrasupercritical environments for power production, In: *ASM Handbook*, Vol. 13C. OH: ASM International, Materials Park (2006).

[3] Shreir L.L., Jarman R.A. and Burnstein G.T., *Principles of Corrosion and Oxidation*, Vol. 1. Oxford: Butterworth-Heinemann (1994).

[4] Geethamani P. and Kasthuri P.K., The inhibitory action of expired asthalin drug on the corrosion of mild steel in acidic media: A comparative study, *Journal of the Taiwan Institute of Chemical Engineers*, 63(4) (2016) 490–499.

[5] Goni L. K.M.O. and Mazumder M.A.J., Green corrosion inhibitors. In: Singh A. (ed.), *Corrosion Inhibitors*. IntechOpen (2019). http://dx.doi.org/10.5772/intechopen.81376

[6] Geethamani P., Kasthuri P.K. and Aejitha S., Mitigation of mild steel corrosion in 1 M sulphuric acid medium by Croton sparciflorus—A green inhibitor, *Chemical Science Review and Letters*, 2(6) (2014) 507–516.

[7] Geethamani P., Kasthuri P.K. and Aejitha S., A new pharmaceutically expired drug acts as corrosion inhibitor for mild steel in acidic medium, *Elixir Corrosion & Dye*, 76 (2014) 28406–28410.

[8] Bentiss F., Traisnel M., Lagrenee M., The substituted 1,3,4-oxadiazoles: A new class of corrosion inhibitors of mild steel in acidic media, *Corrosion Science*, 42 (2000) 127.

[9] Sharma S. and Chaudhary R.S., Inhibitive action of methyl red towards corrosion of mild steel in acids, *Bulletin of Electrochemistry*, 16 (2000) 267.

[10] Lichtenstein A.G., The silver bridge collapse recounted, *Journal of Performance of Constructed Facilities*, 7 (1994) 249–261.

[11] Bowonder B., The Bhopal accident, *Technological Forecasting and Social Change*, 32(2) (1987) 169–182.

[12] Stress Corrosion Cracking Failure [Internet]. (2018). Available from: https://corrosion-doctors.org/Forms-SCC/swimming.htm

[13] Revie R.W. and Uhlig H.H., *Corrosion and Corrosion Control: An Introduction to Corrosion Science and Engineering*, 4th ed. New Jersey: Wiley (2008) 512. ISBN: 978-0-471-73279-2

[14] Raja P.B., Ismail M., Ghoreishiamiri S., Mirza J., Ismail M.C., Kakooei S. et al., Reviews on corrosion inhibitors: A short view, *Chemical Engineering Communications*, 203 (2016) 1145–1156.

[15] Javaherdashti R., How corrosion affects industry and life, *Anti-Corrosion Methods and Materials*, 47 (2000) 30–34.

[16] Uhlig H.H., The cost of corrosion to the United States, *Corrosion*, 6 (1950) 29–33.

[17] Koch G., Varney J., Thompson N., Moghissi O., Gould M. and Payer J., *International Measures of Prevention, Application, and Economics of Corrosion Technologies Stud*, Houston: NACE International (2016) 216.

[18] Hoar T.P., Review lecture: Corrosion of metals: Its cost and control, Proceedings of the Royal Society of London, *Series A: Mathematical and Physical Sciences*, 348 (1976) 1–18.

[19] Bennett L.H., Kruger J., Parker R.L., Passaglia E., Reimann C., Ruff A.W. and Yakowitz, H., Economic Effects of Metallic Corrosion in the United States: A Report to the Congress by the National Bureau of Standards, National Bureau of Standards. Washington: U. S. Government Printing Office. Publication No. 511-1, 65.

[20] Peterson M.J., Bhopal Plant Disaster, International Dimensions of Ethics Education in Science and Engineerings (2009). Available from: www.umass.edu/sts/ethics. (PDF) The Impact of Corrosion on Society. Available from: www.researchgate.net/publication/226696397

[21] NTSB: Aircraft Accident Report—Aloha Airlines Flight 243, Boe-ing 737-200, National Transportation Safety Board Bureau of Accident Investigation, Washington (1989). (PDF) The Impact of Corrosion on Society. Available from: www.researchgate.net/publication/226696397

[22] OMAFRA: Corrosion of Roof Truss Gusset Plates in Farm Buildings [cited 20 July 2010]; available from: www.omafra.gov.on.ca/english/engineer/facts/94-035.htm#Intro. (PDF) The Impact of Corrosion on Society. Available from: www.researchgate.net/publication/226696397

[23] Mercer A.D., *Corrosion Inhibition: Principles and Practice*. Oxford: Butterworths Heinemann (1994).

[24] Aejitha S., Kasthuri P.K. and Geethamani P., Inhibitory action of acalypha indica extract on corrosion of mild steel in hydrochloric acid medium, *Chemical Science Review and Letters*, 2(7) (2014) 566–573.

[25] Taghavikish M., Dutta N.K. and Choudhury N.R., Emerging corrosion inhibitors for interfacial coating, *Coatings*, 7(12) (2017) 217–245.

[26] Sowmyashree A.S., Somya A., Kumar C.B.P., and Rao S., Novel nano corrosion inhibitor, integrated zinc titanate nano particles: Synthesis, characterization, thermodynamic and electrochemical studies, *Surfaces and Interfaces*, 22 (2021) 100812. https://doi.org/10.1016/j.surfin.2020.100812

[27] Sowmyashree A.S., Somya A., Kumar S., Kumar P. and Rao S., Discotic anthraquinones as novel corrosion inhibitor for mild steel surface, *Journal of Molecular Liquids*, 347 (2022) 118194. http://dx.doi.org/10.1016/j.molliq.2021.118194

[28] Arthur D.E., Jonathan A., Ameh P.O. and Anya C., A review on the assessment of polymeric materials used as corrosion inhibitor of metals and alloys, *International Journal of Industrial Chemistry*, 4(2) (2013) 1–9. DOI: 10.1186/2228-5547-4-2

[29] Palou R.M., Olivares-Xomelt O. and Likhanova N.V., Environmentally friendly corrosion inhibitors, In: Sastri V.S. (ed.), *Green Corrosion Inhibitors: Theory and Practice*, 1st ed. New Jersey: Wiley (2011), pp. 257–303. DOI: 10.1002/9781118015438.ch7

[30] Sastri V.S. (ed.), *Green Corrosion Inhibitors: Theory and Practice*, 1st ed. New Jersey: Wiley (2011), p. 310. DOI: 10.1002/9781118015438

[31] Somya A., Hybrid ion exchangers, In: *Hybrid Nanomaterials-Flexible Electronics Materials*. London, UK: IntechOpen (2020), Apr 25, ISBN: 978-1-83880-338. DOI: http://dx.doi.org/10.5772/intechopen.92116

[32] Somya A., Metal phosphates: Their role as ion exchangers in water purification. In: Gupta R.K. (ed.), *Metal Phosphates and Phosphonates. Engineering Materials*. Cham: Springer (2023). DOI: https://doi.org/10.1007/978-3-031-27062-8_19

[33] Miralrio A. and Espinoza Vázquez A., Plant extracts as green corrosion inhibitors for different metal surfaces and corrosive media: A review, *Processes,* 8(8) (2020) 942. https://doi.org/10.3390/pr8080942

[34] Sowmyashree A.S., Somya A., Rao S., Pradeep Kumar C.B., Al-Romaizan A.N., Hussein M.A., Khan A., Marwani H.M. and Asiri A.M., Potential sustainable electrochemical corrosion inhibition study of *Citrus limetta* on mild steel surface in aggressive acidic media, *Journal of Materials Research and Technology*, 24 (2023) 984–994. https://doi.org/10.1016/j.jmrt.2023.02.039

[35] Inamuddin, Ahamed M.I., Luqman M., Altalhi T. (eds.), *Sustainable Corrosion Inhibitors*, Materials Research Forum LLC, 2021 (2021) 107. ISBN: 164490148X, 9781644901489.

[36] Gacitua W., Ballerini A. and Zhang J., Polymer nanocomposites: Synthetic and natural fillers a review, Maderas. *Ciencia y tecnología*, 7(3) (2005) 159–178.

[37] Anandhan S. and Bandyopadhyay S., Polymer nanocomposites: From synthesis to applications, *Nanocomposites and Polymers with Analytical Methods*, 1 (2011) 1–28.

[38] Shi X., Nguyen T.A., Suo Z., Liu Y. and Avci R., Effect of nanoparticles on the anticorrosion and mechanical properties of epoxy coating, *Surface and Coatings Technology*, 204(3) (2009) 237–245, https://doi.org/10.1016/j.surfcoat.2009.06.048.

[39] Mahmoodi A. and Ebrahimi M., Role of a hybrid dye-clay nano-pigment (DCNP) on corrosion resistance of epoxy coatings, *Progress in Organic Coatings*, 114(2018) 223–232, https://doi.org/10.1016/j.porgcoat.2017.10.022.

[40] Nazari M.H., Zhang Y., Mahmoodi A., Xu G., Yu J., Wu J. and Shi X., Nanocomposite organic coatings for corrosion protection of metals: A review of recent advances, *Progress in Organic Coatings*, 162 (2022) 106573, https://doi.org/10.1016/j.porgcoat.2021.106573

[41] Huang W.F. et al., Super-hydrophobic polyaniline-TiO2 hierarchical nanocomposite as anticorrosion coating, *Material Letters*, 258(2020) 126822. https://doi.org/10.1016/j.matlet.2019.126822.

[42] Selim M.S., Yang H., Wang F.Q., Fatthallah N.A., Huang Y. and Kuga S., Silicone/ZnO nanorod composite coating as a marine antifouling surface, *Applications of Surface Science*, 466(2019) 40–50, https://doi.org/10.1016/j.apsusc.2018.10.004

[43] Verma S., Mohanty S. and Nayak S.K., Preparation of hydrophobic epoxy–polydimethylsiloxane–graphene oxide nanocomposite coatings for antifouling application, *Soft Matter*, 16(5) (2020) 1211–1226. https://doi. org/10.1039/C9SM01952A

[44] An K. et al., Large-scale preparation of superhydrophobic cerium dioxide nanocomposite coating with UV resistance, mechanical robustness, and anticorrosion properties, *Surface of Coatings Technology*, 384(2020) 125312. https://doi. org/10.1016/j.surfcoat.2019.125312

[45] Xiong L., Yu M., Li Y., Kong X., Li S. and Liu J., Modified salicylaldehyde@ZIF-8/ graphene oxide for enhancing epoxy coating corrosion protection property on AA2024-T3, *Progress in Organic Coatings*, 142(2020) 105562. https://doi.org/10.1016/ j.porgcoat.2020.105562

[46] Xiong L., Yu M., Li Y., Kong X., Li S. and Liu J., Modified salicylaldehyde@ZIF-8/ graphene oxide for enhancing epoxy coating corrosion protection property on AA2024-T3, *Progress in Organic Coatings*, 142(2020) 105562. https://doi.org/10.1016/ j.porgcoat.2020.105562.

[47] Cao X., Huang F., Huang C., Liu J. and Cheng Y.F., Preparation of graphene nanoplate added zinc-rich epoxy coatings for enhanced sacrificial anode-based corrosion protection, *Corrosion Science*, 159 (2019) 108120. https://doi.org/10.1016/j. corsci.2019.108120.

[48] Cheng L. et al., Effect of graphene on corrosion resistance of waterborne inorganic zinc-rich coatings, *Journal of Alloys and Compounds*, 774 (2019) 255–264. https://doi.org/10.1016/j.jallcom.2018.09.315

[49] Li C.-C., Lai T.-Y. and Fang T.-H., Corrosion resistant coatings based on zinc nanoparticles, epoxy and silicone resins, *Journal of Nanoscience Nanotechnology*, 20(10) (2020) 6389–6395. https://doi.org/10.1166/jnn.2020.18709.

[50] Niroumandrad S., Rostami M. and Ramezanzadeh B., Effects of combined surface treatments of aluminium nanoparticle on its corrosion resistance before and after inclusion into an epoxy coating, *Progress in Organic Coatings*, 101 (2016) 486–501. https://doi.org/10.1016/j.porgcoat.2016.09.010.

[51] Fang F.F., Choi H.J. and Joo J., Conducting polymer/clay nanocomposites and their applications, *Journal of Nanoscience and Nanotechnology*, 8 (4) (2008) 1559–1581, https://doi. org/10.1166/jnn.2008.036

[52] Fu S., Sun Z., Huang P., Li Y. and Hu N., Some basic aspects of polymer nanocomposites: A critical review, *Nano Material Science*, 1(1) (2019) 2–30. https://doi.org/10.1016/j.nanoms.2019.02.006

[53] Lam C.K. and Lau K.T., Localized elastic modulus distribution of nanoclay/epoxy composites by using nanoindentation, *Composite Structures*, 75 (2006) 553–558. https://doi.org/10.1016/j.compstruct.2006.04.045

[54] Hartwig A., Sebald M., Pütz D. and Aberle L., Preparation, characterisation and properties of nanocomposites based on epoxy resins — An overview, *Macromolecular Symposia*, 221 (1) (2005) 127–136, https://doi.org/10.1002/masy.200550313

[55] Shi G., Zhang M.Q., Rong M.Z., Wetzel B. and Friedrich K., Friction and wear of low nanometer Si3N4 filled epoxy composites, *Wear*, 254 (7–8) (2003) 784–796. https://doi.org/10.1016/S0043-1648(03)00190-X

[56] Becker O., Varley R. and Simon G., Morphology, thermal relaxations and mechanical properties of layered silicate nanocomposites based upon high-functionality epoxy resins, *Polymer*, 43 (16) (2002) 4365–4373. https://doi.org/10.1016/ S0032-3861(02)00269-0

[57] Crosky A., Kelly D., Li R., Legrand X., Huong N. and Ujjin R., Improvement of bearing strength of laminated composites, *Composite Structures*, 76 (3) (2006) 260–271. https://doi.org/10.1016/j.compstruct.2006.06.036

[58] Pandey Y.N., Papakonstantopoulos G.J. and Doxastakis M., Polymer/nanoparticle interactions: bridging the gap, *Macromolecules*, 46 (13) (2013) 5097–5106. https://doi.org/10.1021/ma400444w

[59] He X. and Shi X., Self-repairing coating for corrosion protection of aluminum alloys, *Progress in Organic Coating*, 65(1) (2009) 37–43. https://doi.org/10.1016/j. porgcoat.2008.09.003

[60] Montemor M.F., Functional and smart coatings for corrosion protection: a review of recent advances, *Surface and Coatings Technology*, 258 (2014) 17–37. https://doi.org/ 10.1016/j.surfcoat.2014.06.031.

[61] He X. and Shi X., Self-repairing coating for corrosion protection of aluminum alloys, *Progress in Organic Coatings*, 65 (1) (2009) 37–43. https://doi.org/10.1016/j.porgcoat.2008.09.003

[62] Wei H. et al., Advanced micro/nanocapsules for self-healing smart anticorrosion coatings, *Journal of Materials Chemistry A*, 3(2) (2014) 469–480. https://doi.org/10.1039/ C4TA04791E

[63] Shi X., On the use of nanotechnology to manage steel corrosion, *Recent Patents on Engineering*, 4 (1) (2010) 44–50. https://doi.org/10.2174/187221210790244730

[64] Stankiewicz A., Szczygieł I. and Szczygieł B., Self-healing coatings in anti-corrosion applications, *Journal of Materials Science*, 48 (23) (2013) 8041–8051. https://doi.org/ 10.1007/s10853-013-7616-y

[65] Le Y., Hou P., Wang J. and Chen J.-F., Controlled release active antimicrobial corrosion coatings with ag/SiO2 core–shell nanoparticles, *Materials Chemistry and Physical*, 120 (2–3) (2010) 351–355. https://doi.org/10.1016/j. matchemphys.2009.11.020

[66] Abdullayev E., Price R., Shchukin D., Lvov Y., Halloysite tubes as nanocontainers for anticorrosion coating with benzotriazole, *ACS Applied Materials & Interfaces*, 1(7) (2009) 1437–1443. https://doi.org/10.1021/am9002028

[67] Yeganeh M., Omidi M., Mortazavi S.H.H., Etemad A., Nazari M.H. and Marashi S.M., Application of mesoporous silica as the nanocontainer of corrosion inhibitor, In: Rajendran S., Nguyen T.A., Kakooei S., Yeganeh M., Li Y. (eds.), *Corrosion Protection at the Nanoscale*, Elsevier (2020) 275–294. https://doi. org/10.1016/B978-0-12-819359-4.00015-5

[68] Ward C.J., DeWitt M. and Davis E.W., Halloysite nanoclay for controlled release applications, In: Nagarajan R. (ed.), *ACS Symposium Series Vol. 1119*, American Chemical Society, Washington, DC (2012) 209–238. https://doi.org/10.1021/ bk-2012-1119.ch010

[69] Khajouei A., Jamalizadeh E. and Hosseini S.M.A., Corrosion protection of coatings doped with inhibitor-loaded nanocapsules, *Anti-Corrosion Methods and Materials*, 62 (2) (2015) 88–94. https://doi.org/10.1108/ACMM-11-2013-1317

[70] Verma C., Quraishi M.A., Ebenso E.E. and Hussain C.M., Recent advancements in corrosion inhibitor systems through carbon allotropes: Past, present, and future, *Nanoselect*, 2 (12) (2021) 2237–2255. https://doi.org/10.1002/nano.202100039

[71] Norizan M.N., Moklis M.H., Ngah Demon S.Z., Halim N.A., Samsuri A., Mohamad I. S., Knight V.F. and Abdullah N., Carbon nanotubes: Functionalisation and their application in chemical sensors, *RSC Advances*, 10 (71) (2020) 43704. https://doi.org/10.1039/D0RA09438B

[72] Zare E.N., Lakouraj M.M. and Ramezani A., Efficient sorption of Pb(ii) from an aqueous solution using a poly(aniline-co-3-aminobenzoic acid)-based magnetic core–shell nanocomposite, *New Journal of Chemistry*, 40 (2016) 2521.

[73] Baghayeri M., Zare E.N. and Lakouraj M.M., Monitoring of hydrogen peroxide using a glassy carbon electrode modified with hemoglobin and a polypyrrole-based nanocomposite, *Microchimica Acta*, 182 (2015) 771.

[74] Dresselhaus G., Dresselhaus M.S. and Saito R., *Physical Properties of Carbon Nanotubes*, London: Imperial College Press (1998). ISBN: 9781783262410, 1783262419

[75] Ioniţă M. and Prună A., Polypyrrole/carbon nanotube composites: Molecular modeling and experimental investigation as anti-corrosive coating, *Progress in Organic Coatings*, 72 (2011) 647.

[76] Kumar A.M. and Gasem Z.M., In situ electrochemical synthesis of polyaniline/*f*-MWCNT nanocomposite coatings on mild steel for corrosion protection in 3.5% NaCl solution, *Progress in Organic Coatings*, 78 (2015) 387.

[77] Farag A.A., Kabel K.I., Elnaggar E.M. and Al-Gamal A.G., Influence of polyaniline/multiwalled carbon nanotube composites on alkyd coatings against the corrosion of carbon steel alloy, *Corrosion Review*, 35(2) (2017) 85.

[78] Qiu G., Zhu A. and Zhang C., Hierarchically structured carbon nanotube–polyaniline nanobrushes for corrosion protection over a wide pH range, *RSC Advances*, 7 (2017) 35330.

[79] Salvetat J.-P., Bonard J.-M., Thomson N., Kulik A., Forro L., Benoit W. and Zuppiroli L., Mechanical properties of carbon nanotubes, *Applied Physics A*, 69 (1999) 255.

[80] Sivaraj D. and Vijayalakshmi K., Novel synthesis of bioactive hydroxyapatite/f-multiwalled carbon nanotube composite coating on 316L SS implant for substantial corrosion resistance and antibacterial activity, *Journal of Alloys and Compounds*, 777 (2019) 1340.

[81] Dong F., Yang X., Guo L., Qian Y., Sun P., Huang Z., Xu X. and Liu H., A tough, healable, and recyclable conductive polyurethane/carbon nanotube composite, *Journal of Colloid and Interface Science B*, 631 (2023) 239–248, https://doi.org/10.1016/j.jcis.2022.11.045.

[82] Zare E. N., Lakouraj M. M., Ghasemi S. and Moosavi E., Emulsion polymerization for the fabrication of poly(o-phenylenediamine)@multi-walled carbon nanotubes nanocomposites: characterization and their application in the corrosion protection of 316L SS, *RSC Advances*, 5 (2015) 68788.

[83] Ruan Z., Tian Y., Ruan J., Cui G., Iqbal K., Iqbal A., Ye H., Yang Z. and Yan S., Synthesis of hydroxyapatite/multi-walled carbon nanotubes for the removal of fluoride ions from solution, *Applied Surface Science*, 412 (2017) 578–590. https://doi.org/10.1016/j.apsusc.2017.03.215

[84] Sivaraj D. and Vijayalakshmi K., Enhanced antibacterial and corrosion resistance properties of Ag substituted hydroxyapatite/functionalized multiwall carbon nanotube nanocomposite coating on 316L stainless steel for biomedical application, *Ultrasonics Sonochemistry*, 59 (2019) 104730.

[85] Prasannakumar R., Chukwuike V., Bhakyaraj K., Mohan S. and Barik R.C., Electrochemical and hydrodynamic flow characterization of corrosion protection persistence of nickel/multiwalled carbon nanotubes composite coating, *Applied Surface Science*, 507 (2020) 145073.

[86] Rajyalakshmi T., Pasha A., Khasim S., Lakshmi M., Murugendrappa M. and Badi N., Enhanced charge transport and corrosion protection properties of polyaniline–carbon nanotube composite coatings on mild steel, *Journal of Electronic Materials*, 49 (2020) 341.

[87] Zhang Y., Zhu H., Zhuang C., Chen S., Wang L., Dong L. and Yin Y., TiO2 coated multi-wall carbon nanotube as a corrosion inhibitor for improving the corrosion resistance of BTESPT coatings, *Materials Chemistry and Physics*, 179 (2016) 80–91. https://doi.org/10.1016/j.matchemphys.2016.05.012

[88] Palaniappan N., Cole I. S., Kuznetsov A. E. and Justin Thomas K. R., Experimental and DFT studies of carbon nanotubes covalently functionalized with an imidazole derivative for electrochemical stability and green corrosion inhibition as a barrier layer on the nickel alloy surface in a sulphuric acidic medium, *RSC Advances*, 9 (2019) 38677–38686. https://doi.org/10.1039/C9RA08123B

[89] Palaniappan N., Cole I., Kuznetsov A. and Thomas K. J, Experimental and DFT studies of carbon nanotubes covalently functionalized with an imidazole derivative for electrochemical stability and green corrosion inhibition as a barrier layer on the nickel alloy surface in a sulphuric acidic medium, *RSC Advances*, 9 (2019) 38677.

[90] Somya, A. and Varshney, A. P., 2023. Functionalized thin film coatings for automotive coatings. In: Kumar, A., Verma, C. & Thakur, A. (eds.), *Corrosion Mitigation Coatings: Functionalized Thin Film Fundamentals and Applications*. Walter de Gruyter GmbH & Co KG. DOI: https://doi.org/10.1515/9783111016160-016

[91] Peter, A. and Somya, A., Anticorrosion nano coatings and potential role of simulation in nano coatings, *Journal of Applied Optics,* (2024) 113. https://appliedopticsjournal.net/index.php/JAO/article/view/91

10 Porous Nano/Micro-Carriers of Corrosion Inhibitors for Smart Coatings

Hedie Kazemi, Reza Goudarzi, and Bahram Ramezanzadeh

10.1 INTRODUCTION

10.1.1 Corrosion

The term "corrosion" is used for the description of metals' degradation that occurs as a result of chemical or electrochemical reactions during environmental exposure [1]. The process of metal corrosion is caused by irreversible chemical reactions that convert a pure metal into a chemically more stable form, such as sulfides, oxides, hydroxides, etc. Corrosive environments can be in solid, liquid, or gaseous forms. In general, these environments are referred to as electrolytes. Anodic and cathodic reactions are observed on the metal surface in these electrolytes during ion transfer. Corrosion occurs at the surface of materials in several ways. As a first step, there is no attack across the entire surface; rather, only a small area is affected. A difference in corrosion resistance may also result in the formation of this condition along grain boundaries and other weak points. In industrial machinery and metallic equipment, corrosion damages the machine over time and reduces the overall value of the product [2].

10.1.1.1 Factors Contributing to Corrosion

Except for noble metals such as gold and platinum, most metals found in nature are oxide compounds. The reason for this is the thermodynamic stability of metals in their compound state which is much higher than their stability in the elemental state. It takes a lot of energy to extract metals from oxides into free states since they exist in oxide form. In the presence of external environmental conditions such as moisture and oxygen, they can revert to their combined state through this supplied energy. As an example, the corrosion of iron that results from exposure to external conditions causes brown-colored hydrated ferrous oxide to form when the iron is exposed to external conditions [2]. There are two main factors that determine the rate of corrosion: the type of the metal and the nature of the environment.

10.1.2 Methods for Prevention of Corrosion

There are many accidents caused by corrosion in pipelines. Corrosion can be minimized in many ways. The use of anticorrosion coatings, inhibitors, and cathodic protection are some of these methods. Researchers have studied different aspects of corrosion. Temperature and humidity in the processing area are two important parameters in addition to the properties of the metal and the chemicals that will be used during the process [3]. Various techniques are used to minimize the

DOI: 10.1201/9781032677255-10

corrosion of materials to a level that is acceptable for them to reach their natural life expectancy. It is possible to eliminate corrosion in some circumstances through corrosion control methods. The most common methods used to deal with corrosion are the selection of suitable materials, environmental control, cathodic protection, anodic protection, corrosion inhibitors, and protective coatings [4].

10.1.3 Using Corrosion Inhibitors as Part of Anticorrosion Coatings

Corrosion inhibitors can be used directly and indirectly (encapsulated) in anticorrosion coatings. The advantages and disadvantages of each method are disused in detail.

10.1.3.1 Direct Use of Inhibitors

In this case, corrosion inhibitors are directly incorporated into the coating bed and they form a passive film as they are adsorbed onto the metal surface. Consequently, corrosive substances cannot penetrate into the metal substrate, thereby preventing corrosion. It is important to note that although corrosion inhibitors are directly added to the coating matrix, they can still have a considerable anticorrosion effect on the metal surfaces. It has, however, been contested by many researchers. As an example, some believe that adding benzotriazole (BTA) directly to the coating will be ineffective due to its water solubility. Furthermore, it may result in the formation of holes in the coating, which may significantly reduce its barrier performance [5].

10.1.3.2 Indirect Use of Inhibitors

Organic or inorganic nanocarriers can be used in smart anticorrosion coatings that have the ability to carry and release active substances (such as inhibitors) to protect surfaces. Corrosion inhibitors can also be used indirectly by encapsulating them in nanocarriers or microcapsules. By using nanocarriers, corrosive factors such as moisture, oxygen, and other corrosive factors can be prevented from reaching the metal surface, which means that the coating can self-repair. Nanocarriers release corrosion inhibitors under specific conditions and in a controlled manner. For example, by changing the pH, a nanocontainer can release its inhibitor and prevent the progress of corrosion [5, 6].

10.1.4 Smart Anticorrosive Coating

Coatings that are smart are those that sense their environment and respond appropriately to it. There are several types of smart coatings, including antimicrobial, conductive, antifouling, self-healing, and self-cleaning coatings.

Each surrounding material is designed to achieve its unique performance in a particular application. In recent years, there has been an increased focus on developing advanced materials that prevent cracks and provide long-term protection performance. Materials scientists are challenged with the task of developing new materials that offer greater safety and longer lifetimes and require fewer repairs. Self-healing is considered to be one of the most promising methods since it seems that it is not possible in reality to develop a substance that will never die.

Self-healing is defined as the process by which a material is able to repair or heal damaged material without external intervention or in a natural way. Materials that are self-healing are those that are capable of recovering their properties independently in a manner that allows them to last longer than similar materials without the capability of self-healing. The use of self-healing coatings is necessary for the full or partial repair of coated areas that have been damaged due to aging or unexpected aggression [5]. Self-healing anticorrosive smart coatings are divided into intrinsic and extrinsic categories as discussed below.

10.1.4.1 Extrinsic Self-Healing Smart Anticorrosive Coating

Self-healing smart coatings are capable of healing cracks when the local environment is altered or if an external force impacts the active surface [7]. In extrinsic self-healing anticorrosion coatings, the factors that repair the coating and prevent corrosion are the materials added to the coating matrix (such as healing agents and corrosion inhibitors). Molecular migration speed is an important factor in these coatings. Adding inhibitors and healing agents directly to the coating may result in the deficiency of these materials due to their low solubility within the coating matrix, and if these materials are highly soluble in the coating matrix, the metal will be protected, but the preservation period will be short because the inhibiting substances and healing agents are consumed rapidly.

The most commonly used approach to controlling healing agent release is that which encloses solids, liquids, and gases in an inert shell, isolating and protecting these particles from the environment through an inert shell. Microencapsulation is a process of enclosing microparticles in inert shells. For this purpose, organic and inorganic nanocarriers with different capacities and structures can be used. The best materials for micro/nanocontainers are those with porous, hollow, or layer structures. The structure and investigation of organic and inorganic nanocarriers and their comparison will be done in the next section.

Usually, when it is decided to inject the inhibitors with a small dose into the capsules, inorganic capsules that have a small internal space can be used. When it is intended to inject inhibitors with a high dose into capsules, organic capsules with a large internal space should be selected. As the coating cracks, the nanocapsules slowly release the liquid inside, repairing the crack and preventing corrosion.

The key point in using these nanocarriers is to be able to prepare and select a nanocapsule based on the application of the coating, whose inhibitor loading and release process is well under control. In this way, it is possible to expand the life of the coating for a longer time [5].

10.1.4.2 Intrinsic Self-Healing Smart Anticorrosion Coating

In intrinsic self-healing anticorrosion coatings, the polymer itself is responsible for performing coating repair. A smart anticorrosion coating with extrinsic self-healing properties can limit the number of repairs. Clear coatings can become cloudy if nanocontainers are added. Further, when the nanocontainer releases the healing agent/inhibitor, it creates a new gap in the coating. As a result, the corrosive material is exposed to a new path of penetration, which in turn affects the overall properties of the coating. In contrast, the ability of a smart anticorrosion coating to self-heal depends on the physical and chemical structures of the polymer matrix. Polymer chains can undergo chemical and physical reactions when heated, illuminated, or subjected to other external stimuli. In comparison with external self-healing intelligent anticorrosion coatings, one of the most obvious advantages is that the repair process can be repeated several times without additional healing agents required. Self-healing anticorrosion coatings are both chemically and physically capable of healing themselves.

10.1.4.2.1 Chemical Performance of Intrinsic Self-Healing Coatings

It is well known that the initial mobility of the polymer molecular chains increases with energy supply. There is a contact between a polymer chain on the side wall of a crack and another polymer chain on the opposite wall, which reduces the distance between the chains on the two sides. The energy provided by this process enables reversible covalent or non-covalent bonds to be formed between polymer chains. This results in the restoration of the coating's network structure and its properties. Several types of healing reactions are based on reversible covalent bonds, most notably cycloaddition reactions, chain exchange reactions, and free radical reactions.

10.1.4.2.2 Physical Performance of Intrinsic Self-Healing Coatings

The molecular chains of the polymer penetrate the scratch as the scratch temperature rises above the polymer glass transition or melting temperature. In some cases, corrosive factors can be found within the coating scratch during the repair process. To prevent the presence of these corrosive factors from reducing the adhesion of the metal and the coating, it is prudent to add corrosion inhibitors to these coatings [5].

10.1.5 Nanocarriers in Smart Anticorrosion Coating

In general, micro/nanocarriers should have the following characteristics:

- Chemical and mechanical stability
- High loading capacity
- Compatibility with the coating matrix
- Suitable size and structure
- Ability to recognize damage or corrosion
- Release ability regulated by damage development rate [5, 8].

Anticorrosion coatings that dynamically react to corrosive environments offer a new solution to corrosion problems. Nanocarriers are often used as components of these coatings in order to encapsulate corrosion inhibitors and provide targeted corrosion protection. By designing and preparing containers with a variety of functionalities, it is possible to develop insights into creating a new generation of smart coatings that have self-healing properties in response to a variety of stimuli, including pH, electrochemical potential, redox, aggressive corrosion ions, light, heat, magnetic fields, and mechanical forces. There are two main types of nanocarriers: organic and inorganic. In this section, the properties of organic and inorganic nanocarriers will be discussed [9].

10.1.5.1 Organic Nanocarriers

Organic nanocarriers are often composed of polymers that have the inherent advantage of high versatility and a wide scope for functionalization. These properties allow them to be tuned toward specific encapsulation and release features, helping development of the intelligent and precisely controlled anti-corrosion systems [10].

The inherent flexibility of organic polymers leads to the production of firm nanocontainers, which are less likely to break due to mechanical stress but compared to the inorganic-based coating they are weaker in mechanical properties. In addition, the biodegradability of certain polymers can contribute to environmentally friendly anticorrosion solutions.

However, organic nanocarriers are associated with a series of challenges. The potential for degradation under certain corrosive conditions may limit their scope of application. Also, organic nanocarriers may be susceptible to ultraviolet (UV) damage, reducing their lifespan in outdoor applications. Another concern is the possible leaching of organic components that can introduce impurities into the coating system [5].

10.1.5.1.1 Nano Polymer Containers

As an example, the use of microcontainers containing polyurea formaldehyde (PUF) and polymerizable agents has been presented as one of the first strategies to develop self-healing coatings with repair capabilities [5, 11].

10.1.5.1.2 Micellar Nanocarriers

Another form of organic nanocontainer in smart coatings comes from micelles. In one study, block copolymer micelles were loaded with corrosion inhibitors, demonstrating a self-healing effect when incorporated into a coating on aluminum alloy. Micelles are particularly interesting due to their capacity to encapsulate both hydrophilic and hydrophobic substances in their core and shell, respectively. This dual encapsulation activity allows the use of a combination of inhibitors within a single nanocontainer, potentially synergizing their effects [12].

10.1.5.1.3 Layer-by-Layer Assembled Nanocarriers

Layer-by-Layer (LbL) assembly offers another technique for creating organic nanocarriers. Researchers have used this method to build polymeric shells around corrosion inhibitor cores, creating nanocarriers that can be integrated into a coating system [13, 14].

10.1.5.2 Inorganic Nanocontainers

Mineral nanocontainers, which are often made of materials such as silica or alumina, have different characteristics. They are inherently more resistant to UV rays and high-temperature conditions, increasing their suitability for harsh environments. Mineral nanocontainers often exhibit high mechanical strength and provide durability to the coating system.

In terms of encapsulation, mesoporous silica nanoparticles (MSNs) provide an excellent example. The well-defined pore structure enables precise control over the encapsulation and release of inhibitors.

Mineral nanocarriers can be mesoporous silica or titania, ion exchange nanoclays, and halloysite nanotubes. Although inorganic nanocarriers increase the strength of the coating, they can also lead to mechanical vulnerability under certain stress conditions. The process of making inorganic nanocontainers can be more complex and energy-efficient than their organic counterparts. Finally, disposal issues and potential environmental impacts of minerals need careful consideration. In this section, several key examples of mineral nanocarriers, their performance, and their transformative effect on anticorrosion coatings will be discussed [15].

10.1.5.2.1 Carbon Nanotubes

Carbon nanotubes (CNTs) possess cylindrical structures comprised of carbon atoms arranged in a hexagonal lattice. Their tube-like morphology at the nanoscale, coupled with their exceptional mechanical and thermal properties, has resulted in their utilization for corrosion protection when combined with active substances. He et al. [16] performed the synthesis of multi-walled CNTs using a chemical approach. These functional nanoreservoirs were filled with the environmentally friendly inhibitor and dispersed within the epoxy resin, which was subsequently applied as a coating onto steel substrates [16, 17].

10.1.5.2.2 Halloysite Nanotubes

Halloysite nanotubes (HNTs) are naturally occurring aluminosilicate minerals with hollow tubular structures. HNTs possess a unique composition characterized by two distinct basal surfaces. The outer surface is composed of layers of tetrahedral silicate, exhibiting a relatively weak negative charge, while the inner lumen is made up of gibbsite octahedral layers of aluminum oxide (Al_2O_3), providing a strong positive charge. This electrostatic contrast is advantageous for loading negatively charged compounds into the HNT, as it facilitates the repulsion of the outer surface while enabling the incorporation of these compounds into the inner lumen. It is worth noting that the outer surface is hydrophilic, whereas the inner lumen demonstrates hydrophobic properties. In contrast to CNTs, which tend to form Van der Waals bonds with other nanotubes, resulting in their dispersion in solution, HNTs exhibit relatively low interactions with their own nanotubes [18].

These nanotubes can encapsulate corrosion inhibitors within their lumen, acting as effective nanocarriers, offering the capability to accommodate and selectively release a wide range of active agents, including inorganic and organic compounds. For instance, Manasa et al. successfully loaded HNTs with cationic inhibitors Ce^{3+}/Zr^{4+} by etching the lumen of the tubes. The effectiveness of the HNTs-dispersed coatings in self-healing was evaluated using scanning vibrating electrode technique (SVET) after exposure to NaCl solution for 24 hours [19].

Additionally, Patra et al. demonstrated that coatings based on HNTs capped with polymeric nanocapsules containing BTA exhibited improved corrosion resistance on mild steel compared to uncapped HNTs [20].

10.1.5.2.3 Layered Double Hydroxides

Layered double hydroxides (LDHs) represent another class of inorganic nanocarriers. These are layered materials with a sandwich-like structure, wherein metal hydroxide layers encapsulate anionic species.

LDH is primarily suitable for corrosion inhibition in neutral or alkaline systems. This is because LDH, being a layered material, tends to dissolve in acidic environments. The corrosion inhibition properties of LDH are mainly attributed to its structural and chemical characteristics, which include the formation of a physical protective film and the adsorption of chloride ions through anion exchange processes [21].

Various types of LDHs can be synthesized by altering the combination of two metal cations and the anions inserted between the layers, as shown in Figure 10.1. LDHs have been loaded with corrosion inhibitors like vanadate or phosphonate anions and integrated into coatings for aluminum alloys. Alibakhshi et al. investigated the corrosion prevention properties of phosphate-intercalated Mg-Al LDHs, top-coated with epoxy-polyamide. The results showed significant active corrosion protection [22]. Hang et al. demonstrated that Mg-Al LDHs loaded with 2-benzothiazolylthiosuccinic acid could control the release of the inhibitor when incorporated into an epoxy coating [23].

These smart coatings displayed controlled inhibitor release, responsive to pH changes characteristic of corrosion onset, demonstrating effective protection [24, 25].

10.1.5.2.4 Zeolites

Zeolites, microporous aluminosilicate minerals, offer another avenue for inorganic nanocontainer development. Their high thermal stability, mechanical strength, and ion exchange capacity make them suitable for encapsulating corrosion inhibitors [27].

10.1.5.3 Metal-Organic Frameworks

Metal-organic frameworks (MOFs) are crystalline materials composed of metal ions interconnected by organic linkers, forming a highly porous structure. Like MSNs, MOFs have a large surface area and tunable pore sizes but offer an added benefit in terms of their functionality. The organic linkers within MOFs can be tailored to deliver specific functions, including the selective absorption and

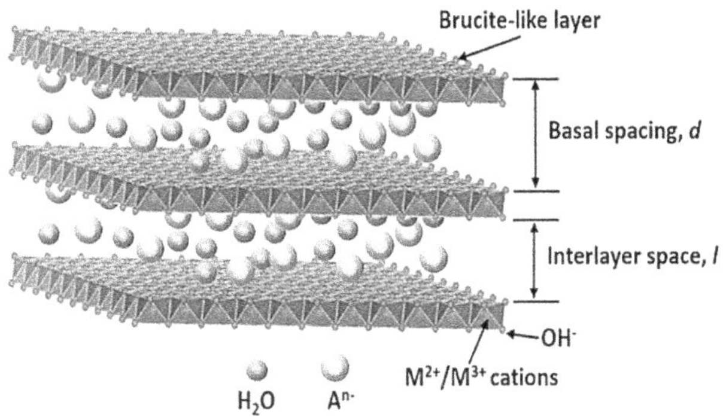

FIGURE 10.1 Structure of the LDHs [26].

triggered release of encapsulated substances. In the context of anti-corrosive coatings, this means MOFs can provide superior control over inhibitor release [28].

10.2 POROUS MATERIALS AS NANOCARRIERS IN SMART COATINGS

Among different types of carriers, nano/microporous carriers have attracted considerable attention due to their unique properties. Nano/microporous carriers have several advantages over their nonporous counterparts. They have a high surface-to-volume ratio, which allows them to hold a large amount of corrosion inhibitor. In addition, pores allow for the controlled release of inhibitors, which can be released from the pores in response to changes in the environment. In the following sections, we will introduce a number of porous carriers.

10.2.1 MOFs AS ANTICORROSION AGENTS

10.2.1.1 Structure and Morphology

MOFs are compounds that consist of ions or clusters of metal ions coordinated by organic ligands to produce one-dimensional, two-dimensional (2D), or three-dimensional (3D) structures. Due to their porous structure, they can be considered a type of porous material. Gases and drugs can be stored, delivered, or held in these pores. The overall structure of MOFs is presented in Figure 10.2 [29, 30].

The MOF complexes in some cases have antimicrobial properties, self-cleaning properties, as well as easy recovery properties. The MOFs are different from other porous nanoparticles in that they are capable of adapting to changes in the microstructure by manipulating the type and amount of metal ions and organic linkers. It is possible to synthesize MOFs in a variety of shapes, including linear, square, planar, and triangular, with a variety of properties. A MOF structure can be composed of di-, tri-, and tetravalent metal ions. In order to optimize MOFs for high-tech applications, it is necessary to carefully select metal ions and organic ligands [29]. There are several characteristics of MOFs that make them particularly suitable for application in anticorrosive coatings. First, they have a high surface area-to-volume ratio, which allows them to hold large amounts of corrosion inhibitors. Second, the size and shape of the pores in MOFs can be adjusted by changing the metal ions or organic ligands, allowing for the tuning of the release characteristics

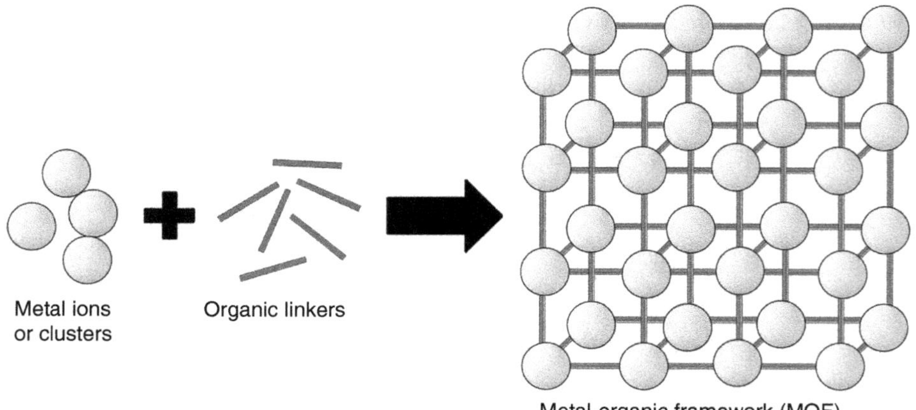

FIGURE 10.2 Schematic structure of the MOF.

of the inhibitors. Third, MOFs can be made from a wide range of materials, allowing for the selection of materials resistant to a specific corrosive environment [31].

10.2.1.2 Anticorrosion Applications of MOFs

10.2.1.2.1 Corrosion Inhibitor

Some MOFs can act in corrosive environments just like corrosion inhibitors and delay corrosion. For example, Cu-MOFs form a hydrophobic layer on the metal surface in an acidic environment, which slows down the anodic and cathodic reactions. In another research, ZIF-7 nanoparticles were used as corrosion inhibitors for the polymer coating system to increase the corrosion resistance of carbon steel in an acidic environment. Also, ZIF-67 nanoparticles (ZIF-67@APS) have shown obvious inhibitory properties for mild steel in saline solution [30].

10.2.1.2.2 Anticorrosion Nanofillers

The structure of MOFs is compatible with the polymer matrix and can act as anticorrosion nanofillers. For this application, MOFs with a more compact structure, such as Ni-MOFs, are more suitable.

10.2.1.2.3 Nanocarriers for Corrosion Inhibitors

A sensitive container (or carrier) used to load inhibitors has proved to be the most effective method to maintain the controlled release of inhibitors. Since MOFs are porous and have many active sites, we conclude that these materials have high adsorption capacity and can be used as nanocontainers containing corrosion inhibitors in smart anticorrosion coatings. In recent years, MOFs have emerged as promising materials for the development of smart anticorrosive coatings. The porous nature of MOFs allows them to act as nanocontainers, capable of storing corrosion inhibitors and releasing them when needed. As a consequence of the instability of the coordination bonds, local pH changes with the initiation of corrosion reactions can affect the stability of a MOF structure in an aqueous environment, which is why most MOFs have a pH dependence. This provides a dynamic and responsive method of corrosion protection that can adapt to changing conditions and provide long-term protection. Table 10.1 summarizes some research conducted in the field of using MOFs as anticorrosion nanocarriers.

In the following, the way of loading the inhibitor inside the MOFs and how they are released (when corrosion occurs) by the MOFs will be further examined [28].

10.2.1.3 Loading Corrosion Inhibitor into MOFs

The loading mechanism primarily depends on the properties of the MOF and the corrosion inhibitor. The MOF structure plays a significant role in the loading mechanism. Because MOFs have a large surface area and high porosity, they have a high loading capacity. Moreover, the chemical functionality of the MOF can be tuned to enhance the interaction with the corrosion inhibitor, thereby improving the loading efficiency.

The properties of the corrosion inhibitor also influence the loading mechanism. For instance, the size of the inhibitor molecules must be compatible with the pore size of the MOF. Furthermore, the chemical compatibility between the inhibitor and the MOF is crucial for effective loading. The process of loading inhibitors into MOFs involves several steps. First, the MOF is synthesized using a suitable metal ion and an organic ligand. The choice of the metal ion and ligand can be tailored to the specific requirements of the application, such as the desired pore size and chemical stability. Once the MOF is synthesized, the corrosion inhibitor can be loaded into the pores of the MOF through a process known as impregnation. This involves soaking the MOF in a solution of the inhibitor, allowing the inhibitor molecules to diffuse into the pores of the MOF. In order to prevent corrosion on metal surfaces, the loaded MOF can be incorporated into a coating formulation [29].

TABLE 10.1
The research was conducted on the use of MOFs as nanocarriers for delivering corrosion inhibitors to smart coatings

MOF	Corrosion Inhibitor	Substrate	Corrosive Media	The Study's Major Findings	Reference
ZIF-8	ATT	Mild steel	0.5 M NaCl solution	It has been shown that the composite ATT@ZIF-8 exhibits superior corrosion protection for mild steel, with an inhibition efficiency of over 97%.	[32]
ZIF-7	BTA	Carbon steel Q235	0.1 M HCl solution	It presented a superior barrier performance, which was 99.4% efficient in inhibiting.	[33]
Ce-MOF	BTA	Steel sheet	NaCl solution	Ce-MOF@TEOS composite material demonstrates excellent dispersibility in epoxy coatings and improved anticorrosion properties.	[34]
Zn-BTC	BTA	Aluminum	NaCl solution	BTA inhibitors were demonstrated to have 98.8% inhibition efficiency in preventing aluminum from corroding after immersion in corrosive solutions for three days. The released inhibitors form a dense barrier in the corrosion area, which decreases corrosion rates.	[35]
ZIF-8	BTA	Carbon steel	NaCl solution	Due to its excellent compatibility, ZBT exhibits high corrosion resistance, as do hydrophobic materials and corrosion inhibitors, as well as ferric tannate's ability to form protective films during the immersion process.	[36]
ZIF-8	Zinc Gluconate (ZnG)	Magnesium alloy AZ31B	NaCl solution	A corrosion inhibitor, ZnG@ZIF-8, was incorporated into the epoxy and it was not found to be corroding after 40 days of immersion in corrosive media.	[37]
ZIF-8	HMSN-BTA	Carbon steel Q235	NaCl solution	It was observed that the material was highly corrosion-resistant and self-healing in a significant way.	[38]

10.2.1.4 Synthesis of MOFs

The synthesis of MOFs typically involves the reaction of a metal ion or cluster with an organic ligand under suitable conditions. The choice of the metal ion, organic ligand, and reaction conditions can be used to control the structure and properties of the resulting MOF.

MOFs are typically synthesized by mixing the metal salt solution with the ligand solution or by adding solvent to a solid salt and ligand mixture. Solvents can be selected according to various aspects such as reactivity, solubility, redox potential, stability constant, etc.

The solvents affect the thermodynamics and activation energy of a reaction. It has also been attempted to synthesize MOFs in solid phase, in addition to liquid phase, because solid-phase synthesis is more rapid and straightforward. In addition to difficulties in obtaining single crystals, solid-state synthesis is usually unable to determine the product structure, which is otherwise quite straightforward during solution phase reactions. For the synthesis of MOFs, various methods have been used, including microwave-assisted, electrochemical, mechanochemical, and sonochemical methods [39]. Common methods for the synthesis of MOFs are shown in Figure 10.3.

10.2.2 MSNs

MSNs are a class of nanomaterials composed of silica and have a unique porous structure. These nanoparticles possess pores with sizes ranging from 2 to 50 nm. The presence of these pores imparts MSNs with a large surface area, which can be modified by attaching different chemical groups.

10.2.2.1 Structure and Morphology

In the early 1990s, a group of 15 researchers at Mobil Oil Company, along with Kuroda and colleagues, made a significant discovery known as mesoporous silicas. This breakthrough was driven by the need to expand the applications of zeolites. Materials containing mesoporous silica, such as MCM-41 [40], MCM-48 [41], and SBA-15 [42], feature fascinating 2D hexagonal and 3D cubic structures. These MSNs possess extraordinary characteristics, such as controlled particle size, porosity, morphology, and exceptional chemical stability. Consequently, they have gained immense attention as carriers for drugs, catalysts for diagnostic purposes, and for separation and sensing applications. Researchers have conducted extensive studies on MSNs, exploring their potential in diverse areas including energy storage, water and air purification, drug delivery, and smart release of active agents. Moreover, their value extends to coating applications, making them highly versatile and sought-after materials [43, 44].

MSN offers numerous advantages, including its remarkable chemical stability, low reactivity, and compatibility with various matrices. The synthesis of MSN is relatively straightforward, adding to its appeal. MSN is an ideal adsorption and desorption material due to its high surface area and mesoporous structure. Effective adsorbents require a significant specific surface area, and MSN fulfills this criterion. Depending on the amount of porosity in the adsorbent, the adsorption properties will differ, whether it involves long-range interactions (physisorption) or chemical interactions such as covalent and ionic bonds (chemisorption). Physical adsorption, especially, thrives in pores that are comparable in size to the molecular radius of the adsorptive species. Thus, materials with micropores or small mesopores are considered to be suitable for adsorption.

In high-performance coatings, MSN can be used as nanocontainers for anticorrosive agents. It is important to carefully examine both the loading of active agents and their controlled release in these coatings. In order to maximize corrosion inhibitor loading into porous nanocarriers, different strategies can be employed. Factors such as the dimensions and chemistry of the inhibitors need to be taken into account. The porosity of the nanocarriers should be tailored to accommodate an adequate quantity of active agents and enable the inhibitor to penetrate the porous structure effectively. The loading capacity plays a significant role in ensuring the anticorrosive performance of the smart system. It is also important that the chemical interactions between the nanocarrier and the inhibitor are carefully designed in order to facilitate the prompt release of the inhibitor. To be effective, nanocarriers must be able to retain the inhibitor loaded inside for a suitable period of time, and their chemical structure must remain stable under the specific environmental conditions where they are used.

Mesoporous silica materials (MSMs) have been successfully synthesized across a wide range of sizes, from the nanoscale to the microscale, exhibiting high surface areas (700–1000 $m^2.g^{-1}$) and significant pore volumes (0.6–1 $cm^2.g^{-1}$). In MSMs, surface silanol groups determine whether the molecule is hydrophobic or hydrophilic. This characteristic allows for their functionalization and loading with various substances, making them highly versatile for numerous applications [45].

MSNs exhibit different mesostructures (Figure 10.4), some of which are named after their respective developers. These mesostructures include:

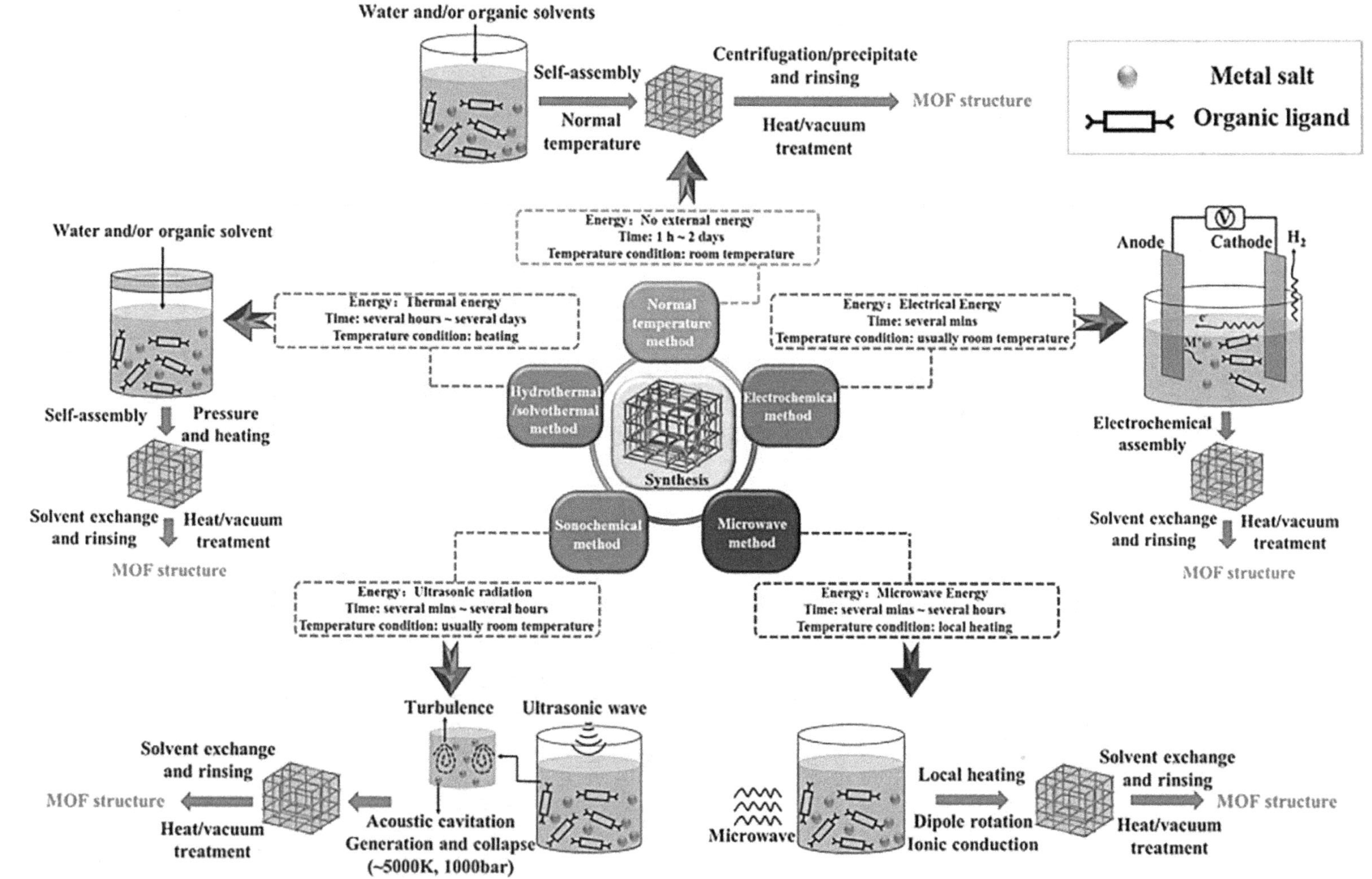

FIGURE 10.3 Typical fabrication approaches for MOFs for corrosion protection [28].

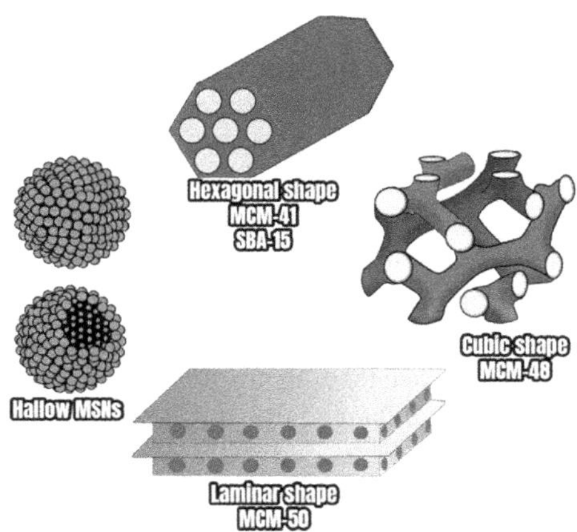

FIGURE 10.4 Mesostructures of mesoporous silica nanoparticles (MSNs).

1. Hexagonal structures: Examples include "Mobil Composition of Matter number 41" (MCM-41) [40], and "Santa Barbara Amorphous type number 15" (SBA-15) [42].
2. Cubic structures: One notable example is MCM-48 [41].
3. Lamellar structures: Represented by MCM-50 [46].
4. Other random structures: This category includes "Technische Universiteit Delft number 1" (TUD-1) [47], "Hiroshima Mesoporous Material number 33" (HMM-33) [48], and "Michigan State University type number 1" (MSU-1) [49].

The presence of these distinct mesostructures contributes to the diverse properties and functionalities exhibited by MSMs. These mesoporous silica materials find applications in various scientific and technological fields, owing to their tunable characteristics and ability to accommodate different substances within their porous structures [50].

10.2.2.2 Anticorrosion Applications of MSNs

In the field of self-healing corrosion protective coatings, MSNs are extensively utilized as carriers for inhibitors due to their beneficial properties, such as biocompatibility, high stability, large specific surface area (700–1000 m^2/g), high pore volume (0.6–1 cm^2/g), and the presence of organic functional groups. MSNs can effectively load various types of inhibitors, both organic and inorganic, for efficient anticorrosion applications. Examples of organic inhibitors include BTA, imidazole, 8-hydroxyquinoline, and sulfamethazine, while inorganic inhibitors encompass cerium, phosphate, zinc phosphate, molybdates, copper, and others [51].

Synthesized mesoporous silica particles were successfully used to encapsulate dodecylamine. Additionally, the incorporation of SBA-15 mesoporous silica loaded with dodecylamine into an alkyd primer has shown promising results. It is interesting to note that while the addition of mesoporous silica alone improved the barrier properties of the alkyd primer, the addition of silica-containing dodecylamine had a significant effect on their physical barrier performance. This enhancement effectively blocks defects and facilitates the release of the encapsulated inhibitor, resulting in improved active anticorrosion properties and self-healing capabilities. Techniques such as electrochemical impedance spectroscopy and SVET have confirmed these advancements [52, 53].

In a separate study conducted by Keyvani et al. [54], mesoporous silica nanocontainers were employed as intelligent hosts for molybdate corrosion inhibitors embedded in epoxy-coated steel. Compared with the coatings containing only mesoporous silica, epoxy coatings containing molybdate inhibitors demonstrated superior corrosion resistance. A significant aspect of this improvement is the controlled release of the corrosion inhibitor from the mesoporous silica resulting from the changes in pH within the corrosion zone causing the corrosion inhibitor to release. Additionally, the coating with inhibitor-loaded MSNs exhibited a more intact surface appearance after two months of immersion in chloride media, in contrast to the coating with MSNs alone. The corrosion inhibitor is released into the corrosion zone as a result of this, which further enhances the coating's protective capabilities [54].

Successful encapsulation of calcium phosphate into MSN particles has also been achieved by Lamprakou et al. [52] under pH 2 conditions. The release of calcium phosphate from the MSNs is pH-dependent, with an increase in pH leading to the degradation of the MSN structure. The barrier properties of the coating are enhanced by the inclusion of inhibitory pigment within the MSNs. As an example, MSN coatings loaded with BTA molecules have been designed and functionalized so that the molecules can be released under controlled conditions in acidic environments. A nanocarrier, loaded with BTA molecules, provides protection against photodegradation, and preserves the anticorrosion properties. According to the accelerated corrosion tests performed on coated copper-based alloy discs exposed to hydrochloric acid water vapor, the resulting polymer nanocomposite coatings have had reliable and long-term protective properties [52, 55].

10.2.2.3 Loading Corrosion Inhibitor into MSN

Scientists have conducted experiments in which corrosion inhibitors were introduced into the pores of mesoporous silica without modifying the silica itself. Several types of inhibitors have been loaded into mesoporous silica, including both organic and inorganic compounds. Among organic inhibitors BTA, 8-HQ, MBT, Rhodamine 6G, dodecylamine, polyvinylpyrrolidone, and 1-hydroxybenzotriazole can be mentioned. A variety of inorganic inhibitors have also been utilized, including $Mo_{12}Na_3O_{40}P$ and $Ce(NO_3)_3.6H_2O$, in addition to other substances, including quaternary ammonium salts and zinc gluconate. Among these inhibitors, organic molecules are most frequently used [45].

Typically, mesoporous silica suspension is mixed with an aqueous or organic inhibitor solution under reduced pressure. Vacuuming allows the inhibitor to penetrate the mesoporous cavities by removing oxygen and moisture from the pores. However, in certain instances, the vacuum step has been omitted and mesoporous silica has been mixed with corrosion inhibitor solution for a specific period. Using this approach may pose a risk of premature corrosion inhibitor leakage from the pores without requiring any external stimulation [45].

In the pursuit of synthesizing mesoporous silica nanocarriers, a group of researchers employed in-situ techniques in a one-step procedure. For example, Maia et al. utilized a method where mesoporous silica was prepared by introducing an Ammonia solution to the CTAB solution. In the following step, MBT has been added to the initial solution, and dissolved in ethyl ether. Finally, TEOS was incorporated while vigorously stirring the mixture, which was then placed in a sealed container for 24 hours [56]. This particular synthesis approach facilitated the diffusion of corrosion inhibitors throughout the entire pore structure of the mesoporous materials, offering a convenient and efficient method of synthesis. A single-stage polymerization process utilizing microemulsion oil-in-water microemulsions was employed to encapsulate the MBT within the mesoporous silica shells in the study mentioned above [56]. In spite of this, it is important to note that the presence of the inhibitor during the formation of the mesoporous structure could compromise its structural integrity [45].

10.2.2.4 Synthesis of MSNs

Mesoporous silicas have been synthesized with exceptional properties using the templating method since the first mesoporous silicas were discovered using cationic surfactants as templates. Surface areas of these materials are high, pore sizes are customizable, pore volumes are significant, and morphological characteristics are diverse. Surfactant-templated mesoporous silicas can be precisely tailored by carefully adjusting reaction conditions such as temperature, pH value, surfactant concentration, and silica source to exhibit a variety of mesostructures (such as disordered, wormhole-like, hexagonal, cubic, and lamellar mesophases), morphologies (such as spheres, hollow spheres, fibers, tubules, gyroids, helical fibers, crystals, and hierarchical structures), and dimensions ranging from nanometers to centimeters [57].

The interaction of surfactants and silica species has been extensively studied with a view to synthesizing specific mesoporous silicas. It has been established that the formation of mesostructured surfactant-silica nanocomposites arises from favorable interactions between the organic and inorganic components. In addition, the kinetics of sol-gel chemistry, such as reaction temperatures, water contents, and pH values, have a significant impact on the morphologies and dimensions of the resulting materials. By carefully controlling the self-assembly process and silica condensation rate, precise engineering of the sizes, mesostructures, and morphologies of mesoporous silicas becomes feasible.

Several key characteristics are desired in an ideal synthesis of MSNs in order to maximize their versatility. In addition, it is capable of maintaining a well-suspended stable solution, controlling pore size, assuring a uniform particle size distribution, and exhibiting a large pore volume. Two conditions must be met in order to synthesize mesoporous silicas. First, it is essential to ensure that MSNs grow uniformly from 30 to 300 nm in size through well-controlled nucleation and growth rates. Second, it is important to prevent the MSNs from adhering to each other during subsequent processing steps. A variety of strategies have been explored to produce nanosized mesoporous silicas during the early stages of synthesis development. A variety of cosolvents, bases, surfactants, and polymer protection strategies were employed in these strategies [58].

One notable modification to the Stöber synthesis of monodisperse silica spheres was introduced by Grün et al. [59]. They incorporated a cationic surfactant into the reaction mixture, leading to the formation of submicrometer-scaled MCM-41 particles [59]. Additionally, Cai et al.[9] discovered that maintaining a highly dilute surfactant condition was crucial in obtaining MCM-41 silicas with sizes around 100 nm. Another successful approach achieved mesostructured silica nanoparticles with sizes below 50 nm using a double surfactant system.

It is important to note that all the mentioned strategies played a significant role in tailoring the synthesis of mesoporous silicas and addressing the challenges associated with achieving desirable particle sizes and preventing agglomeration. These findings have paved the way for advancements in the synthesis techniques and the production of mesoporous silicas with enhanced properties and controlled characteristics. He et al. achieved the successful synthesis of uniform MSNs ranging from 40 to 600 nm by employing a phosphate buffer solution as the reaction medium and adjusting various parameters such as temperature, structure-directing agent, cosurfactant, and cosolvent [58].

During the synthesis process, it is very important to pay close attention to the processes of nucleation, growth, and aggregation caused by the silica condensation reaction. The goal is to achieve a well-controlled process that ensures a brief and uniform nucleation period while avoiding any form of aggregation. This is usually done using very dilute alkaline conditions. As a result of these conditions, a negatively charged and dense surface is created, which significantly reduces the chance of aggregation between particles [58]. Similarly, He et al. demonstrated the importance of precise parameter adjustments during the synthesis process. By carefully selecting the reaction medium and controlling various factors, they achieved uniform MSNs with sizes ranging from 40 to 600 nm, catering to specific application requirements [58].

The advancements highlighted by these studies have greatly contributed to the synthesis of MSNs for various fields, particularly in biomedicine. The ability to achieve better control over the particle size and suspension stability opens up new possibilities for tailored applications and improved performance in diverse areas of research and technology. With the rapid advancement of various applications, there is an increasing demand for MSNs with unique structures and performance characteristics. Nanomaterials of hollow and rattle types are gaining interest due to their potential applications in enzyme immobilization, confined-space catalysis, as well as acoustic, thermal, and electrical insulation. Consequently, the development of novel fabrication methods for hollow and rattle-type MSNs has become a prominent topic in the field of nanotechnology [60].

Traditionally, hollow-type MSNs are produced using a dual template method. A hard or soft template is used to construct a hollow interior, while a soft template induces mesopores in the shell by acting as a pore-forming agent. The pore templates are then coated with a silica matrix by the sol-gel process, and the templates can be removed by thermal calcination and/or solvent extraction.

A multi-step coating process is employed on the core particles when fabricating rattle-type particles to form a removable middle layer [42]. This method can be laborious, expensive, and challenging, particularly when targeting small-sized nanoparticles with complex structures. Consequently, researchers are actively exploring new methods that offer simplicity, controllability, and scalability in fabrication. It has been possible to synthesize different types of hollow and rattle-type nanoparticles using a variety of approaches, including soft template synthesis, selective etching, and self-template synthesis [60].

These innovative fabrication methods hold great promise for tailoring the structures and properties of MSNs, enabling their application in various fields. The continuous exploration and refinement of these methods will undoubtedly contribute to the advancement of nanotechnology and pave the way for novel applications of MSNs.

10.2.3 Zeolites

Zeolites are a class of minerals that can be found naturally or created through artificial synthesis. They possess a distinctive crystalline structure consisting of aluminosilicate compounds, which form a 3D porous framework. These porous structures are the key to the exceptional properties exhibited by zeolites, making them immensely valuable across a wide range of applications.

10.2.3.1 Structure and Morphology

Zeolites are crystalline aluminosilicate minerals renowned for their microporous structure. These minerals consist of SiO_4 and AlO_4 tetrahedra as their fundamental building blocks, with silicon or aluminum atoms at the centers and oxygen atoms at the vertices, linking adjacent tetrahedra. These building blocks form a 3D structure with interconnected channels and cavities, resulting in a crystal with a substantial specific surface area [61].

Zeolites find a wide range of applications as adsorbents, catalysts, and ion exchangers. As a nontoxic material, they are even used in the food and pharmaceutical industries. Zeolite coatings can be grown as continuous films on diverse supports, including organic materials like plastics and wood, as well as inorganic materials such as ceramics, metals, and glass. The adhesion of zeolite crystals to the surface is primarily attributed to electrostatic interactions rather than covalent bonding [61].

Within the interconnected network of zeolites, oxygen atoms occupy the corners, while silicon atoms or aluminum atoms are located in the center. In general, zeolites can be categorized as either synthetic or natural. In the production of synthetic zeolites, rapid formation processes take place involving specific combinations of substances under alkaline and hydrothermal conditions (100 to 200°C). However, natural zeolites are formed from volcanic ash and sea salt over millions of years, and they possess larger particle sizes compared to their synthetic counterparts [62].

Zeolites consist of two integral atoms: silicon and aluminum. Furthermore, zeolites can be classified based on their structural adaptations. Synthetic zeolites possess approximately 200 types, while natural zeolites have 10 different types. Depending on the structure type, one or more specific zeolites may be incorporated, each with a special physical and chemical composition [62].

10.2.3.2 Zeolites as Corrosion Inhibitor Nanocarriers

Coatings based on zeolites can provide both barrier and active protection, thereby enhancing their anticorrosion properties. In the paint industry, chromate inhibitors are used widely to provide a high level of durability and self-healing capabilities. However, in an effort to offer a greener and more sustainable alternative that eliminates the need for chromates, the development of nanoporous and mesoporous ion-exchange fillers can be utilized to provide nanocarriers for inhibitors. Zeolites, as ion-exchange materials, can be employed to create active pigments by substituting sodium cations with self-healing properties [63].

An inhibitor agent can be doped within the zeolite filler, which can inhibit corrosion when released during corrosion. Zeolites possess a negatively charged surface, which is balanced by loosely bonded cations within their framework. Cationic species, such as Ce^{3+} and La^{3+} ions, can be trapped within zeolite particle cages via cation exchange. By releasing cationic species (Mn^{+} and H^{+}) and altering pH levels, these containers release the entrapped inhibitors. The substrate can be protected for a long time because the inhibiting ions are selectively transferred to the damaged site, strengthening the oxide layer. Zeolite containers function as active capsules. When the coating is damaged locally, the active capsule releases the inhibitor stored in its structure, thus initiating the healing process. This imparts an additional functional property to the coating, wherein the inhibiting action can effectively make the coating self-healing. For instance, in chromate coatings, when a scratch occurs, the active surface undergoes spontaneous repassivation. To achieve similar healing capabilities, a highly effective, active, environmentally friendly, and safe pigment must be incorporated to provide the same inhibitory capabilities as chromates. As a result, ion exchange agents are effective and suitable compounds that can be incorporated into zeolite structures, providing inhibitory activity [64].

The research conducted by Zhang and colleagues [65] focused on the successful loading of Ce^{3+} ions onto zeolite microparticles (NaX) with a high capacity. The zeolite used had a Si/Al atomic ratio of 1.3%. In the presence of surrounding cations, such as Mg^{2+} and Na^{+}, the release of Ce ions was controlled. Among these cations, Na^{+} was more effective in stimulating the release of Ce compared to Mg^{2+}. Surprisingly, the amount of Ce release did not show significant variation with changes in the pH of the solution. The researchers also discovered that an epoxy resin doped with CeNaX and coated on a AZ31 surface treated with plasma electrolytic oxidation exhibited excellent self-healing ability against corrosion [65].

10.2.3.3 Loading Corrosion Inhibitor into Zeolites

Loading involves impregnating the zeolites with the corrosion inhibitor. A variety of techniques are available for achieving this, including vacuum impregnation, incipient wetness impregnation, and solvent evaporation. The goal is to ensure maximum uptake of the inhibitor while safeguarding the structural characteristics of the zeolites.

10.2.3.4 Synthesis of Zeolites

Zeolite-based films can be synthesized by two different methods: ex situ or in situ crystallization. With the in situ method, the zeolite film is directly grown on the substrate under hydrothermal conditions from a clear synthesis solution. This approach enables the formation of a fully covered and well-oriented zeolite crystal layer. The in situ crystallization occurs directly at the interface between the metal substrate and the liquid, making it suitable for coating complex geometries and narrow or confined spaces. There are various in situ crystallization techniques that can be used, such as hydrothermal, ionothermal, and dry-gel conversion [66].

On the other hand, the ex situ synthesis method involves immersing a support material in a suspension containing pre-formed zeolite crystals together with a solvent and binder. The substrate is then brought into contact with the zeolite seeds using techniques like dip, spin, or wash coatings. This approach is less time-consuming than the in situ method, does not require extensive pretreatment, and provides control over eventual final film thickness by adjusting the number of coats applied [66].

10.2.4 Covalent Organic Frameworks

Covalent organic frameworks (COFs) possess distinct properties due to their well-defined internal order, which sets them apart from conventional covalent polymers lacking such structural regularity.

10.2.4.1 Structure and Morphology

The COF is a special type of covalent organic polymer that is porous and crystalline. They consist of organic molecules that are connected in a regular pattern, forming well-defined coordination networks and adjustable pores. These structures possess unique properties such as high crystalline structure, low density, a porous matrix, and a large specific surface area. COFs can be precisely tailored at the molecular and atomic levels, allowing for versatile functions and active sites. This makes them highly promising for various biomedical applications including drug delivery, diagnostic imaging, and disease therapy. The performance of COFs in these applications is influenced by factors such as the composition, dimensions, and loading of guest molecules [67].

In 2005, pure 2D organic frameworks known as COF-1 and COF-5 were synthesized. These frameworks were formed by connecting organic building blocks by forming strong covalent bonds. Since then, a variety of COFs have been developed, including those that are hydrazone-linked, imine-linked, and keto-enol-linked. On the basis of their structural properties and building units, COFs can be classified into three categories: imine-based COFs, boron-containing COFs, and triazine-based COFs. Further classification of boron-containing COFs can be made according to the process of their formation, which can involve the co-condensation of multiple units or the self-condensation of single units. Brunauer-Emmett-Teller measurements reveal that boron-containing COFs have high surface areas, low densities, and excellent thermal stability [68, 69].

10.2.4.2 Synthesis

A COF is synthesized through reversible covalent bond formation, such as condensation to generate boroxines or imines. The reversibility of these reactions enables the formation of crystalline COF structures. However, structural order is not solely determined by reversibility. To achieve regular pore structures in COF materials, additional factors such as structural rigidity, symmetric multi-connectivity, and the building unit's geometry and angle play a crucial role in the reversible formation of covalent bonds [70].

Following the pioneering work of Yaghi and his colleagues on the solvothermal synthesis of COF powders, numerous research groups have explored various strategies to expand the possibilities of COF synthesis. Different methods have been employed, including solvothermal, ionothermal, microwave, and room-temperature methods. Among these approaches, solvothermal synthesis is the most commonly used. Analogous to the autoclave synthesis of inorganic zeolites, solvothermal synthesis of COF powders typically involves heating in a sealed container for a period ranging from two to nine days [69].

10.2.4.3 Application of COFs as Carriers of Corrosion Inhibitors

2D covalent organic frameworks (2D-COFs) have garnered considerable attention in current research due to their remarkable chemical stability, large specific surface area, and nanoporous structure. Unlike other 2D nanomaterials, 2D-COFs typically have weaker interlayer interactions, making them more compatible and dispersible in organic anticorrosion coatings. One specific example is the TpPa-1 monomer (Tp: 1,3,5-Triformylphloroglucinol, Pa-1: 1,4-phenylenediamine) used for synthesizing the TpPa-1 2D-COF. This monomer contains amino groups that can crosslink with the epoxy bonds in the epoxy resin during the curing process. This property allows the TpPa-1

2D-COF to disperse well in the epoxy coatings, reducing defects and enhancing their physical shielding effect [71].

Additionally, COFs with pore diameters of approximately 2 nm are ideal for loading inhibitors into nanocarriers due to their nanoporous structure. It has been shown that surface modification with COFs improves compatibility and dispersion, thereby improving the filling role of COF-grown MoS2 nanosheets in epoxy coatings. This modification has resulted in a significant improvement of approximately 50% in adhesion strength, as demonstrated by Najmi et al. [72].

10.2.5 Activated Carbon

Activated carbon is a type of amorphous carbon material known for its finely crafted and intricate porous structure. It has a long history and is widely used as an absorbent material in various applications. One of its primary functions is the removal of heavy metals and toxic gases. Activated carbon is also employed in air and water filtration processes, where it effectively removes impurities. Additionally, it finds utility in electrochemical systems and energy storage devices, particularly as a component in supercapacitors. Its unique porous structure allows for high surface area and adsorption capacity, making it a versatile material for a range of practical applications [73]. An amorphous or disordered arrangement of carbon atoms characterizes activated carbon. Carbon atoms form a 3D network that creates micropores, mesopores, and macropores. These pores contribute to the extensive internal surface area of activated carbon, which can range from hundreds to thousands of square meters per gram. It is this remarkable surface area that gives activated carbon its exceptional adsorption capacity [74, 75].

10.2.5.1 Synthesis

Activated carbons are synthesized using two main methods: physical activation and chemical activation. In physical activation, the process involves two distinct thermal steps. As a first step, the carbonaceous precursor is carbonized or pyrolyzed at extreme temperatures, typically 700–900 °C, in an inert atmosphere in order to avoid combustion. During this step, heteroatoms are removed, and volatiles are released, resulting in char containing a high proportion of carbon (the higher the carbonization temperature, the higher the carbon content), but with limited pore development. Gasification is the second step in the physical activation procedure. This step selectively removes the most reactive carbon atoms through controlled gasification reactions, leading to the creation of the characteristic porosity in activated carbons. The temperature at which gasification occurs differs depending on the gasification agent employed, which can include water vapor, CO_2, or O_2 (air) [76].

Chemical activation is another method used to synthesize the activated carbons, involving a single thermal step that consists of three distinct stages. In the first stage, the carbonaceous precursor is impregnated with the activating agent using various techniques. Some researchers use an aqueous solution of the activating agent, which is brought into contact with the carbonaceous precursor. Precursors and activating agents can also be formulated directly through a simplified and direct physical blend. The impregnation ratio or ratio of activating agents to carbonaceous precursors is one of the most important synthesis parameters that affect the textural characteristics of activated carbon. Chemical activation proceeds by subjecting the materials to the inert atmospheric thermal treatment after impregnation. The specific temperature depends on the chosen activating agent. During this stage, depolymerization, dehydration, and condensation reactions occur; thus, higher carbon yields are achieved compared to physical activation. This is due to the limitations on tar and volatile formation [76].

The final stage of the chemical activation process involves washing the materials to remove residual byproducts that could obstruct the newly formed porosity. This is followed by drying. In cases where "soft" carbon precursors, such as biomass waste, are subjected to NaOH or KOH activation, an initial carbonization step is typically necessary to prevent the organic matter of the precursor from dissolving. This step ensures that subsequent activation can be achieved [77].

10.2.6 Mineral Clays

Clay minerals, such as montmorillonite (MMT) and bentonite, have gained significant attention due to their excellent stability, cost-effectiveness, properties of cation exchange, as well as their specific surface area. Among these minerals, MMT stands out as an environmentally friendly mineral with a 2D and 3D interlayer. The hydrated cations present in the MMT structure, such as Na^+ and Ca^{2+}, can be exchanged with other compounds, including organic and inorganic corrosion inhibitors. As a result, MMT clay can act as a container of inhibitors, allowing for rapid and long-term release performance [78, 79].

Numerous studies have investigated the use of MMT-based corrosion inhibitor nanocarriers. For example, Ghazi et al. synthesized benzimidazole intercalated MMT nanoclays, which were used to provide active protection in the epoxy coatings [80].

It is important to note that a typical clay mineral like MMT has a structure composed of two tetrahedral sheets and one octahedral sheet arranged in a sandwich-like configuration. Layered materials like smectic clays, including Na^+-montmorillonite (Na^+-MMT), have gained popularity in the preparation of polymer-clay nanocomposites in recent decades. The crystal structure of MMT comprises three sublayers: two tetrahedral silica sheets bonded to an edge-shared octahedral sheet of alumina. The interlayer space in MMT can accommodate hydrated cations like Na+, effectively neutralizing negative charges [79].

As noted previously, MMT is a type of mineral that consists of stacked layers. The thickness of each layer is approximately 1 nm. Each layer consists of an O-Al(Mg)-O octahedral sheet sandwiched between two O-Si-O tetrahedral sheets. Due to isomorphous substitution, the layer carries a positive charge, resulting in the presence of cations in the interlayer space of MMT. The primary particles of clay are formed by the neighboring layers being held together by van der Waals and electrostatic forces.

Adding MMT increases the diffusion pathway length for water, oxygen, and aggressive ions. By exchanging cationic corrosion inhibitors with cations of MMT, corrosion can be effectively suppressed. For example, an epoxy nanocomposite coating was prepared by intercalating cerium nitrate into Na-MMT and a low corrosion rate was achieved on the carbon steel substrates and a corrosion rate reduction was observed in the area of the scribe.

10.2.7 Other Porous Nano/Micro-Carriers

10.2.7.1 Diatomaceous Earth

Diatomaceous earth (DE), or diatomite, is a naturally occurring mineral that has been extracted from nearly 30 countries across the globe for quite some time. Diatomite possesses distinctive properties, including high porosity, lightweight nature, substantial surface area, small particle size, chemical stability, and poor thermal conductivity, thereby opening up a multitude of applications. Its visual appearance commonly resembles that of a pale-colored powder, with a reported particle size range of approximately 10–200 mm. Regardless of its origin, diatomite predominantly consists of SiO_2 as the main phase, accompanied by other phases such as CaO, MgO, Fe_2O_3, and Al_2O_3. Notably, there has been a recent surge in literature dedicated to DE [81].

Diatomite can be classified into three primary groups related to energy-related studies. These groups are based on the specific ways diatomite is utilized in experimental investigations. The first group involves the synthesis of materials using raw DE. The second group utilizes DE frustules as sacrificial materials or templates. In the third group, DE is used as a precursor for silicon-based materials [81].

DE can be used as a precursor to creating novel functional materials by preserving and modifying its nanoscale morphology. In some studies, raw diatomite is used as a substrate material to host active compounds, particularly in thermal energy storage applications. Diatomite also finds applications in fireproof cement, and insulation materials, and serves as an absorbent in the manufacturing of explosives due to its exceptional resistance to heat and chemical actions [82].

2,5-dimercaptothiadiazole and cerium nitrate have been loaded onto DE particles. These loaded particles were then incorporated into epoxy-amine coatings to evaluate their performance in terms of corrosion inhibition. An in situ real-time optical electrochemical setup and local electrochemistry were used to test the coatings under continuous immersion and cyclic (wet/dry) conditions. The results showed that DE, which is a naturally occurring nanoporous and hollow silica microparticle, can effectively serve as a container for corrosion inhibitors like cerium nitrate ($Ce(NO_3)_3$). Cerium nitrate is known for its ability to inhibit corrosion in Cu-rich aluminum alloys. When the loaded DE particles were combined with epoxy-amine, a corrosion-resistant coating with significant inhibitory properties was obtained. Importantly, the occurrence of unwanted side reactions between the cerium inhibitor and the matrix was reduced compared to the direct addition of cerium salts to epoxy/amine coatings. This facilitated the diffusion of an adequate amount of Ce^{3+} ions to the metallic substrate, especially at scribe locations, providing protection to the metal in the presence of corrosive electrolytes [83].

10.2.7.2 Inorganic Oxides

Inorganic oxides are widely used as carriers for corrosion inhibitors due to their beneficial properties. Zinc oxide (ZnO), aluminum oxide (Al_2O_3), cerium oxide (CeO_2), titanium dioxide (TiO_2), and silica (SiO_2) are notable examples of these oxides.

Zinc oxide is valued for its barrier properties and its ability to release inhibiting ions. On metal surfaces, it forms a protective layer that prevents corrosive agents from penetrating. Aluminum oxide is known for its chemical stability and resistance to corrosion. As a carrier, it provides a protective barrier that shields the metal substrate from aggressive substances.

The self-healing properties of cerium oxide make it particularly advantageous. Repairing cracks and defects in the oxide layer maintains the protective barrier and extends the metal's life. On the other hand, titanium dioxide forms a dense and stable oxide layer on metal surfaces, protecting them from corrosion-causing environmental factors.

Silica, a commonly used inorganic oxide, is chemically inert and capable of forming a stable barrier layer. It is often incorporated into the coatings or applied as a surface treatment to enhance corrosion resistance.

These inorganic oxides exhibit versatility as carriers for corrosion inhibitors, offering different protective mechanisms and compatibility with various metal substrates. They play a crucial role in preventing or slowing down the corrosion process and find applications in industries such as aerospace, automotive, marine, and infrastructure [84].

10.2.7.3 Hydrogels as Carriers for Corrosion Inhibitors

Hydrogels are 3D networks of hydrophilic polymers known for their exceptional water-absorbing capacity and ability to retain aqueous solutions. These unique properties make them highly suitable for various applications, including serving as carriers for corrosion inhibitors in smart coatings. In a 2015, a study conducted wherein chitosan-based hydrogels were used with silver nanoparticles, which possess antimicrobial properties. The research demonstrated the effective adsorption of silver nanoparticles within the chitosan hydrogel and their subsequent controlled release. This study revealed the promising potential of hydrogels as carriers for active agents, including corrosion inhibitors. Using hydrogels as carriers for corrosion inhibitors offers several advantages. Their high water-absorbing capacity enables efficient loading and retaining of the inhibitors. Additionally, hydrogels provide controlled release capabilities, allowing for prolonged and targeted corrosion protection. Their compatibility with aqueous environments and ability to adhere to various substrates make them well-suited for coating applications. The integration of hydrogels as carrier systems in corrosion protection strategies opens up new possibilities for the development of smart coatings that can actively respond to corrosion conditions. Ongoing research and advancements in this field are expected to enhance the effectiveness and practicality of hydrogel-based corrosion inhibitor delivery systems [85].

10.3 CONCLUSION

Encapsulating corrosion inhibitors within the micro/nanocarriers for controlled release has demonstrated enhanced durability and efficiency in anticorrosion coatings. Various porous micro/nanocarrier materials, such as MSNs, MOFs, and Zeolites, have been employed in combination with organic/inorganic inhibitors to investigate their corrosion behavior and release mechanisms in coatings.

These carriers play a pivotal role in protecting metals against different types of corrosion. In this chapter, numerous carriers have been introduced, and their loading and release mechanisms of inhibitors have been studied. The anticorrosion behavior of each system is closely associated with the control of inhibitor release rate and the provision of active-passive corrosion protection. However, several aspects remain unexplored and present challenges in the field.

Designing carriers with optimal properties, ensuring compatibility with the coating matrix, addressing stability issues, and improving loading efficiency are among the unresolved challenges. To achieve future advancements, it is essential to address these challenges and explore new pathways for the development of highly effective corrosion protection systems.

By addressing these existing challenges and exploring new approaches, the field of micro/nanocarriers for controlled release inhibitors can contribute significantly to the advancement of corrosion protection technologies.

REFERENCES

(1) K.P. Balan. Chapter nine – Corrosion. In *Metallurgical Failure Analysis*, Balan, K. P. Ed.; Elsevier, 2018; pp. 155–178.
(2) S. Harsimran, S. Kumar, K. Rakesh, Overview of corrosion and its control: A critical review, *Proceedings on Engineering Sciences. 3*(2021) 13–24. DOI: 10.24874/PES03.01.002.
(3) S. Zehra, M. Mobin, R. Aslam. Chapter 2 – Corrosion prevention and protection methods. In *Eco-Friendly Corrosion Inhibitors*, Guo, L., Verma, C., Zhang, D. Eds.; Elsevier, 2022; pp. 13–26.
(4) S.J. Kulkarni, A review on studies and research on corrosion and its prevention, *International Journal of Research & Review. 2*(2015) 5.
(5) G. Cui, Z. Bi, S. Wang, J. Liu, X. Xing, Z. Li, B. Wang, A comprehensive review on smart anti-corrosive coatings, *Progress in Organic Coatings. 148*(2020) 105821. DOI: https://doi.org/10.1016/j.porgcoat.2020.105821
(6) P.B. Raja, M.A. Assad, M. Ismail. Chapter 6 – Inhibitor-encapsulated smart nanocontainers for the controlled release of corrosion inhibitors. In *Corrosion Protection at the Nanoscale*, Rajendran, S., Nguyen, T. A. N. H., Kakooei, S., Yeganeh, M., Li, Y. Eds.; Elsevier, 2020; pp. 91–105.
(7) H. Wei, Y. Wang, J. Guo, N.Z. Shen, D. Jiang, X. Zhang, X. Yan, J. Zhu, Q. Wang, L. Shao, et al., Advanced micro/nanocapsules for self-healing smart anticorrosion coatings, *Journal of Materials Chemistry A. 3*(2015) 469–480. DOI: 10.1039/C4TA04791E
(8) T. Liu, L. Ma, X. Wang, J. Wang, H. Qian, D. Zhang, X. Li, Self-healing corrosion protective coatings based on micro/nanocarriers: A review, *Corrosion Communications. 1*(2021) 18–25. DOI: https://doi.org/10.1016/j.corcom.2021.05.004
(9) H. Cai, P. Wang, D. Zhang, Smart anticorrosion coating based on stimuli-responsive micro/nanocontainer: A review, *Journal of Oceanology and Limnology. 38*(2020) 1045–1063. DOI: 10.1007/s00343-020-0058-x

(10) R. Rodriguez, D.G. Bekas, S. Flórez, M. Kosarli, A.S. Paipetis, Development of self-contained microcapsules for optimised catalyst position in self-healing materials, *Polymer. 187*(2020) 122084. DOI: https://doi.org/10.1016/j.polymer.2019.122084
(11) H. Pulikkalparambil, S. Siengchin, J. Parameswaranpillai, Corrosion protective self-healing epoxy resin coatings based on inhibitor and polymeric healing agents encapsulated in organic and inorganic micro and nanocontainers, *Nano-Structures & Nano-Objects. 16*(2018) 381–395. DOI: https://doi.org/10.1016/j.nanoso.2018.09.010
(12) J.-B. Xu, Y.-Q. Cao, L. Fang, J.-M. Hu, A one-step preparation of inhibitor-loaded silica nanocontainers for self-healing coatings, *Corrosion Science. 140*(2018) 349–362. DOI: https://doi.org/10.1016/j.corsci.2018.05.030
(13) S.H. Sonawane, B.A. Bhanvase, A.A. Jamali, S.K. Dubey, S.S. Kale, D.V. Pinjari, R.D. Kulkarni, P.R. Gogate, A.B. Pandit, Improved active anticorrosion coatings using layer-by-layer assembled ZnO nanocontainers with benzotriazole, *Chemical Engineering Journal. 189–190*(2012) 464–472. DOI: https://doi.org/10.1016/j.cej.2012.02.076
(14) D.V. Andreeva, D. Fix, H. Möhwald, D.G. Shchukin, Self-healing anticorrosion coatings based on pH-sensitive polyelectrolyte/inhibitor sandwichlike nanostructures, *Advanced Materials. 20*(2008) 2789–2794.
(15) E. Shchukina, H. Wang, D.G. Shchukin, Nanocontainer-based self-healing coatings: current progress and future perspectives, *Chemical Communications. 55*(2019) 3859–3867. DOI: 10.1039/C8CC09982K
(16) Y. He, C. Zhang, F. Wu, Z. Xu, Fabrication study of a new anticorrosion coating based on supramolecular nanocontainer, *Synthetic Metals. 212*(2016) 186–194. DOI: https://doi.org/10.1016/j.synthmet.2015.10.022
(17) R. Subasri, S.H. Adsul, S. Manasa, Smart nanocontainers for anticorrosion applications, *Smart Nanocontainers. 1*(2020) 399–412.
(18) K.A. Zahidah, S. Kakooei, M.C. Ismail, P.B. Raja, Halloysite nanotubes as nanocontainer for smart coating application: A review, *Progress in Organic Coatings. 111*(2017) 175–185.
(19) S. Manasa, A. Jyothirmayi, T. Siva, S. Sathiyanarayanan, K.V. Gobi, R. Subasri, Effect of inhibitor loading into nanocontainer additives of self-healing corrosion protection coatings on aluminum alloy A356.0, *Journal of Alloys and Compounds. 726*(2017) 969–977. DOI: https://doi.org/10.1016/j.jallcom.2017.08.037
(20) R. Patra, A. Gautam, K. Gobi, R. Subasri, Hybrid silane coatings based on benzotriazole loaded aluminosilicate nanotubes for corrosion protection of mild steel, *Silicon. 15*(2023) 1–16.
(21) C. Jing, B. Dong, A. Raza, T. Zhang, Y. Zhang, Corrosion inhibition of layered double hydroxides for metal-based systems, *Nano Materials Science. 3*(2021) 47–67. DOI: https://doi.org/10.1016/j.nanoms.2020.12.001
(22) E. Alibakhshi, E. Ghasemi, M. Mahdavian, B. Ramezanzadeh, S. Farashi, Active corrosion protection of Mg-Al-PO43– LDH nanoparticle in silane primer coated with epoxy on mild steel, *Journal of the Taiwan Institute of Chemical Engineers. 75*(2017) 248–262. DOI: https://doi.org/10.1016/j.jtice.2017.03.010
(23) T.T.X. Hang, T.A. Truc, N.T. Duong, N. Pébère, M.-G. Olivier, Layered double hydroxides as containers of inhibitors in organic coatings for corrosion protection of carbon steel, *Progress in Organic Coatings. 74*(2012) 343–348. DOI: https://doi.org/10.1016/j.porgcoat.2011.10.020
(24) M.L. Zheludkevich, S.K. Poznyak, L.M. Rodrigues, D. Raps, T. Hack, L.F. Dick, T. Nunes, M.G.S. Ferreira, Active protection coatings with layered double hydroxide nanocontainers of corrosion inhibitor, *Corrosion Science. 52*(2010) 602–611. DOI: https://doi.org/10.1016/j.corsci.2009.10.020
(25) M. Tabish, G. Yasin, M.J. Anjum, M.U. Malik, J. Zhao, Q. Yang, S. Manzoor, H. Murtaza, W.Q. Khan, Reviewing the current status of layered double hydroxide-based smart nanocontainers for corrosion inhibiting applications, *Journal of Materials Research and Technology. 10*(2021) 390–421. DOI: https://doi.org/10.1016/j.jmrt.2020.12.025
(26) H. Boumeriame, E.S. Da Silva, A.S. Cherevan, T. Chafik, J.L. Faria, D. Eder, Layered double hydroxide (LDH)-based materials: A mini-review on strategies to improve the performance for photocatalytic water splitting, *Journal of Energy Chemistry. 64*(2022) 406–431. DOI: https://doi.org/10.1016/j.jechem.2021.04.050

(27) L. Rassouli, R. Naderi, M. Mahdavain, The role of micro/nano zeolites doped with zinc cations in the active protection of epoxy ester coating, *Applied Surface Science. 423*(2017) 571–583. DOI: https://doi.org/10.1016/j.apsusc.2017.06.245

(28) L. Jiang, Y. Dong, Y. Yuan, X. Zhou, Y. Liu, X. Meng, Recent advances of metal–organic frameworks in corrosion protection: From synthesis to applications, *Chemical Engineering Journal. 430*(2022) 132823. DOI: https://doi.org/10.1016/j.cej.2021.132823

(29) F. Seidi, M. Jouyandeh, M. Taghizadeh, A. Taghizadeh, H. Vahabi, S. Habibzadeh, K. Formela, M.R. Saeb, Metal-organic framework (MOF)/epoxy coatings: A review, *Materials. 13*(2020) 2881.

(30) M. Ramezanzadeh, B. Ramezanzadeh, M. Mahdavian, G. Bahlakeh, Development of metal-organic framework (MOF) decorated graphene oxide nanoplatforms for anti-corrosion epoxy coatings, *Carbon. 161*(2020) 231–251. DOI: https://doi.org/10.1016/j.carbon.2020.01.082

(31) S. Rauf, M.T. Vijjapu, M.A. Andrés, I. Gascón, O. Roubeau, M. Eddaoudi, K.N. Salama, Highly selective metal–organic framework textile humidity sensor, *ACS Applied Materials & Interfaces. 12*(2020) 29999–30006.

(32) H. Tian, W. Li, A. Liu, X. Gao, P. Han, R. Ding, C. Yang, D. Wang, Controlled delivery of multi-substituted triazole by metal-organic framework for efficient inhibition of mild steel corrosion in neutral chloride solution, *Corrosion Science. 131*(2018) 1–16. DOI: https://doi.org/10.1016/j.corsci.2017.11.010

(33) Y. Guo, J. Wang, D. Zhang, T. Qi, G.L. Li, pH-responsive self-healing anticorrosion coatings based on benzotriazole-containing zeolitic imidazole framework, *Colloids and Surfaces A: Physicochemical and Engineering Aspects. 561*(2019) 1–8. DOI: https://doi.org/10.1016/j.colsurfa.2018.10.044

(34) K. Cao, Z. Yu, D. Yin, Preparation of Ce-MOF@TEOS to enhance the anti-corrosion properties of epoxy coatings, *Progress in Organic Coatings. 135*(2019) 613–621. DOI: https://doi.org/10.1016/j.porgcoat.2019.06.015

(35) Z. Mohammadpour, H.R. Zare, Fabrication of a pH-sensitive epoxy nanocomposite coating based on a Zn-BTC metal–organic framework containing benzotriazole as a smart corrosion inhibitor, *Crystal Growth & Design. 21*(2021) 3954–3966. DOI: 10.1021/acs.cgd.1c00284

(36) C. Yang, W. Xu, X. Meng, X. Shi, L. Shao, X. Zeng, Z. Yang, S. Li, Y. Liu, X. Xia, A pH-responsive hydrophilic controlled release system based on ZIF-8 for self-healing anticorrosion application, *Chemical Engineering Journal. 415*(2021) 128985. DOI: https://doi.org/10.1016/j.cej.2021.128985

(37) B. Ren, Y. Chen, Y. Li, W. Li, S. Gao, H. Li, R. Cao, Rational design of metallic anti-corrosion coatings based on zinc gluconate@ZIF-8, *Chemical Engineering Journal. 384*(2020) 123389. DOI: https://doi.org/10.1016/j.cej.2019.123389

(38) C. Zhou, Z. Li, J. Li, T. Yuan, B. Chen, X. Ma, D. Jiang, X. Luo, D. Chen, Y. Liu, Epoxy composite coating with excellent anticorrosion and self-healing performances based on multifunctional zeolitic imidazolate framework derived nanocontainers, *Chemical Engineering Journal. 385*(2020) 123835. DOI: https://doi.org/10.1016/j.cej.2019.123835

(39) C. Dey, T. Kundu, B.P. Biswal, A. Mallick, R. Banerjee, Crystalline metal-organic frameworks (MOFs): Synthesis, structure and function, *Acta Crystallographica Section B: Structural Science, Crystal Engineering and Materials. 70*(2014) 3–10.

(40) K. Kermannezhad, A. Chermahini, M. Momeni, B. Rezaei, Application of amine-functionalized MCM-41 as pH-sensitive nano container for controlled release of 2-mercaptobenzoxazole corrosion inhibitor, *Chemical Engineering Journal. 306*(2016). DOI: 10.1016/j.cej.2016.08.004

(41) T.-W. Kim, P.-W. Chung, V.S.Y. Lin, Facile synthesis of monodisperse spherical MCM-48 mesoporous silica nanoparticles with controlled particle size, *Chemistry of Materials. 22*(2010) 5093–5104. DOI: 10.1021/cm1017344.

(42) M. Amini, R. Naderi, M. Mahdavian, A. Badiei, Effect of piperazine functionalization of mesoporous silica type SBA-15 on the loading efficiency of 2-mercaptobenzothiazole corrosion inhibitor, *Industrial & Engineering Chemistry Research. 59*(2020) 3394–3404. DOI: 10.1021/acs.iecr.9b05261

(43) I.I. Slowing, J.L. Vivero-Escoto, B.G. Trewyn, V.S.Y. Lin, Mesoporous silica nanoparticles: Structural design and applications, *Journal of Materials Chemistry. 20*(2010) 7924–7937. DOI: 10.1039/C0JM00554A

(44) A. Mehmood, H. Ghafar, S. Yaqoob, D. Gohar, B. Ahmad, Mesoporous silica nanoparticles: A review, *Journal of Developing Drugs. 06*(2017). DOI: 10.4172/2329-6631.1000174

(45) M. Yeganeh, M. Omidi, S.H.H. Mortazavi, A. Etemad, M.H. Nazari, S.M. Marashi. Chapter 15 – Application of mesoporous silica as the nanocontainer of corrosion inhibitor. In *Corrosion Protection at the Nanoscale*, Rajendran, S., Nguyen, T. A. N. H., Kakooei, S., Yeganeh, M., Li, Y. Eds.; Elsevier, 2020; pp. 275–294.

(46) S. Wang, H. Li, Structure directed reversible adsorption of organic dye on mesoporous silica in aqueous solution, *Microporous and Mesoporous Materials. 97*(2006) 21–26. DOI: 10.1016/j.micromeso.2006.08.005

(47) T. Heikkilä, J. Salonen, J. Tuura, M. Hamdy, G. Mul, N.a. Kumar, T. Salmi, D.Y. Murzin, L. Laitinen, A.M. Kaukonen, Mesoporous silica material TUD-1 as a drug delivery system, *International Journal of Pharmaceutics. 331*(2007) 133–138.

(48) A.B.D. Nandiyanto, S.-G. Kim, F. Iskandar, K. Okuyama, Synthesis of spherical mesoporous silica nanoparticles with nanometer-size controllable pores and outer diameters, *Microporous and Mesoporous Materials. 120*(2009) 447–453.

(49) C. Boissiere, M. Kümmel, M. Persin, A. Larbot, E. Prouzet, Spherical MSU-1 mesoporous silica particles tuned for HPLC, *Advanced Functional Materials. 11*(2001) 129–135.

(50) F. Olivieri, R. Castaldo, M. Cocca, G. Gentile, M. Lavorgna, Mesoporous silica nanoparticles as carriers of active agents for smart anticorrosive organic coatings: A critical review, *Nanoscale. 13*(2021) 9091–9111. DOI: 10.1039/D1NR01899J

(51) M. Yeganeh, N. Asadi, M. Omidi, M. Mahdavian, An investigation on the corrosion behavior of the epoxy coating embedded with mesoporous silica nanocontainer loaded by sulfamethazine inhibitor, *Progress in Organic Coatings. 128*(2019) 75–81. DOI: https://doi.org/10.1016/j.porgcoat.2018.12.022

(52) Z. Lamprakou, H. Bi, C.E. Weinell, S. Tortajada, K. Dam-Johansen, Smart epoxy coating with mesoporous silica nanoparticles loaded with calcium phosphate for corrosion protection, *Progress in Organic Coatings. 165*(2022) 106740. DOI: https://doi.org/10.1016/j.porgcoat.2022.106740

(53) J.M. Falcón, L.M. Otubo, I.V. Aoki, Highly ordered mesoporous silica loaded with dodecylamine for smart anticorrosion coatings, *Surface and Coatings Technology. 303*(2016) 319–329. DOI: https://doi.org/10.1016/j.surfcoat.2015.11.029

(54) A. Keyvani, M. Yeganeh, H. Rezaeyan, Application of mesoporous silica nanocontainers as an intelligent host of molybdate corrosion inhibitor embedded in the epoxy coated steel, *Progress in Natural Science: Materials International. 27*(2017) 261–267. DOI: https://doi.org/10.1016/j.pnsc.2017.02.005

(55) R. Castaldo, M.S. de Luna, C. Siviello, G. Gentile, M. Lavorgna, E. Amendola, M. Cocca, On the acid-responsive release of benzotriazole from engineered mesoporous silica nanoparticles for corrosion protection of metal surfaces, *Journal of Cultural Heritage. 44*(2020) 317–324. DOI: https://doi.org/10.1016/j.culher.2020.01.016

(56) F. Maia, J. Tedim, A.D. Lisenkov, A.N. Salak, M.L. Zheludkevich, M.G. Ferreira, Silica nanocontainers for active corrosion protection, *Nanoscale. 4*(2012) 1287–1298.

(57) S.-H. Wu, C.-Y. Mou, H.-P. Lin, Synthesis of mesoporous silica nanoparticles, *Chemical Society Reviews. 42*(2013) 3862–3875. DOI: 10.1039/C3CS35405A.

(58) S.-H. Wu, Y. Hung, C.-Y. Mou, Mesoporous silica nanoparticles as nanocarriers, *Chemical Communications. 47*(2011) 9972–9985. DOI: 10.1039/C1CC11760B.

(59) M. Grün, I. Lauer, K.K. Unger, The synthesis of micrometer-and submicrometer-size spheres of ordered mesoporous oxide MCM-41, *Advanced Materials. 9*(1997) 254–257.

(60) F. Tang, L. Li, D. Chen, Mesoporous silica nanoparticles: Synthesis, biocompatibility and drug delivery, *Advanced Materials. 24*(2012) 1504–1534.

(61) L. Calabrese, E. Proverbio, A brief overview on the anticorrosion performances of sol-gel zeolite coatings, *Coatings. 9*(2019) 409.

(62) H. Serati-Nouri, A. Jafari, L. Roshangar, M. Dadashpour, Y. Pilehvar-Soltanahmadi, N. Zarghami, Biomedical applications of zeolite-based materials: A review, *Materials Science and Engineering: C. 116*(2020) 111225. DOI: https://doi.org/10.1016/j.msec.2020.111225

(63) L. Calabrese, Anticorrosion behavior of zeolite coatings obtained by in situ crystallization: A critical review, *Materials. 12*(2019) 59.

(64) A. Gavrilović-Wohlmuther, A. Laskos, E. Kny. Chapter 12 – Corrosion inhibitor–loaded smart nanocontainers. In *Corrosion Protection at the Nanoscale*, Rajendran, S., Nguyen, T. A. N. H., Kakooei, S., Yeganeh, M., Li, Y. Eds.; Elsevier, 2020; pp. 203–223.

(65) G. Zhang, L. Wu, A. Tang, X. Ding, B. Jiang, A. Atrens, F. Pan, Smart epoxy coating containing zeolites loaded with Ce on a plasma electrolytic oxidation coating on Mg alloy AZ31 for active corrosion protection, *Progress in Organic Coatings. 132*(2019) 144–147. DOI: https://doi.org/10.1016/j.porgcoat.2019.03.046

(66) M.N. Katariya, A.K. Jana, P.A. Parikh, Corrosion inhibition effectiveness of zeolite ZSM-5 coating on mild steel against various organic acids and its antimicrobial activity, *Journal of Industrial and Engineering Chemistry. 19*(2013) 286–291. DOI: https://doi.org/10.1016/j.jiec.2012.08.013

(67) S. Yao, Z. Liu, L. Li, Recent progress in nanoscale covalent organic frameworks for cancer diagnosis and therapy, *Nano-Micro Letters. 13*(2021) 176.

(68) H. Wang, Z. Zeng, P. Xu, L. Li, G. Zeng, R. Xiao, Z. Tang, D. Huang, L. Tang, C. Lai, Recent progress in covalent organic framework thin films: Fabrications, applications and perspectives, *Chemical Society Reviews. 48*(2019) 488–516.

(69) X. Liu, D. Huang, C. Lai, G. Zeng, L. Qin, H. Wang, H. Yi, B. Li, S. Liu, M. Zhang, Recent advances in covalent organic frameworks (COFs) as a smart sensing material, *Chemical Society Reviews. 48*(2019) 5266–5302.

(70) J.L. Segura, M.J. Mancheño, F. Zamora, Covalent organic frameworks based on Schiff-base chemistry: Synthesis, properties and potential applications, *Chemical Society Reviews. 45*(2016) 5635–5671.

(71) T. Liu, W. Li, C. Zhang, W. Wang, W. Dou, S. Chen, Preparation of highly efficient self-healing anticorrosion epoxy coating by integration of benzotriazole corrosion inhibitor loaded 2D-COF, *Journal of Industrial and Engineering Chemistry. 97*(2021) 560–573. DOI: https://doi.org/10.1016/j.jiec.2021.03.012

(72) P. Najmi, N. Keshmiri, M. Ramezanzadeh, B. Ramezanzadeh, M. Arjmand, Design of nacre-inspired 2D-MoS2 nanosheets assembled with mesoporous covalent organic frameworks (COFs) for smart coatings, *ACS Applied Materials & Interfaces. 14*(2022) 54141–54156.

(73) J. Lan, B. Wang, C. Bo, B. Gong, J. Ou, Progress on fabrication and application of activated carbon sphere in recent decade, *Journal of Industrial and Engineering Chemistry. 120*(2023) 47–72. DOI: https://doi.org/10.1016/j.jiec.2022.12.045

(74) T.-H. Liou, Development of mesoporous structure and high adsorption capacity of biomass-based activated carbon by phosphoric acid and zinc chloride activation, *Chemical Engineering Journal. 158*(2010) 129–142. DOI: https://doi.org/10.1016/j.cej.2009.12.016

(75) H. Behloul, H. Ferkous, N. Bougdah, S. Djellali, M. Alam, C. Djilani, A. Sedik, D. Lerari, B.-H. Jeon, Y. Benguerba, New insights on the adsorption of CI-Reactive Red 141 dye using activated carbon prepared from the ZnCl2-treated waste cotton fibers: Statistical physics, DFT, COSMO-RS, and AIM studies, *Journal of Molecular Liquids. 364*(2022) 119956. DOI: https://doi.org/10.1016/j.molliq.2022.119956

(76) J. Bedia, M. Peñas-Garzón, A. Gómez-Avilés, J.J. Rodriguez, C. Belver, Review on activated carbons by chemical activation with FeCl3, *C. 6*(2020) 21.

(77) C. Demarchi, B. Michel, N. Nedelko, A. Slawska-Waniewska, P. Dluzewski, A. Kaleta, R. Minikayev, T. Strachowski, L. Lipińska, J. Dalmagro, et al., Preparation, characterization, and application of magnetic activated carbon from termite feces for the adsorption of Cr(VI) from aqueous solutions, *Powder Technology. 354*(2019) 432–441. DOI: 10.1016/j.powtec.2019.06.020

(78) W. Wang, A. Wang. Nanoscale clay minerals for functional ecomaterials: Fabrication, applications, and future trends. In *Handbook of Ecomaterials*, Martínez, L. M. T., Kharissova, O. V., Kharisov, B. I. Eds.; Springer International Publishing, 2019; pp. 2409–2490.

(79) M. Izadi, T. Shahrabi, I. Mohammadi, B. Ramezanzadeh, Synthesis of impregnated Na+-montmorillonite as an eco-friendly inhibitive carrier and its subsequent protective effect on silane coated mild steel, *Progress in Organic Coatings. 135*(2019) 135–147. DOI: https://doi.org/10.1016/j.porgcoat.2019.05.037

(80) A. Ghazi, E. Ghasemi, M. Mahdavian, B. Ramezanzadeh, M. Rostami, The application of benzimidazole and zinc cations intercalated sodium montmorillonite as smart ion exchange inhibiting pigments in the epoxy ester coating, *Corrosion Science. 94*(2015) 207–217. DOI: https://doi.org/10.1016/j.corsci.2015.02.007

(81) P. Aggrey, M. Nartey, Y. Kan, J. Cvjetinovic, A. Andrews, A.I. Salimon, K.I. Dragnevski, A.M. Korsunsky, On the diatomite-based nanostructure-preserving material synthesis for energy applications, *RSC Advances. 11*(2021) 31884–31922.
(82) J.-Y. Wang, N. De Belie, W. Verstraete, Diatomaceous earth as a protective vehicle for bacteria applied for self-healing concrete, *Journal of Industrial Microbiology and Biotechnology. 39*(2012) 567–577.
(83) P.J. Denissen, V. Shkirskiy, P. Volovitch, S.J. Garcia, Corrosion inhibition at scribed locations in coated AA2024-T3 by cerium- and DMTD-loaded natural silica microparticles under continuous immersion and wet/dry cyclic exposure, *ACS Applied Materials & Interfaces. 12*(2020) 23417–23431. DOI: 10.1021/acsami.0c03368
(84) M.L. Zheludkevich, R. Serra, M.F. Montemor, M.G. Ferreira, Oxide nanoparticle reservoirs for storage and prolonged release of the corrosion inhibitors, *Electrochemistry Communications. 7*(2005) 836–840.
(85) L.R.L. Santos, C.E.B. Marino, I.C. Riegel-Vidotti, Silica/chitosan hybrid particles for smart release of the corrosion inhibitor benzotriazole, *European Polymer Journal. 115*(2019) 86–98. DOI: https://doi.org/10.1016/j.eurpolymj.2019.03.008

11 Utilization of Encapsulated Corrosion Inhibitors in the Biomedical Industry

Khasan Berdimuradov, Elyor Berdimurodov, El Ibrahimi Brahim, Burak Tuzun, Hicham Es-soufi, Sadhucharan Mallick, and Bakhtiyor Borikhonov

11.1 INTRODUCTION

11.1.1 Brief Overview of the Importance of Corrosion Inhibition in the Biomedical Industry

The biomedical industry is responsible for developing and manufacturing a diverse range of medical devices and equipment designed to improve the health and well-being of patients. A critical aspect of ensuring the safety and effectiveness of these devices is the prevention of corrosion, as it can lead to material degradation, loss of mechanical integrity, and potential harm to patients. Therefore, corrosion inhibition in the biomedical industry is of paramount importance [1–3]. Corrosion is a naturally occurring electrochemical process that affects various metallic materials used in biomedical applications, such as stainless steel, titanium, and cobalt-chromium alloys. These materials are widely utilized due to their excellent mechanical properties, biocompatibility, and corrosion resistance. However, the harsh physiological environment within the human body, including varying pH levels, the presence of aggressive ions, and the dynamic loading conditions, can lead to localized or general corrosion of these materials over time [4, 5]. To mitigate the detrimental effects of corrosion, researchers have explored different methods to protect metallic materials in biomedical applications. One promising approach is the use of encapsulated corrosion inhibitors. These inhibitors are substances that, when released from their encapsulation matrix, can interact with the metal surface or the surrounding environment to slow down or prevent the corrosion process. Encapsulation offers several advantages over traditional corrosion inhibition methods, such as improved stability, targeted delivery, and controlled release kinetics [6, 7].

In this chapter, we will discuss the utilization of encapsulated corrosion inhibitors in the biomedical industry, focusing on their applications, encapsulation techniques, types of inhibitors, encapsulation materials, release mechanisms, and the associated advantages and challenges. By understanding these critical aspects, we aim to highlight the potential of encapsulated corrosion inhibitors in enhancing the performance and longevity of medical devices and equipment, ultimately benefiting patients and healthcare providers.

11.1.2 The Role of Encapsulated Corrosion Inhibitors in Biomedical Applications

Encapsulated corrosion inhibitors play a vital role in protecting medical devices and equipment from the detrimental effects of corrosion in the demanding environment of the human body. They offer several advantages over traditional corrosion protection methods, including improved stability, targeted delivery, and controlled release kinetics. These properties are critical in addressing

DOI: 10.1201/9781032677255-11

the unique challenges associated with corrosion in biomedical applications, such as the need for biocompatibility, long-term performance, and minimal adverse effects on the surrounding tissues [8, 9]. In biomedical applications, encapsulated corrosion inhibitors can be integrated into various types of implantable devices, including orthopedic implants, cardiovascular stents, and dental implants, as well as surgical instruments and equipment. By incorporating these corrosion inhibitors, the service life and performance of medical devices can be significantly enhanced, reducing the risk of device failure, infection, and other complications. The role of encapsulated corrosion inhibitors in biomedical applications can be summarized through the following key aspects [10–12]:

- Enhanced corrosion protection: Encapsulated corrosion inhibitors provide a barrier between the metallic surface and the corrosive environment, effectively reducing the corrosion rate and extending the service life of the devices.
- Targeted delivery: The encapsulation of corrosion inhibitors allows for localized delivery at the exact location where corrosion protection is required, minimizing the inhibitor's systemic exposure and reducing the risk of potential side effects on patients.
- Controlled release: The release of corrosion inhibitors can be precisely controlled through the design of the encapsulation system, ensuring a consistent and sustained release of inhibitors over time. This feature is particularly beneficial in biomedical applications, where long-term protection is often required.
- Biocompatibility: Encapsulation materials used for corrosion inhibitors can be selected based on their biocompatibility, ensuring that the corrosion protection system does not induce any adverse effects on the surrounding tissues or the patient's overall health.
- Adaptability: Encapsulated corrosion inhibitors can be tailored to suit specific biomedical applications and requirements by adjusting the encapsulation materials, inhibitor types, and release mechanisms. This adaptability allows for the development of customized corrosion protection solutions that can effectively address the unique challenges associated with each application.

By fulfilling these critical roles, encapsulated corrosion inhibitors have the potential to revolutionize the way corrosion protection is approached in the biomedical industry, leading to safer, more reliable, and longer-lasting medical devices and equipment for the benefit of patients and healthcare providers alike. For example, the antibacterial activity assessment indicated that adding carbon nanotubes (CNTs) to the magnesium (Mg) matrix effectively inhibited the growth of *Escherichia coli* (*E. coli*) and *Staphylococcus aureus* (*S. aureus*). Overall, the study shows that CNTs serve as an effective reinforcement for Mg–2.5Zn–0.5Zr/CNTs biocomposites, resulting in enhanced mechanical, degradation, and antibacterial performance (Figure 11.1) [13].

11.2 ENCAPSULATION TECHNIQUES FOR CORROSION INHIBITORS

11.2.1 Overview of Encapsulation Methods

Encapsulation is a process by which active agents, such as corrosion inhibitors, are entrapped within a protective coating or matrix. The encapsulation methods can be categorized into physical and chemical processes, each having its unique advantages and disadvantages. The choice of the encapsulation method depends on factors such as the desired release properties, the nature of the corrosion inhibitor, and the requirements of the specific biomedical application. It indicated an overview of common encapsulation methods used for corrosion inhibitors [14, 15].

a. Physical Encapsulation Methods
    - Spray drying: A solution or suspension containing the corrosion inhibitor and the encapsulation material is atomized into fine droplets and dried in a stream of hot gas. The resulting particles contain the corrosion inhibitor entrapped within the encapsulation

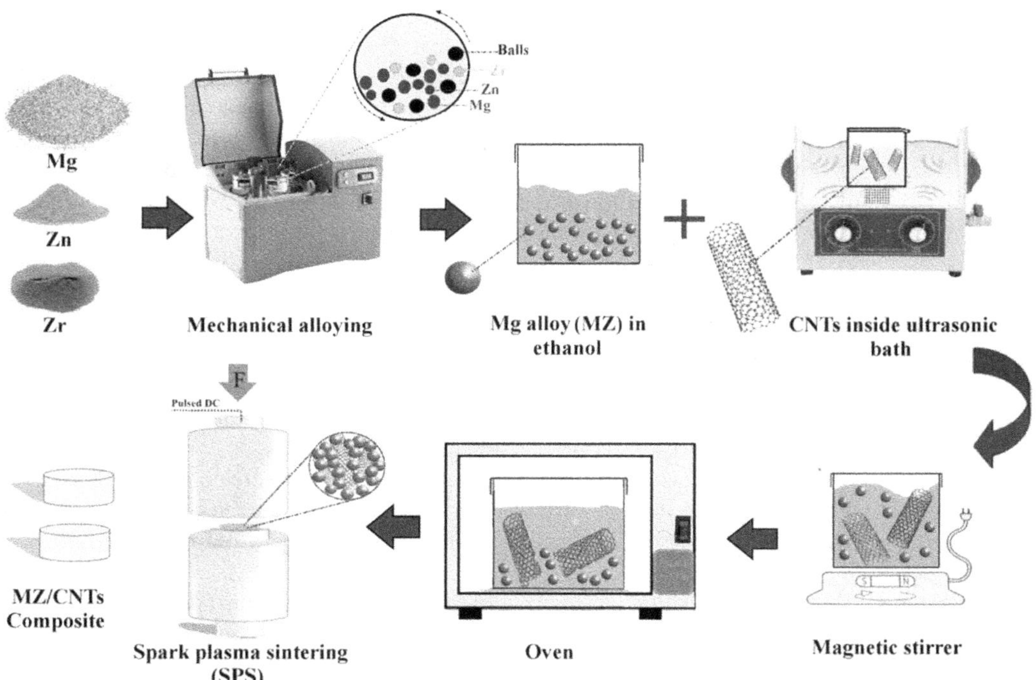

FIGURE 11.1 Representation of the process of MZ/CNTs composite preparation using SPM and SPS methods [13].

material. Spray drying is a versatile and scalable method, suitable for both heat-sensitive and heat-stable corrosion inhibitors.

- Coacervation: This method involves the phase separation of a polymer solution by changing the solvent conditions, such as pH or temperature, leading to the formation of a dense coacervate phase that entraps the corrosion inhibitor. Coacervation is suitable for encapsulating heat-sensitive corrosion inhibitors and offers good control over particle size and morphology.
- Solvent evaporation: The corrosion inhibitor and encapsulation material are dissolved in a suitable solvent, which is then evaporated, leaving behind solid particles of encapsulated corrosion inhibitor. This method is suitable for encapsulating hydrophobic corrosion inhibitors and offers good control over the particle size and composition.
- Electrospraying: A high voltage is applied to a solution or suspension containing the corrosion inhibitor and encapsulation material, causing the formation of fine droplets that solidify as they are collected on a grounded substrate. This method offers excellent control over particle size and morphology and is suitable for a wide range of corrosion inhibitors and encapsulation materials.

b. Chemical Encapsulation Methods

- Polymerization: The corrosion inhibitor is entrapped within a polymer matrix formed by the in situ polymerization of monomers or prepolymers. This method offers excellent control over the release properties and can be tailored to suit specific biomedical applications. However, the polymerization conditions may need to be carefully controlled to avoid adverse effects on the corrosion inhibitor.

- Interfacial polymerization: A solution containing the corrosion inhibitor is brought into contact with a second solution containing a complementary reactive species, leading to the formation of a polymer film at the interface between the two solutions. This film encapsulates the corrosion inhibitor and can be collected by subsequent processing steps. Interfacial polymerization offers good control over film thickness and composition and is suitable for a wide range of corrosion inhibitors and encapsulation materials.
- Mineralization: The corrosion inhibitor is encapsulated within an inorganic matrix, such as silica or hydroxyapatite, formed by the controlled precipitation of ions from a solution. Mineralization offers excellent chemical and thermal stability, as well as biocompatibility, making it suitable for biomedical applications. However, the encapsulation efficiency may be lower compared to other methods, and the release properties may be more challenging to control.

Each encapsulation method has its unique advantages and challenges, and the choice of the most suitable method depends on the specific requirements of the biomedical application, the nature of the corrosion inhibitor, and the desired release properties. By selecting the appropriate encapsulation method, it is possible to develop effective and tailored corrosion protection solutions for various biomedical applications.

11.2.2 Selection Criteria for Encapsulation Techniques in Biomedical Applications

The choice of an appropriate encapsulation technique for corrosion inhibitors in biomedical applications depends on various factors, including the properties of the corrosion inhibitor, the desired release profile, the biocompatibility of the encapsulation materials, and the specific requirements of the application. The key selection criteria are reported here for encapsulation techniques in biomedical applications [16–18]:

- Corrosion inhibitor properties: The encapsulation method should be compatible with the physical and chemical properties of the corrosion inhibitor. Factors to consider include the inhibitor's solubility, stability, and sensitivity to heat, light, or chemical reactions. For example, heat-sensitive inhibitors may require a low-temperature process like coacervation, while hydrophobic inhibitors may be more suitable for solvent evaporation.
- Release profile: The desired release profile of the corrosion inhibitor, such as immediate, sustained, or stimuli-responsive release, could be achieved with the chosen encapsulation method. For example, interfacial polymerization can offer a controlled release by adjusting the film thickness, while electrospraying allows for the incorporation of stimuli-responsive materials that can trigger the release upon specific environmental changes.
- Biocompatibility: The encapsulation method should allow for the use of biocompatible encapsulation materials that do not induce adverse effects on the surrounding tissues or the patient's overall health. Additionally, the process should not involve the use of toxic solvents or reagents that could compromise the biocompatibility of the final product.
- Particle size and morphology: The encapsulation method should offer control over the particle size and morphology of the encapsulated corrosion inhibitor, as these factors can influence the release properties, bioavailability, and performance of the corrosion protection system. For instance, electrospraying and spray drying methods provide excellent control over particle size and morphology.
- Scalability: The encapsulation method should be suitable for scaling up from laboratory to industrial production levels without significant changes in the properties or performance of the encapsulated corrosion inhibitor. Spray drying and solvent evaporation are examples of scalable methods.

- Cost-effectiveness: The encapsulation method should offer a balance between the desired performance and economic feasibility. Factors to consider include the cost of raw materials, equipment, and energy, as well as the process efficiency and yield.
- Adaptability: The encapsulation method should be adaptable to suit the specific requirements of different biomedical applications, such as the incorporation of additional functional components, the use of alternative encapsulation materials, or the adjustment of release properties.

By considering these selection criteria, it is possible to identify the most suitable encapsulation technique for corrosion inhibitors in biomedical applications, ensuring effective corrosion protection while meeting the specific requirements and constraints of the application. Figure 11.2 demonstrates that bacterial growth was halted around the MZ/CNTs nanocomposites containing 0.3–0.9 wt% CNTs, while bacterial proliferation persisted around the MZ matrix. Furthermore, the images reveal a broader zone of inhibition around the composite MC3 (3.3 mm) on the agar plate compared to the composite MC1 (2.05 mm), which has a lower CNT content. Consequently, the antibacterial properties of the nanocomposite samples are related to their CNT content, meaning that increasing the amount of CNTs in the nanocomposite corresponds to larger inhibition zones. The inhibition zones for *E. coli* and *S. aureus* ranged from 0.24 to 3.1 mm and 0.33 to 3.3 mm, respectively [13].

The antibacterial activity of carbon nanomaterials (CNMs) depends on factors such as their composition, target microorganisms, surface modification, and reaction environment. The antibacterial mechanism of CNMs involves attacking the membrane/wall of microbial cells, damaging cellular structures, and disrupting the physical mechanisms associated with the biological separation of microbial cells from their surroundings. CNMs also generate oxidative stress conditions by producing toxic substances like reactive oxygen species (ROS) and exert chemical antimicrobial

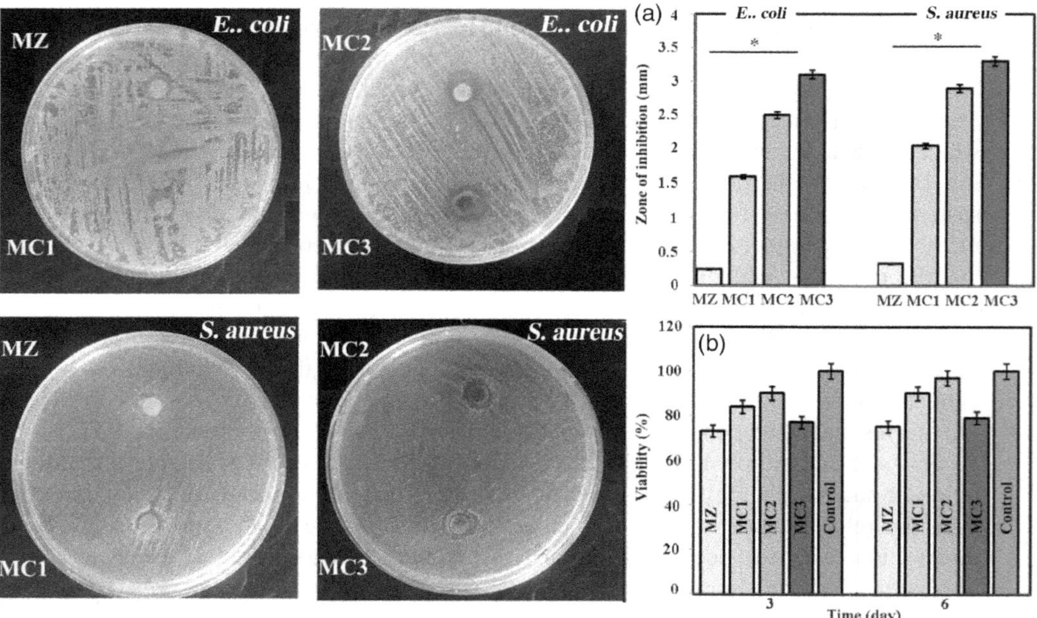

FIGURE 11.2 (a) Antimicrobial properties evaluated through the disk diffusion assay against both gram-positive (*S. aureus*) and gram-negative (*E. coli*) bacteria, and (b) cell viability of MG63 cells cultivated for different durations on the MZ, MC1, MC2, and MC3 biocomposite samples [13].

effects based on the interaction of microorganisms with CNMs [13]. Interactions between CNMs and microbes facilitate electron transfer, promoting ROS-independent oxidative stress by removing electrons from the microbial surface, ultimately leading to cell death.

11.3 BIOMEDICAL APPLICATIONS OF ENCAPSULATED CORROSION INHIBITORS

11.3.1 Implantable Medical Devices

Implantable medical devices play a crucial role in the treatment and management of various medical conditions, including orthopedic, dental, and cardiovascular disorders. These devices are typically made from metallic materials, such as stainless steel, titanium, and cobalt-chromium alloys, due to their excellent mechanical properties, biocompatibility, and corrosion resistance. However, the demanding physiological environment in the human body can lead to corrosion and degradation of these materials over time. Encapsulated corrosion inhibitors offer a promising approach to enhance the corrosion protection of implantable medical devices, thus extending their service life and reducing the risk of device-related complications. Some examples of implantable medical devices where encapsulated corrosion inhibitors can be applied are discussed [19–21]:

- Orthopedic implants: These devices, such as joint replacements, bone plates, and screws, are used to repair or replace damaged bones and joints. Encapsulated corrosion inhibitors can be incorporated into the implant coatings or integrated within the bulk material to provide long-term corrosion protection, reducing the risk of implant failure, inflammation, and infection.
- Cardiovascular stents: Stents are used to restore blood flow in blocked or narrowed blood vessels. The incorporation of encapsulated corrosion inhibitors into the stent coatings can help prevent corrosion-induced degradation of the metallic framework, reducing the risk of in-stent restenosis and prolonging the stent's functional life.
- Dental implants: These devices are used to replace missing teeth by providing a support structure for dental prosthetics. Encapsulated corrosion inhibitors can be applied to the implant surface or incorporated into the bulk material to enhance the corrosion resistance of the dental implant, ensuring long-term stability and minimizing the risk of peri-implantitis.
- Cochlear implants: These devices are used to restore hearing in patients with severe hearing loss. The incorporation of encapsulated corrosion inhibitors into the implant components can help prevent corrosion-induced device failure and extend the service life of the implant, ensuring consistent and reliable performance.
- Pacemakers and defibrillators: These electronic devices are implanted to regulate heart rhythm and prevent arrhythmias. Encapsulated corrosion inhibitors can be integrated into the device's metallic components and coatings to provide corrosion protection, reducing the risk of device failure and ensuring the long-term reliability of these life-saving devices.

The use of encapsulated corrosion inhibitors in implantable medical devices offers significant benefits, such as targeted delivery, controlled release, and biocompatibility, which can contribute to improved device performance, longevity, and patient safety. By addressing the unique challenges associated with corrosion in the physiological environment, encapsulated corrosion inhibitors hold great potential for revolutionizing the design and development of implantable medical devices across various biomedical applications.

11.3.2 Orthopedic Implants

Orthopedic implants are medical devices used to replace or support damaged bones and joints, helping patients regain mobility and alleviate pain. Examples of orthopedic implants include joint replacements (e.g., hip, knee, and shoulder), bone plates, screws, and rods. These implants are typically made from metallic materials, such as stainless steel, titanium, and cobalt-chromium alloys, which offer excellent mechanical properties, biocompatibility, and corrosion resistance. However, in the demanding physiological environment, corrosion and degradation of these materials can occur, leading to implant failure, inflammation, and infection. Encapsulated corrosion inhibitors can play a critical role in extending the service life of orthopedic implants and reducing the risk of complications. There are several ways encapsulated corrosion inhibitors can be applied to orthopedic implants [22–25].

- Implant coatings: Encapsulated corrosion inhibitors can be integrated into the coatings applied to the implant surface. These coatings can be made from various materials, such as polymers, ceramics, or bioactive glass. The encapsulated inhibitors provide localized corrosion protection, releasing corrosion inhibitors at a controlled rate when the coating is exposed to corrosive conditions. This targeted delivery helps prevent implant degradation while minimizing the risk of adverse effects on the surrounding tissue.
- Bulk material modification: Encapsulated corrosion inhibitors can be incorporated into the bulk material of the orthopedic implant. This approach helps provide corrosion protection throughout the entire structure of the implant, ensuring long-term stability and reducing the risk of mechanical failure. The incorporation of inhibitors can be achieved through techniques such as powder metallurgy or additive manufacturing.
- Self-healing systems: Encapsulated corrosion inhibitors can be combined with self-healing materials to develop advanced orthopedic implant systems. These systems release the encapsulated inhibitors when the implant is subjected to mechanical stresses or corrosive conditions, providing active corrosion protection and allowing the implant to "heal" itself, thereby extending its service life and improving its performance.
- Stimuli-responsive release: Encapsulated corrosion inhibitors can be designed to release their contents in response to specific environmental triggers, such as changes in pH, temperature, or mechanical stress. This approach allows for a more targeted and efficient delivery of corrosion inhibitors, providing protection when and where it is needed most.

Incorporating encapsulated corrosion inhibitors into orthopedic implants can significantly enhance their corrosion resistance, leading to improved implant performance, longevity, and patient outcomes. By addressing the unique challenges associated with corrosion in the physiological environment, encapsulated corrosion inhibitors hold great potential for revolutionizing the design and development of orthopedic implants for various applications. Mg-based alloys have great potential as temporary orthopedic implants due to their biodegradability and favorable mechanical properties. However, their rapid corrosion rate in the physiological environment significantly limits their clinical application. In this study, the researchers developed an anticorrosive polymeric coating using polycaprolactone (PCL) and lawsone to improve the corrosion resistance of AZ31 Mg alloy (Figure 11.3) [26].

Lawsone, a natural compound, was incorporated into the PCL coating to enhance its passive corrosion protection ability. Lawsone achieves this by strongly chelating with dissolving Mg2+ ions. The anticorrosion properties of the PCL-lawsone coatings were evaluated through electrochemical techniques and in vitro immersion tests. The results demonstrated a significant improvement in the corrosion resistance of the bare Mg alloy. In addition to improving corrosion resistance, the incorporation of lawsone also provided the coating with strong antibacterial activity against *E. coli* and *S. aureus* bacterial strains. The fabricated coating exhibited cytocompatibility, with a viability of over 85% toward human fetal osteoblast cells. These findings highlight the potential of using lawsone as a natural corrosion inhibitor to create corrosion-protective, antibacterial, and biocompatible coatings on Mg-based biodegradable implants. This development could open new

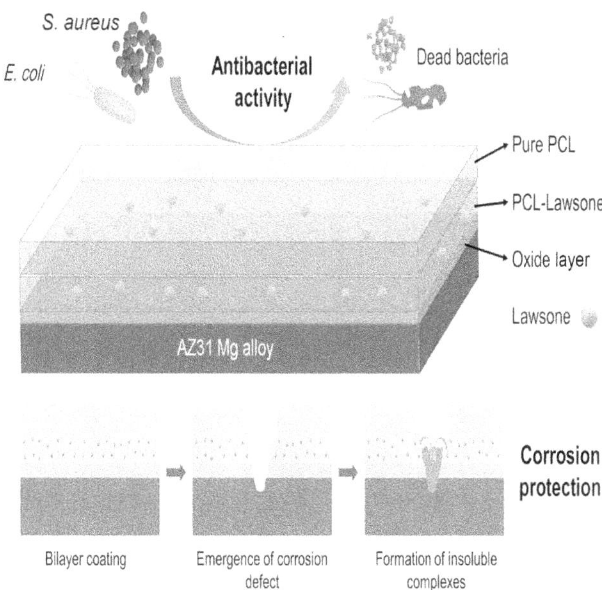

FIGURE 11.3 Anticorrosive polymeric coating based on polycaprolactone (PCL) for temporary orthopedic implants [26].

possibilities for the use of Mg-based alloys in temporary orthopedic implants, improving their performance and safety in clinical applications.

11.3.3 Cardiovascular Stents

Cardiovascular stents are tubular mesh-like devices used to restore blood flow in blocked or narrowed blood vessels, particularly in the coronary and peripheral arteries. They are typically made from metallic materials, such as stainless steel, cobalt-chromium alloys, or shape-memory alloys like nitinol, due to their excellent mechanical properties, biocompatibility, and corrosion resistance. However, the physiological environment in the human body can lead to corrosion and degradation of these materials over time. Encapsulated corrosion inhibitors offer a promising approach to enhance the corrosion protection of cardiovascular stents, thus extending their service life and reducing the risk of device-related complications, such as in-stent restenosis or stent thrombosis. Some ways through which encapsulated corrosion inhibitors can be applied to cardiovascular stents are discussed here [27–29]:

- Stent coatings: Encapsulated corrosion inhibitors can be integrated into the stent coatings, which can be made from various materials, such as polymers or bioabsorbable materials. The encapsulated inhibitors provide localized corrosion protection, releasing corrosion inhibitors at a controlled rate when the coating is exposed to corrosive conditions. This targeted delivery helps prevent stent degradation while minimizing the risk of adverse effects on the surrounding tissue and reducing the likelihood of complications.
- Drug-eluting stents: In addition to corrosion inhibitors, other therapeutic agents, such as antiproliferative drugs or anti-inflammatory agents, can be co-encapsulated with the inhibitors. This approach combines the benefits of corrosion protection with the controlled release of therapeutic agents, which can help prevent in-stent restenosis or stent thrombosis.

- Stimuli-responsive release: Encapsulated corrosion inhibitors can be designed to release their contents in response to specific environmental triggers, such as changes in pH, temperature, or mechanical stress. This approach allows for a more targeted and efficient delivery of corrosion inhibitors, providing protection when and where it is needed most.
- Self-healing systems: Encapsulated corrosion inhibitors can be combined with self-healing materials to develop advanced stent systems. These systems release the encapsulated inhibitors when the stent is subjected to mechanical stresses or corrosive conditions, providing active corrosion protection and allowing the stent to "heal" itself, thereby extending its service life and improving its performance.

The use of encapsulated corrosion inhibitors in cardiovascular stents offers significant benefits, such as targeted delivery, controlled release, and biocompatibility, which can contribute to improved stent performance, longevity, and patient outcomes. By addressing the unique challenges associated with corrosion in the physiological environment, encapsulated corrosion inhibitors hold great potential for advancing the design and development of cardiovascular stents across various applications.

11.3.4 Dental Implants

Dental implants are medical devices used to replace missing teeth by providing a support structure for dental prosthetics such as crowns, bridges, or dentures. They consist of a screw-like titanium or titanium alloy post that is surgically inserted into the jawbone, acting as an artificial tooth root. While titanium and its alloys are known for their excellent biocompatibility, mechanical properties, and corrosion resistance, the demanding oral environment can still lead to corrosion and degradation over time. Encapsulated corrosion inhibitors can help enhance the corrosion protection of dental implants, ensuring long-term stability and minimizing the risk of complications such as peri-implantitis. Some ways on how encapsulated corrosion inhibitors can be applied to dental implants are suggested [23, 30, 31]:

- Implant coatings: Encapsulated corrosion inhibitors can be integrated into the coatings applied to the dental implant surface. These coatings can be made from various materials, such as polymers, ceramics, or bioactive glass. The encapsulated inhibitors provide localized corrosion protection, releasing corrosion inhibitors at a controlled rate when the coating is exposed to corrosive conditions. This targeted delivery helps prevent implant degradation while minimizing the risk of adverse effects on the surrounding tissue and bone.
- Bulk material modification: Encapsulated corrosion inhibitors can be incorporated into the bulk material of the dental implant. This approach helps provide corrosion protection throughout the entire structure of the implant, ensuring long-term stability and reducing the risk of mechanical failure. The incorporation of inhibitors can be achieved through techniques such as powder metallurgy or additive manufacturing.
- Stimuli-responsive release: Encapsulated corrosion inhibitors can be designed to release their contents in response to specific environmental triggers, such as changes in pH or mechanical stress. This approach allows for a more targeted and efficient delivery of corrosion inhibitors, providing protection when and where it is needed most.
- Combination with antimicrobial agents: In addition to encapsulated corrosion inhibitors, antimicrobial agents can be co-encapsulated or incorporated separately into the implant coatings or bulk material. This combination helps protect the dental implant from both corrosion and bacterial colonization, reducing the risk of peri-implantitis and other implant-related infections.

Incorporating encapsulated corrosion inhibitors into dental implants can significantly enhance their corrosion resistance, leading to improved implant performance, longevity, and patient outcomes. By addressing the unique challenges associated with corrosion in the oral environment, encapsulated corrosion inhibitors hold great potential for advancing the design and development of dental implants for various applications.

11.3.5 Surgical Instruments and Equipment

Surgical instruments and equipment play a crucial role in various medical procedures, including diagnostics, treatment, and surgery. These devices are typically made from metallic materials, such as stainless steel or titanium, due to their excellent mechanical properties, corrosion resistance, and biocompatibility. However, the demanding conditions in the clinical environment, including exposure to bodily fluids, sterilization processes, and frequent cleaning, can lead to corrosion and degradation of these materials over time. Encapsulated corrosion inhibitors offer a promising approach to enhance the corrosion protection of surgical instruments and equipment, thus extending their service life and reducing the risk of device-related complications. Some ways through which encapsulated corrosion inhibitors can be applied to surgical instruments and equipment are discussed [1, 2, 6].

- Coatings: Encapsulated corrosion inhibitors can be integrated into the coatings applied to the surface of surgical instruments and equipment. These coatings can be made from various materials, such as polymers, ceramics, or hybrid materials. The encapsulated inhibitors provide localized corrosion protection, releasing corrosion inhibitors at a controlled rate when the coating is exposed to corrosive conditions. This targeted delivery helps prevent instrument degradation while maintaining their functionality and performance.
- Bulk material modification: Encapsulated corrosion inhibitors can be incorporated into the bulk material of surgical instruments and equipment. This approach helps provide corrosion protection throughout the entire structure of the device, ensuring long-term durability and reducing the risk of mechanical failure. The incorporation of inhibitors can be achieved through techniques such as powder metallurgy or additive manufacturing.
- Stimuli-responsive release: Encapsulated corrosion inhibitors can be designed to release their contents in response to specific environmental triggers, such as changes in pH, temperature, or mechanical stress. This approach allows for a more targeted and efficient delivery of corrosion inhibitors, providing protection when and where it is needed most.
- Combination with antimicrobial agents: In addition to encapsulated corrosion inhibitors, antimicrobial agents can be co-encapsulated or incorporated separately into the coatings or bulk material. This combination helps protect surgical instruments and equipment from both corrosion and bacterial colonization, reducing the risk of healthcare-associated infections and contributing to a safer clinical environment.

Incorporating encapsulated corrosion inhibitors into surgical instruments and equipment can significantly enhance their corrosion resistance, leading to improved device performance, longevity, and patient safety. By addressing the unique challenges associated with corrosion in the clinical environment, encapsulated corrosion inhibitors hold great potential for advancing the design and development of surgical instruments and equipment across various applications.

11.4 TYPES OF ENCAPSULATED CORROSION INHIBITORS FOR BIOMEDICAL USE

Encapsulated corrosion inhibitors can be grouped into three main categories based on their chemical composition: organic corrosion inhibitors, inorganic corrosion inhibitors, and hybrid corrosion inhibitors. Each type offers its unique advantages and potential applications in the biomedical field.

11.4.1 Organic Corrosion Inhibitors

Organic corrosion inhibitors are organic compounds that form a protective film on the surface of a metal, reducing the rate of corrosion. These inhibitors often work through adsorption processes, physical or chemical interactions with the metal surface, or by disrupting the electrochemical processes that contribute to corrosion. Examples of organic corrosion inhibitors include the following [8, 9, 18]:

- Amines: These are nitrogen-containing organic compounds that can form a protective film on metal surfaces, often due to their affinity for the metal surface and their ability to displace water molecules.
- Azoles: These are a class of organic compounds containing nitrogen and/or sulfur heterocyclic rings, which can form a protective film on metal surfaces by interacting with the metal's oxide layer.
- Phosphonates: These are organic compounds containing phosphorus, which can adsorb onto metal surfaces and chelate metal ions, reducing the metal's susceptibility to corrosion.

11.4.2 Inorganic Corrosion Inhibitors

Inorganic corrosion inhibitors are inorganic compounds that can protect metals from corrosion by forming a passive oxide film on the metal surface, by precipitation of insoluble compounds, or by interfering with the electrochemical processes that lead to corrosion. Examples of inorganic corrosion inhibitors include the following [11, 17]:

- Silicates: These compounds can form a protective film on the metal surface by reacting with metal ions to create insoluble silicate compounds, which act as a barrier against corrosive species.
- Phosphates: Inorganic phosphates can form insoluble precipitates that protect the metal surface from corrosive species or can act as anodic inhibitors, affecting the electrochemical processes leading to corrosion.
- Molybdates: These compounds can form a passivating oxide film on the metal surface, reducing the rate of corrosion.

Self-repairing illustrative diagrams of abraded dicalcium phosphate dihydrate (DCPD) coatings in Hank's (a–c) and normal saline (NS) (d–f) solutions are presented in Figure 11.4 [32].

The self-healing process of scratched DCPD coatings in both Hank's and NS solutions is depicted in a series of illustrative diagrams. DCPD coatings are commonly used to improve the biocompatibility and bioactivity of implant materials. These coatings can self-repair when subjected to damage, such as scratches, through a series of transformations involving different calcium phosphate phases. The diagrams are divided into two sets, representing two different types of solutions:

- Hank's solution (a–c): Hank's balanced salt solution is a standard solution used to simulate the physiological environment. The diagrams display the self-healing process of the DCPD coating in this solution.
- NS solution (d–f): NS represents an alternative solution, which might have different properties compared to Hank's solution. The diagrams similarly depict the self-repairing process of the DCPD coating in this solution.

The self-healing process involves a series of transformations occurring in the scratched DCPD coating:

- ACP (amorphous calcium phosphate): Upon exposure to the solution, the damaged DCPD coating starts to transform into an ACP phase. This phase serves as an intermediate during the self-healing process.

- HA (hydroxyapatite): As the self-healing process continues, the ACP phase further converts into HA. HA is a stable, biocompatible, and bioactive calcium phosphate phase that is chemically similar to the mineral component of bone.

These diagrams help visualize the self-repairing mechanism of DCPD coatings when exposed to different solutions. This self-healing property is beneficial for the long-term performance and durability of implant materials, as it can help maintain the bioactivity and biocompatibility of the coatings even when subjected to mechanical damage.

11.4.3 Hybrid Corrosion Inhibitors

Hybrid corrosion inhibitors are developed by combining the properties of both organic and inorganic corrosion inhibitors. This approach enables the creation of multifunctional corrosion inhibitors that can offer enhanced corrosion protection, often through synergistic effects. Examples of hybrid corrosion inhibitors include the following [8, 9, 15]:

- Organosilanes: These compounds contain both organic and inorganic components, with the organic part providing adsorption onto the metal surface and the inorganic part forming a protective oxide film.
- Organic–inorganic hybrid coatings: These coatings consist of a combination of organic polymers and inorganic materials, such as nanoparticles or sol-gel-derived materials, which can provide a combination of adsorption, barrier, and passivation protection mechanisms.

The choice of encapsulated corrosion inhibitor type depends on the specific requirements of the biomedical application, such as biocompatibility, corrosion protection performance, and release kinetics. By selecting the appropriate type of encapsulated corrosion inhibitor, it is possible to tailor the corrosion protection properties of biomedical devices and materials, leading to improved performance, longevity, and patient outcomes.

11.5 RELEASE MECHANISMS AND CONTROLLED RELEASE OF CORROSION INHIBITORS

The release of corrosion inhibitors from encapsulation systems plays a key role in their performance and efficacy. The release mechanisms can be broadly categorized into passive release mechanisms and stimuli-responsive release mechanisms. Understanding these mechanisms and the factors that affect their release is essential for designing efficient encapsulated corrosion inhibitors for biomedical applications [16–18].

11.5.1 Passive Release Mechanisms

Passive release mechanisms involve the gradual release of corrosion inhibitors from the encapsulation system without the need for any external stimuli. These mechanisms often rely on processes like diffusion, degradation, or swelling of the encapsulation material.

Examples of passive release mechanisms include the following:

- Diffusion: The corrosion inhibitor molecules diffuse through the encapsulation material due to a concentration gradient. The release rate is determined by the diffusion coefficient of the inhibitor in the encapsulation material and the thickness of the encapsulation barrier.
- Degradation: The encapsulation material degrades over time through physical, chemical, or enzymatic processes, leading to the release of the corrosion inhibitor. The release rate is determined by the degradation rate of the encapsulation material.

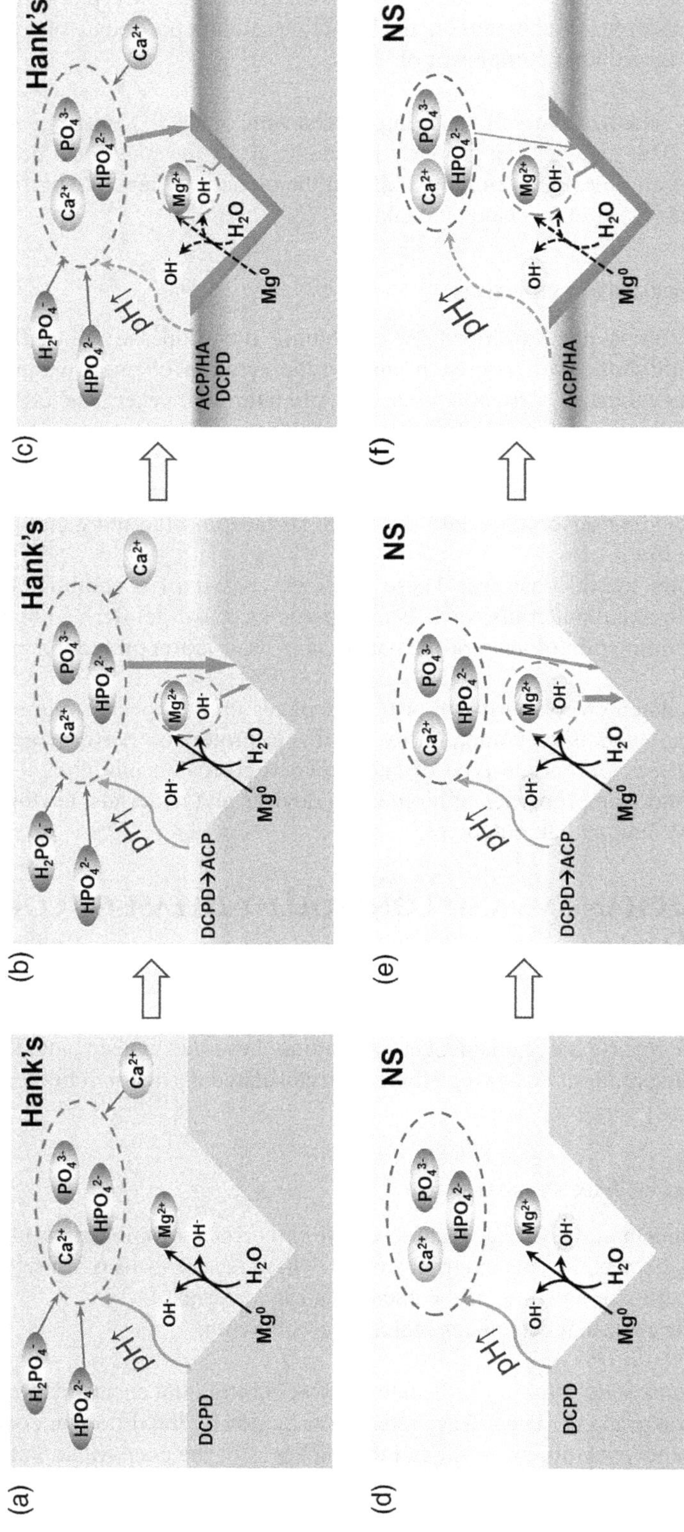

FIGURE 11.4 Self-repairing illustrative diagrams of abraded DCPD coatings in Hank's (a–c) and NS (d–f) solutions (DCPD: dicalcium phosphate dihydrate coating; ACP: amorphous calcium phosphate; HA: hydroxyapatite) [32].

- Swelling: In some cases, encapsulation materials can swell in the presence of water or other solvents, leading to an increase in the free volume within the material and the release of the corrosion inhibitor. The release rate depends on the swelling behavior of the encapsulation material.

11.5.2 Stimuli-Responsive Release Mechanisms

Stimuli-responsive release mechanisms involve the release of corrosion inhibitors from the encapsulation system in response to specific environmental triggers. These mechanisms enable a more targeted and efficient delivery of corrosion inhibitors, providing protection when and where it is needed most. Examples of stimuli-responsive release mechanisms include the following [30, 31]:

- pH-sensitive: Some encapsulation materials can change their properties, such as swelling or degradation behavior, in response to changes in pH. This can be exploited to release corrosion inhibitors in environments with a specific pH, such as acidic conditions in the vicinity of corrosion sites.
- Temperature-sensitive: Certain encapsulation materials exhibit temperature-dependent properties, such as a lower critical solution temperature or an upper critical solution temperature, which can be used to trigger the release of corrosion inhibitors at specific temperature ranges.
- Mechano-responsive: Encapsulation systems can be designed to release corrosion inhibitors in response to mechanical stress, such as strain or impact, providing active corrosion protection when the device is subjected to mechanical stresses.

A self-healing coating system (silk-polymerized amine [PA]) with pH-responsive properties and osteogenic activity was developed and applied to an Mg-1Ca alloy. The coating system consisted of three layers: a pure silk fibroin coating on top, an inhibitor-containing silk fibroin coating with PA, and a fluoride film at the bottom. The silk-PA coating significantly improved the corrosion resistance of the bare Mg-1Ca alloy even after 14 days of exposure to a corrosive medium (Figure 11.5) [33].

11.5.3 Factors Affecting the Release of Corrosion Inhibitors from Encapsulation Systems

Several factors can influence the release of corrosion inhibitors from encapsulation systems, including:

- Encapsulation material properties: The chemical structure, molecular weight, and crosslinking density of the encapsulation material can affect the release mechanisms and kinetics of corrosion inhibitors.
- Corrosion inhibitor properties: The molecular weight, solubility, and affinity of the corrosion inhibitor for the encapsulation material can impact the release behavior.
- Encapsulation method: The choice of encapsulation method, such as spray coating, layer-by-layer assembly, or electrospinning, can influence the release kinetics by affecting the encapsulation material's structure and morphology.
- Environmental conditions: Factors such as temperature, pH, and mechanical stress can impact the release mechanisms and kinetics of corrosion inhibitors from encapsulation systems.

Understanding the release mechanisms and the factors that affect the release of corrosion inhibitors from encapsulation systems is crucial for designing efficient encapsulated corrosion inhibitors with tailored release profiles for various biomedical applications. This knowledge can help optimize the corrosion protection performance, longevity, and biocompatibility of the encapsulated corrosion inhibitor systems.

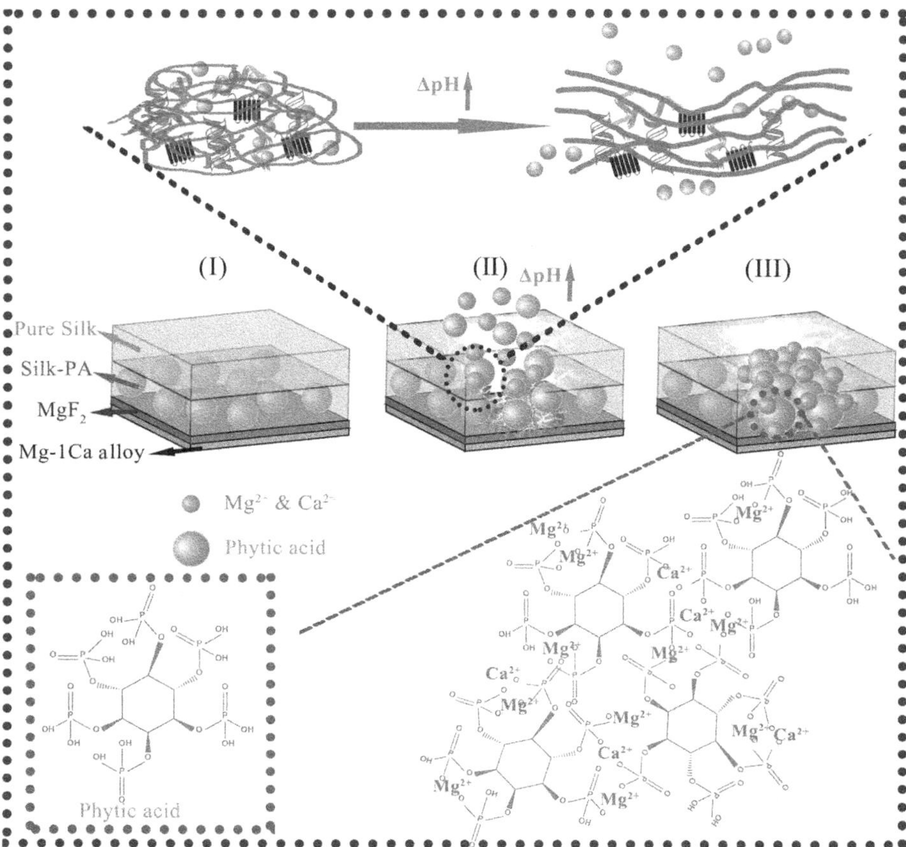

FIGURE 11.5 Illustration of the self-repairing process when silk-PA is exposed to the submersion environment [33].

11.6 ADVANTAGES AND CHALLENGES OF ENCAPSULATED CORROSION INHIBITORS IN BIOMEDICAL APPLICATIONS

The use of encapsulated corrosion inhibitors in biomedical applications offers several advantages, but it also comes with some challenges and limitations. Here, the key benefits and potential drawbacks of using encapsulated corrosion inhibitors in biomedical applications are discussed [29, 31].

11.6.1 Enhanced Corrosion Protection

Encapsulated corrosion inhibitors can provide superior corrosion protection compared to traditional inhibitors. The encapsulation process helps protect the active corrosion inhibitors from premature degradation, ensuring their effectiveness over a longer period. Additionally, the controlled release of corrosion inhibitors from the encapsulation system ensures that the desired concentration of inhibitors is maintained at the corrosion site, resulting in enhanced corrosion protection.

11.6.2 Controlled Release and Targeted Delivery

One of the main advantages of encapsulated corrosion inhibitors is their ability to provide controlled release and targeted delivery of corrosion inhibitors. Encapsulation systems can be designed to release corrosion inhibitors in response to specific environmental triggers, such as changes in pH, temperature, or mechanical stress. This enables a more targeted and efficient delivery of corrosion inhibitors, providing corrosion protection when and where it is needed most [29–31].

11.6.3 Improved Biocompatibility

Encapsulated corrosion inhibitors can potentially improve the biocompatibility of corrosion inhibitors used in biomedical applications. The encapsulation process can help minimize the exposure of surrounding tissues and cells to potentially harmful corrosion inhibitors. Furthermore, the controlled release of corrosion inhibitors can reduce the overall dosage required for effective corrosion protection, minimizing the potential for adverse biological effects.

11.6.4 Challenges and Limitations

Despite the numerous advantages of encapsulated corrosion inhibitors, there are also some challenges and limitations associated with their use in biomedical applications:

- Design and optimization: The design and optimization of encapsulation systems can be complex, as it requires a thorough understanding of the release mechanisms, environmental triggers, and the interaction between the corrosion inhibitor and the encapsulation material.
- Cost and manufacturing: Encapsulation techniques can be more expensive and time-consuming compared to traditional inhibitor application methods. The scalability and cost-effectiveness of encapsulation processes need to be considered for large-scale biomedical applications.
- Long-term stability: The long-term stability of encapsulated corrosion inhibitors under physiological conditions and sterilization processes is an essential consideration for their use in biomedical applications. The encapsulation materials and corrosion inhibitors must maintain their integrity and functionality over extended periods.
- Regulatory approval: The use of encapsulated corrosion inhibitors in biomedical devices may require additional regulatory approvals and extensive testing to ensure their safety and efficacy.

Despite these challenges, encapsulated corrosion inhibitors hold great promise for improving the corrosion protection, service life, and biocompatibility of biomedical devices and materials. Continued research and development efforts are needed to overcome these challenges and fully realize the potential of encapsulated corrosion inhibitors in biomedical applications.

11.7 FUTURE PERSPECTIVES AND RESEARCH DIRECTIONS

Encapsulated corrosion inhibitors have shown great potential in enhancing the corrosion protection of biomedical devices and materials. As research continues to advance in this field, emerging trends and technologies are likely to shape the future landscape of encapsulated corrosion inhibitors for biomedical applications. In this section, some of these emerging trends and potential research areas are discussed [8–10].

11.7.1 Emerging Trends and Technologies in Encapsulated Corrosion Inhibitors for Biomedical Applications

- Nanotechnology: The use of nanotechnology in encapsulating corrosion inhibitors opens up new possibilities for creating more efficient and targeted delivery systems. Nanoparticles, nanocapsules, and nanofibers can provide unique release characteristics and enhanced surface interactions with the target materials, resulting in improved corrosion protection performance.

- Smart materials: The development of smart materials that can respond to specific environmental stimuli, such as pH, temperature, or mechanical stress, can lead to more efficient and targeted release of corrosion inhibitors. These materials can potentially provide corrosion protection only when needed, reducing the overall dosage of inhibitors and minimizing the risk of adverse biological effects.
- Multifunctional coatings: Combining corrosion inhibitors with other functional components, such as antimicrobial agents, anti-inflammatory drugs, or bioactive molecules, can lead to the development of multifunctional coatings that simultaneously address various challenges in biomedical applications, such as corrosion, infection, and inflammation.

11.7.2 Potential Research Areas and Opportunities for Further Development

Research into developing new biocompatible encapsulation materials and optimizing their properties for specific biomedical applications is essential to improve the performance and safety of encapsulated corrosion inhibitors.

- In vitro and in vivo testing: Rigorous in vitro and in vivo testing of encapsulated corrosion inhibitors is necessary to evaluate their corrosion protection performance, biocompatibility, and potential side effects. The development of standardized testing protocols and models for evaluating encapsulated corrosion inhibitors in a biomedical context is crucial.
- Synergistic effects: Investigating the synergistic effects between different types of corrosion inhibitors, encapsulation materials, and other functional components can lead to the development of more effective and efficient corrosion protection systems.
- Personalized medicine: The use of encapsulated corrosion inhibitors can potentially contribute to the development of personalized medical devices, where the corrosion protection is tailored to the specific needs of individual patients based on factors like their physiology, the type of implant, or the local environment within the body.

The future of encapsulated corrosion inhibitors for biomedical applications holds great promise as researchers continue to explore new materials, techniques, and applications. By addressing the challenges and capitalizing on the opportunities outlined above, encapsulated corrosion inhibitors can significantly improve the performance, longevity, and biocompatibility of biomedical devices and materials, ultimately benefiting patients and healthcare providers alike.

11.8 CONCLUSION

Encapsulated corrosion inhibitors have emerged as a promising approach to enhance the corrosion protection of biomedical devices and materials. By encapsulating corrosion inhibitors within protective matrices, these systems offer several advantages over traditional corrosion inhibitors, including enhanced corrosion protection, controlled release and targeted delivery, and improved biocompatibility. The development of encapsulated corrosion inhibitors relies on a thorough understanding of release mechanisms, such as passive and stimuli-responsive mechanisms, and the factors that influence the release of inhibitors from encapsulation systems. When designing encapsulated corrosion inhibitors, consideration must be given to the properties of both the corrosion inhibitor and the encapsulation material. Despite the numerous advantages, there are also challenges and limitations associated with the use of encapsulated corrosion inhibitors in biomedical applications. These include the complexity of design and optimization, cost and manufacturing concerns, long-term stability, and regulatory approval. Emerging trends and technologies, such as nanotechnology and smart materials, along with potential research areas like biocompatible encapsulation materials and personalized medicine, offer exciting opportunities for the development of next-generation corrosion inhibitors for biomedical applications.

In conclusion, encapsulated corrosion inhibitors hold great promise for improving the corrosion protection, service life, and biocompatibility of biomedical devices and materials. Continued research and development efforts are needed to overcome existing challenges and fully realize the potential of encapsulated corrosion inhibitors in biomedical applications, ultimately benefiting patients and healthcare providers alike.

REFERENCES

1. Anjum, M.J., et al., Green corrosion inhibitors intercalated Mg: Al layered double hydroxide coatings to protect Mg alloy. *Rare Metals*, 2021. 40: p. 2254–2265.
2. Bakhsheshi-Rad, H.R., et al., In-vitro corrosion inhibition mechanism of fluorine-doped hydroxyapatite and brushite coated Mg–Ca alloys for biomedical applications. *Ceramics International*, 2014. 40(6): p. 7971–7982.
3. Bakhsheshi-Rad, H.R., et al., Synthesis of a novel nanostructured zinc oxide/baghdadite coating on Mg alloy for biomedical application: In-vitro degradation behavior and antibacterial activities. *Ceramics International*, 2017. 43(17): p. 14842–14850.
4. Chen, L., et al., A layer-by-layer assembled coating for improved stress corrosion cracking on biomedical magnesium alloy in cell culture medium. *Surface and Coatings Technology*, 2020. 403: p. 126427.
5. Cheng-hao, M.O.U.Z.-q.L., Effect of different artificial body fluids and their pH on corrosion of biomedical metallic materials. *Journal of Chinese Society for Corrosion and Protection*, 2009. 18(2): p. 126–130.
6. Cuartas-Marulanda, D., et al., Natural coatings and surface modifications on magnesium alloys for biomedical applications. *Polymers*, 2022. 14(23): p. 5297.
7. Dong, Q., et al., Dual self-healing inorganic-organic hybrid coating on biomedical Mg. *Corrosion Science*, 2022. 200: p. 110230.
8. Giger, E.V., B. Castagner, and J.-C. Leroux, Biomedical applications of bisphosphonates. *Journal of Controlled Release*, 2013. 167(2): p. 175–188.
9. Gu, Y., et al., Long-term corrosion inhibition mechanism of microarc oxidation coated AZ31 Mg alloys for biomedical applications. *Materials & Design (1980–2015)*, 2013. 46: p. 66–75.
10. Hassan, N., et al., Electrochemical evaluation of synthesized s-triazine derivatives for improving 316L stainless steel for biomedical applications. *Monatshefte für Chemie-Chemical Monthly*, 2019. 150: p. 1761–1771.
11. Heimann, R.B., Osseoconductive and corrosion-inhibiting plasma-sprayed calcium phosphate coatings for metallic medical implants. *Metals*, 2017. 7(11): p. 468.
12. Heimann, R.B., Magnesium alloys for biomedical application: Advanced corrosion control through surface coating. *Surface and Coatings Technology*, 2021. 405: p. 126521.
13. Zhao, J., et al., Carbon nanotube (CNT) encapsulated magnesium-based nanocomposites to improve mechanical, degradation and antibacterial performances for biomedical device applications. *Coatings*, 2022. 12: p. 1589. DOI: 10.3390/coatings12101589
14. Herenda, S., et al., In vitro biomedical corrosion and enzyme activity inhibition on modified Cu-Zn-Al bioalloy. *Corrosion Reviews*, 2023. 41(4): pp. 443–454.
15. Indira, K., U.K. Mudali, and N. Rajendran, Corrosion behavior of electrochemically assembled nanoporous titania for biomedical applications. *Ceramics International*, 2013. 39(2): p. 959–967.
16. Kausar, A., Potential of polymer/fullerene nanocomposites for anticorrosion applications in the biomedical field. *Journal of Composites Science*, 2022. 6(12): p. 394.
17. Li, J., et al., A silicate-loaded MgAl LDH self-healing coating on biomedical Mg alloys for corrosion retardation and cytocompatibility enhancement. *Surface and Coatings Technology*, 2022. 439: p. 128442.

18. Li, P., et al., In vitro corrosion inhibition on biomedical shape memory alloy by plasma-polymerized allylamine film. *Materials Letters*, 2012. 89: p. 51–54.
19. Liu, C., et al., In vitro corrosion behavior of multilayered Ti/TiN coating on biomedical AISI 316L stainless steel. *Surface and Coatings Technology*, 2006. 200(12–13): p. 4011–4016.
20. Majidi, H.J., et al., Fabrication and characterization of graphene oxide-chitosan-zinc oxide ternary nano-hybrids for the corrosion inhibition of mild steel. *International Journal of Biological Macromolecules*, 2020. 148: p. 1190–1200.
21. Manivasagam, G., D. Dhinasekaran, and A. Rajamanickam, Biomedical implants: Corrosion and its prevention-A review. *Recent Patents on Corrosion Science*, 2010. 2(1): p. 40–54.
22. Martínez, A.L., et al., Immobilization of Zn species in a polypyrrole matrix to prevent corrosion and microbial growth on Ti-6Al-4V alloy for biomedical applications. *Progress in Organic Coatings*, 2020. 144: p. 105650.
23. Nady, H., et al., Assessment of H2O2/albumin and glucose on the biomedical iron alloys corrosion in simulated body fluid: Experimental, surface, and computational investigations. *Journal of Molecular Liquids*, 2021. 339: p. 116823.
24. Pragathiswaran, C., et al., Experimental investigation and electrochemical characterization of titanium coated nanocomposite materials for biomedical applications. *Journal of Molecular Structure*, 2021. 1231: p. 129932–129932.
25. Rodríguez-Alonso, L., et al., Hybrid sol-gel coatings doped with non-toxic corrosion inhibitors for corrosion protection on AZ61 magnesium alloy. *Gels*, 2022. 8(1): p. 34–34.
26. Asadi, H., et al., A multifunctional polymeric coating incorporating lawsone with corrosion resistance and antibacterial activity for biomedical Mg alloys. *Progress in Organic Coatings*, 2021. 153: p. 106157.
27. Sheng, Y., et al., A compound Schiff base coating on biomedical magnesium alloy for enhanced corrosion resistance and biocompatibility. *Smart Materials in Manufacturing*, 2023. 1: p. 100003–100003.
28. Tan, J.K.E., P. Balan, and N. Birbilis, Advances in LDH coatings on Mg alloys for biomedical applications: A corrosion perspective. *Applied Clay Science*, 2021. 202: p. 105948–105948.
29. Uddin, M.N., S.S. Ahmed, and S.M.R. Alam, Biomedical applications of Schiff base metal complexes. *Journal of Coordination Chemistry*, 2020. 73(23): p. 3109–3149.
30. Xiong, P., et al., A pH-sensitive self-healing coating for biodegradable magnesium implants. *Acta Biomaterialia*, 2019. 98: p. 160–173.
31. Zhou, X., et al., Smart corrosion inhibitors for controlled release: A review. *Corrosion Engineering, Science and Technology*, 2023. 58(2): p. 190–204.
32. Dong, Q., et al., Insights into self-healing behavior and mechanism of dicalcium phosphate dihydrate coating on biomedical Mg. *Bioactive Materials*, 2021. 6(1): p. 158–168.
33. Xiong, P., et al., Osteogenic and pH stimuli-responsive self-healing coating on biomedical Mg-1Ca alloy. *Acta Biomaterialia*, 2019. 92: p. 336–350.

12 Coating Characterization with Electrochemical Techniques

Athira Ajayan, Vismaya Joseph, Anila Paul, and Abraham Joseph

12.1 INTRODUCTION

Corrosion is a relentlessly vital challenge of the utmost relevance in the engineering and industrial domains, stipulated to the financial and safety concerns. Even though corrosion is a spontaneous natural phenomenon that cannot be ceased, it can be delayed using several mitigation strategies, with coating being the overarching one. Coatings are a versatile approach as they can be potentially employed on a wide range of materials, are simple to cohere, are non-intrusive, and can enhance the aesthetics of structures. They can also be geared to meet environmental prerequisites as well as specific application demands, simultaneously offering flexibility and adaptability. While no single option is sufficient in all scenarios, coating mitigation stipulates a convenient and efficient approach for inhibiting corrosion.

Coatings refrain from corrosion by cathodic protection, anodic passivation, electrolytic inhibition, or active corrosion inhibition mechanisms. Withal, 'smart coatings' considered as the forthcoming phase of coating currently reside in vying prominence [1]. Materials that are accustomed to dynamically adapting their characteristics in response to external stimuli are referred to as 'smart', and the notion of 'smart coating' refers to those coatings that possess a knack for sensing their surroundings and initiate an appropriate response to the stimulus [2]. Beyond stimuli responsiveness, these coatings possess exclusive characteristics comprising self-healing capabilities, antifouling and antimicrobial characteristics, self-cleaning properties, environmental adaptability, sensing as well as monitoring capabilities, active release of functional agents, and energy efficiency [3–8]. Smart coatings thus impart dynamic control, superior performance, and greater sustainability by incorporating the aforementioned attributes.

Electrochemical characterization techniques are more pertinent and often used to study the corrosion characteristics and the mitigation performance of coatings owing to the corrosion process being overwhelmingly electrochemical in origin. The electrical parameters related to coatings are acquired and analyzed by electrochemical approaches. It has the potential to gather key information about coating characteristics, particularly corrosion resistance, adhesion, barrier characteristics, ion diffusion, as well as other electrochemical processes occurring over the coating-substrate interface, by applying controlled potentials or currents. Electrochemical techniques for coating characterization lend real-time information on the electrochemical behavior of coatings, facilitating the detection of localized corrosion, monitoring coating deterioration, and the assessment of protective properties. This real-time monitoring capability is especially beneficial for evaluating coating performance in dynamic and fluctuating environments. Electrochemical techniques also furnish quantitative measurements of coating aspects such as coating thickness, porosity, and adhesion strength, which provide substantial information for quality control and performance

DOI: 10.1201/9781032677255-12

analysis. Furthermore, by evaluating coating performance in specific environmental scenarios and simulating real-world situations, these approaches assist in fabricating tailored coating that meets specific applications.

In particular, two strategies—global and local—are often used to sort electrochemical characterization strategies. Global electrochemical coating characterization includes exploring the holistic electrochemical behavior of the coating system as an ensemble, thereby giving a macroscopic glimpse into its performance. Here, the response to a perturbation corresponds to a surface-averaged measurement accountable to the behavior of the entire surface. In contrast, the primary emphasis of local electrochemical coating characterization is on exploring specific areas or localized regions of the coating. It enables the detection and mapping of local variations in electrochemical activity, spotting areas of potential vulnerability and imperfections. Local characterization delivers in-depth insights into localized corrosion behavior and coating heterogeneity, whereas global characterization offers a peripheral evaluation of the coating [9].

In global coating characterization, as intact unperturbed segments are also taken into account during the normalization operation, values obtained by large area measurements may not be an exact characteristic of the coating, as corrosion may only occur at some minuscule defect areas. However, it is an innately intricate endeavor to translate global electrochemical techniques into local strategies. Besides soaring computation demands, there is an enormous upsurge in mechanical instrumentation, such as in the positioning capabilities of a probe, microscope, or micro-electrochemical cell, often with the sub-micron resolution required [10]. Further, local processes can be time-consuming and have limited coverage as they only assess specific regions and a meticulous selection of the specific region is required to assure representativeness.

In a nutshell, global methods provide a broad assessment of the overall performance of coating, whereas local techniques provide detailed information concerning localized behavior. The choice between the two approaches is determined by the specific objectives of the study, the extent of information required, and the obstacles of each strategy. The combination of global and local approaches enables an in-depth comprehension of the nature and features of coating, strengthening its efficiency and robustness.

In this chapter global and local electrochemical coating characterization techniques are dealt with in sections 12.2 and 12.3, respectively. The global characterization techniques discussed are electrochemical impedance spectroscopy (EIS), potentiodynamic polarization (PDP), and electrochemical noise method (ENM) while the local coating characterization section covers scanning vibrating electrode technique (SVET), localized electrochemical impedance spectroscopy (LEIS), scanning electrochemical microscopy (SECM), scanning kelvin probe technique (SKPT), and scanning ion-selective electrode technique (SIET).

12.2 GLOBAL ELECTROCHEMICAL TECHNIQUES

12.2.1 Electrochemical Impedance Spectroscopy

In EIS, an electrochemical system is stimulated by sinusoidal, small-amplitude perturbations of potential or current, and the response to these perturbations in terms of current or potential is monitored over a wide range of frequencies (μH to MHz). The stimulus and response data are utilized to determine the transfer function, which is electrochemical impedance in the case of an electrochemical system. Either applying a potential perturbation and gauging the current response in the potentiostatic mode or applying a current perturbation and gauging the potential response in the galvanostatic mode could potentially be utilized to determine the EIS data. Contingent to the nature of the electrochemical system, either potentiostatic or galvanostatic mode can be applied. The potentiostatic mode is generally preferred for electrochemical systems with higher impedance, while galvanostatic impedance is usually better suited for electrochemical systems with low impedance and higher current levels [11, 12].

Speculate the case wherein a marginal sinusoidal potential is applied to an electrochemical system so that nonlinear effects can be omitted, as shown by equation (12.1).

$$V(t) = V_0 \sin(\omega t) \tag{12.1}$$

where V_0 is the potential amplitude, ω is the angular frequency, t is the time, and the term ωt represents the phase of the waveform. The ensuing current response has a distinct amplitude (I_0) and a phase shift (φ), as given by equation (12.2).

$$I(t) = I_0 \sin(\omega t + \varphi) \tag{12.2}$$

The Fourier transform of sinusoidal voltage and current signals converts the data from the time domain into the frequency domain, giving voltage and current phasors ($\bar{V}$ and $\bar{I}$)respectively (equations 12.3 and 12.4), and thus enables the determination of electrochemical impedance.

$$\bar{V} = V_0 e^{-j\omega t} \tag{12.3}$$

$$\bar{I} = I_0 e^{-j[\omega t + \varphi(\omega)]} \tag{12.4}$$

The impedance (Z) can be defined as the ratio of the voltage and current phasors,

$$Z = \frac{V_0}{I_0} e^{j\varphi(\omega)} = Z_0 e^{j\varphi(\omega)} \tag{12.5}$$

The magnitude of impedance gives the level of resistance to the flow of alternative current (AC), while the phase angle provides the extent of phase shift between voltage and current. It can also be represented as equation (12.6).

$$Z = Z_0 \left[cos\varphi(\omega) + jsin\varphi(\omega)\right] = Z' + jZ'' \tag{12.6}$$

where $j = \sqrt{-1}$; Z' is the real part and Z'' is the imaginary part which embodies, respectively, frequency-dependent resistance and reactance. The resistive or ohmic component represents the energy represented as heat, while the reactive component gives the energy stored and released as an electromagnetic field. At higher frequencies, the reactive component is inductive and is directly correlated to frequency; however, when operating at lower frequencies, it possesses a capacitive characteristic and is inversely correlated to frequency. In the EIS technique, the transfer function, impedance is measured as a function of frequency in a linear, causal, and time-invariant system.

The EIS instrumentation consists of a signal generator to produce the perturbation signal, a potentiostat/galvanostat to regulate the applied potential or current, an electrochemical cell, and a response assessing system. The working electrode, reference electrode, and counter electrode are encompassed within the electrochemical cell, and an impedance measurement system tracks the response system. EIS data can be assessed through various strategies, including alternating-current bridges, Lissajous curves, phase-sensitive detection and lock-in amplifiers, frequency-response analyzers (FRAs), Laplace transforms, and wavelet transforms [13]. Of these, the dominant approach belongs to the FRA category, which relies on the orthogonality of sines and cosines to determine the real and imaginary parts of the complex impedance at a given frequency. A computer with specialized software stewards the instrumentation set and acquires data and performs analysis.

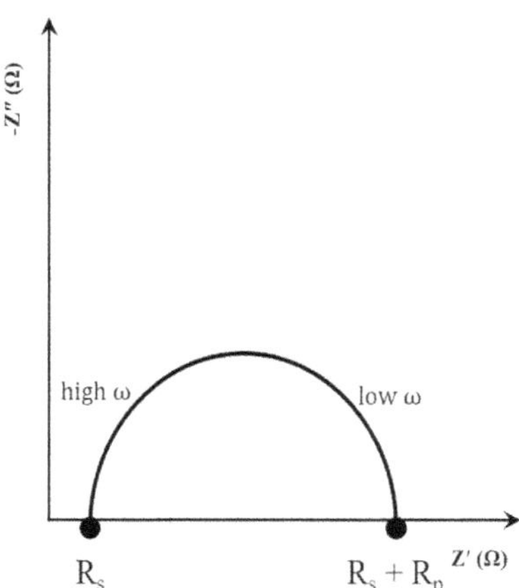

FIGURE 12.1 A general representation of the EIS Nyquist plot in Cartesian coordinates.

The Nyquist plot and the Bode plot are the two most popular avenues to depict EIS data, despite there being other approaches as well [14].

Nyquist Plot: The Nyquist plot is a graph of the imaginary component of the impedance (-*Z*'') versus the real component of the impedance (*Z*'). The data is often represented on a graph as a series of data points, each of which embodies the impedance at that specific frequency. The resistance and any frequency-dependent behavior like inductive or capacitive behavior of the material are readily apparent using the Nyquist plot. The gestalt of the plot unveils the electrochemical mechanism of the system. Rudimentarily, the intercept on the real axis and diameter of the semicircle in the Nyquist plot portray, respectively, solution resistance (R_S) and charge transfer or polarization resistance (R_p) as embodied in Figure 12.1 [13, 15].

Bode Plot: The Bode plot is composed of two distinct plots: videlicet magnitude versus frequency and phase versus frequency as illustrated in Figure 12.2. The Bode plot is adept at illustrating frequency-dependent system behavior, such as resonance, relaxation, or diffusion-controlled processes. In a Bode plot, a peak in the magnitude can verify the existence of a resonance or relaxation process, whereas a plateau in the phase can attest to a diffusion-controlled process [15, 16].

Impedance data can be simulated either via deploying analogs or physical models [17]. An analog devours mathematical functions or equivalent electrical circuits to explain experimental impedances with the intent to verify an adequate fit between calculated and observed impedances with the minimal possible parameters. Electrical equivalent circuits (EECs) are widely employed analogs that primarily consist of a combination of resistors, inductors, and capacitors. A resistor is used to model the resistance of the system to current flow and emulates the ohmic resistance of the electrolyte and electrodes.

As the impedance of a resistor is equivalent to real resistance and frequency-independent, it doesn't cause a phase shift in the voltage and current signals. The charge storage capacity of the electrochemical system is simulated using a capacitor, which represents the double-layer capacitance at the electrode-electrolyte interface. The impedance of a capacitor is frequency-dependent and purely imaginary, meaning that it causes a 90° phase shift between the voltage and current signals. An inductor can be employed to mimic the energy storage in a magnetic field that occurs when an electric current passes through an electrochemical system.

Electrochemical or faradaic elements such as constant phase elements (CPE, Q) and Warburg (W) define secondary circuit elements. Constant phase elements are electrical circuit elements that are used to simulate the conduct of non-ideal capacitors and are portrayed as a parallel combination of capacitors and resistors. A broad range of nonideal capacitive behavior can be described by the

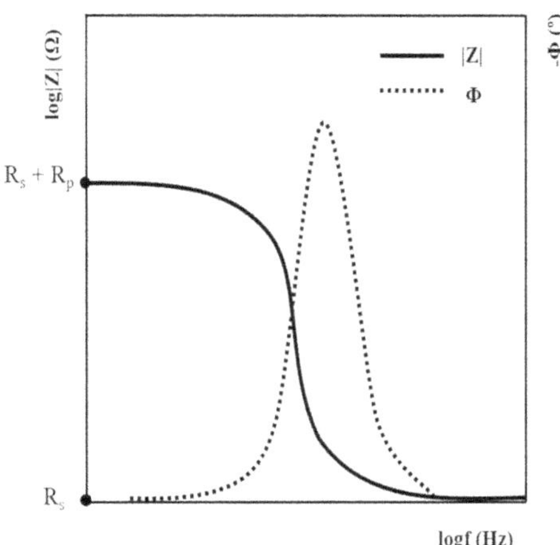

FIGURE 12.2 A general representation of the EIS Bode plot.

TABLE 12.1
Mathematical description of the impedance of common equivalent circuit elements

Electrical Element	Impedance	Terms
Resistor	$Z_R = R$	R—Resistance in Ohm [Ω]
Capacitor	$Z_C = \frac{1}{j\omega C}$	C—Capacitance in Siemen [S]
Inductor	$Z_L = j\omega L$	L—Inductance in Henry [H]
Constant phase element	$Z_{CPE} = \frac{1}{Y_0(j\omega)^{\alpha}}$	Y_0—Magnitude of the admittance at zero frequency; α—Power-law exponent
Warburg element	$Z_W = \frac{Y_0}{(j\omega)^{1/2}}$	Y_0—Magnitude of the admittance at zero frequency

CPE, which is a generalization of the ideal capacitor. The power-law exponent (α) is a gauge of the deviation from ideal behavior, with a value of 1 indicating ideal capacitive behavior and values of $\alpha > 1$ or $\alpha < 1$suggesting nonideal capacitive behavior. A Warburg element is frequently used to model the demeanor of diffusion-limited processes at an electrode-electrolyte interface and is considered to be akin to a series-connected resistor and capacitor equivalent circuit.

The impedance of each of the aforementioned electrical elements is defined mathematically in Table 12.1. Each element of the circuit corresponds to a physical process and contributes to the total impedance of the system. The values of these circuit elements are adjusted to fit the experimental results obtained from EIS measurements and the parameters are determined. The chi-squared value determines the fidelity of the data modeling to a given electrical equivalent circuit. However, it should be emphasized that due to the fact certain circuits are mathematically identical, simulating an impedance spectrum with more than one circuit is possible. Furthermore, they do not directly offer insights into the physical processes occurring in the electrochemical system. On the other hand, physical models endeavor to account for the mechanism of the processes occurring at the interface in terms of physico-electrochemically valid concepts, in addition to reproducing the phenomenon of interest. These models can offer a more accurate and comprehensive understanding of the system, albeit they are often more complicated and demand more in-depth knowledge of the system under investigation.

The Kramers–Kronig (K&K) relations can be leveraged to assess the validity of the impedimetric data [15]. They enable us to deduce the imaginary part of the impedance from the real part, and vice versa since they are related by the Hilbert transform [18]. The aforementioned fundamental condition on the real and imaginary components of the impedance spectrum is required to be satiated in order for the measured data to be deemed consistent. Major discrepancies from the K&K relationship imply that the measurement is not accurate, the circuit model used to fit the data is illicit for the system under investigation, or the electrochemical system exhibits nonideal behaviors that the circuit model is unfit to reimburse.

EIS is a versatile, noninvasive, in-situ coating characterization technique that can provide information on the electrochemical behavior and corrosion resistance of the coating [19, 20]. The corrosion impedance/resistance can be assessed by monitoring changes in the impedance values over time in a corrosive environment. A high impedance may characterize better coating performance and a decrease in the impedance values may indicate corrosion of the substrate or the coating itself. Besides, EIS can be used to evaluate the integrity as well as the durability of the coating by monitoring the impedance values over time, and if there is a diminution in impedance values it is possibly an indicator of coating deterioration, such as pinholes or cracks [21]. Further, by grabbing cognizance of the characteristic impedance values, coating thickness can be determined. The high-frequency impedance values correspond to the coating capacitance and are inversely proportional to the coating thickness [22]. The impedance of the coating–substrate interface can be utilized to gauge the adhesion of the coating as well [23]. Good adhesion is characterized by a high impedance value, while delamination of the coating or poor adhesion is detected by a low impedance value.

The processing of coating impedance data often involves the usage of EECs [24]. As a way to depict the behavior of coatings, Beaunier et al. suggested a very rudimentary electrical model in Figure 12.3, in which R_s, R_c, C_c, C_{dl}, and R_p refer to solution resistance, coating resistance, coating capacitance, double-layer capacitance, and charge transfer or polarization resistance, respectively [23, 24]. The electrical characteristics and phenomena affiliated with the coating including conductivity, insulating properties, charge storage capacity, polarization behavior, film thickness, barrier performance, and the presence of defects or barriers can be delimited by these elements and their numerical values can be obtained.

12.2.2 Electrochemical Noise Method

ENM is an atomic-level long-term corrosion diagnosing technique. The term 'electrochemical noise' is because it measures the fluctuations of current and potential of low frequency and

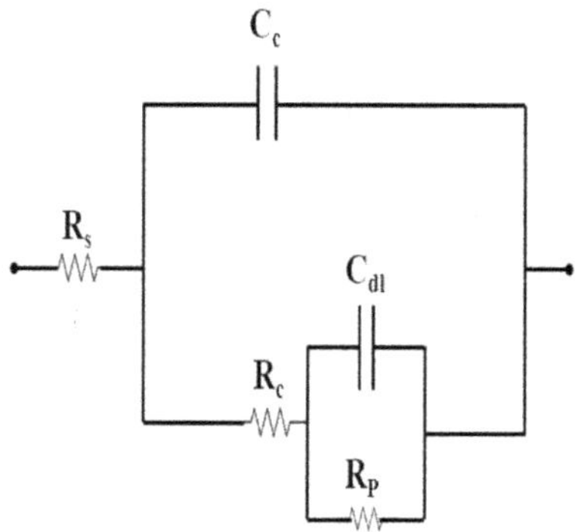

FIGURE 12.3 Simplified equivalent circuit model for coated metal.

low amplitude associated with the corroding metal structures. In contrast to other new electrochemical techniques like electrochemical impedance spectroscopy, scanning reference electrode technique, SVET, and wire beam electrode techniques, the ENM detects the current in the metal–solution interface rather than the solution phase [25]. Iverson and Tyagai used sound as a source of information in the 1960s and 1970s, and Iverson recorded the first electrochemical potential noise (EPN). This discovery paved a new path that correlated the frequency and amplitude of electrode potential fluctuations with corrosion processes. Setwart et al. supported this method by recording the fluctuations in current associated with the phenomenon called electrochemical current noise (ECN), of metastable pitting for the first time. Eden et al. modified this method by introducing a three-electrode system for the measurement of ECN [26]. One of the most important applications of the ENM is the determination of general corrosion rates using electrochemical noise resistance (R_n), which is calculated by dividing potential noise with current noise as standard deviation. The benefits of the ENM include the rapid monitoring of the corrosion process with the omission of applied external signals for the collection of experimental data. Due to the simplicity of the ENM experimental setup, online corrosion monitoring is also possible [27]. Still, the ENM is a challenging one to apply practically because it needs the identification of the type of corrosion and the difficulty in the interpretation of electrochemical noise data. ENM analysis is excellent in early detection and warning of corrosion, especially localized corrosion.

Iverson developed the first EN measuring device as a simple device containing a working electrode (WE) as corroding material and a low-noise reference electrode (RE) to detect open-circuit potential fluctuations using a sensitive high-impedance voltmeter. It helps to identify the type of corrosion by giving electrode voltage fluctuations characteristic to each type of corrosion and thereby helping scientists to calculate the rate of corrosion. Then a more sensitive double RE device was developed by Hladky and Dawson. Later they modified this system to a more simpler form by introducing two identical WEs instead of using a working and a reference electrode [26]. A voltmeter is used to record the potential difference between these two WEs as potential noise. Generally, the weight loss method is considered the gold standard for identifying the corrosion rate. But the discovery of these two working electrode systems attracted more attention due to the ability of their system to the early detection of corrosion initiation and identify the type of corrosion without taking much more time as compared to the gravimetric method. One limitation of this method lies in identifying which of the two electrodes is the source of the noise. Another type of noise measurement is via ECN. This method uses two identical electrodes to measure the current or a single electrode held at a fixed potential to measure the current. Comparing the two, the former is easy to handle as it avoids the difficulty of holding the WE at a fixed potential [28]. A method of overcoming these difficulties is to use EPN and ECN simultaneously for EN measurement. The combination of electrochemical potential and current noise is more powerful than the individual measurements. Eden et al. were probably the first to use a device consisting of a three-electrode electrochemical cell as a noise detection device where ECN is measured as the current between two identical WEs under potentiostatic control using a zero resistance ammeter (ZRA) or EPN is measured as the potential between two identical WEs under galvanostatic control concerning a third WE as a reference electrode [29]. Notable advantages of this device are that it uses no applied external polarization signal and that it collects potential and current noise data simultaneously. This device is now effectively the standard apparatus used to measure electrochemical noise.

In potential measurement, the amplitude of EPN is less than 1 mV. Then the average potential may be several hundred mV by using a true reference electrode. So, the system must have good sensitivity. There are two methods to offset the average direct current (DC) level. The first method is to subtract predefined values. The second method involves the removal of dc by using a very low-frequency high-pass filter [30]. Generally current is measured according to Ohm's law, I=VR. In ECN, the resistor placed in the current path leads to a voltage difference between two WEs. So, the use of a current amplifier produces a voltage output that is proportional to a current input, with no voltage drop between the input terminals [31]. Therefore, the device is called a ZRA.

The Rn is mathematically represented as:

$$R_n = \frac{\sigma_E A}{\sigma_I} \tag{12.7}$$

where σ_E and σ_I are the standard deviations of potential and current, respectively, and A is the area of the sample. According to the Stern–Geary equation, corrosion current density can also be calculated with the use of R_n:

$$i_{corr} = \frac{B}{R_n} \tag{12.8}$$

where i_{corr} is the corrosion current density and B is the Stern–Geary coefficient. Significant advances have been made over the past decades in electrochemical noise analysis for sensing corrosion, especially localized corrosion [30]. Various forms of potential and current noise detection devices have been designed for different applications. Noise signatures and noise resistance techniques have been developed to provide a practical means of early corrosion warning.

12.2.3 Potentiodynamic Polarization

PDP is a prominent electrochemical technique that uses DC and has the potential to find out the corrosion rate of a coated metal specimen. Polarization is an extremely important corrosion parameter, which enables one to derive the rate of corrosion processes from measurements of the relation between polarization and current density [29]. This electrode kinetic method is also called the Tafel-plot technique which exhibits a curve showing the relationship between electrode potential (ΔE) and polarization current density (log i) as shown in Figure 12.4 [32]. This plot consists of anodic and cathodic polarization curves. An essential corrosion parameter called corrosion current density (i_{corr}) can be determined from this plot by extending the linear area of the two polarization curves to intersect a point [33].

This point of intersection corresponding to the x-axis gives i_{corr} value, whereas that corresponding to the y-axis denotes E_{corr} value. The slope of the anodic curve is called the anodic Tafel slope, whereas the slope of the cathodic curve is called the cathodic Tafel slope, which are represented as β_a and β_c, respectively, and are known as mechanistic indicators of the corrosion process. Wagner and Traud derived the current-density and polarization relationship using mixed potential theory as given below:

$$i = i_{corr}\left[exp\left(\frac{2.3\Delta E}{\beta_a}\right) - exp\left(\frac{-2.3\Delta E}{\beta_c}\right)\right] \tag{12.9}$$

where $\Delta E = E - E_{corr}$ and $\beta_a = \dfrac{2.3RT}{F(\alpha_+)a}, \beta_c = \dfrac{2.3RT}{F(\alpha_-)c}$.

Normally the instrumental setup of this voltage-regulated system uses a three-electrode setup, which is polarized by giving a range of potentials starting from –250 mV to +250 mV with a 1mV per second scan rate from cathodic to the anodic region [34]. Then the current is produced in response to the applied potential cause's anodic and cathodic potentials to find out the kinetics of

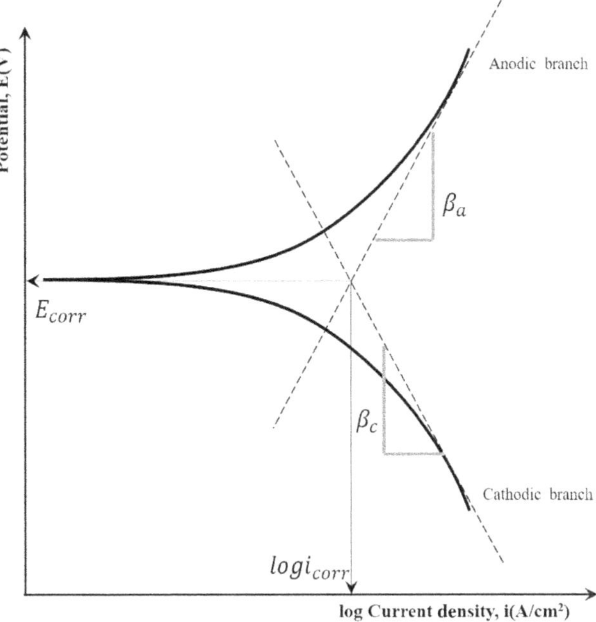

FIGURE 12.4 A general illustration of Tafel's plot.

anodic and cathodic processes. The protection efficiency (η) of the coating can be calculated from i_{corr} using the following equation:

$$H = \frac{i^*_{corr} - i_{corr}}{i^*_{corr}} \tag{12.10}$$

where i^*_{corr} is the corrosion current density of the bare metal and i_{corr} is the corrosion current density of the coated metal. The Stern–Geary equation utilizes an important parameter known as polarization resistance (Rp), which underscores the effectiveness of the anticorrosive coating. The subscript denotes the protective efficiency of the coating. The Rp value of a coated specimen can be calculated using the equation involving $icorr$, βa, and βc:

$$R_p = \frac{\beta_a \beta_c}{2.303(\beta_a + \beta_c) i_{corr}} \tag{12.11}$$

The higher the R_p value, the lowerthe i_{corr}, and corrosion rate confirms the better anticorrosive properties of the coating material, whereas a higher E_{corr} value indicates the corrosion protection behavior of that coating material [35].

The rapid detection of corrosion, especially for localized types of corrosion, with an easy determination of η, R_p, and corrosion rate is the main advantage of PDP. PDP measurement needs a large amount of current density. The usage of an excessive amount of current density will damage the surface coating of the metal specimen, thereby introducing errors in the results and also affecting the uncompensated solution resistance [36]. Another disadvantage of this method is the use of a wide potential range with the measurement of current density which may also alter the results in the case of adsorbed species being potential-dependent [30].

12.3 LOCAL ELECTROCHEMICAL TECHNIQUES

12.3.1 Scanning Electrochemical Microscopy

SECM is a versatile technique that combines the principles of electrochemistry and scanning probe microscopy in tandem to study local electrochemical processes with high spatial resolution. SECM enables the investigation of various electrochemical phenomena, including corrosion, electrocatalysis, and ion transport, making it a valuable tool for coating characterization [37, 38].

The SECM instrumentation is composed of an electrochemical cell with a three-electrode configuration (ultramicroelectrode [UME], reference electrode, and counter electrode), a potentiostat, a feedback system, a scanning system, and a data acquisition/control system. While the reference and counter electrodes complete the electrochemical circuit, the UME monitors the local electrochemical activity. A UME probe (or tip) is usually utilized by SECM to induce chemical changes and gather electrochemical responses while scanning the substrate, which acts as the working electrode [37]. A multitude of UMEs, such as microband electrodes, cylindrical electrodes, micro rings, disk-shaped electrodes, and hemispherical electrodes, frequently serve as SECM probes. Of these, disk-shaped geometry is preferred since this configuration provides significant benefits for electroanalytical applications, including a significantly reduced ohmic potential drop in solution and double-layer charging current, the ability to reach a steady state in seconds or milliseconds, and a small dimension facilitating microanalysis. Micro-dimensioned probes, on the other hand, are being upgraded to nanopipette-based probes, for niche multifaceted applications. A potentiostat manages the potential difference between the microelectrode and the reference electrode alongside a feedback system that maintains the distance between the tip and the substrate constant while scanning. The ultramicroelectrode is swept across the sample surface by the scanning system, which is powered by motors or piezoelectric actuators. The electrochemical data are captured and archived by a data acquisition and control system. The versatility of SECM can be augmented with advanced setups encompassing imaging systems, spectroscopic techniques, or environmental control systems. These integrated components facilitate precise control, scanning, and data acquisition for high-resolution electrochemical imaging in coating characterization and other applications [39].

Feedback mode, tip generation/substrate collection, substrate generation/tip collection, penetration mode, and ion transfer feedback mode comprise several of the possible SECM operation or measurement modes [40]. Further, SECM can be categorized into amperometric SECM, potentiometric SECM (commonly referred to as SIET which is discussed separately in Section 12.3.5), and AC-SECM based on the mode of detection [9].

With SECM, processes and structural features at the substrate of interest are investigated as the electrode tip moves near the surface of a conductive, semiconductive, or insulating substrate which is immersed in a solution. The electrochemical activity and/or topography of the substrate can be retrieved by moving the probe in the x- or y-directions, while moving the probe in the z-direction of the substrate reveals information on the kinetics of electron transfer at the substrate surface.

The quantitative theory of SECM leverages mathematical modeling of electrochemical processes and mass transport to acquire quantitative information about the local electrochemical aspects of a sample [41]. It incorporates principles comprising Fick's laws of diffusion for describing species transport, boundary conditions for electroactive species at the electrode surface, feedback equations relating current, tip-substrate distance, and electrochemical properties, as well as numerical simulations to solve complex equations and anticipate experimental findings. The analysis entails fitting experimental data, interpreting experimental data, extracting quantitative parameters, and optimizing experimental conditions. However, due to the complexity of electrochemical processes and influencing factors, SECM quantitative analysis might be difficult. The theory aids in characterizing coatings and other surfaces with high spatial precision and provides a more profound understanding of electrochemical behavior.

SECM is an efficient means for coating characterization owing to its dexterity to elicit and monitor local changes in chemical composition and electrochemical characteristics with a high spatial resolution. It further enables the assessment of coating thickness, uniformity in adhesion, and corrosion resistance. In order to identify defects, differences in corrosion resistance, and areas of coating degradation, SECM scans a microelectrode over the coating surface and measures the electrochemical response. Electrochemical parameters, such as corrosion rates, can be quantitatively analyzed using feedback mode, where the microelectrode is maintained at a fixed distance from the coating surface. To acquire a thorough understanding of coating behavior, SECM can be used in conjunction with other approaches [41, 42].

12.3.2 Scanning Vibrating Electrode Technique

SVET, a genre of scanning probe microscopy, is employed to investigate localized corrosion and electrochemical processes on conducting materials. At the outset, SVET is used to detect ionic currents generated in biological processes like electrophysiology, tissue regeneration, and cell differentiation. However, in the inception phase, nonvibrating reference electrodes were utilized to measure the potential distribution in solutions. This scanning reference electrode approach, comprising parallel electrodes in an electrochemical cell, measures the current density from the potential difference with an ohmic drop in solution using equation 12.12.

$$I = \kappa E = \frac{-\kappa \times \Delta V}{\Delta r \times l} \tag{12.12}$$

Where κ is the solution conductivity, E is the electric field in solution, and ΔV is the potential difference between two points separated by the distance Δr in the direction of current flow. The current density obtained is noise-prone; meanwhile, it can be dwindled by vibrating electrodes leading to enhanced sensitivity [43].

A mechanically vibrating probe and lock-in amplifier are implemented in SVET to enhance signal recovery and improve sensitivity by about an order of 6. An electrometer receives the signal and noise response from the vibrating probe in the instrumentation setup, and a lock-in amplifier significantly reduces the noise component. The output is further filtered as the probe traverses over the surface of interest, and following a series of measurements, the data map is developed. The genesis of a vibrating electrode and the modulation of signals by a lock-in amplifier enhanced the signal-to-noise ratio and, as ensued, the output of SVET to better in agreement with the theoretical value obtained [43].

The current density distribution over the surface of a conducting material immersed in a solution is determined using the SVET. Recognizing that the conductivity of the electrolytic solution above the surface is substantially lower than that of the metal, ionic current causes a potential gradient to forge in the electrolytic solution. This potential gradient is measured by the vibrating probe by converting the DC field to an AC signal. Further, the AC signal obtained is converted to ionic current density by calibration procedure [9]. With a vibrating conductive probe that can monitor the potential difference between the probe and electrolyte in contact with corroding metal, SVET can perceive the local cathodic and anodic reactions. Contingent on the measurement as well as instrument category and the approximate distance of the probe from the surface, the probe may vibrate along multiple axes or merely a single axis. The current at the desired location is determined by measuring the potential difference in the solution built up by the flow of electrons as well as ions and then comparing the known vibrational amplitude to the resistivity of the solution.

$$I = -\left(\frac{1}{\rho}\right)\left(\frac{\Delta V}{2d}\right) \tag{12.13}$$

where I is the current, ΔV is the voltage difference, d is the amplitude of vibration, and ρ is the resistivity of the electrolyte. The process of obtaining the current density is repeated over a selected surface. Scanning in a plane parallel to the sample surface using the probe will give the potential difference at each point providing a map of the current distribution from the grid of points [10].

For accurate values the calibration is done by placing the probe at a fixed distance from a point current source of known current (I); the current density (i) at distance r from the point source is given by

$$i = \frac{I}{4\pi r^2} \tag{12.14}$$

The results are influenced by numerous factors like current density differences between the area of measurement and its surroundings, the distance of the vibrating probe from the sample surface, vibrating amplitude, the sensitivity of the detector, and the resistivity of the solution.

The degradation or deterioration of corrosive protective coatings can be evaluated using different electrochemical methods. The lifetime of the protective film can be attributed to the resistance of the film (R_f), which is often determined using the EIS method, and to the current density measured using SVET [44].

SVET measures the potential difference to determine the current density (i) in relation to the immersion time (t). It demonstrates that the current density increases with an increase in immersion time, and a greater value of current density is associated with a high degree of corrosion in relation to the ion diffusion coefficient. These findings can be related to R_f or t_s determined by EIS, which stands for the resistance and longevity of a protective film, respectively. R_f values are inversely proportional to the current density and decrease with increasing immersion time. If the current density measured from SVET is higher, the lifetime of the coating will decrease, indicating its degradation [44].

Localized detection of coating by SVET helps in investigating the self-healing or repairing property by current density values. The conventional coating exhibits an explicit anodic activity promptly after being immersed in the corrosive medium, and this response increases as the immersion period extends. However, the self-healing coating has neither anodic activity nor corrosion products; therefore, the SVET technique can be utilized to analyze the self-healing capabilities by monitoring the anodic activity. Nevertheless, many of the repairing processes are triggered only after the corrosion; therefore, initially the anodic activity is observed and then the absence of flux indicates the anticorrosive and self-healing ability of coating [44].

12.3.3 Local Electrochemical Impedance Spectroscopy

The theory of LEIS is similar to that of EIS but with the added dimension of localized measurements. In lieu of applying small sinusoidal AC voltage perturbation to the entire surface of the material, LEIS entails applying it to a microelectrode in contact with a localized zone or point, and the resulting current response is measured [45]. The measurement of the local current density in close proximity to the interface under examination is a vital requirement for the LEIS device. Among several possibilities, the dual probe system promises to be an excellent fit as it is possible to make resilient metallic microelectrodes (UME). Further, both the inter-microelectrode distance and the UME dimension should, in preliminary approximation, influence the resolution of the probe [46].

A probe consisting of reference microelectrodes embedded in close proximity to the interface of interest is used to gauge the potential difference $\left(\Delta V_{probe}\right)$ in solution, from which the local current density (i_{loc})is obtained.

$$i_{loc}(\omega) = \frac{\Delta V_{probe}(\omega) \times \kappa}{d} \quad (12.15)$$

where κ is the electrolyte conductivity and d is the distance between probes. Therefore, the local impedance (z) can be derived from the ratio of the global AC potential perturbation $(\bar{V})$ and the local AC solution current response measured with the dual-electrode according to Ohm's law.

$$z = \frac{\bar{V}(\omega)}{i_{loc}(\omega)} = \frac{\bar{V}(\omega) \times d}{\Delta V_{probe}(\omega) \times \kappa} \quad (12.16)$$

Impedance measurements could be conducted on one point at a multitude of frequencies or by mapping the surface at a fixed frequency, enabling the recognition of electrochemically active zones [47].

Akin to EIS, the impedance spectrum can be demonstrated as a Nyquist plot or a Bode plot, and equivalent circuit models can be used to derive information regarding the electrochemical characteristics of the localized area or point of interest.

LEIS is a potent means for coating characterization, offering pertinent information on the adhesion, corrosion resistance, and barrier properties of the coatings [48]. On account of the localized nature of measurement, LEIS reveals the electrochemical traits of specific regions or points on the coating surface, which enable the mapping of localized defects or fluctuations in the properties of the coatings [49].

LEIS can be utilized to figure out the thickness of coatings by monitoring the impedance over different frequencies, which is further handy for grading the homogeneity of the coating or locating changes in thickness that potentially hint at defects or irregularities. Further, merely by measuring the impedance at the coating substrate interface, LEIS can also be utilized to discern the extent of adherence. The long-term performance and degradation of coatings in various environments can be investigated by measuring the impedance of the coating at various intervals using LEIS [50].

12.3.4 Scanning Kelvin Probe Technique

SKPT is a nondestructive and noninvasive approach for assessing corrosion potential that utilizes a metallic microelectrode probe termed the Kelvin probe as a reference electrode. In this case, a parallel plate capacitor is developed with air as the dielectric medium, and it is externally connected to establish a Volta potential difference (VDP).

The vibration of the probe produces an AC that is annulled by applying a DC voltage. The voltage applied to rescind the current can be recognized as VDP, and the potential of the sample surface can be calculated using the known potential of the reference probe [9]. SKPT requires neither electrolytic contact nor a conducting channel between the probe and the sample; in turn, it can be measured in a vacuum, open air, humid air, or with a single drop of electrolyte. According to the difference in work function between the probe and the substrate, a potential formed at the conductor surface is brought into electrical contact, ensuring that electrons move from lower work function material to higher work function material. This configuration of the probe and substrate establishes a parallel plate capacitor.

Kelvin probe and sample surfaces are delegated as capacitor plates with potentials of E_{F1} and E_{F2} and work functions of Φ_1 and Φ_2, respectively. If the conductors are not connected, there is no potential difference; on the other hand, if they are connected, the charge flow will generate the potential difference by the equalization of Fermi levels. Figure 12.5 illustrates this diagrammatically. The potential difference (V_c) related to the work function is given by equation 12.17.

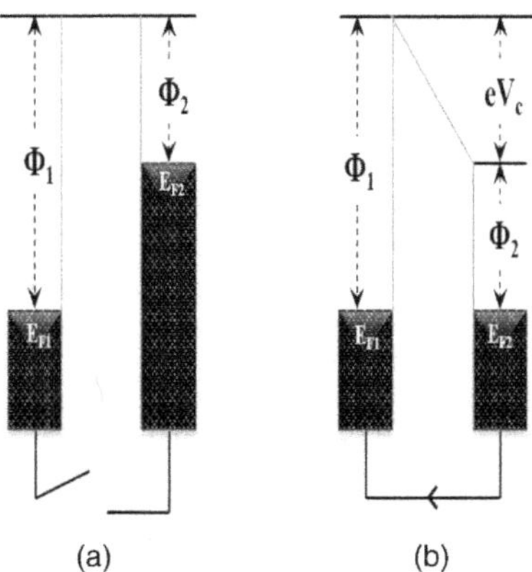

FIGURE 12.5 Charge and energy level when plates are (a) not electrically connected and (b) electrically connected.

$$-eV_c = \Phi_1 - \Phi_2 \tag{12.17}$$

The instrumentation provides the equal and opposite potential to nullify the generated potential difference, from which the corrosion potential (E_{corr}) can be directly calculated using equation 12.18.

$$E_{corr} = K + \frac{(\Phi_1 - \Phi_2)}{e} \tag{12.18}$$

where K is a constant obtained from the calibration of a given probe using different materials such as copper, zinc, and iron.

The potential difference between the sample substrate and Kelvin probe is directly proportional to the corrosion potential. Therefore, SKP can be used to investigate the location of defects in the coatings and to measure the delamination of the coating. The bare substrate shows negative potentials, whereas the coated surface has high potential, indicating inhibition of corrosion. The shift of positive to negative potentials represents the defected area and the delamination gives the potentials of bare substrate. The measurement of the potential difference between two interfaces provides the rate of delamination. The mechanism of delamination along with the influence of air, N_2, and CO_2 in coating delamination and the effect of pigment addition can be investigated by SKPT. The characterization of the defect can be done by this technique by visualizing the distribution of current densities. At the advanced stage, the anodic potential is observed and the formed rust can act as a cathode which cyclically creates another anode beneath the coating. Current mapping can give a complete picture of the electrochemical behaviors beneath the coating layer. SKPT was successively used to study the corrosion protection mechanism disbandment and the effect of oxygen partial pressure, local geometry, and thickness of electrolyte beneath the disbanded coating [9, 51].

The pretreatment of the surface before the application of coatings to improve anticorrosive coatings can protect the corrosion reaction by blocking cathodic, anodic, or both reactions. The protective nature and mechanistic details of pretreatment can be investigated by SKPT. The shift in negative and positive potentials with respect to the nontreated surfaces is used to understand the mechanism. Metal processing steps can induce mechanical stress and cause changes in defects characterized by the work function measurements. Trace quantities of elements in metal alloy composition can affect the overall properties, including corrosion performance, which causes an impact on work function. Such changes produced by intermetallic particles or inclusion can be measured by SKPT. Hydrogen inclusion, evolution, permeation, diffusion, and adsorption can be studied along with the diffusion of oxygen known as the rate-limiting factor in corrosion [10]. However, submicroscopic distinctions are not possible with SKPT; hence, this technique is often combined with atomic force microscopy (AFM) by using an AFM tip for potential measurement. The combined technique known as scanning Kelvin probe force microscopy (SKPFM) can measure surface topography and potential distributions with a resolution of 0.1μm ranging up to nanometers. The mismatching results on rough surfaces can be corrected by capacitive height potential measurements and can be related to anticorrosion studies [10].

Coupling of SKP with AFM will result in topographic imaging and high-resolution measurement of surface potential. In recent studies the use of SKPFM provided both morphology and surface potential details. In accordance with the thickness of the coating, the morphology also changes, and the role of intermetallic particles in activation and its effect on changing the directional propagation of filament head can be related to delamination.

12.3.5 Scanning Ion-Selective Electrode Technique

SIET is used in electrochemical analysis to map the distribution of specific ions in a solution or across a surface. It combines the principles of ion-selective electrodes with spatial scanning capabilities. Ion-selective electrodes are electrochemical sensors that respond selectively to specific ions in a solution and entail a membrane that selectively interacts with the target ion and generates an electrical potential proportional to its concentration. pH electrodes, potassium-selective electrodes, and chloride-selective electrodes are some common examples of ion-selective electrodes. SIET is essentially potentiometric scanning electrochemical microscopy, which measures the pH and the presence of ions on the electrolyte surface.

The SIET setup comprises a reference electrode, often an Ag/AgCl reference electrode, and a microelectrode which is a micropipette packed with an ionophore that can detect ions or pH. The measurements are conducted under zero current conditions using a two-electrode galvanic cell setup in which the reference electrode and microelectrode are both tethered to a three-dimensional motorized stage or robert arm. A computer was integrated to control the motion, speed, and area that the micropipette was intended to defray while scanning over the surface. The sample is immersed in the electrolyte, and electrodes are positioned in close proximity so that an optical camera embedded within the instrument can determine the position of the probe with respect to the sample, its distance from the sample, the calibration of microelectrode, and image of the sample [10]. The calibration step is required to be conducted beforehand as well as after the sample measurement, and it has to be recorded as potential versus time dependence with an increase in activity.

The potentiometric control mode monitors the potential difference induced by variations in chemical potential due to ionic concentration variation using an ion-selective membrane which permits only specific ions [10]. The potential response will be of the Nernstian genre, which essentially means that it will be a linear function of the logarithm of ionic activity or concentration. The ion selectivity of the membrane can also be detected by a modified separate solution method along with the response time. A dual drop cell can measure the response time; however, other properties such as potential stability, low detection limit, and functional pH region should be validated for application studies [52].

Ion-selective electrodes can detect variations in the concentration of hydrogen ions, hydroxide ions, and metal ions in the electrolyte prompted by cathodic and anodic corrosion processes. The oxidation of metal at the anode gives hydrated cations, which subsequently undergo hydrolysis, leading to metal hydroxyl complexes as described by equations 12.19 and 12.20.

$$M \rightarrow M^{n+} + ne^- \quad (12.19)$$

$$M^{n+} + H_2O \rightarrow M(OH)^{(n-1)+} + H^+ \quad (12.20)$$

The rate of hydrolysis can be determined by using equation 12.20, and it largely depends on the stability of the hydroxyl complex formed. The acidity may be determined by using pH calculations. These stability constants and pH values can be used for corrosion studies. It is possible to determine the rate of corrosion, ionic diffusion, and the presence of interfering ions from pH measurements by juxtaposing the findings with established metrics. By monitoring the pH values of anodic sites and comparing them with known values, it is possible to evaluate the inhibitor coating and impediment of anodic dissolution. This is because anodic dissolution can lead to significant acidification, and the inhibitor can prevent this by altering pH measurements. The self-healing systems are also worth investigating by monitoring the pH, where the corrosion reaction may lead to a local pH change that can trigger the release of an inhibitor whenever a certain pH is required for the anodic and cathodic reactions to take place. The total quantity of cations released by anodic dissolution can be measured using cation-selective microelectrodes to demonstrate the rate at which anodic reactions proceed. In corrosive environments like NaCl, as Cl^- migrates toward the positive zone, Na^+ migrates in the opposite direction, ending up with a diminution in sodium ions and an upsurge in chloride ions. The selective detection of these ions can also be utilized to investigate corrosion behavior.

12.4 CONCLUSIONS

Electrochemical characterization techniques are preferred for studying the corrosion behavior and effectiveness of coatings over other strategies due to their dominant electrochemical nature. Major global and local electrochemical techniques that offer acuity at electrochemical mechanisms and monitor the performance of coatings in corrosion mitigation are discussed in this chapter in detail and are summarized in Table 12.2 along with their pros and cons.

With their efficacy and the macro-scale perceptions, global methods enable a comprehensive tangential assessment of the electrochemical behavior of the coating. However, they lack spatial resolution and might gloss over localized electrochemical phenomena. Local techniques, on the other hand, offer a high spatial resolution, allowing for in-depth analysis of specific areas and detecting localized corrosion even at the nanoscale. Nonetheless, local techniques can be time-consuming and require precise positioning and scanning, while data interpretation may be more complex due to the localized nature of the measurements. With global techniques offering an overall understanding and local techniques offering detailed insights, the choice between them depends on the objectives of the analysis. The combination of both is capable of providing a complete comprehension of the electrochemical behavior and performance.

TABLE 12.2
Overview of global and electrochemical characterization techniques

Technique	Working Principle	Application in Coating Characterization	Advantages	Disadvantages
		Global Electrochemical Techniques		
EIS	Impedance, which is a transfer function, is measured as a function of frequency	Provide information on electrochemical behavior; integrity, thickness, defects, degradation, adherence, durability, and performance of coating	• Versatile and nondestructive • High sensitivity • Analyze a wide range of coating systems	• Invasive nature • Time-consuming • Complex data interpretation • Limited spatial resolution and frequency range
ENM	Measures spontaneous fluctuations in electrical current or potential of low frequency and amplitude over a specific period	Assess the corrosion behavior, degradation, and performance of coatings	• In situ monitoring • Nondestructive technique • Provides real-time information • Minimal sample requirement	• Invasive nature • Lack of detailed mechanistic information • Limited sensitivity to early-stage degradation • Interpretation challenges • Limited spatial resolution • Need for comparative analysis
PDP	Measures current response while continuously scanning the potential of working electrode	Evaluates coating performance and integrity, identify coating defects, compare different coating systems or formulations, evaluate the effect of environmental variables, and optimize coating formulation.	• Versatile and nondestructive • Quantitative and rapid measurements • Comparative analysis • Real-time monitoring • Sensitivity to coating defects	• Invasive • Potential disturbance to the coating system • Coating thickness effect • Limited spatial resolution • Interpretation challenges
		Local Electrochemical Techniques		
SCEM	Monitor current response at probe tip while scanning surface	Detects and characterizes coating defects, delamination, or porosity by mapping the local electrochemical activity at the coating-substrate interface; access coating thickness, uniformity, and adhesion	• High spatial resolution • Real-time monitoring • Surface sensitivity • Quantitative analysis • High chemical selectivity	• Requires redox-active mediator • Limited depth profiling capabilities • Interpretation challenges
SVET	Measure local current density by scanning a vibrating microelectrode probe over a sample surface	Detection and mapping of defects, evaluate uniformity, thickness, adhesion, and delamination of coating; Identification of localized corrosion and corrosion under coatings.	• High spatial resolution • Real-time monitoring • Surface sensitivity • Quantitative analysis • Ability to study coating-substrate interactions.	• Surface roughness effects • Limited depth profiling capabilities • Potential electrode polarization effects • Interpretation challenges

(continued)

TABLE 12.2 (Continued)
Overview of global and electrochemical characterization techniques

Technique	Working Principle	Application in Coating Characterization	Advantages	Disadvantages
LEIS	Impedance, a localized surface area is measured as a function of frequency	Detects localized coating defects, thickness, adherence, and degradation mechanism; provides information regarding corrosion initiation, progression, and the effectiveness of the coating as a protective barrier;	• High spatial resolution • Surface sensitivity • Comprehensive analysis of resistive and capacitive components • Rapid data acquisition for efficient mapping or multiple location analysis • Potential for in situ measurements	• Stray inductance • Limited depth profiling capability • Interpretation challenges • Influence of probe–sample interaction. • Limitations for Large-Scale Studies
SKPT	Measures work function and surface potential by analyzing the contact potential difference between a vibrating probe and a sample surface	Evaluate adhesion, surface charge distribution, coating degradation, localized defects, performance in corrosive environments, and analysis of surface modifications	• High spatial resolution • Quantitative measurements • Real-time monitoring • Surface-sensitive • Quality control and optimization of coating processes	• Limited depth profiling capabilities • Interpretation challenges • Sensitivity to probe–sample distance and surface roughness • Limited chemical specificity • Restricted applicability to conductive coatings
SIET	Measures potential generated in response to the concentration of a specific ion while scanning an ion-selective microelectrode across the surface	Enable spatial mapping and analysis of specific ions, providing insights into ion transport, diffusion, and distribution within coatings, and assisting in the evaluation of coating performance and properties	• Selective detection and mapping of specific ions within coatings. • High spatial resolution • Real-time monitoring • Quantitative measurements of ion concentrations. • Surface specific	• Limited to specific ions • Complex electrode fabrication process • Interference from other ions or coating matrix • Surface roughness effects on ion detection • Limited depth profiling capabilities

REFERENCES

1. Nazeer AA & Madkour M (2018) Potential use of smart coatings for corrosion protection of metals and alloys: A review. *Journal of Molecular Liquids* 253:11–22.
2. Baghdachi J (2009) *Smart Coatings.* (ACS Publications).
3. Hughes AE, Cole IS, Muster TH, & Varley RJ (2010) Designing green, self-healing coatings for metal protection. *NPG Asia Materials* 2(4):143–151.
4. Ali A, *et al.* (2023) Eco-friendly biodegradable polyurethane based coating for antibacterial and anti-fouling performance. *Chinese Journal of Chemical Engineering* 54:80–88.
5. Zhang F, *et al.* (2018) Self-healing mechanisms in smart protective coatings: A review. *Corrosion Science* 144:74–88.
6. Xiao H, *et al.* (2023) Smart sensing coatings for early warning of degradations: A review. *Progress in Organic Coatings* 177:107418.
7. Montemor MF (2014) Functional and smart coatings for corrosion protection: A review of recent advances. *Surface and Coatings Technology* 258:17–37.
8. Wei H, *et al.* (2015) Advanced micro/nanocapsules for self-healing smart anticorrosion coatings. *Journal of Materials Chemistry A* 3(2):469–480.

9. Upadhyay V & Battocchi D (2016) Localized electrochemical characterization of organic coatings: A brief review. *Progress in Organic Coatings* 99:365–377.
10. Jadhav N & Gelling VJ (2019) The use of localized electrochemical techniques for corrosion studies. *Journal of the Electrochemical Society* 166(11):C3461.
11. Srinivasan R & Fasmin F (2021) *An Introduction to Electrochemical Impedance Spectroscopy* (CRC Press).
12. Orazem ME & Tribollet B (2009) Electrochemical impedance spectroscopy. *Angewandte Chemie International Edition* 48:1532–1534.
13. Lasia A (2002) *Electrochemical Impedance Spectroscopy and Its Applications* (Springer).
14. Wang S, *et al.* (2021) Electrochemical impedance spectroscopy. *Nature Reviews Methods Primers* 1(1):41.
15. Lazanas AC & Prodromidis MI (2023) Electrochemical impedance spectroscopy—A tutorial. *ACS Measurement Science Au* 3:162–193.
16. Chang B-Y & Park S-M (2010) Electrochemical impedance spectroscopy. *Annual Review of Analytical Chemistry* 3:207–229.
17. Macdonald DD (2006) Reflections on the history of electrochemical impedance spectroscopy. *Electrochimica Acta* 51(8–9):1376–1388.
18. Szekeres KJ, Vesztergom S, Ujvári M, & Láng GG (2021) Methods for the determination of valid impedance spectra in non-stationary electrochemical systems: Concepts and techniques of practical importance. *ChemElectroChem* 8(7):1233–1250.
19. Sykes J, Whyte E, Yu X, & Sahir ZS (2017) Does "coating resistance" control corrosion? *Progress in Organic Coatings* 102:82–87.
20. Murray JN (1997) Electrochemical test methods for evaluating organic coatings on metals: An update. Part III: Multiple test parameter measurements. *Progress in Organic Coatings* 31(4):375–391.
21. Szociński M & Darowicki K (2023) Sensing the onset of epoxy coating degradation with combined Raman spectroscopy/atomic force microscopy/electrochemical impedance spectroscopy. *Progress in Organic Coatings* 174:107302.
22. Murray JN (1997) Electrochemical test methods for evaluating organic coatings on metals: An update. Part II: Single test parameter measurements. *Progress in Organic Coatings* 31(3):255–264.
23. Margarit-Mattos I (2020) EIS and organic coatings performance: Revisiting some key points. *Electrochimica Acta* 354:136725.
24. Murray JN (1997) Electrochemical test methods for evaluating organic coatings on metals: An update. Part I. Introduction and generalities regarding electrochemical testing of organic coatings. *Progress in Organic Coatings* 30(4):225–233.
25. Xia D-H, *et al.* (2020) Electrochemical noise applied in corrosion science: Theoretical and mathematical models towards quantitative analysis. *Journal of the Electrochemical Society* 167(8):081507.
26. Tan Y (2009) Sensing localised corrosion by means of electrochemical noise detection and analysis. *Sensors and Actuators B: Chemical* 139(2):688–698.
27. Kearns JR, Scully JR, Roberge PR, Reichert DL, & Dawson JL (1996) *Electrochemical Noise Measurement for Corrosion Applications* (ASTM International).
28. Xia D-H, Song S-Z, & Behnamian Y (2016) Detection of corrosion degradation using electrochemical noise (EN): Review of signal processing methods for identifying corrosion forms. *Corrosion Engineering, Science and Technology* 51(7):527–544.
29. Roberge PR (2007) *Corrosion Inspection and Monitoring* (John Wiley & Sons).
30. Yang L (2020) *Techniques for Corrosion Monitoring* (Woodhead Publishing).
31. Cottis R (2001) Interpretation of electrochemical noise data. *Corrosion* 57(03):265–298.
32. Hsissou R (2021) Review on epoxy polymers and its composites as a potential anticorrosive coatings for carbon steel in 3.5% NaCl solution: Computational approaches. *Journal of Molecular Liquids* 336:116307.
33. Jones DA (1996) Principles and prevention. *Corrosion* 2:168.
34. Ahmad Z (2006) *Principles of Corrosion Engineering and Corrosion Control* (Elsevier).
35. Ghali E, Sastri VS, & Elboujdaini M (2007) *Corrosion Prevention and Protection: Practical Solutions* (John Wiley & Sons).
36. Sliem M, *et al.* (2021) Monitoring of under deposit corrosion for the oil and gas industry: A review. *Journal of Petroleum Science and Engineering* 204:108752.

37. Bard AJ, Fan FRF, Kwak J, & Lev O (1989) Scanning electrochemical microscopy. Introduction and principles. *Analytical Chemistry* 61(2):132–138.
38. Polcari D, Dauphin-Ducharme P, & Mauzeroll J (2016) Scanning electrochemical microscopy: A comprehensive review of experimental parameters from 1989 to 2015. *Chemical Review* 116(22):13234–13278.
39. Amemiya S, Bard AJ, Fan F-RF, Mirkin MV, & Unwin PR (2008) Scanning electrochemical microscopy. *Annual Review of Analaytical Chemistry* 1:95–131.
40. Sun P, Laforge FO, & Mirkin MV (2007) Scanning electrochemical microscopy in the 21st century. *Physical Chemistry Chemical Physics* 9(7):802–823.
41. Mirkin MV, Nogala W, Velmurugan J, & Wang Y (2011) Scanning electrochemical microscopy in the 21st century. Update 1: Five years after. *Physical Chemistry Chemical Physics* 13(48):21196–21212.
42. Santana JJ, Izquierdo J, & Souto RM (2022) Uses of scanning electrochemical microscopy (SECM) for the characterization with spatial and chemical resolution of thin surface layers and coating systems applied on metals: A review. *Coatings* 12(5):637.
43. Bastos A, Quevedo M, Karavai O, & Ferreira M (2017) On the application of the scanning vibrating electrode technique (SVET) to corrosion research. *Journal of the Electrochemical Society* 164(14):C973.
44. Sekine I (1997) Recent evaluation of corrosion protective paint films by electrochemical methods. *Progress in Organic Coatings* 31(1–2):73–80.
45. Lillard R, Moran P, & Isaacs H (1992) A novel method for generating quantitative local electrochemical impedance spectroscopy. *Journal of the Electrochemical Society* 139(4):1007.
46. Huang VM, *et al.* (2011) Local electrochemical impedance spectroscopy: A review and some recent developments. *Electrochimica Acta* 56(23):8048–8057.
47. Gharbi O, Ngo K, Turmine M, & Vivier V (2020) Local electrochemical impedance spectroscopy: A window into heterogeneous interfaces. *Current Opinion in Electrochemistry* 20:1–7.
48. Lillard R, Kruger J, Tait W, & Moran P (1995) Using local electrochemical impedance spectroscopy to examine coating failure. *Corrosion* 51(4):251–259.
49. Wittmann M, Leggat R, & Taylor S (1999) The detection and mapping of defects in organic coatings using local electrochemical impedance methods. *Journal of the Electrochemical Society* 146(11):4071.
50. Zou F & Thierry D (1997) Localized electrochemical impedance spectroscopy for studying the degradation of organic coatings. *Electrochimica Acta* 42(20–22):3293–3301.
51. Nazarov A & Thierry D (2019) Application of scanning Kelvin probe in the study of protective paints. *Frontiers in Materials* 6:192.
52. Lamaka S, Souto RM, & Ferreira MG (2010) In-situ visualization of local corrosion by scanning ion-selective electrode technique (SIET). *Microscopy: Science, Technology, Applications and Education* 3:2162–2173.

13 Coating Characterization with Surface Morphological Techniques

Nam Nguyen Dang, Lai Xuan Bach, Kim Long Duong Ngo, Cam Tu Hoang Ngoc, and Thi-Bich-Ngoc Dao

13.1 INTRODUCTION

The images that provide the final morphology, topography, and related distribution of the metallic surfaces after using corrosion inhibitors are indispensable for corrosion inhibitor characterizations. The basic is a light microscope result but it is not enough to show full surface information of morphology and topography, as well as the related distribution. Up to now, scanning electron microscopy (SEM) has been one of the best choices for characterizing surface morphology. Most qualitative reports can combine SEM results with other spectroscopic results with the view that this combination is more powerful than the combination of SEM and probe analyzer results. However, the best way that can fully support morphology, topography, and distribution is to provide all information on the final metallic surfaces worked in inhibited systems [1–5]. It is noticed that all surface morphological techniques are localized and/or selective measurements; therefore, it is important how to provide the reproducibility of surface morphological results. Optical microscopy (OM) is first used for evaluating the post-surface morphology. However, OM can perform at low magnification and presents a poor surface view, while the inhibited surface morphology requires more information and obvious images. Immediately, SEM has been powerfully used till now and provides more evident information and clear images of the post-surface morphology which can be used to persuade all readers. The mainly powerful information provided by SEM could be counted as a wide range of magnification with high resolution and the traceable standard for magnification. Figure 13.1 compares the post-metallic surfaces observed by SEM and OM showing the full advantages and disadvantages of these techniques.

Post-surface topography after using a corrosion inhibitor is also needed to characterize the surface's profile related to its roughness using a profilometer and atomic force microscopy (AFM). Atomic force microscopy can provide more surface features and topography, such as roughness, pits, and cracks, at a very high resolution in comparison with a profilometer [6]. Both techniques can be used for characterizing surface roughness and/or surface finish, as well as burrs and other microscopic shapes within a very short time. The results are presented by the peak, one-dimensional scan, and two-dimensional (2D) map images with varying height, depth, and spacing that can verify satisfaction of the requirements for the used inhibitors [7–9] as shown in Figure 13.2. Furthermore, the use of line scan in profilometer technique can measure the film thickness, depending on the experiment setup [7, 10]. However, there are very few studies that have used this technique for verifying the film thickness. Manh and colleagues [10] have used this technique to determine the film thickness via the image of a line scan, however, need to show the peak belonging to this image to easily identify the film thickness [7]. Another technique that provides the exact film thickness without any effects and must be counted is focused ion beam scanning

DOI: 10.1201/9781032677255-13

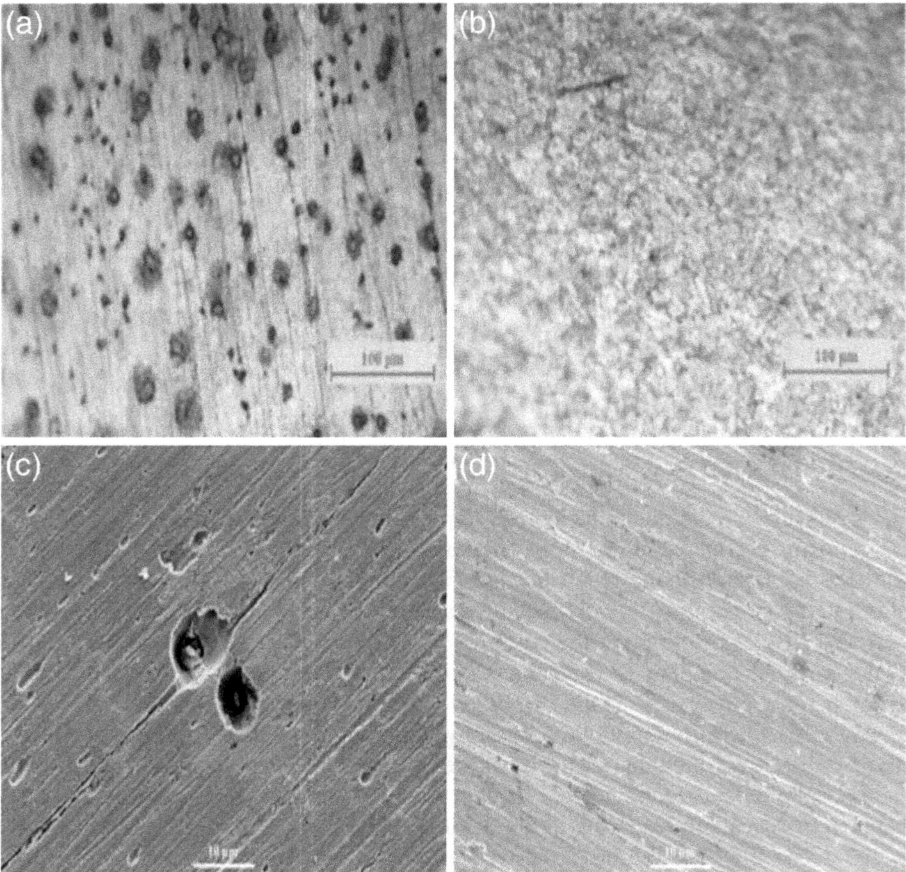

FIGURE 13.1 Comparison of uninhibited and inhibited surface morphologies observed by OM (a and b) and SEM (c and d).

electron microscopy (FIB-SEM). Shedrack and colleagues [11] used FIB-SEM for characterizing film thickness, and localized deposition of inhibitors on intermetallic particles in the alloy surface. Importantly, FIB-SEM can be used to fabricate the specific specimens which will be used to characterize by other techniques such as transmission electron microscopy/high-resolution transmission electron microscopy (TEM/HRTEM), and EDS of cross-section related to the film thickness. Wang and colleagues [12] used FIB-SEM to determine the properties and structure of three layers on the weathering steel surface, as well as fabricate the specific specimens used for HRTEM and scanning transmission electron microscopy with energy-dispersive X-ray spectroscopy (STEM-EDS) measurements. Furthermore, FIB-SEM can produce prototyping 2D and three-dimensional (3D) structures which are extremely interesting in pitting corrosion studies, such as pit shapes and depths. These properties cannot be observed by OM or SEM techniques. Lillard and colleagues [13] used FIB-SEM to reveal the details of the pit propagation at the grain boundary. They reported the inclusions and pitting initiation as shown in Figure 13.3, supporting the important information on pitting corrosion. The most attractive of the FIB-SEM technique is the ion (Xe) species which is inert with both conducted (metallic) and semiconductor materials, avoiding any influences on measured specimens such as doping [14]. It also produces large area cross-section 3D volumes with high accuracy and a wide dimension of a maximum of around 500 μm for each direction

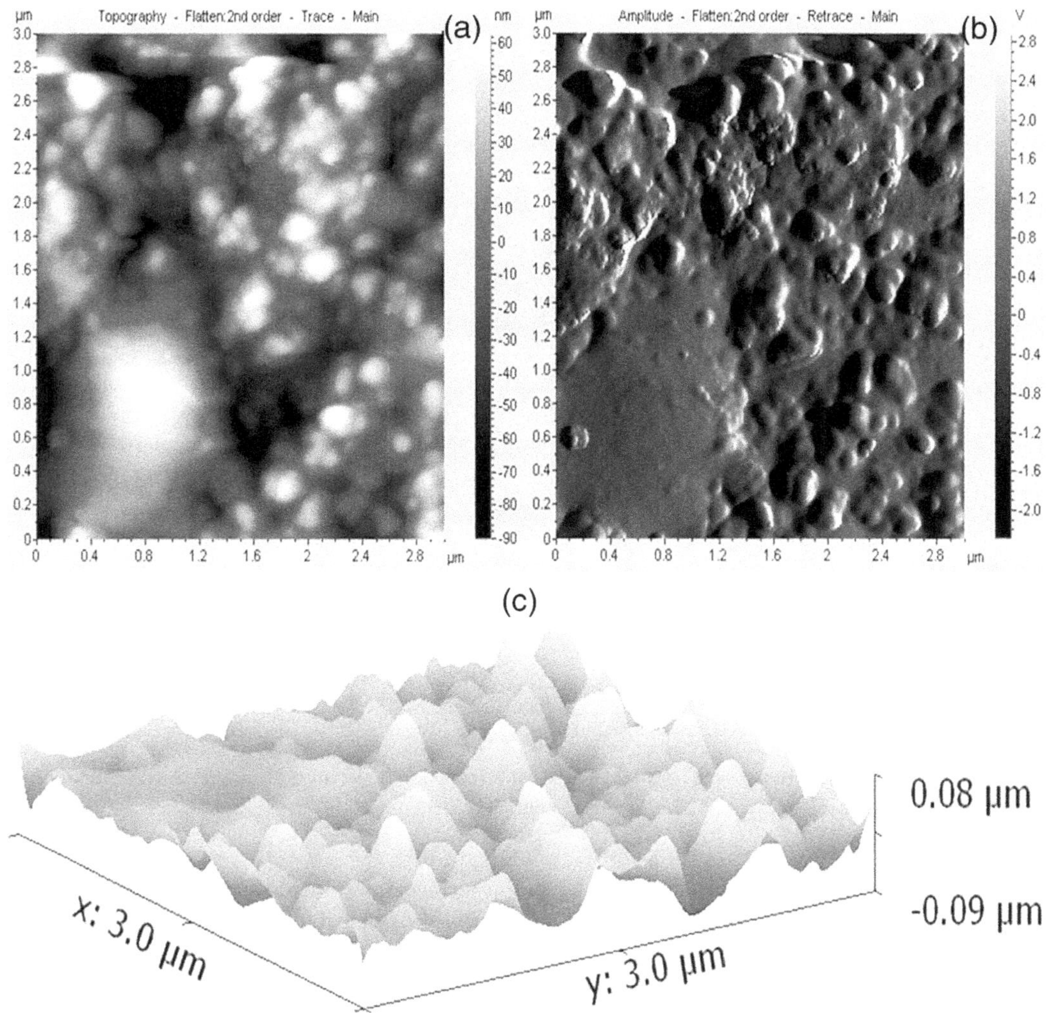

FIGURE 13.2 Surface topography of inhibited surface observed by AFM 2D (a and b) and 3D (c).

at very low current and short time. However, the cost and availability of this technique are still questionable.

To achieve a complete study in corrosion inhibitors, the element distributions that contribute to the morphology and topography of the inhibited surfaces also need to be characterized by various techniques. Raman mapping can detect both inorganic and organic compounds in the presence of peaks and mapping. Zhang and colleagues [15] reported the FeO, Fe_3O_4, and γ-FeOOH formed on carbon steel surfaces in both spectra and distribution maps. They also reported both spectra and distribution maps for mussel adhesive protein (Mefp-1)-related compound, γ-FeOOH, and Fe_3O_4 when carbon steel was coated. That is interesting information to interrupt the film formation mechanism that is related to the inner and outer layer for the surface finish of coating. In addition, the typical Raman map of the spalled oxides was also used to investigate the individual contribution of each element product supporting the corrosion product layer that resists the corrosion of steel [16]. Interestingly, Kharitonov and colleagues [17] used Raman techniques for characterizing the localized surface chemical events contributed to the AA6063-T5 alloy by vanadate corrosion

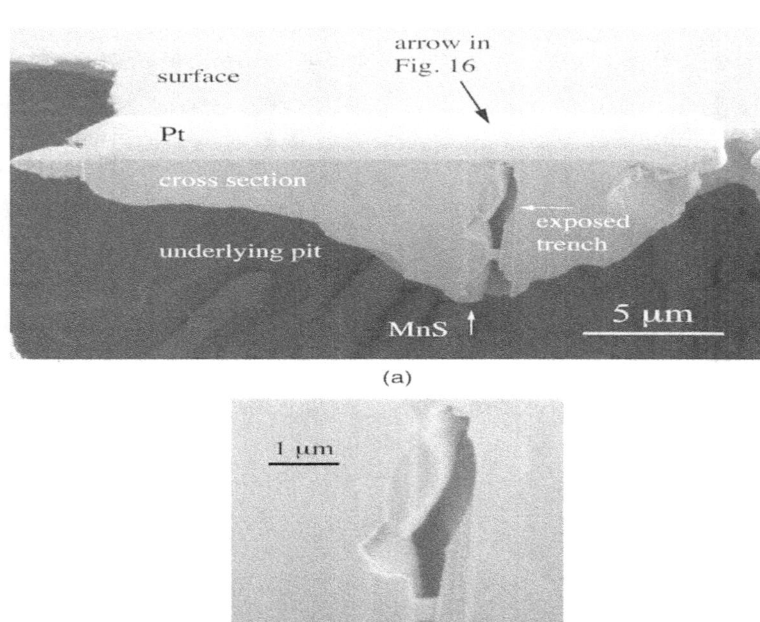

FIGURE 13.3 FIB-SEM presents (a) pitting and (b) MnS inclusion in the alloy structure. Surface topography of inhibited surface observed by AFM 2D (a and b) and 3D (c).

inhibitors. In this case, Raman was used to detect the surface components and localized surface, as well as depth in the form of mapping, indicating components distribution constructed the surface finish of the inhibited AA6063-T5 alloy surface. It is noticed that Raman analysis is not influenced by water and is done without damaging the specimens, indicating a suitable use to examine the inhibited surface finish. However, the resolution of the images is still questionable. Furthermore, the popular measurement is energy dispersive X-ray (spectroscopy) mapping due to its availability, cost saving, and familiar operation. The qualitative results can be represented by both element peak and mapping images that indicate the occurrence of each element distributed to the morphology and topography. Nam and colleagues [18] reported the clear element distribution on the inhibited AS1020 and X65 surfaces (Figure 13.4). The results indicated that the different contribution of Fe, C, Pr, and O elements to the inhibited morphology and topography surfaces is due to the localized deposition of corrosion inhibitor. The authors also indicated that the surface was blocked by the particle's shadow, resulting in the occurrence of some dark areas, but this did not influence the accuracy of EDS mapping results. With the same purpose, Li and colleagues [19] presented the uniform distribution of Ce, C, and O on the inhibited metal surface and concluded that the Ce^{3+} ion complexed with tannic acid during the absorption processes. Furthermore, Ma and colleagues [20] successfully investigated the S, N, and S occurrences to ensure the encapsulation of benzotriazole and 2-mercaptobenzothiazole inhibitors co-incorporated in nanocontainers. However, EDS measurement is unable to detect the elements that are lighter than the boron element, hardly identify the peak standard due to the different properties from the individual constituent elements when a combination of elements is analyzed, and heavily distinguish the element peaks with the background noise.

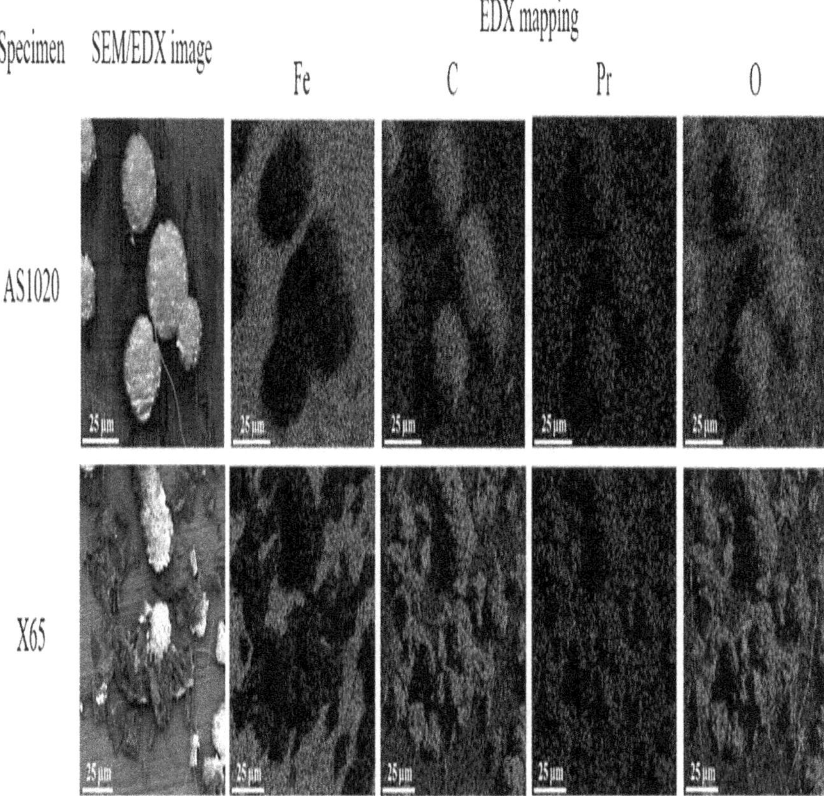

FIGURE 13.4 Surface topography and related distribution of the inhibited AS1020 and X65 steel surfaces.

To solve the above limitations, time-of-flight secondary ion mass spectrometry (ToF-SIMS), electron probe micro-analyzer (EPMA), and Auger electron spectroscopy (AES) are recommended. EPMA, ToF-SIMS, and AES offer element and isotope mapping for bulk surfaces with high resolution with very small feature sizes, but some of them are still unable to detect the hydrogen element. EPMA can produce the quantitative chemical analysis of the surface and internal compositional structures at small spatial scales, high sensitivity, and wide crystal range. It is normally used to characterize the alloy microstructures such as element distribution [21–27] and has recently extended to apply to corrosion inhibitor studies [28]. Li and colleagues [28] reported the influence of sodium dodecyl sulfate on the morphology and atomic composition of the layer formed on the AM50 Mg alloy. An interesting conclusion was exposed as almost all corrosion products were concentrated around the Al-Mn intermetallic and S-containing corrosion inhibitor, implying adsorption located at the top of intermetallic and the surrounding α-Mg. It means that sodium dodecyl sulfate acted as a localized inhibition related to the micro-galvanic corrosion mitigation of AM50 Mg alloy. Besides these advantages for corrosion inhibitor studies, EPMA meets some limitations such as the inability to detect lithium, helium, and hydrogen elements, overlapping peak positions, oxides (not as cations), and the inability to distinguish the exact valence states of Fe. ToF-SIMS is also a powerful technique to characterize element distribution in surface morphologies and provides an imaging mass spectrometry of isotopic, elemental, and molecular distribution information. ToF-SIMS provides the mapping elements with high resolution, particularly favored with organic molecule distributions on solid surfaces. This technique can be used for parallel observations of all compounds formed on the surface and provides the chemical images

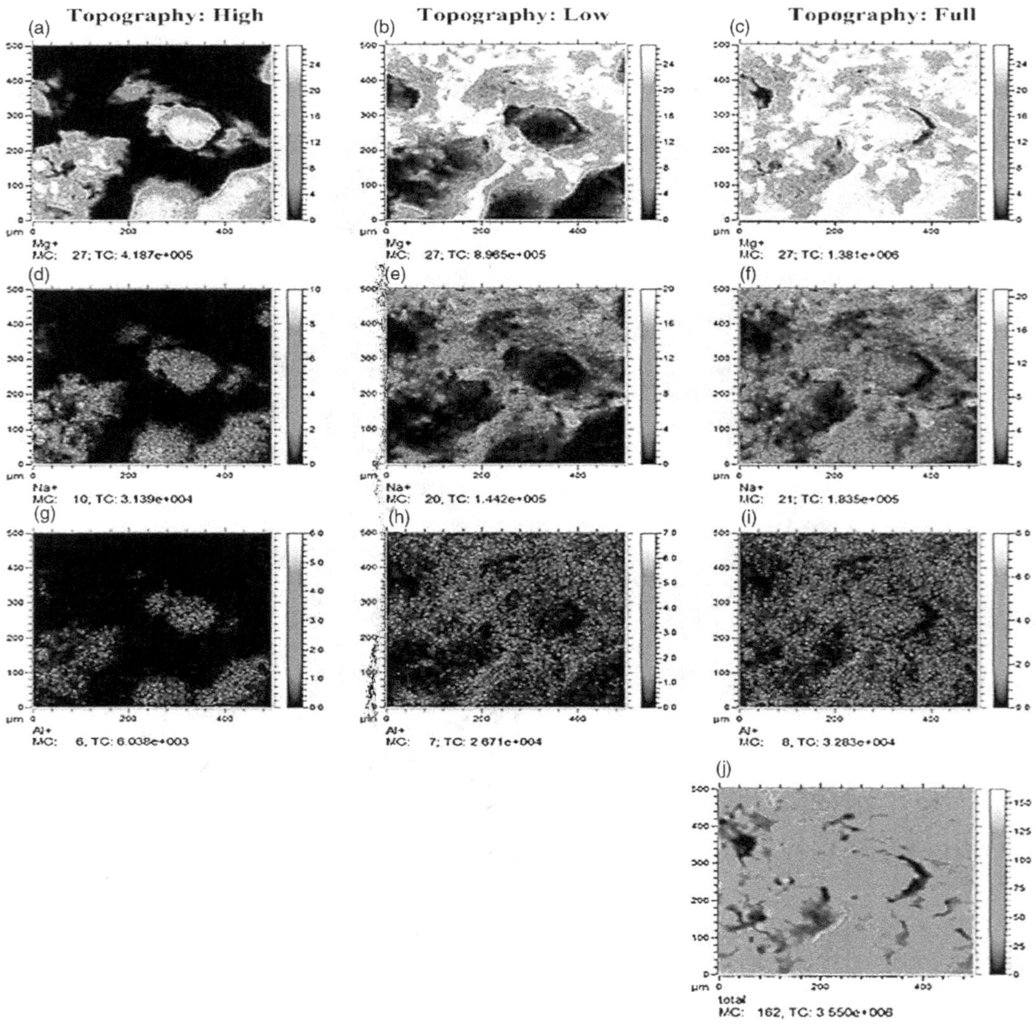

FIGURE 13.5 Surface topography observed by ToF-SIMS with positive polarity: (a, b, and c) Mg^+, (d, e, and f) Na^+, and (g, h, and k) Al^+ (left to right columns assigned to ion signals of high, low, and full topography range).

related to surface topography [29]. Esmaily and colleagues [30] reported the full ToF-SIMS scene and SIMS maps of the surface topography of AM50 alloy before and after corrosion performed by both positive and negative polarities as shown in Figures 13.5 and 13.6. The use of ToF-SIMS analysis concluded that the distribution of ionic species selectively concentrated nearby the cathodic and anodic locations. In addition, ToF-SIMS also provides the depth profile characterization and presents the 3D distribution of surface topography [29]. It is noticed in the operational process of ToF-SIMS measurement that the ion beam must remain as low as possible to release the results without the chemical damage that easily happens with organic molecules. It is also hard to produce quantitative analyses and sometimes shift the image due to changing and takes a long time to completely analyze a single data set. Other surface topography-sensitive analytical techniques that provide the mapping of the element distribution on the solid surface could use AES. AES can analyze mapping, quantity, and chemical stratigraphy, as well as depth profiles which can be used to replace a combination of XPS, ToF-SIMS, EPMA, and EDS analyses. More interestingly, AES

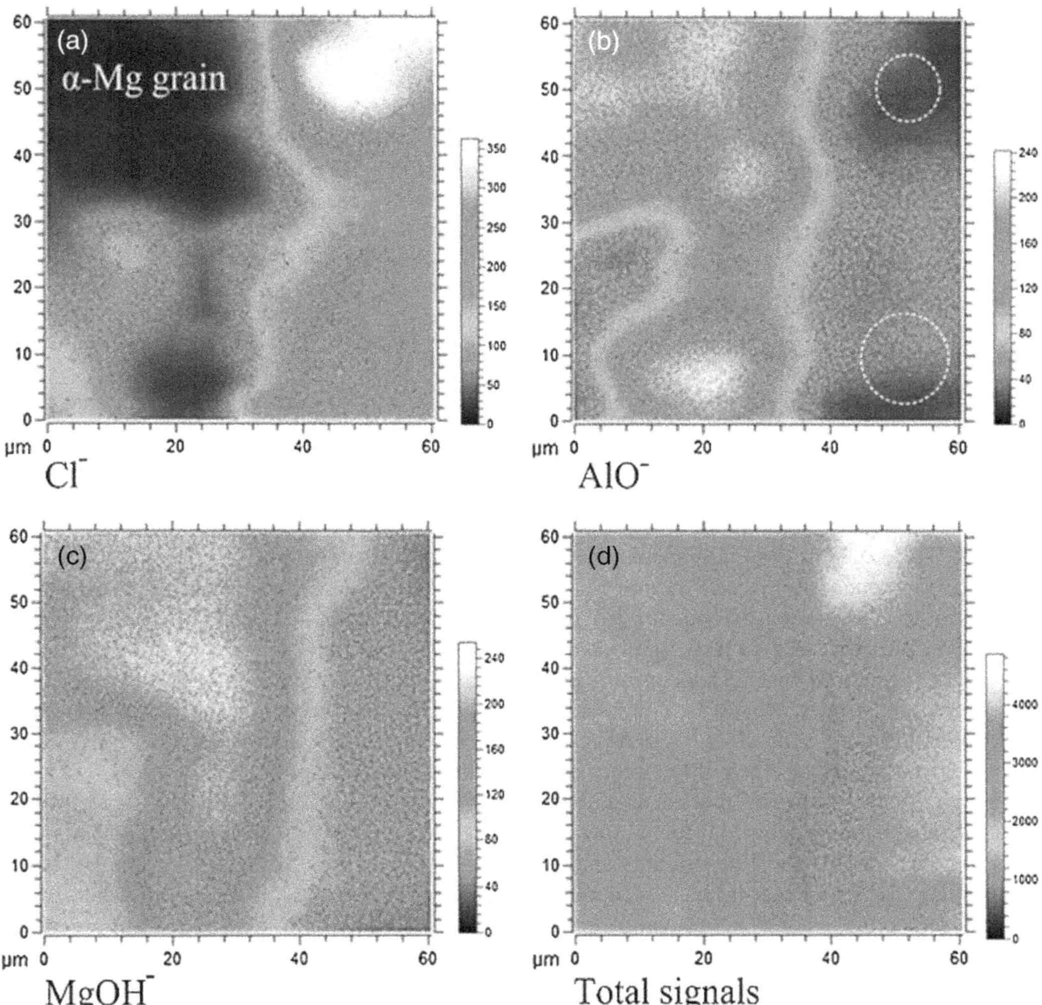

FIGURE 13.6 Negative ion images of surface topography observed by ToF-SIMS: (a) Cl^- ions, (b) AlO^- ions, (c) $MgOH^-$ ions, and (d) total signals map.

can be used for characterizing the top few atomic monolayers, a wide range of elements from light (such as lithium) to heavier elements, homogeneous and heterogeneous components on the surfaces of the materials, and thickness of thin film layer, etc. Liang and colleagues [31] reported the copper topography after sputtering graphene via C_{KLL}, $Cu_{LMM,}$ O_{KLL}, and S_{KLL} Auger electron maps. Based on the results of Auger electron maps and supported DFT calculations, they invented graphene nucleation growth on the copper substrate. Furthermore, Bellhouse and Mcdermid [32] reported the surface oxide morphology of steel using AES via secondary electron image and some steel element maps such as Fe, Mn, Si, and O maps. The investigated results provide evidence of the influence of oxygen partial pressure on steel regarding the selective oxidation and reactive wetting phenomena. Therefore, AES performs extremely well in studying surface morphology and distribution and is capable of coating characterization with morphology, topography, and related distribution. However, it is noticed that AES operates at an ultrahigh vacuum, detects all small amounts of surface contaminations, and hardly characterizes the insulating specimens.

Hopefully, this chapter can provide an overview approach that can suggest the way to choose the most suitable techniques for the morphology, topography, and related distribution characterization of the inhibited metallic surfaces. In addition, the individual techniques and their possible applications to inhibited surfaces have also been described and it is suggested that a combination of several techniques including morphology (OM or SEM), topography (profilometer or AFM or FIB-SEM), and related distribution (EDS, Raman, ToF-SIMS, AES, or EPMA) could be the best choice for morphology, topography, and related distribution characterization of inhibited surfaces in finding new corrosion inhibitor system.

13.2 LIGHT MICROSCOPE

13.2.1 Optical Microscopy

The OM system provides image magnification for small objects. It utilizes visible light and a lens system for magnification and is the most fundamental and popular microscope invented in the 17th century. Traditional OM only allowed for direct observation of images, but contemporary versions are also equipped with charged-coupled device (CCD) cameras, enabling magnified images to be saved. The operating principle of the optical microscope uses a positive lens system. When an object is placed on the microscope slide, visible light from the light source interacts with it via scattering, transmitting, etc., before reaching the objective lens (focal length f_1), resulting in an intermediate image that is opposite in direction and larger than the object. By using an ocular lens of focal length, the researcher can observe this image as a virtual image that is 1,000 times bigger than the actual object. This common optical microscope magnification can be adjusted using the adjustment bracket. Nowadays, an observed image can be saved in the same direction as the object by employing a camera with an image flip system. The camera acts as a lens of focal length, transforming the virtual image into an intermediate image on the diaphragm [33].

The process of preparing a material sample for OM involves several steps that depend on the material being observed. Various techniques can be used to prepare materials, including slicing and mounting the sample on a slide, staining and/or embedding the sample, and dehydrating the sample to eliminate any water that may cause damage. Additionally, before viewing the sample with an optical microscope, it needs to be coated with a special immersion oil that acts as a mounting media. These immersion oils are designed to control refraction when the light passes through different environments, enabling the maximum collection of light and optimal magnification and resolution of the surface morphology of the material [34]. Compared to electron microscopy measures, OM provides several advantages. This technique is less complicated and costly as it does not necessitate specialized equipment. Additionally, optical microscopes offer the ability to observe live samples without causing any damage, while electron microscopes often need samples to be treated and prepared before viewing. In the study of corrosion, OM finds widespread application. OM enables the detection and morphology study of coatings formed from corrosion inhibitors in the study of corrosion and protection of metallic materials. In a study about the epoxy coating on the surface of cold rolled steel (CRS), some additives, such as organically modified montmorillonite (OMMT) clay or nano-glass flake (GF), were included to improve the mechanical performance and effectively inhibit the corrosion of the new coating [35]. Nematollahi and colleagues [35] conducted a study on adding OMMT and nano-GF to the epoxy resin mixture (at 1 and 3 wt.%) in other ways. The synthesized nanocomposites were applied to CRS. The properties and morphology of the coating after adding additives were preliminary observed by OM. Based on the analysis of the OM results, the researcher came to some quite interesting conclusions. For the sample with OMMT added, when mechanically stirred for 2 h, the agglomerates appeared quite large (about 25 μm – 1% OMMT and 40 μm – 3% OMMT). When sonicated, the agglomerates became significantly smaller in size and fewer in number. After 1 h of sonication, the agglomerates were completely absent, reflecting a sonication procedure suitable for the synthesis of a non-agglomerated nanomaterial. In addition, the micrograph showed no agglomeration in the GF–epoxy mixture even after mechanical agitation. From there, the researchers drew the best conclusions about the process of synthesizing the two-epoxy nano-types mentioned above and then performed experiments on their ability to inhibit corrosion to CRS in 5 wt.% NaCl. Thus, OM results can give clear information on the properties of nanocoating as well as the size of nanoparticles in composites, supporting studies to test the feasibility of synthetic coatings to protect materials from corrosion.

Up to now, OM has become an important technique for the wide research fields. Loc and colleagues [36] used OM to realize the graphene growth on the copper substrate at large scales via the selective oxidation of the copper substrate. OM is also used for characterizing the morphology and minor pits, as well as the amount of corrosion on the Co surface in DI water containing corrosion inhibitors [37]. Interestingly, the surface states including local surface defects, natural anode and/or cathode of an electrochemical reaction in identified zones, and location of bubble occurrence can be qualitatively identified [38–41]. Particularly, the use of OM can determine the local electrochemical impedance of the surface (Figure 13.7) [41, 42]. In general, OM is usually

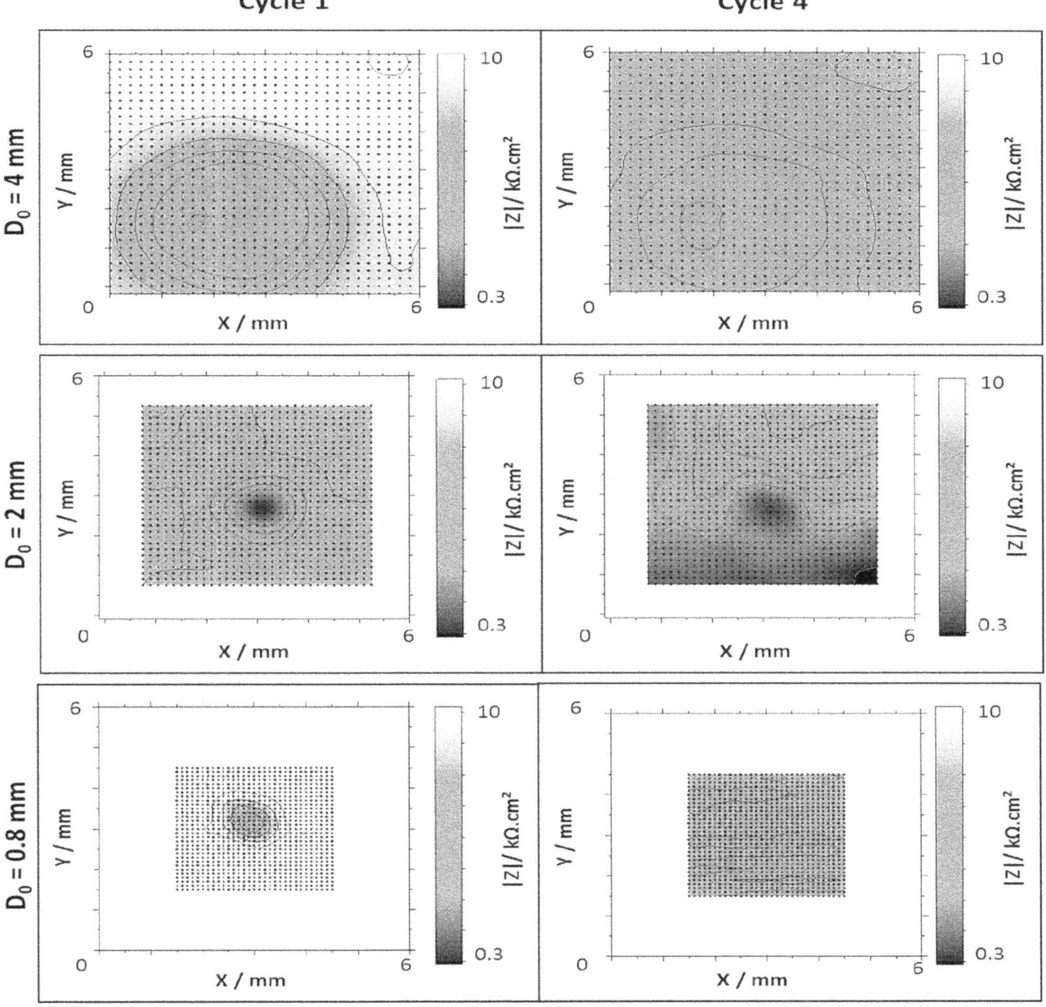

FIGURE 13.7 Local impedance modulus and Grad |Z| in the vicinity of the defects with diameters of 4.0, 2.0, and 0.8 mm after the first and fourth degradation cycles.

used for characterizing surface morphologies including uniform and pitting corrosion [43], surface topographies including drop distribution [44–46], pits [43], and trenching [47], as well as supporting inhibition performance [48, 49], filiform corrosion [50], and coating thickness via cross-section measurement. Therefore, OM is a useful tool for observing the coatings on the surface of various materials. The technique is particularly effective in visualizing the morphology and thickness of the coating and observing any changes that may occur due to external factors such as corrosion. OM is also ideal for identifying the type of coating on a sample using staining or other specialized techniques. However, it is important to note that OM has certain limitations. It has relatively poor resolution compared to other microscopy techniques such as electron microscopy. Additionally, OM has difficulty penetrating opaque coatings and materials, which can hinder the visibility of certain types of coatings. Despite these limitations, OM remains a widely used method for the visual examination of surface coatings due to its nondestructive nature, ease of use, and cost-effectiveness.

13.2.2 Profilometer

A 3D optical profiler is a highly sophisticated type of optical microscope that accurately measures the 3D surface topography of an object by analyzing the light reflected from its surface. This instrument is particularly valuable in the study of corrosive materials due to its ability to identify even the slightest changes on the material's surfaces, including surface roughness and the presence of the coating on the surface related to corrosion inhibition [51]. The main difference between OM and a 3D optical profiler lies in their measurement techniques. While OM relies on image exaggeration and resolution, a 3D optical profiler employs digital rationality to measure the surface of every portion of the object, quick scan, either using contacting or non-contacting, as well as great precision and can be measured outdoors. This allows the 3D optical profiler to provide clear and precise images, overcoming the limitations of OM.

The 3D optical profilometer provides high-resolution surface observation and can identify the elevation of post-corrosion sites, making it useful in detecting and observing the coating characterization on one material. Sreekanth and colleagues [52] proposed the protection of oxide-coated magnesium alloy, via a plasma electrolytic oxidation system in an aqueous solution, using $Na_2B_4O_7.10H_2O$ and K_2TiF_6 as additives. The alloy was intended for use as a biological implant, and samples with and without 2 g/L of each additive were observed in a simulated body fluid using profilometry imaging as shown in Figure 13.8. The images revealed the pitting corrosion locations and corresponding depth profiles on the surface. The nonadditive samples displayed pitting corrosion sites that were around 20 μm lower than the material surface, demonstrating the limitation of plasma electrolytic oxidation (PEO) coating for ZM21 material. The pore depth profiles of the samples present of $Na_2B_4O_7.10H_2O$ or K_2TiF_6 were reduced significantly. Simultaneously, the 3D optical profilometer result confirmed the limit of the surface region of the coating at about 5–7 μm thickness on the surface. Nevertheless, it is unclear which additive is a more effective inhibitor if the study does not combine other measures. According to the research results, $Na_2B_4O_7.10H_2O$ more effectively inhibits corrosion than K_2TiF_6. Furthermore, the typical 3D profilometer can detect pitting corrosion and determine the maximum pit depth and width (Figure 13.9), which can be used for evaluating the pit penetration rate [53]. Mandal and colleagues [54] used the 2D and 3D surface profiles to characterize the surface topography and roughness of steel after immersing in an aggressive solution without and with inhibitor addition. Based on the measured average (S_a) and root mean square roughness (S_q) and surface topography, as well as some supported results, they concluded that the barrier layer was formed on the steel surface immersed in a solution containing corrosion inhibitor, resulting in higher corrosion resistance. Several studies [51–57] also used a profilometer to characterize the surface topography of steel after immersion in 2 M HCl solutions

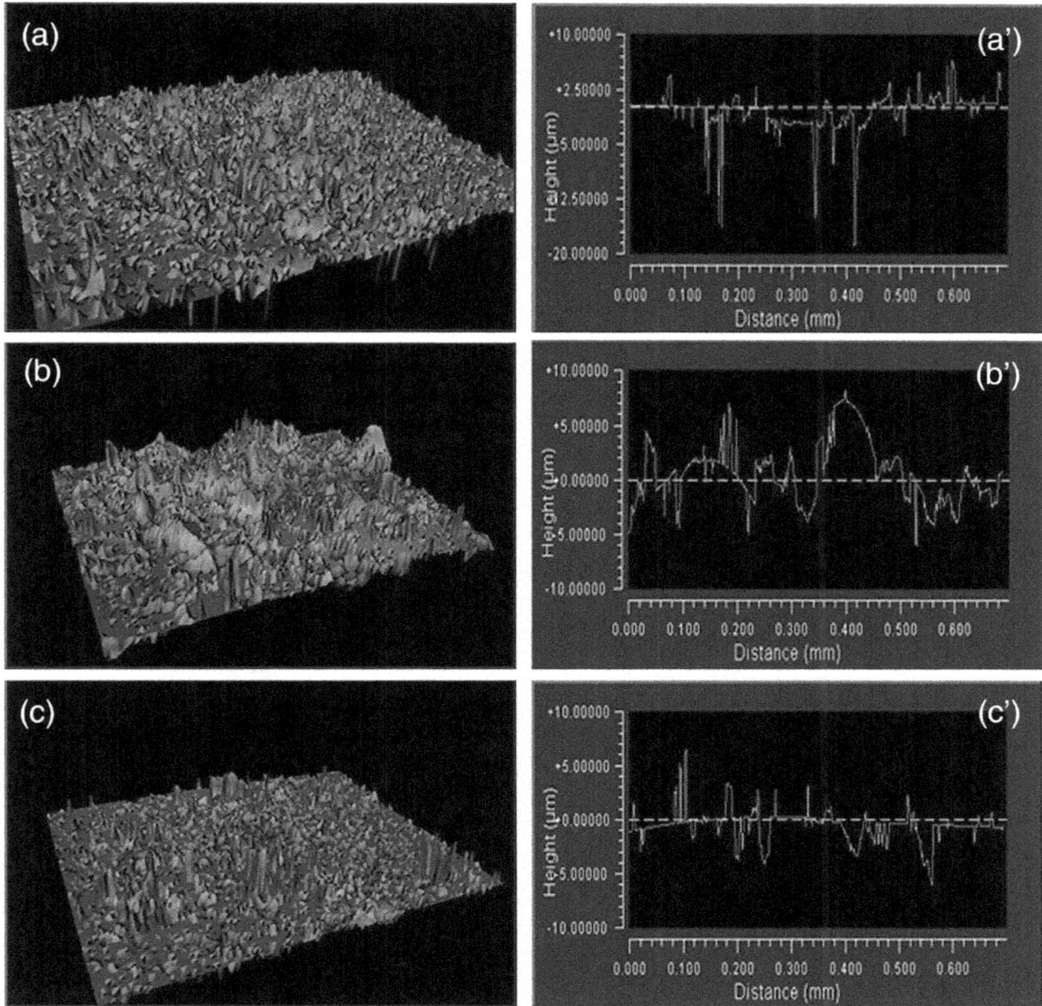

FIGURE 13.8 Surface morphology and topography images observed by optical profilometry: (a–c) three-dimensional surfaces and (a'–c') depth profiles.

without and with sulfonated chitosan addition. Based on the 3D images, the surface topography was characterized by maximum height, surface roughness, and maximum peak height, which were indirectly evaluated after corrosion happened with steel in investigated solutions. Hence, a profilometer can provide valid information such as the average and root mean square surface roughness, and maximum peak-to-valley, as well as the average of the ten greatest peak-to-valley separations for surface topography of post-surface [43–57]. Importantly, a profilometer can be used to measure coating thickness [10] and pitting sizes [58].

Therefore, a profilometer could be a powerful technique to characterize the surface topography of post-surfaces. It can be used for scanning the surfaces closed to the structure, resulting in reliable data and being an effective technique for evaluating surface roughness under certain conditions. However, it is noticed that the profilometer produces low-density areas and several atypical locations of investigated areas [59, 60].

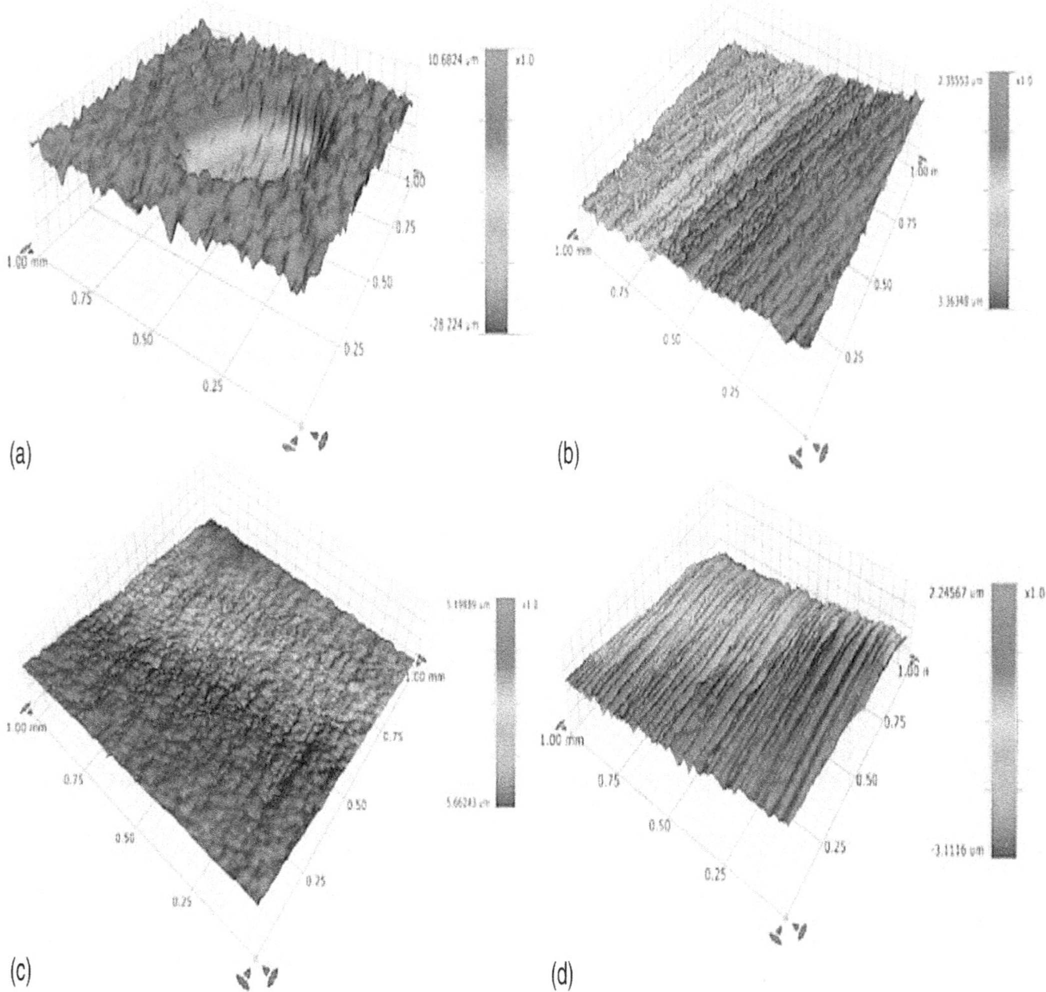

FIGURE 13.9 3D profile images showing (a) the size of the pit on the uninhibited steel surface and surface roughness of inhibited steel surfaces in aggressive solutions containing (b) N-acryloylmorpholine, (c) N-acryloylpyrrolidine, and (d) N-vinylcaprolactam.

13.3 ELECTRON MICROSCOPE

To formulate new materials with specific functions for different applications, a comprehensive understanding of a material's surface, interface, and interior structure is pivotal. Analyzing the morphology and structure of coatings is made possible through the use of advanced electron microscopy techniques such as SEM, FIB-SEM, and TEM. Coatings often rely on their surface characteristics to perform specific functions in various applications. By exploring the morphology of coatings at different length scales, their relationship with functional properties can be comprehended. The use of advanced electron microscopy in coating characterization has resulted in better-designed coatings that exhibit higher performance and an extended lifespan. This advancement in coating characterization has had a transformational impact in numerous fields, including aerospace, automotive, and healthcare applications. Therefore, examining the coating surface finish and how it influences functional qualities using electron microscopy has now become essential in designing the next generation of engineering materials. In this section, we will discuss the fundamental principles, advantages, and limitations of these advanced electron microscopy techniques for characterizing the morphology of coatings.

13.3.1 Scanning Electron Microscopy

When compared to other electron microscopy methods, SEM has a number of distinct advantages. When an electron beam scans a surface, it stimulates the production of secondary electrons, which are then detected and reconstructed into the image of the surface of the object. The electron beam is created by a gun and is focused by the electromagnetic lenses into a fine point on the sample. The sample is held and positioned by the sample stage, which has precise 3D movement capability. The detectors detect and measure various signals produced by the interaction between the primary electrons and the sample, such as secondary electron detectors detecting low-energy electrons emitted from the sample surface, while backscattered electron detectors measure the electrons that are reflected from the sample, or X-ray detectors can measure X-rays generated. The imaging system processes these signals detected by the detectors, generating an image that is displayed on the viewing screen, as well as performing digital processing to improve the clarity and sharpness of the image. The control system regulates the electron gun, lenses, detectors, sample stage, and imaging system to ensure optimal settings for imaging. The computer and software are used for data acquisition, storage, processing, and analysis.

After the material has been coated to inhibit corrosion, SEM will be used for evaluating the morphology structure of the coating, including visualizing surface roughness, cracks, pitting, and other features [61–66]. Additionally, a number of elements have an impact on coatings' efficacy. A beam of high-energy electrons is produced by an electron cannon inside the SEM after the specimen is secured inside. The beam is focused onto the surface of the specimen by a series of electromagnetic lenses, producing secondary electrons, backscattered electrons, and other types of signals depending on the SEM settings. These signals are then detected by the detectors, which measure the signal intensity and provide valuable insight into the morphology and composition of the sample's surface. The imaging system processes these signals using computer algorithms to generate an image that is displayed on the SEM's viewing screen. The operator can adjust the SEM settings to optimize imaging quality and acquire high-resolution images and analytical data on the microstructure and composition of the specimen.

SEM provides several benefits, including high resolution, expansive magnification range, and specific imaging capabilities. Its resolution capabilities range from 1 to 10 nm, allowing for clear imaging of nanoscale structures. The SEM also has a high magnification range, up to 400,000x, providing detailed images of a sample surface. It is a selective tool for identifying materials by their unique surface characteristics, such as texture, shape, and composition. Besides, there are still some drawbacks of this analysis tool. The primary limitations of this imaging technique are the cost of the equipment and the need for highly skilled operators. Additionally, the instrument is sensitive to the parameters of operation, such as optimizing the electron beam voltage, current, and focus to achieve the proper contrast and resolution.

Since both the environment and the substance of the substrate play a significant role, not all materials demonstrate high anticorrosion performance from coatings [67]. According to SEM conclusions, Jun and colleagues [68] observed that utilizing PEO coating for Mg alloys results in a porous surface with numerous minuscule and heterogeneous pores. The potential for surface corrosion and harm to the coating has been caused by thickness. However, the anticorrosion performance increases dramatically when the PEO coating has been combined with polyurethane on the top layer, with the coating uniformity increasing and the number of defects substantially lowering. For aluminum alloys, a coating technique that includes an anodized layer with further deposition of a superhydrophobic sol-gel coating has also been investigated and is found

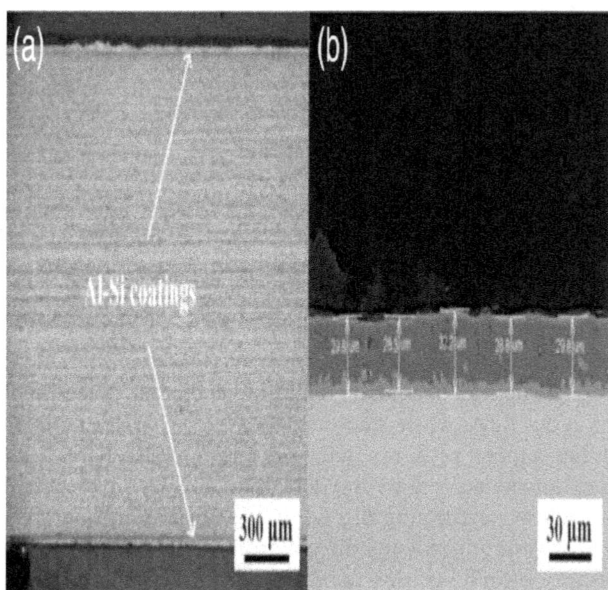

FIGURE 13.10 Coating thickness observed by (a) OM and (b) SEM.

to be capable of replacing chromate-transformed coatings [69]. SEM was utilized to measure the number of holes generated on the substrate surface based on the surface morphology of the coated and untreated samples. By affecting the surface with cavitation pulses, the mantle is consequently rapidly shattered, resulting in the development of craters and minor pits. In comparison to samples with low porosity, samples with higher porosity have lower resistance to corrosion and erosion. Therefore, the best way to present SEM results for the surface finish after using a corrosion inhibitor could be to use both the lowest and highest magnifications with clear images. It is noticed that SEM is one of the most important measurements and must be required for all coating studies. SEM can provide a very high resolution with a wide range of magnifications to characterize the surface morphology and topography [59–69], the forms of corrosion such as general and localized corrosions [8, 55, 56], as well as coating thickness via cross-section measurement [70, 71]. Figure 13.10 shows the coating thickness observed by both SEM and OM cross-sections [72]. In addition, SEM can observe the defects and microcracks in the coating, coating properties at specific locations such as top, within, and bottom coating [61, 73], coating pores, inner pores, pore structures [74–77], or pitting width [58], and the morphology of uninhibited and inhibited surfaces [78–82]. This information indicates that SEM is an integral technique in coating characterization.

13.3.2 Focused Ion Beam Scanning Electron Microscopy

FIB-SEM technique combines two different technologies: a focused beam of ions to sputter and remove materials from a measured surface and a scanning electron microscope to generate high-resolution images of the exposed surface. The FIB-SEM instrument consists of several key components, including an ion source, an electron source, a vacuum system, and a sample stage. The ion source generates a beam of ions typically composed of gallium (Ga^+), which is focused on and directed toward the surface of the sample [83]. The ion beam is controlled by a series of lenses and apertures, which ensure that the beam can be precisely controlled in terms of current and spot size, allowing for fine-scale removal of material and imaging of the sample. The sample stage holds the sample and allows for precise movements in three dimensions, such that the ion beam can scan across the sample surface with micrometer-scale precision. The electron source, located at a right angle to the ion beam, generates a beam of electrons that is also carefully focused and directed toward the surface. When the electrons interact with the sample surface, they produce high-resolution images of the exposed surface and its nanoscale features. Scanning system controls the movements of the ion and electron beams across the sample surface. With nanometer-scale precision, the scanning system can produce images of and/or remove material from the sample surface. To avoid any sample contamination and to ensure the stability of the ion and electron beams, the entire apparatus is assembled inside a high vacuum chamber.

In coating research, FIB-SEM can be used to study the morphology and composition of coatings and to evaluate the effectiveness of coatings in terms of their ability to resist wear, corrosion, and other forms of degradation. For example, FIB-SEM can be used to study the adhesion and failure mechanisms of coatings on various substrates [84, 85], to analyze the morphology of coatings deposited using different techniques [86], and to evaluate the effect of different environmental conditions on coating performance. FIB-SEM is a powerful tool for materials analysis, device fabrication, and biological imaging, but it is not without its drawbacks. One of the main advantages of FIB-SEM is its ability to generate high-resolution images and perform precise modifications to samples with nanometer-scale accuracy. It is helpful in pitting corrosion study including surface morphology and topography, as well as the accuracy of shape, width and depth of the pits. In addition, FIB-SEM can be used to create 3D images of complex structures [83] and to analyze the composition of materials at the nanoscale and film thickness [86–89]. The most interesting prospect of FIB-SEM measurement is the high-resolution images of the exact depth and width of the pits (Figure 13.11) [58], each coating layer, grain growth, interface behaviors in the multilayer

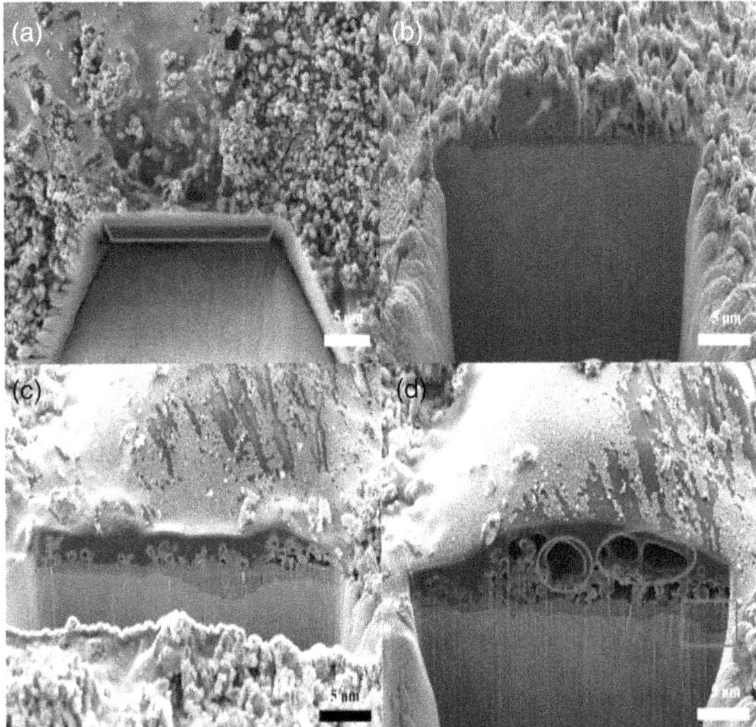

FIGURE 13.11 FIB-SEM images showing (a) FIB milling location, (b) no pitting content, (c) pits containing area, and (d) pits containing area with small milling current.

coating system [90], 3D pore spaces and sizes at nanoscales, as well as degradation within the coating [91]. It can also be used to characterize 3D pore structures at different scales [92], multiple material phases and repositioning artifacts [93], sampling thin (measured TEM specimen), fragile, and complex surface layers of cultural materials [94], degradation of the passive film [95], open porosity of spheroidized materials [96], and providing cryogenic volumetric imaging [97]. In addition, the defects and microcracks within the coating, coating properties at specific locations such as top, within, and bottom coating, as well as coating interfaces become possible to characterize using FIB-SEM. These specific benefits indicate that FIB-SEM is also an important technique for characterizing battery electrode materials, catalysis, cement, electrochemistry, etc. However, FIB-SEM is a time-intensive process that requires extensive training and expertise to operate effectively. In addition, the ion beam is destructive to the surface of the sample, leading to artifacts in the images and limiting the accuracy of the analysis. Furthermore, FIB-SEM tends to be expensive to operate and maintain, making it inaccessible to many researchers. Nonetheless, its versatility and precision continue to make it an invaluable tool for materials science and engineering research, particularly in the study of surface finish after using corrosion inhibitors.

13.3.3 Transmission Electron Microscopy

The material's nanoscale characteristics were examined using transmission AFM. Especially with different analytical capabilities and imaging modes, it has greatly aided in the study of chemical properties and nanostructures [98, 99]. The information is crucial for understanding the coating mechanism and how to create coating materials because the coating material is made up of a variety of atoms and may comprise nanoparticles. The examinations of materials using TEM have been started with the advancement of science to ease the sample preparation process in the evaluation analysis. The primary goal of TEM is to gather data on morphology, structure, and nanochemistry at the nanometer or even sub-nanometer level. These data were gathered in a very good spatial location, making correlation easy [98].

The basic principle behind TEM is similar to that of a conventional optical microscope [100], but instead of using visible light to image the sample, a beam of electrons is used. The electrons are generated by an electron gun, which emits a beam of electrons that is focused onto the sample using electromagnetic lenses. As the electrons pass through the measured sample, they interact with the atoms and electrons in the sample, which causes them to scatter and diffract. Some of the electrons are absorbed by the sample, while others continue to pass through and are collected on the other side of the sample by a detector. The detector records the intensity of the electrons that have passed through the sample, which is used to generate a reconstruction of the sample. The image is created by using a computer to convert the measured electron intensities into a grayscale image. The resolution of a TEM image depends on a number of factors, including the wavelength of the electrons, the quality of the electron optics, and the stability of the instrument. In modern TEMs, it is possible to achieve resolutions down to a few picometers, which allows for the visualization of individual atoms and the study of nanoscale materials.

However, as TEM is a destructive technique, it has a significant drawback in that it necessitates cutting the materials into incredibly thin slices, frequently only a few micrometers thick. In the case of coatings, this entails taking the sample out of the coating and embedding it in a resin, which may change the coating structure. The preparation site needs to be carefully chosen because the studied area is only a few square micrometers, which does not permit prolonged observations. Additionally, this means that no single result can be used to reflect all of the coating's features. The drawback is that, but by TEM results, the cross-section of HAp coating biologically treated on SS 316L substrate has been studied [101]. Thereby, the development of crystallinity and morphological changes in HAp crystals were investigated. The elements influencing the coating development process were assessed concurrently. TEM is frequently used in conjunction with SEM/

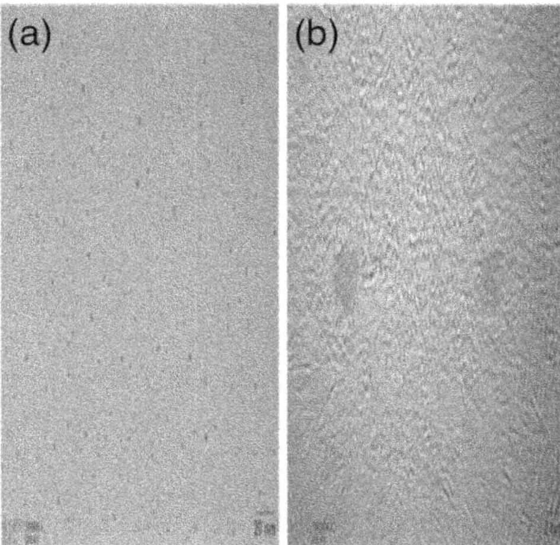

FIGURE 13.12 Different resolution TEM images of the inhibited copper surface.

EDS and Raman to overcome this deficit or to acquire a more detailed picture of the properties of the sample than with any one technique. Kovalov and colleagues [102] combined TEM with EDS mapping and some electrochemical techniques to assert that the majority of MnS inclusion was not dissolved and insignificantly effect to 304L stainless steel in solutions containing chloride ions. While TEM provides high-resolution images of a sample and can reveal details of its structure and composition at the nanoscale, SEM/EDS provides information on surface morphology and composition of sample elements, which can be useful for characterizing coatings or surface treatments. Alternatively, SEM imaging can be performed on samples that are not as thin as is required for TEM. Although TEM is a powerful technique for coating characterization, very few studies have used TEM for characterizing the surface finish after using corrosion inhibitor [103, 104]; the inhibited surface is given in Figure 13.12.

The combination of SEM, FIB-SEM, and TEM provides a comprehensive understanding of coating morphology and structure. These advanced electron microscopy techniques can facilitate the optimization of coating compositions and deposition processes, leading to better coating performance. Coatings with well-characterized morphology and structure are vital to applications in numerous fields, such as energy, aerospace, and bioengineering.

13.4 PROBE ANALYZER

Probe analyzer is a powerful technique used to analyze data from different kinds of scanning probe microscopy (SPM) techniques, such as AFM and EPMA. SPM is a type of microscopy that is used to create high-resolution images and perform measurements on surfaces at the nanoscale level. The technique uses a very sensitive probe that comes into close contact with the surface being studied. Measurements are made as the probe interacts with and scans across the surface. Probe analyzer is an essential tool for researchers working in the field of nanotechnology and surface science.

13.4.1 Atomic Force Microscopy

AFM or scanning force microscopy is one of the techniques grouped under the larger category of scanning probe microscopies. The standard system comprises the main components: a sharp tip, a cantilever in charge of horizontal position, a Piezoelectric (PZT) actuator in charge of vertical position, a laser, and a position-sensitive photodetector. All movements are regulated by a controller which reads the feedback of laser deflecting of the end point of the cantilever. The AFM probe is the combination of a micro-cantilever with a sharp tip. The probe's tip, possessing a monomolecular point, facilitates nanometer-resolution imaging, while the micro-cantilever functions as a force sensor capable of detecting even the slightest sample deformation. As a result, the AFM enables high-sensitivity force measurements. The principal operation is based on scanning the tip over the measured surface with mechanisms that provide feedback enabling the PZT scanners to stabilize the tip's constant force or constant height over the sample surface. During scanning, as the tip moves up and down with the sample surface, the cantilever deflects the laser beam. The deflection provides intensity variation between the upper and lower photodetectors. Feedback from the photodiode difference signal, controlled through software from the computer, regulates the tip to maintain a constant force or constant height above the sample. The real-time height deviation is observed by the PZT transducer in the constant force mode, whereas the deflection force on the sample is recorded in the constant height mode.

The output of SPM is typically displayed as a matrix consisting of the measured topography or some other signal, such as the phase, as the x-y position on the surface. The process of data description involves mapping the heights, obtained from the z-controller, to gray levels in an image for an optimal choice. In order to improve image quality, image processing techniques are utilized to eliminate image artifacts like noises, resulting in a refined representation of the data. The tip in the contact mode means the distance between the probe tip and the sample surface remains constant during the scanning process. A topographic image is generated by dragging the probe's tip over the surface. However, this operational mode has a significant disadvantage in that it damages both the probe tip and the measured surface. Additionally, the imaging process is strongly affected by any adsorbent layer present on the sample surface. During the noncontact mode, the cantilever tip is positioned approximately 50–150 Å upon the measured surface to probe the attractive van der Waals forces between the sample surface and the tip. The frequency, amplitude, and phase of the oscillation will be affected by the interaction between the sample and the probe, thus more information about modulated samples in the signal. Moreover, the noncontact mode is the method used for the first high-resolution photograph captured using AFM in a high vacuum. The tapping mode is essentially an advancement of the noncontact dynamic mode. The cantilever is vibrated by the piezoelectric oscillator mounted on it, with amplitudes ranging from 100 to 200 nm and frequency close to the private oscillation frequency. The advantage of this mode is that it prevents the transducer from dragging on the specimen's surface, which otherwise can damage the sample, and eliminates any adhesive forces between the probe and the sample, thereby avoiding any interference in the image. It also helps avoid imaging interference due to the appearance of liquid layers on the sample surface. The AFM can apply both adherent and suspended cells while simultaneously gaining mechanical and structural information under nanometer scales and the imaging with native conditions will require no staining, labeling, or fixation necessary, and other benefits such as assessment of multiple physical, high signal-to-noise ratio, biological, and chemical information. However, it limits to less than 1 cell/10 min and the cell activities may be affected by the mechanical poking of the AFM tip, and it also restricts to surfaces.

With the capability of direct measurement of intermolecular forces alongside atomic-resolution characterization, the AFM tool may be employed in applications across various engineering domains. AFM is also capable of in situ imaging through nanoindentation, which excludes the requirement to move the sample, switch tips, utilize a separate instrument for imaging the indentation, or relocate the scanning area [105–109]. The aim of Yang and colleagues' study was to investigate the performance of a sol-gel coating, and its combination with a polyurethane unicoat, on aluminum alloy during exposition in aqueous Harrison's solution. The coating was composed of ZrO_2 and SiO_2 with a ratio of 1:3.4. The AFM technique was employed to monitor the evolution of the coating system for aluminum corrosion protection during immersion [110]. Employing metallic surface pretreatment by sol-gels was a contemporary strategy in surface science and

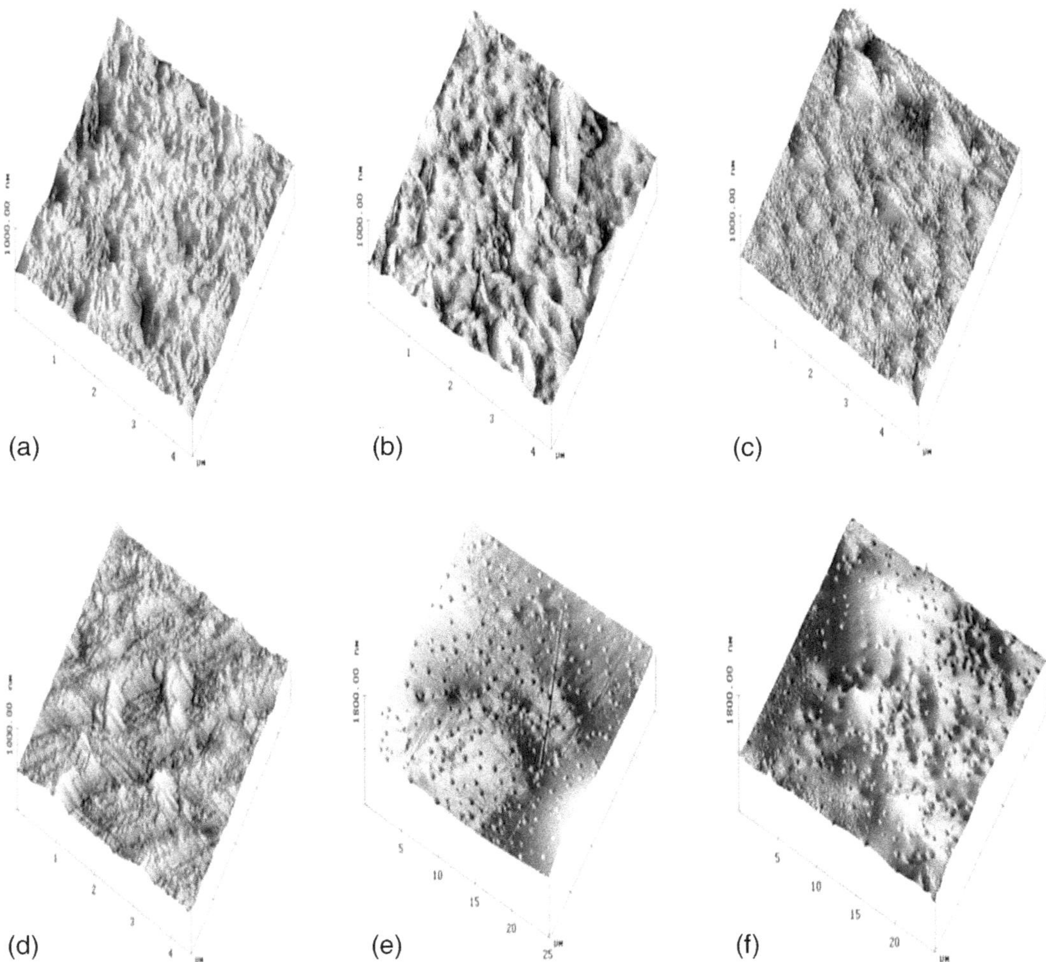

FIGURE 13.13 AFM images of (a) surface before corrosion, (b) surface after two days corrosion, (c) alloy surface after polishing, (d) alloy surface after corrosion, (e) topcoat surface, and (f) blisters under-coating of a sol-gel conversion coating after corrosion.

safeguarding against electrochemical corrosion. In this study, AFM was utilized in either contact or tapping modes using silicon nitride or sharp silicon tips. The AFM included 13.5 μm E and 125 μm J scanners. The scan range of the images varied from 125 μm to 500 nm. The rough surface of a sol-gel coating before corrosion is shown in Figure 13.13(a). Due to the roughness of the aluminum surface and the thinness of the sol-gel film, the surface topography reflected the underlying roughness in that image, and according to the z-range set. To gain a more profound comprehension of sol-gel functioning on the aluminum substrate, the electrochemical properties of the coating on the aluminum alloy were characterized in Harrison's solution. After a 2-day immersion period in the investigated solution, the surface produced some visible blisters as illustrated in Figure 13.13(b). As shown in Figure 13.13(c), an initially polished surface exposed the small spots of the grains of the alloy surface, which may be oxide particles that are quickly created after polishing. The Al surface underwent significant changes after a three-day immersion period, as depicted in Figure 13.13(d). Within the first one- or two-day exposition, several discernible pits occurred on the alloy surface, and pitting corrosion swiftly developed. An investigation was carried out to assess the efficacy of the investigated coating and polyurethane TT-P-2756 self-priming topcoat. Figure 13.13(e) presents an AFM image of the investigated coating before corrosion, displaying the parameters including 250 and 30–50 nm of the diameter and depth. This crater formation was created during the coated process, but their depth was negligible compared to the thickness of the coating (80–100 μm), indicating that they likely do not influence the coating performance. However, after four weeks of exposition, the surface showed a rise in surface roughness. It was attributed to the coating degradation and the blisters formed under-coating (Figure 13.13(f)). The biggest blisters observed through AFM measurements were around 13 μm and 338 nm in diameter and height. Even after 28 weeks of exposed time, the blisters remained around 10.5 μm and 420 nm in diameter and height. AFM results ensure that the blisters were insignificantly affected by immersion time. The AFM results strengthened the understanding of the sol-gel coating's behavior on Al metal surfaces during immersion in Harrison's solution. It was presumed that a stable barrier layer of the mixed oxides consisting of aluminum oxide and silicon oxide was initially formed at the interface during the corrosion process, preventing further pit development. It indicates that AFM is a unique technique used to visualize the surface topography of the inhibited surface finish [111–113, 118, 119]. In addition, AFM can detect the topography and corresponding bearing histograms of the surface, and identify the scratch and grinding directions, as well as the details of the formation of ditch and ridges on the surface finish [104], plastic deformation [105], and surface roughness [110–117]. Importantly, AFM can provide both qualitative and quantitative information on surface finish with simple specimen preparation and viability in both air and liquid medium [118-120]. Some individual features can be detected by AFM at nanoscale levels, even of a few nanometers [121]. The morphology changes observed by AFM images can be used for determining selective passivation [122]. Bunch and colleagues [123] observed the air leakage over time via AFM line profile measurements of inward bubble deflection. Gu and colleagues [124] used AFM measurements to monitor the surface morphology changes of the amine-cured epoxy coating under progressed ultraviolet exposition by measuring unexposed air/coating surface, unexposed coating bulk, unexposed coating/substrate interface, and field exposed air/coating surface. Interestingly, AFM can be easily used in combination with other techniques such as SEM, infrared (IR) analysis, and electrochemistry for characterizing surface morphologies and properties. Firstly, a combination of AFM with SEM and IR was used to characterize the heterogeneous internal morphology and chemical heterogeneity of epoxy phenolic resins as shown in Figure 13.14, where IR was mapped by line-by-line to get only localized fluctuations [125]. Furthermore, the combination of AFM and Raman analysis operated in an electrochemical cell was also used to evaluate the activation of passive titanium [126]. The detailed study indicated that the dissolution of the upper TiO_2 portion of the passive film resulted in a titanium monoxide thin layer formed on the electrode surface due to the cathodic electrification of Ti electrodes. Furthermore, high-speed AFM has also been developed to capture the hierarchical assembly of septins [127], μs-dynamics of unlabeled biomolecules [128], self-assembly of short pieces of the Tetrahymena telomeric DNA sequence d[G4T2G4] in physiologically relevant aqueous solution [129], antibiotic action in bacterial membranes [130], cancer detection [131], and contact failures in nanoscale semiconductor device [132]. Therefore, it indicates that AFM can be applied to broad applications including materials and manufacturing, coatings and thin films, semiconductors, biomaterials, electronics, biology, polymers, composites, and contaminant detections. In particular, AFM could be an ideal method to investigate the modified surfaces for corrosion protection.

13.4.2 Electron Probe Micro-Analyzer

EPMA is an advanced microbeam instrument that is utilized for nondestructive, determining the chemical composition of small volumes of solid materials and can be done in situ. When operating

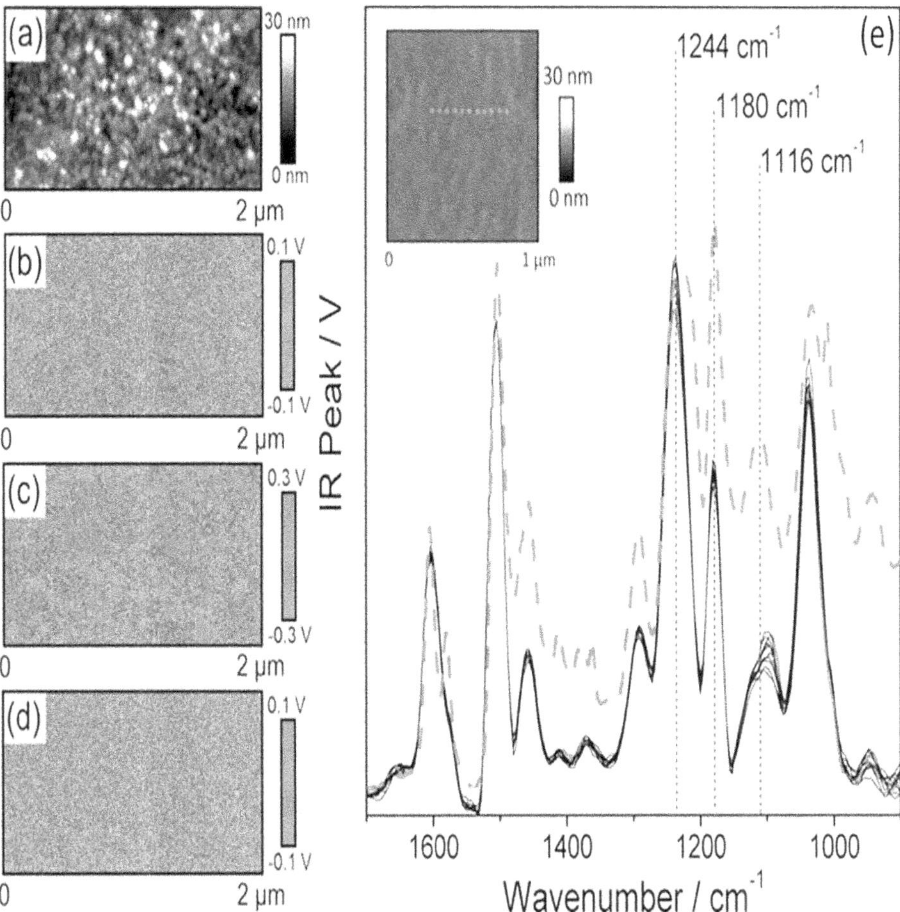

FIGURE 13.14 AFM-IR images: (a) height and following IR-induced deflection at (b) 1116 cm^{-1}, (c) 1244 cm^{-1}, and (d) 1180 cm^{-1}, and (e) AFM-IR fingerprint region spectra.

an electron microprobe, a measured specimen is positioned within a vacuum and exposed to a highly focused beam of high-energy (accelerated) electrons (5–30 keV). Upon bombarding the sample, various electron–atom interactions occur. These interactions occur within an extremely small volume of the sample, ranging from 100 nm^3 to a few μ^3, rendering the microprobe an ideal instrument for carrying out quantitative analysis in situ. These interactions between electrons and the sample also produce heat, derivative electrons, and X-rays. In geological material analysis, secondary and backscattered electrons generated by this interaction are commonly used for imaging the surface or obtaining the material's average composition. Inelastic collisions between the incident electrons and the electrons in the inner shells of atoms in the sample produce X-rays. When an inner-shell electron is ejected from its orbit, a higher-shell electron falls into the vacancy and emits energy as an X-ray, which is characteristic of the element. Electron probe micro-analysis is nondestructive because the X-rays generated accompany loss material, enabling the re-analysis of the same materials more than one time.

An electron probe of EPMA is a main tool to study complex materials at a microscopic scale because of its ability to produce a focused beam of electrons, and the electron probe facilitates the detection and precise measurement of elemental composition in solid materials. With the ability to detect small spatial variations in composition, the in-situ chemical analysis plays a vital role in understanding the characteristics and behavior of complex materials or within chemically zoned materials. The electron probes are commonly equipped with imaging detectors (secondary electron images, backscattered electron detector, and cathodoluminescence) that produce surficial images and compositional structures, providing significant assistance in the analysis process, allowing for more accurate evaluation of the elemental composition and properties of the material. However, it is noticed that the electron probe has limitations as it cannot detect the lightest elements such as H, He, and Li. The different valence states of iron cannot be distinguished by the electron probe, which can pose a challenge in accurately determining the composition of some materials. And certain elements may produce X-rays with peak positions that overlap in both energy and wavelength, requiring separation to accurately distinguish between them.

EPMA could identify and quantify elements in solid samples with great sensitivity and high accuracy [133, 134]. The common technique used for performing compositional characterization of geological materials at small scales is quantitative electron probe microanalysis (EPMA) [135]. In addition to its application in geological samples, EPMA is a commonly used method for analyzing synthetic materials. These include thin films, optical wafers, microcircuits, superconducting ceramics, semiconductors, and element distributions [136–140]. EPMA has been recognized as a powerful analytical technique for identifying and characterizing the composition of these materials. Hao and colleagues [141] used arc spraying techniques to make the FeCrAl coating for carbon steel, followed by high-current pulsed electron beam irradiation to modify the surface structure. Especially, the chemical composition on the cross-section and the modified surface layer of the sample was measured with an EPMA. The FeCrAl coating's cross-section was analyzed using this method in a map scanning mode as shown in Figure 13.15. The backscattered electron image (BEI) for the scanned area was provided, revealing a typical lamellar structure for the investigated coating. The main components of the coating including O, Al, Fe, and Cr were found around the lamellas, suggesting that Al was partial to oxidation during the deposition process. The outer layer of the arc-sprayed coating was abundant in iron oxides. Besides, EPMA measurement indicated that the Fe and Cr distributions were even throughout the surface. The Al and O were enriched at the outer surface. The remelted layer also showed the unchanged distribution of these elements. The EPMA results were used to evaluate the distribution of the chemical composition in the metallic coating before and after high-current pulsed electron beam irradiation. At 20 J/cm^2 of energy density, the surface was entirely continuous with exceptional corrosion resistance. As such, it is recommended that future applications of high-current pulsed electron beam irradiation on the investigated coating focus on pretreatment of the original coating and shorter pulse durations, which help in better corrosion resistance of the steel. Schiffmann and Steinberg [142] suggested that EPMA can determine the chemical composition of solid surfaces with high detection sensitivity and spatial resolution, adjustable depth, and easy and accurate quantification. The study also reported the chemical composition and layer thicknesses of thin multilayer with only one single measurement. In particular, the composition and thickness of a layer buried under one or more other layers can also be determined and the density of thin layers will be identified if the film thickness is provided. In addition, the use of EPMA can confirm the selective desorption and adsorption of the thin film, as well as multilayer formation [143]. EPMA was used to characterize a precursor alloy (Pre. Er-Fe-B) microstructure, which supports the determination of boron-migration-mediated solid-state reaction [144], high accuracy determination of carbon-containing steel [145], the actual Ce content in Ce-$CoSb_3$ skutterudites [146], the Fe, Ni, Cr, Co, and Mo diffusion at the P92 steel interface [147], inclusion glasses with the major element compositions and sulfur contents [148], the mixture of olivine and diopside [149], and segregation of Co, Cr, and W from the dendrite core and Ni, Al, and Ta from the interdendritic regions [150]. Therefore, it indicates that EPMA is an ideal technique to characterize the element distribution that closely relates to the morphology and topography of the post-surfaces.

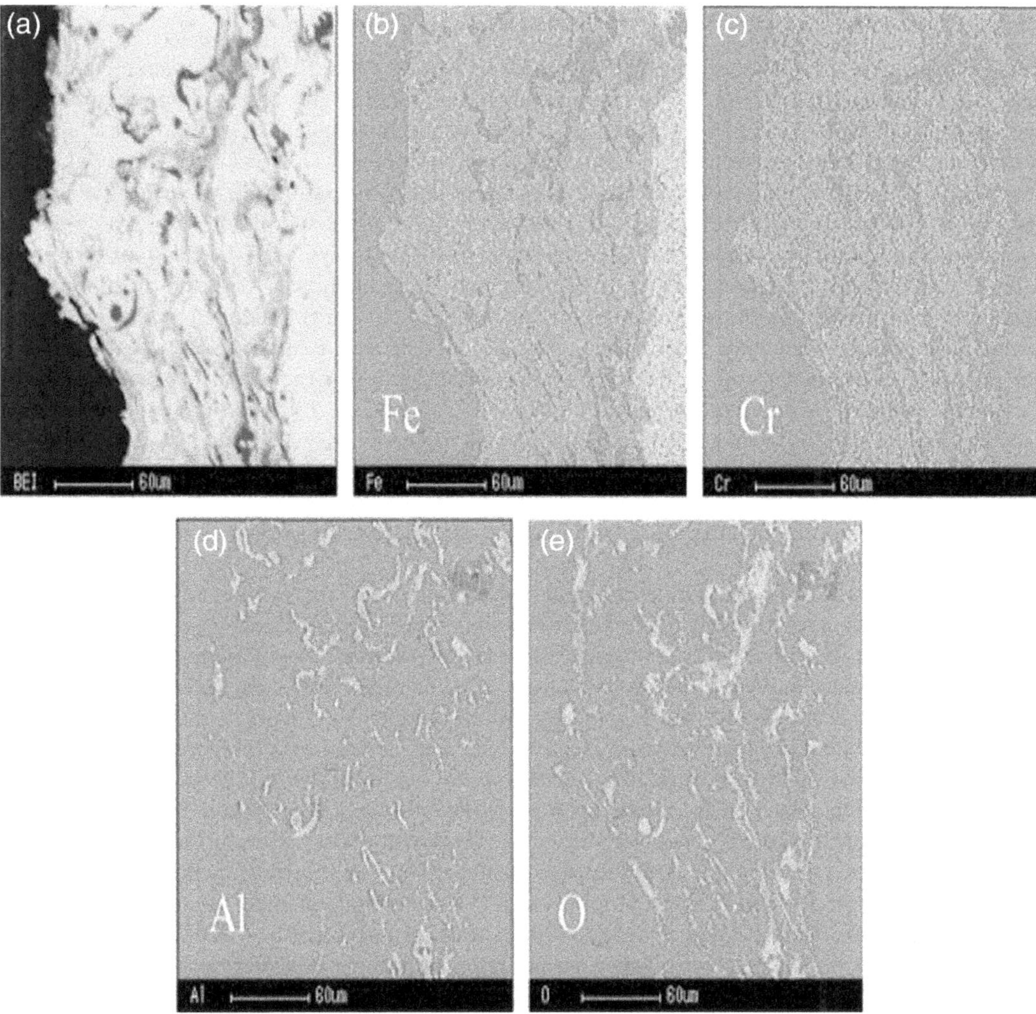

FIGURE 13.15 EPMA maps of (a) BEI image, (b) Fe, (c) Cr, (d) Al, and (e) O distribution on the cross-section of arc-sprayed FeCrAl coating.

13.5 SPECTROSCOPIC TECHNIQUE

A spectrum is a distribution of a physical quantity, such as charge, energy, frequency, time, mass, distance, etc., over another physical quantity for reference. When the referenced quantity is one-dimensional, the data will be called spectral lines. When the referenced quality is position, it is usually 2D and the data will be called spectral images. In general, any characterization technique has two roles: generating the signal and analyzing it. For the context of spectroscopic material characterization, signals are generated via beams of photons, electrons, or ions, usually of high energy and variable intensities. The method of beam generation, although not instructive to the contents of this section, will be mentioned briefly. Beam generation, while not a simple task, is a well-documented process. However, deducing information from the signal is not so straightforward due to the complex interactions between the excitation and the sample. Finally, one should be aware of what effect, depth, and volume these interactions have for each technique.

13.5.1 Energy Dispersive X-Ray Spectroscopy

Energy dispersive X-ray spectroscopy (EDX or EDS) seems to be a SEM technique, meaning that restrictions that apply to SEM, such as sample electrical conductivity and vacuum chamber, will also apply to EDS. It should be nondestructive to conforming samples. In general, there are four components of EDS measurement: an excitation source of e-beam, an X-ray detector, a pulse processor to digitize the signal, and an analyzer to perform data processing. The theory is that sample excitation provides enough energy to release an electron out of the sample, leaving a hole that an outer-shell electron then comes to fill. This process releases X-rays with the energies characteristic of each element. A database is required to identify these elements. As a rule of thumb, EDS generates information sparingly due to the low intensity of the X-ray produced and consequently takes a long time to reduce the high chance of noise. It is suitable to detect heavy metal ions, but not so much for elements with low atomic masses. The depth of generation is around 2 μm and the sample should be mounted properly, taking into account the long measuring time and the vacuum condition [151]. Generated peaks of different elements may also overlap, leading to further dependence on correction and database for better resolution [152]. Generally, the accuracy of EDS allows quantifying the relative abundance of elements on the surface or coating. However, the large depth of 2 μm may include the substrate under the coating, so caution should be taken with the obtained information.

EDS could supply both qualitative and quantitative analysis of the various materials, including coatings as are concerned in this book. However, since the process of generating characteristics, X-ray involves ionizing the related atoms, only elements with more than 3 atomic masses are feasible for measurement, and even then it is not very ideal as more atomically massive elements, or simply those with more density, are favored by this technique. The theory is that as the atoms of the sample are ionized via the excitation electron beam, when the holes are generated in the core shells, only then will X-rays be emitted by electrons moving to fill the hole from outer shells. The energy profile of the signal, such as how many X-ray fluorescence lines there are and their relative intensity among each other, is characteristic of the elemental distribution on the sample surface. Moreover, as said earlier, this signal must be deconvoluted using an existing database to make it presentable. Sample preparation must be strict because the measuring chamber is subjected to a high vacuum. Thorough cleaning, degreasing, and drying of the sample must be followed through to prevent outgassing of organic species. Proper mounting of the sample typically requires double-sided and vacuum-compatible conductive tape to attach the sample to the holder. If the sample conductivity is low, coating (and a tape connecting top to bottom side) or ionic liquid [153] can be used to alleviate the problem. Samples may also be fractured to expose the cross-section view of the coating.

Houk and colleagues [154] used EDS maps to perform the homogeneous distribution of the elemental components of solid-state products, irregularly shaped particles of some zones due to a nonequilibrium product, and the high homogeneity of elemental distribution within particles. $AlPO_4$-coated $LiCoO_2$ particle edges were characterized by EDS mapping which indicated phosphorus-rich clusters closed to oxygen distribution, Al contributed across the thick coating region (Figure 13.16) and the rest of the particle surface (Figure 13.17), and the particle bulk confine of Co [155].

EDS was utilized to characterize various materials and structures. In one study [156], it analyzed erbium-doped SiO2 layers on silicon surfaces, revealing the density of silicon in the spark processed layer and the distribution of erbium within the same region. Additionally, it facilitated atomic-scale chemical distribution and quantification in a nanostructured intermetallic alnico alloy [157], identified a uniform distribution of elements (C, S, and Na) on the electrode surface, resulting in improved electrochemical performance [158], and quantified the heterogeneous distribution of elements (Al, Si, and Fe) in an organic coating [159]. Furthermore, EDS revealed the presence of Sr and Ti in the core region of a TiOx-coated CoOOH/Rh-loaded SrTiO3:Al photocatalyst [160] and the distribution of phosphorus on both surface and grain boundaries of ALD-Li3PO4 coating

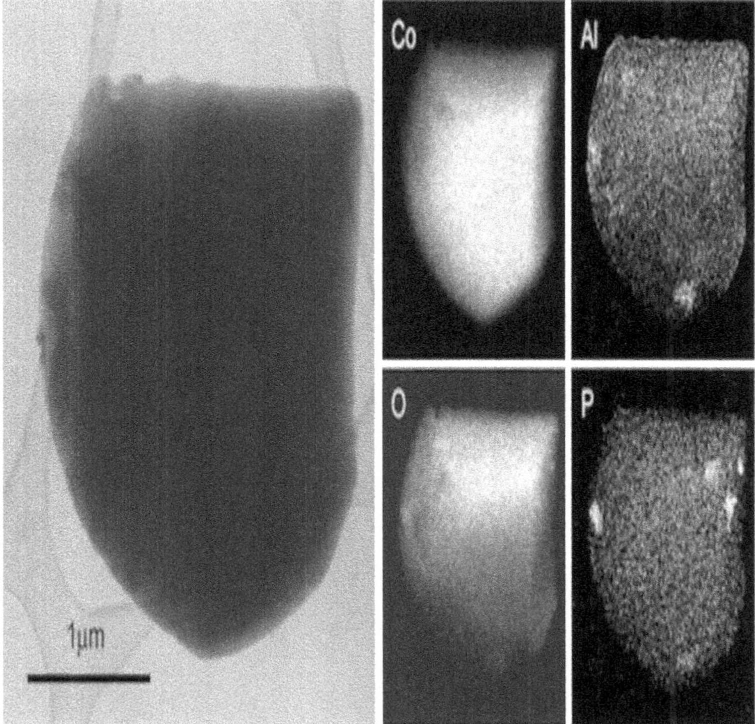

FIGURE 13.16 EDS mapping of the cross-sectional piece of $AlPO_4$-coated $LiCoO_2$.

[161]. Finally, it detected the enrichment of Al(OH)3 on the surface of secondary particles and its presence in the intergranular spaces of primary particles [162]. Regarding elemental distribution related to morphology and topography of post-surface, EDS showed the Fe, O, S, and N distribution maps of Fe, O, S, and N on the steel surface after 24 h of immersion in the sodium chloride solution containing 2-mercaptobenzothiazole-loaded functionalized mesoporous silica and confirmed the release and deposition of the protective film on the steel surface [163]. The C, O, and Na maps showed surface topography and related elemental distribution which enriched C and O elements at 78.63 and 20.84 at.% as a result of the protective film formation via the adsorption of organic species from colophony [164]. The uniform distribution of O, C, N, and Pr maps confirmed the improved corrosion protection of waterborne epoxy-coated iron sheet substrate via halloysite@ polyaniline nanoparticles loaded with the praseodymium (III) cation. The report also concluded that the detected oxygen could be from halloysite nanotubes, the detected C and N were from polyaniline, and the detected uniform Pr distribution suggested the successfully loaded Pr^{3+} on the nanotube surfaces [165]. SEM/EDS mapping also confirmed the corrosion products occurred on the mild steel surface immersed in 15 wt.% HCl solution with the Fe, Cl, and O distributions and its high weight percentage. On the other hand, the decrease in the weight percentage of Fe, O, and Cl, as well as the increase in the weight percentage of C with a uniform distribution and a smooth surface, affirmed the inhibitor species adsorption, formation of the protective film, and minimum surface damage in the 15% HCl solution containing sunflower oil [166]. The maps of Fe, C, La, P, and O distribution on the protective film formed on the steel surface via lanthanum diphenylphosphate compound were also characterized by SEM/EDS to support the dominant cathodic inhibition mechanism [167]. Therefore, EDS could be applied to a wide research field as an ideal technique to characterize the surface topography and related distribution, particularly post-treatment surfaces.

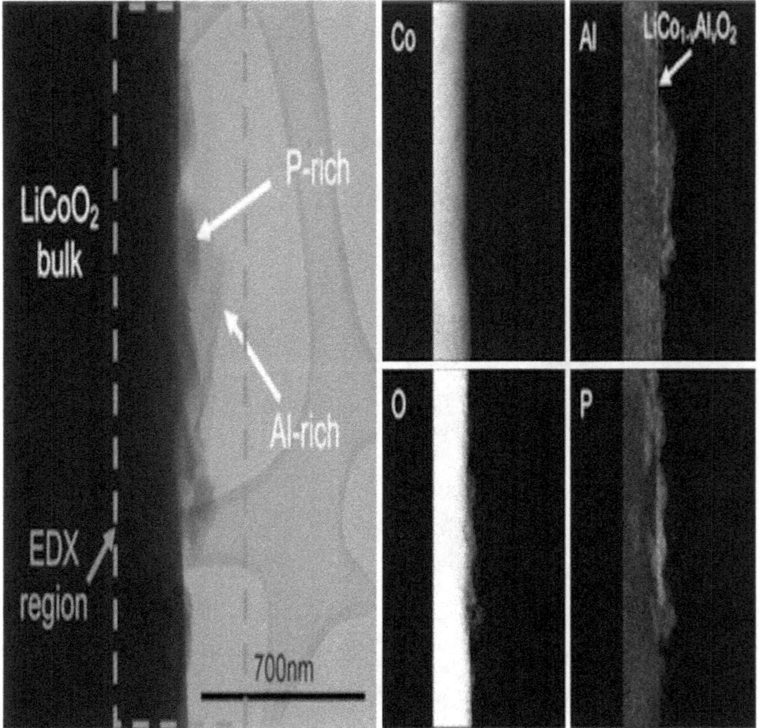

FIGURE 13.17 EDS mapping of $AlPO_4$-coated $LiCoO_2$ particle edge with STEM.

Electron beam generation is the same as that of the SEM technique. However, an EDS spectrum must use software to interpret the detected X-ray signals and reconstruct the surface according to revealed elements. And since the generated signal is different from SEM, the detectors must also have different specifications. An EDS detector can be expected to have (1) a collimator to make sure that generated X-ray signals are authentic; (2) an electron trap to keep out electrons; (3) a window that is transparent to X-rays, even those with low energy; (4) a semiconductor crystal detector that generates charges when hit by X-rays; and (5) electronics to measure those charges and send them to the computer for further processing and analysis.

13.5.2 Raman Spectroscopy

Raman spectroscopy is a light (usually laser) scattering characterization technique due to chemical bond vibrations and their interaction with light [168]. It relies on the polarizability of the molecule and is sensitive to homo-nuclear molecular bonds. Pure metallic bonds do not show on Raman spectroscopy. Instead, only covalent single, double, and triple bonds are detected. The signal is usually detected using CCD detectors, but an interferometer can also be used. Then it calculates relative frequencies in which the sample scatters radiation which depends on the type of vibration and strength of the bond. Functional groups that are susceptible to polarization will introduce strong peaks in the Raman spectrum.

Since it is a bulk technique that scatters light, samples can be measured as-is with a very small volume (less than 1 and 10 μm of the diameter and depth); the identification of species can be represented by the spectra in that volume. However, the cost may be high when compared to that of IR techniques which provide somewhat similar information which is the chemical identification via unique chemical fingerprints and the intensity of scattered light is plotted versus the Raman shift. While scattering does not depend on the wavelength of the light source, Raman shift is performed in the unit of wavenumber which should be inverted to wavelength. The general setup for the Raman spectrometer consists of a laser, microscope, grating, and detector. A computer is an option but also of great value. The light source of a Raman spectrometer is, in general, a laser with a predefined wavelength, which lies between the ultraviolet and near-infrared range depending on the application. The scattered light is focused back to the filter that virtually only lets Raman scatter through. Grating is used to separate the scattered light which will be measured, in most cases, at a CCD detector. Finally, the measurement, which is photocurrent without further processing, is sent to a computer for analysis.

Raman spectroscopy has been recognized as a powerful analysis tool in physical, chemical and biological surface sciences, and so on [169, 170]. Ledinský and colleagues [171] expanded the contactless thickness profiling technology to deposit thin films on rough surfaces with thin film thickness maps. The study also discussed the characteristics applied to single-layer and/or multilayer 2D materials.

Raman mapping, a versatile technique, has been employed across various studies. In one instance [172], it observed edge orientations and stacking orders of bi- and trilayer graphene. Additionally, it demonstrated its capability to detect risk chemicals such as bisphenol, pentachlorophenol, and dichlorophenol with high sensitivity [173]. Moreover, it showcased high surface homogeneity over a wide scale range [174] and provided hyperspectral 2D results, integrating electronic and vibrational spectra with surface topographies [175]. Furthermore, it elucidated the relationship between phase boundaries and giant dielectric responses in a solid-solution system [176] and quantified Si–Si distortion angles and strain-free energy in amorphous silicon structures [177]. Finally, it mapped blended microplastic surfaces, delineated various polymer contributions, and illustrated polymer distributions in surface morphology with quantitative concentrations [178]. Raman mapping has been applied to characterize the corrosion inhibitors as a potential technique for elemental distribution of post-surface related to surface morphology and topography. Ma and colleagues [179] used Raman mapping to obtain benzotriazole molecules' adsorption on AA2024-T3 surface exposed to 3.5 wt.% NaCl solution containing benzotriazole inhibitor. The combination of Raman and other techniques can provide an insight into the inhibition mechanism with the detected surface components and local surface, as well as depth component distribution [180]. Zhang and colleagues [15] reported that Raman maps provided the component distribution maps of Fe_3O_4 and γ-FeOOH on both uninhibited and inhibited surfaces; other iron oxides/hydroxides and Mefp-1 species were also observed on uninhibited and inhibited surfaces as shown in Figures 13.18 and 13.19, respectively. It is noticed that iron hydroxide ($Fe(OH)_3$) as part of corrosion products was not observed by Raman spectroscopy due to its strong fluorescence background. Therefore, Raman spectroscopy can provide the component distribution maps related to surface morphology and topography after corrosion with the evident identification of the enriched molecular vibration information, high efficiency, nondestructiveness, low cost, and easy operation [181]. However, the inelastic Raman scattering of light can happen due to the relatively weak Raman signals at a very small fraction of the photons incident.

13.5.3 Time-of-Flight Secondary Ion Mass Spectrometry

While technically not being a spectroscopy technique, ToF-SIMS has a similar premise. It excites the sample with high-energy ion beams (SIMS) and analyzes the signal via the time-of-flight (ToF) method. The main difference is the consideration of mass for both the excitation and signal. The principle of SIMS is sputtering; as the high energy ion beam (charged particles accelerated by electric field of 1–25 keV) bombards the surface of the sample, collisions are created in "cascades" which eject a part of the topmost layer of the surface. This part can be atoms, molecules, or

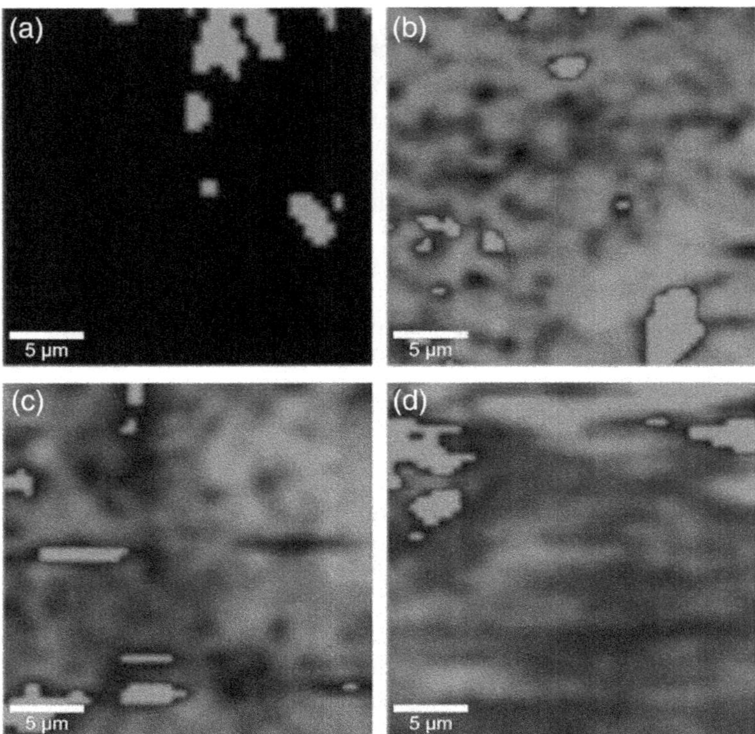

FIGURE 13.18 Component distribution maps observed by Raman technique of uncoated carbon steel surface immersion in 0.1 M NaCl solution after (a) 1 h, (b) 1 day, (c) 2 days, and (d) 4 days.

molecular fractions, some of which may be ionized on the way out. The idea for ToF-SIMS, specifically, is that the ejected material is limited to less than 1% of the top layer due to a low dose of primary ions (less than 10^{13} cm^{-1} and is usually achieved by pulsing of heavy ions) and that these emitted particles (secondary ions) are analyzed by ToF mass spectrometer, hence the name of the technique. In short, the excitation is from a high-energy ion beam, the signal is from secondary ions, and the analyzer is a ToF mass spectrometer [182]. Because the sputtered material is limited to 1% of the first layer of the sample, the depth of characterization differs for each sample depending on the nature of the coating involved and the time required to obtain enough information about the coating can be expected to be long. Furthermore, the identification of peaks from mass spectra requires a database. And unlike SIMS which is a destructive technique capable of characterizing concentrations of impurity at great depth and with high sensitivity, ToF-SIMS is limited to elemental and molecular information on the surface. Moreover, the pulsed ion beam of ToF-SIMS may be focused to obtain very high resolution spectral images. Arguments can be made for mounting an auxiliary sputter source to strike a compromise between quality and depth of characterization [183].

A comprehensive review of ToF-SIMS and the related computation has been compiled by Vickerman [184]. In summary, sputtering is a destructive method and sputter rate can become increasingly more complex with higher involvement of covalent organic matters due to how differently ionized fragments can break off. ToF-SIMS not only measures the yield of elemental carbon, but also gives information on the chemical structure of the destroyed area. Thus, Vickerman promoted the use of damage cross-section σ instead of sputter rate. Ionization happens on the way out of particles emitted from the surface, meaning that the yield of secondary ions is

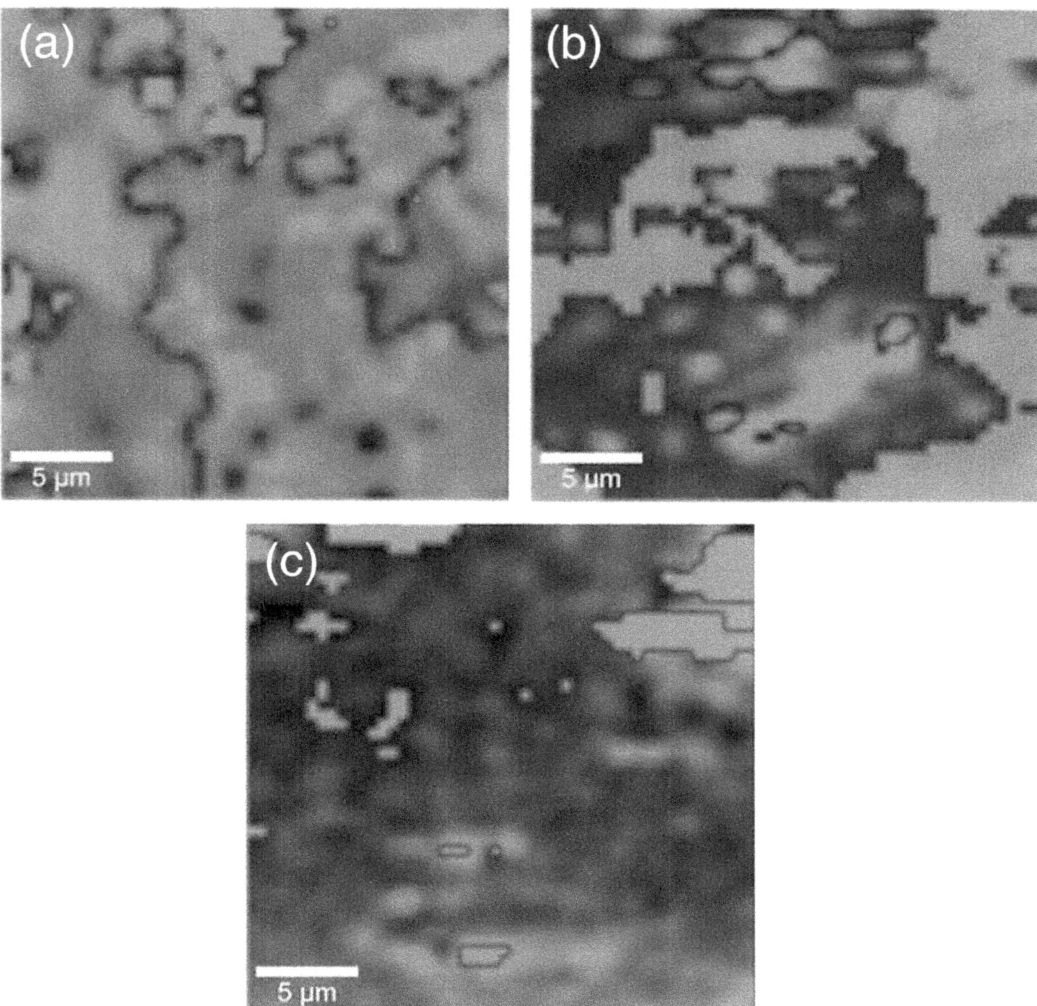

FIGURE 13.19 Component distribution maps observed by Raman technique of the Mefp-1/ceria composite film on steel surface immersion in 0.1 M NaCl solution after (a) 1 h, (b) 1 day, and (c) 2 days.

heavily dependent on the electronic state of the material. In other words, electronic states of the surface decide how secondary ions form. Electron ejection, acid base reactions, and cationization and anionization can all generate ions and they in turn give information about the material being sputtered. Notably, low-mass fragments may reveal energetic recoil within the material. There are four main components for the static SIMS experiment: generating primary particles, forming secondary ions, detecting them, and analyzing them via mass spectrometer. An electron source will also be required for charge compensation. Early static SIMS instruments used quadrupole mass analyzers not unlike dynamic SIMS; however, sensitivity was a problem. In the 1980s, ToF analyzers were adopted and it is now a powerful technique to characterize the surface topography and related distribution of the inhibited surface finish.

Visser and colleagues [185] used ToF-SIMS to investigate the lateral spread of lithium and chemical species of corrosion inhibitor contributing to the protective layers in the defect regions on the AA2024-T3 surface. In this process, nonnegative matrix factorization (Figure 13.20) was used

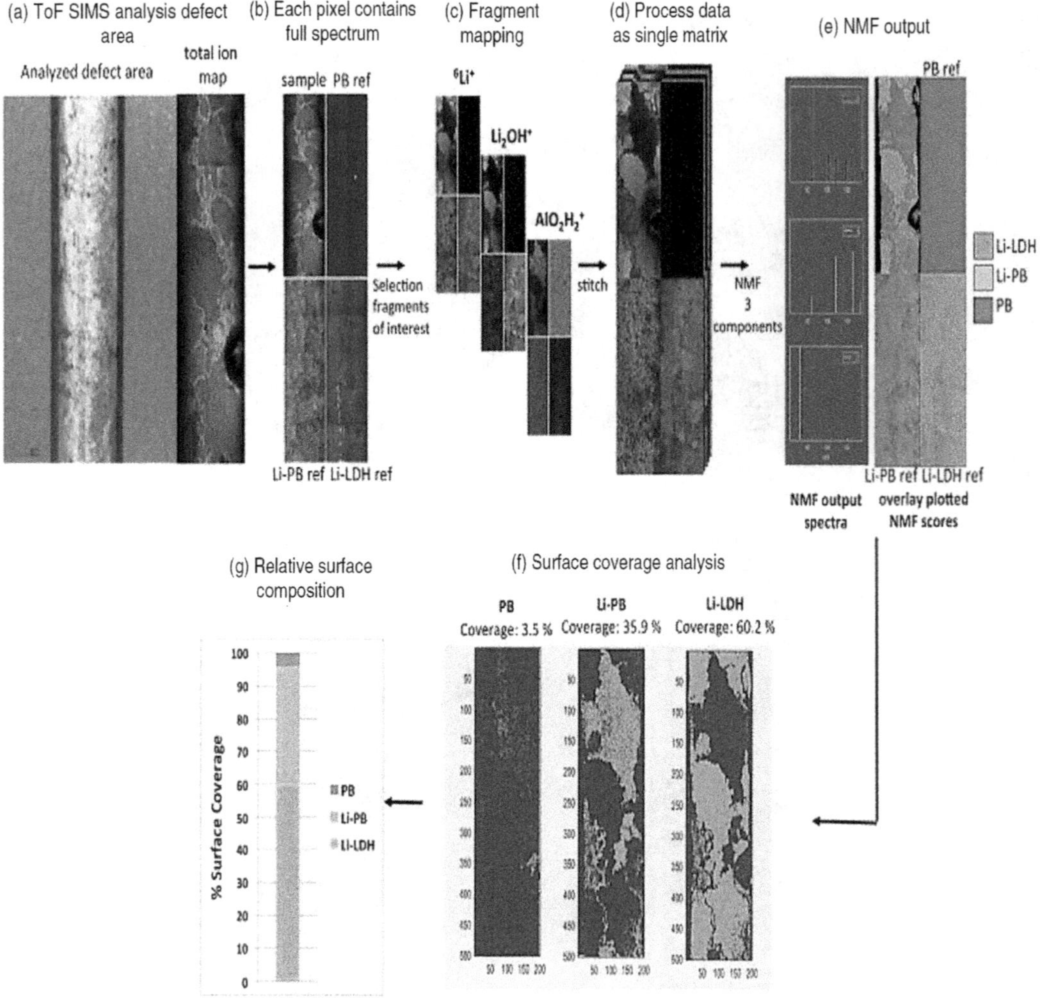

FIGURE 13.20 Non-negative matrix factorization process for the compositional characterization of the protective layer in the defect area.

for detecting the protective layer composition in the defect area. The work focused on the image of the defect area, total ion maps of the defect area, and all maps of each fragment to detect the percentage of surface coverage for each compound and the compositional ratio for all compounds contributed to the protective layer in the defect area. ToF-SIMS also confirmed the composition and growth of the corrosion protective layer on the AA2024-T3 surface by the leachable lithium salts in an organic coating defect [186]. The study also reported that the lateral spread of leaching lithium salts in artificial 1-mm-wide scribes and the distinction between pseudoboehmite, lithium-containing pseudoboehmite, and Al-Li layered double hydroxide layers were also successfully demonstrated. The combination of all ion images indicated that lower Li concentrations concurred with higher Al concentrations and the different morphologies could be attributed to the different compositions of the surface coverage. The use of ToF-SIMS identified the phase segregation in blends as the organic coatings indicated the maps of tapers, reconstruction error, each coating component distribution, phase identification, and morphology characterization via nonnegative matrix factorization analysis [187]. In addition, ToF-SIMS maps were also used for evaluating the molecular characteristics and transport of contaminants to a buried interface of polyaniline films as a conformal aerospace industry coating [188], organic, bioorganic, and biological systems [189–191], active epoxy coating protection for Li_2CO_3-pretreated anodized 2024-T3 alloy [192], polymeric surfaces and interfaces [191], 3D printing of bacterial biofilm resistant composite devices [192], the interfacial chemistry between polymer and a Fe-Cr alloy [193], dimensional stacking of heterostructures [194], heterogeneous oxidation of thermally aged cross-linked polyethylene [195], etc. Therefore, regarding the identification of the protective layer composition of post-surface, ToF-SIMS could be preferred due to its higher sensitivity in comparison with Raman, EDS, and AES.

13.5.4 Auger Electron Spectroscopy

AES is a spectroscopy technique that measures the Auger effect which is one of the consequences of materials being excited with X-rays or other energetic beams. Based on the discovery by Pierre Auger and Lise Meitner in the 1920s electrons can be emitted after a series of internal relaxation events that carry information about the electronic process of the material. The Auger effect is the result of inter- and intrastate transitions of electrons in an excited atom that can be detected and processed to identify the chemical and compositional makeup of the surface. When an atom is excited by a photon or a beam of electrons with energies in the eV and keV range, the Auger event starts as a core state electron is removed due to the said excitation. An unstable hole will be left behind that may be covered by an outer shell electron. When this event happens, another outer electron can be coupled and emitted from the atom if the transferred energy exceeds its binding energy. This is a simplified formula that ignores the effect of ionization on binding energy. Finally, this value of binding energy will provide information on the chemical composition [196].

AES is a unique analysis technique in surface sciences, particularly for the chemical composition of the surface. Prasittichai and colleagues [197] used AES to perform the polyurea film via C and N Auger electron elemental mapping of the self-assembled monolayer and polyurea molecular layer depositions. Their study also reported that polyurea molecular layer selectively deposited on the patterned SiO_2 surface, suggesting the new fabrication method for 3D organic structures. Dong and colleagues [198] used AES to characterize the local corrosion of bimetallic Cu–Sn catalysts using AES element maps of tin and copper before and after the CO_2 reduction reaction. AES detected Sn elements in the patterned region with an enrichment of O elements, and Cu elements in the whole surface, suggesting the evident pieces of local corrosion. In the aging of transition metal dichalcogenide monolayers, AES also supported the detection of gradual oxidation along the grain boundaries and the adsorption of organic contaminants [199]. Combined with the TEM image and the principles of selective deposition of metal on metal, the compositional map of AES results ensured the selective growth of Ru metal thin film on the Mo surface [200]. It is noted that EDS measurements could not provide the accurate composition of element distributions on the substrate surface owing to the resolution limitations and the overlapping of Mo and Ru peaks. In the study of the surface reactivity of Li_2MnO_3 crystals [201], Auger's chemical mapping of Mn, S, and O distribution on the Li_2MnO_3 crystal surface after SO_2 adsorption demonstrated the homogeneous distribution of all of the elements. The methylammonium lead iodide ($MAPbI_3$) perovskite film degraded by heat was analyzed by AES, which indicated uniform C and Pd distribution again and the disappearance of the organic compounds owing to the surface carbonization after long-term heat effect [202]. The nanoscale AES mapping of titanium, manganese, and oxygen of a single particle provided the homogeneous distribution of Mn element in the particle core of microwave-assisted and metal-induced crystallization [203]. Wood and colleagues [204] used AES mapping to determine the solid electrolyte interphase formation and evolution in Li_2S-P_2S_5

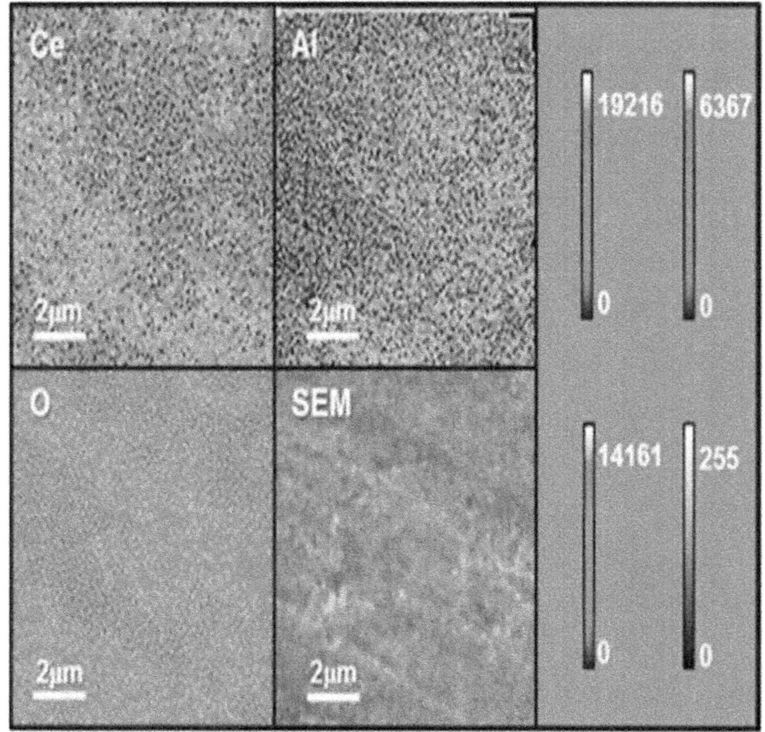

FIGURE 13.21 SAM images of Ce, Al, and O element distribution on cerium conversion coating observed by ToF-SIMS and SEM image of the same area after corrosion.

solid-state electrolytes. The study presented the solid electrolyte interphase heterogeneity and the locations without and with high oxygen concentrations after charging. Arioka and colleagues [205] used AES maps to determine the cause of intergranular penetration by cavity formation. The results showed carbon, nickel, and chromium maps in the fracture surface and oxygen distribution on the intergranular crack surface, as well as a high density of cavities at grain boundaries near the pre-crack and the thick oxide-fully-covered parts of the stress corrosion cracking surface. AES elemental distribution at the hydrogen-induced intergranular crack and the enriched Mg at the cracked grain boundaries suggest the role of Mg in the improved resistance of the hydrogen embrittlement [206]. The dezincification spot on $CuZn_{21}Si_3P$ after 12 months of immersion in tap water was also characterized by AES maps, which indicated high copper, low zinc, and no silicon content in the dezincified area. Oxygen was located between the remaining grain and the dezincification zone where there did not appear to be any silicon signals and the phosphorous was located everywhere in the oxide [207]. Uhart and colleagues [208] used the AES map to characterize the active corrosion protection of cerium for AA2024-T3 alloy in 3.5 wt.% NaCl solution at room temperature. The results indicated good cerium dispersal in the conversion coating deposited on the AA2024-T3 alloy surface. It also suggested that cerium in conversion coating could move to the pit locations during the corrosion event due to its relatively high content in the pit and its center part as shown in Figure 13.21. Other studies also used AES maps to ensure the high corrosion resistance of coating [209, 210].

Advantages of AES include high surface sensitivity due to the fact that the mean free path of emitted electrons, eV to keV in this case, is very short, limiting escape depth to a few nanometers from the surface. However, the low energy of Auger electrons requires ultra-high vacuum conditions for most setups [211]. This also prevents adsorbed gas, which may happen in nonvacuum, from interfering with such a sensitive technique. A regular AES configuration requires a focused electron beam of specific intensities on the specimen and generated electrons are collected via a cylindrical mirror analyzer. The detected signals must be amplified before being sent to data processing due to their low intensity. Therefore, this technique seems to be the most powerful for visualizing the surface topography and related distribution of the inhibited surface finish.

13.6 CONCLUSION

This chapter summarizes the most common techniques used for coating characterization with morphology, topography, and related distribution. Therefore, the morphology, topography, and related distribution of the surface finish after using a corrosion inhibitor should be based on a suitable combination of the above-suggested analysis techniques. The images of coating characterization need to provide both low and high magnifications to fully show the surface morphology and topography. The best consistent combination of coating characterization with full investigation including morphology, topography, and related distribution could choose OM/SEM (prefers SEM) and profilometer/AFM/EPMA and one spectroscopic mapping technique such as EDS (or EDX), Raman, ToF-SIM spectrometry, and AES, which could provide the most fundamental coating characterization.

REFERENCES

1. Carmona-Hernandez A, Campechano-Lira C, Espinoza-Vázquez A, Ramírez-Cano JA, Orozco-Cruz R, Galván-Martínez R (2023) Electrochemical and DFT theoretical evaluation of the Randia monantha Benth extract as an eco-friendly corrosion inhibitor for mild steel in 1 M HCl solution. *J Taiwan Inst Chem Eng* 147:104913.
2. Tran TN, Hoai Vu NS, Tran TT, Jo DS, Huynh TL, Nguyen TTV, Panaitescu C, Nguyen HTT, Nguyen VK, Nguyen Dang N (2024) Precise major compounds in Barringtonia acutangula flower — water extract for mitigating carbon steel corrosion. *J Taiwan Inst Chem Eng* 155:105251.
3. Nguyen Dang N, Tran Van H, Do Thai N, Thai Khac NH (2016) Film formation in Y(4NO2Cin)3 compound on 6061 aluminum alloy to protect against corrosion in chloride ion media. *J Taiwan Inst Chem Eng* 67:495–504.
4. Moumeni O, Mehri M, Kerkour R, Boublia A, Mihoub F, Rebai K, Khan AA, Erto A, Darwish AS, Lemaoui T, Chafai N, Benguerba Y (2023) Experimental and detailed DFT/MD simulation of α-aminophosphonates as promising corrosion inhibitor for XC48 carbon steel in HCl environment. *J Taiwan Inst Chem Eng* 147:104918.
5. Nam ND, Mathesh M, Hinton B, Tan MJY, Forsyth M (2014) Rare earth 4-hydroxycinnamate compounds as carbon dioxide corrosion inhibitors for steel in sodium chloride solution. *J Electrochem Soc* 161:C527–C534.
6. Akhtar S (2022) Applications of atomic force microscopy in corrosion research. In: Toor, I.U. (ed) *Recent Developments in Analytical Techniques for Corrosion Research*. Springer, Cham.
7. Li XY, Cheong BHP, Somers A, Liew OW, Ng TW (2013) Surface-scribed transparency-based microplates. *Langmuir* 29:849–855.
8. Hien PV, Vu NSH, Thu VTH, Sommers A, Nam ND (2017) Study of yttrium 4-nitrocinnamate to promote surface interactions with AS1020 steel. *Appl Surf Sci* 412:464–474.
9. Manh TD, Huynh TL, Thi BV, Lee S, Yi J, Dang NN (2022) Corrosion inhibition of mild steel in hydrochloric acid environments containing Sonneratia Caseolaris leaf extract. *ACS Omega* 7:8874–8886.
10. Manh TD, Hien PV, Nguyen QB, Quyen TN, Hinton BRW, Nam ND (2019) Corrosion inhibition of steel in naturally-aerated chloride solution by rare-earth 4-hydroxycinnamate compound. *J Taiwan Inst Chem Eng* 103:177–189.
11. Gad SM, Emad S, Zhou X, Lyon SB, Jin Zelong, Dagwa IM (2022) Effectiveness of strontium zinc phosphosilicate on the corrosion protection of AA2198-T851 aluminium alloy in sodium chloride solution. *Corros Sci* 209:110725.

12. Wang Y, Li J, Zhang L, Zhang L, Wang Q, Wang T (2020) Structure of the rust layer of weathering steel in A high chloride environment: A detailed characterization via HRTEM, STEM-EDS, and FIB-SEM. *Corros Sci* 177:108997.
13. Lillard RS, Kashfipour MA, Niu W (2016) Pit propagation at the boundary between manganese sulfide inclusions and austenitic stainless steel 303 and the role of copper. *J Electrochem Soc* 163:C440–C451.
14. Bassim N, Scott K, Giannuzzi LA (2014) Recent advances in focused ion beam technology and applications. *MRS Bull* 39:317–325.
15. Zhang F, Brinck T, Brandner BD, Cleasson PM, Dedinaite A, Pan J (2013) In situ confocal Raman micro-spectroscopy and electrochemical studies of mussel adhesive protein and ceria composite film on carbon steel in salt solutions. *Electrochim Acta* 107:276–291.
16. Dubois F, Mendibide C, Pagnier T, Perrard F, Duret C (2008) Raman mapping of corrosion products formed onto spring steels during salt spray experiments. A correlation between the scale composition and the corrosion resistance. *Corros Sci* 50:3401–3409.
17. Kharitonov DS, Sommertune J, Ornek C, Ryl J, Kurilo II, Cleasson PM, Pan J (2019) Corrosion inhibition of aluminium alloy AA6063-T5 by vanadates: Local surface chemical events elucidated by confocal Raman micro-spectroscopy. *Corros Sci* 148:237–250.
18. Nam ND, Panaitescu C, Tan MYJ, Forsyth M, Hinton B (2018) An interaction between praseodymium 4-hydroxycinnamate with AS1020 and X65 steel microstructures in carbon dioxide environment. *J Electrochem Soc* 165:C50–C59.
19. Li C, He Y, Li Z, Li H, Zhao Y (2022) Graphene loaded with corrosion inhibitor cerium (III) cation for enhancing corrosion resistance of waterborne epoxy coating: Physical barrier and self-healing. *Colloids Surf A* 635:128048.
20. Ma L, Wang J, Wang Y, Huo X, Wu S, Fu D, Zhang D (2022) Enhanced active corrosion protection coatings for aluminum alloys with two corrosion inhibitors co-incorporated in nanocontainers. *Corros Sci* 208:110663.
21. Anh NP, Bach LX, Panaitescu C, Sy LV, Dang NN (2022) An investigation of the role of calcium in the microstructure of Mg-5Al-4Sn-based alloys and pitting corrosion resistance. *J Mater Eng Perform* 31:8830–8839.
22. Sy LV, Lal B, Binh PMQ, Nguyen QB, Hung TV, Panaitescu C, Nam ND (2020) The role of alloyed strontium in the microstructures and alkaline electrochemistry of Mg–5Al–4Sn alloys. *RSC Adv* 57:34387–34395.
23. Nam ND, Mathesh M, Le TV, Nguyen HT (2014) Corrosion behavior of Mg-5Al-xZn alloys in 3.5 wt.% NaCl solution. *J Alloys Compd* 616:662–668.
24. Nam ND, Bian MZ, Forsyth M, Seter M, Tan M, Shin KS (2012) Effect of calcium oxide on the corrosion behaviour of AZ91 magnesium alloy. *Corros Sci* 64:263–271.
25. Kim KH, Nam ND, Kim JG, Shin KS, Jung HC (2011) Effect of calcium addition on the corrosion behavior of Mg-5Al alloy. *Intermetallics* 19:1831–1838.
26. Nam ND, Kim WC, Kim JG, Shin KS, Jung HC (2010) Corrosion resistance of Mg-5Al-xSr alloys. *J Alloys Compd* 509:4839–4847.
27. Nam ND, Kim WC, Kim JG, Shin KS, Jung HC (2009) Effect of mischmetal on the corrosion properties of Mg–5Al alloy. *Corros Sci* 51:2942–2949.
28. Li Y, Lu X, Mei D, Zhang T, Wang F (2022) Passivation of corrosion product layer on AM50 Mg by corrosion inhibitor. *J Magnesium Alloys* 10:2563–2573.
29. Fletcher JS, Kotze HL, Armitage EG, Lockyer NP, Vickerman JC (2013) Evaluating the challenges associated with time-of-fight secondary ion mass spectrometry for metabolomics using pure and mixed metabolites. *Metabolomics* 9:535–544.
30. Esmaily M, Malmberg P, Shahabi-Navi M, Svensson JE, Johansson LG (2016) A ToF-SIMS investigation of the corrosion behavior of Mg alloy AM50 in atmospheric environments. *Appl Surf Sci* 360:98–106.
31. Liang T, He G, Huang G, Kong Y, Fu W, Chen H, Wang Q, Iwai H, Fujata D, Liu Y, Xu M (2015) Graphene nucleation preferentially at oxygen-rich Cu sites rather than on pure Cu surface. *Adv Mater* 27:6404–6410.
32. Bellhouse EM, Mcdermid JR (2014) selective oxidation and reactive wetting during galvanizing of a CMnAl TRIP-assisted steel. *Metall Mater Trans A* 42A:2753–2768.

33. Seeger RJ (1996) *Galileo Galilei, His Life and His Works (Men of Physics Series)*, 1st edition, Pergamon Press.
34. Chen Y, Guo Z, Wang X, Qiu C (2008) Sample preparation. *J Chromatogr A* 1184:191–219.
35. Nematollahi M, Heidarian M, Peikari M, Kassiriha SM, Arianpouya N, Esmaeilpour M (2010) Comparison between the effect of nanoglass flake and montmorillonite organoclay on corrosion performance of epoxy coating. *Corros Sci* 52:1809–1817.
36. Duong DL, Han GG, Lee SM et al. (2012) Probing graphene grain boundaries with optical microscopy. *Nature* 490:235–239.
37. Li H, Zhang B, Li Y, Wu P, Wang Y, Xie M (2022) Effect of novel green inhibitor on corrosion and chemical mechanical polishing properties of cobalt in alkaline slurry. *Mater Sci Semicond Process* 146:106691.
38. Kurchavov D, Rustambek U, Ottochian A, Lefèvre G, Seyeux A, Ciofini I, Marcus P, Lair V, Volovitch P (2022) Synergistic effect of ionic liquid (IL) cation and anion inhibits negative difference effect on Mg in water – IL mixtures. *Corros Sci* 209:110723.
39. Li L, Jie W, Yuanzhi W, Dang Nam N (2016) Effect of static annealing on microstructure and texture in extruded Mg-Gd-Y-Zr alloy. *Rare Metal Mate Eng* 45:2263–2268.
40. Sullivan J, Mehraban S, Elvins J (2011) In situ monitoring of the microstructural corrosion mechanisms of zinc-magnesium-aluminium alloys using time lapse microscopy. *Corros Sci* 53:2208–2215.
41. Sanchez T, Kurchavova E, Shkirskiy V, Światowska J, Vivier V, Volovitch P (2021) Detection and quantification of defect evolution at buried metal-oxide-polymer interface on rough substrate by local electrochemical impedance mapping. *Electrochim Acta* 388:138467.
42. Denissen PJ, Shkirskiy V, Volovitch P, Garcia SJ (2020) Corrosion inhibition at scribed locations in coated AA2024-T3 by cerium- and DMTD-loaded natural silica microparticles under continuous immersion and wet/dry cyclic exposure. *ACS Appl Mater Interfaces* 12:23417–23431.
43. Denissen PJ, Homborg AM, Garcia SJ (2019) Interpreting electrochemical noise and monitoring local corrosion by means of highly resolved spatiotemporal real-time optics. *J Electrochem Soc* 166:C3275–C3283.
44. Gateman SM, Gharbi O, Turmine M, Vivier V (2021) Measuring changes in wettability and surface area during micro droplet corrosion measurements. *Electrochim Acta* 399:139402.
45. Zhao W, Babu RP, Chang T, Odnevall I, Hedström P, Johnson CM, Leygraf C (2022) Initial atmospheric corrosion studies of copper from macroscale to nanoscale in a simulated indoor atmospheric environment. *Corros Sci* 195:109995.
46. Schindelholz EJ, Cong H, Jove-Colon CF, Li S, Ohlhausen JA, Moffat HK (2018) Electrochemical aspects of copper atmospheric corrosion in the presence of sodium chloride. *Electrochim Acta* 276:194–206.
47. Olgiati M, Denissen PJ, Garcia SJ (2021) When all intermetallics dealloy in AA2024-T3: Quantifying early stage intermetallic corrosion kinetics under immersion. *Corros Sci* 192:109836.
48. Sullivan J, Cooze N, Gallagher C, Lewis T, Prosek T, Thierry D (2015) In situ monitoring of corrosion mechanisms and phosphate inhibitor surface deposition during corrosion of zinc-magnesium-aluminium (ZMA) alloys using novel time-lapse microscopy. *Faraday Discuss* 180:361–379.
49. Wint N, Ansell P, Edy J, Williams G, McMurray HN (2019) A method for quantifying the synergistic inhibitory effect of corrosion inhibitors when used in combination: A 'chromate generating coating'. *J Electrochem Soc* 166:C580–C588.
50. Kousis C, Keil P, McMurray HN, Williams G (2022) The kinetics and mechanism of filiform corrosion affecting organic coated Mg alloy surfaces. *Corros Sci* 206:110477.
51. Chen F, Brown GM, Song M (2000) Overview of 3-D shape measurement using optical methods. *Opt Eng* 39:10–22.
52. Sreekanth D, Rameshbabu N, Venkateswarlu K, Subrahmanyam C, Krishna LR, Rao KP (2013) Effect of K_2TiF_6 and $Na_2B_4O_7$ as electrolyte additives on pore morphology and corrosion properties of plasma electrolytic oxidation coatings on ZM21 magnesium alloy. *Surf Coatings Technol* 222:31–37.
53. Alanazi NM, Ulhaq MI, Al-Zahrani TY, Saleem Q, Leoni M, Al-Mutahhar FM, Al-Shebil RA, Abeedi TM (2023) Dual-purpose corrosion and kinetic hydrate inhibitors based on 3-acrylamidopropyl trimethyl ammonium copolymers. *React Funct Polym* 190:105626.

54. Mandal S, Bej S, Banerjee P (2023) Insights into the uses of two azine decorated d10-MOFs for corrosion inhibition application on mild steel surface in saline medium: Experimental as well as theoretical investigation. *J Mol Liq* 381:121789.
55. Farhadiana A, Varfolomeeva MA, Shaabanid A, Nasirid S, Vakhitove I, Zaripovab YF, Yarkovoib VV, Sukhov AV (2020) Sulfonated chitosan as green and high cloud point kinetic methane hydrate and corrosion inhibitor: Experimental and theoretical studies. *Carbohydr Polym* 236:116035.
56. Moustafa AHE, Abdel-Rahman HH, Awad MK, Naby AANA, Seleim SM (2022) Molecular dynamic simulation studies and surface characterization of carbon steel corrosion with changing green inhibitors concentrations and temperatures. *Alexandria Eng J* 61:2492–2519.
57. Umoren SA, Suleiman RK, Obot IB, Solomon MM, Adesina AY (2022) Elucidation of corrosion inhibition property of compounds isolated from Butanolic date palm leaves extract for low carbon steel in 15% HCl solution: Experimental and theoretical approaches. *J Mol Liq* 356:119002.
58. Li Y, Feng S, Liu H, Tian X, Xia Y, Kai X, Yu H, Liu Q, Chen C (2020) Bacterial distribution in SRB biofilm affects MIC pitting of carbon steel studied using FIB-SEM. *Corros Sci* 167:108512.
59. Rousseau B, Rivard P, Marache A, Ballivy G, Riss J (2012) Limitations of laser profilometry in measuring surface topography of polycrystalline rocks. *Int J Rock Mech Min Sci* 52:56–60.
60. Nam ND, Bui QV, Mathesh M, Tan MYJ, Forsyth M (2013) A study of 4-carboxyphenylboronic acid as a corrosion inhibitor for steel in carbon dioxide containing environments. *Corros Sci* 76:257–266.
61. Ngoc DTB, Lai LX, Ngo KLD, Manh TD, Dinh TV, Thu XNT, Nguyen DK, Dang NN (2023) Inhibition properties of Vang tea-water extract for carbon steel corrosion in acidic environments. *J Taiwan Inst Chem Eng* 149:104941.
62. Bach LX, Dao TBN, Pham TT, Sporken R, Nguyen TN, Vattikuti SVP, Van NP, Dang NN (2022) Role of SnO_2 nanoparticles for a self-forming barrier layer on a mild steel surface in hydrochloric acid medium containing Piper betle leaf extract. *ACS Omega* 7:38061–38068.
63. Vuong BX, Huynh TL, Tran TQN, Vattikuti SVP, Manh TD, Nguyen-Tri P, Nguyen AT, Hien PV, Dang NN (2022) Corrosion inhibition of carbon steel in hydrochloric acid solution by self-formation of a Malpighia glabra leaf extract-based organic film. *Mater Today Commun* 31:103641.
64. Anh HT, Vu NSH, Huyen LT, Tran NQ, Thu HT, Bach LX, Trinh QT, Vattikuti SVP, Nam ND (2020) Ficus racemosa leaf extract for inhibiting steel corrosion in a hydrochloric acid medium. *Alexandria Eng J* 59:4449–4462.
65. Trung DC, Pham TT, Phan Minh QB, Panaitescu C, Tran NQ, Anh HT, Bach LX, Dang NN (2021) The use of Piper Betle leaf extract for forming a barrier layer on steel surface in hydrochloric acid solution. *Prog Org Coat* 158:106340.
66. Nam ND, Ha PTN, Anh HT, Hoai NT, Hien PV (2019) Role of hydroxyl group in cerium hydroxycinnamate on corrosion inhibition of mild steel in 0.6 M NaCl solution. *J Saudi Chem Soc* 23:30–42.
67. Zhang M, Li C, Wang X, Peng J, Yuan S, Geng H, Zhou Y, Gao Y, Wang H (2021) Ultrahigh anticorrosion performance of polymer-based coating filled with a novel micro network nanofiller. *Corros Sci* 190:109685.
68. Liang J, Liu S, Peng Z, Li R, Wang B (2023) Galvanic corrosion behavior of AZ31 Mg alloy coupled with mild steel: Effect of coatings. *J Mater Res Technol* 24:7745–7755.
69. Hegde M, Mohan J, Mushtaq Warraich MQ, Kavanagh Y, Duffy B, Tobin EF (2023) Cavitation erosion and corrosion resistance of hydrophobic sol-gel coatings on aluminium alloy. *Wear* 524–525:204766.
70. Wu M, Liu Y, Qu W, Guo W, Zhang H, Pei Y, Li S, Gong S (2023) Thickness-related failure behaviors of the thermal barrier coatings under thermal gradient cycling. *Surf Coat Technol* 468:129748.
71. Kim KH, Lee SH, Nam ND, Kim JG (2011) Effect of cobalt on the corrosion resistance of low alloy steel in sulfuric acid solution. *Corros Sci* 53:3576–3587.
72. Chen X, Zhang S, Jiang M, Chen Y, Jiang Y, Wang Z, Jiang N, Chen A, Li B, Lei Z, Chen Y (2023) Microstructure and mechanical properties of laser welded hot-press-formed steel with varying thicknesses of Al–Si coatings cleaned by nanosecond pulsed laser. *J Mater Res Technol* 22:2576–2588.
73. Nam ND, Lee SH, Kim JG, Yi JW, Lee KR (2009) Effect of stress on the passivation of Si-DLC coating as stent materials in simulated body environment. *Diamond Relat Mater* 18:1145–1151.
74. Bach LX, Thuan DV, Thu VTH, Phan TB, Vu NSH, Nam ND (2019) An investigation on titania multilayer coatings for enhanced corrosion resistance of carbon steel in simulated seawater by sol–gel dip coating. *J Mater Res Technol* 8:6400–6406.

75. Lal D, Saputo J, Gildersleeve VEJ, Sampath S (2023) Through thickness changes to stiffness and thermal conductivity in thermal barrier coatings subjected to gradient exposure. *J Eur Ceram Soc* 43:4146–4152.
76. Blanchard F, Kadi MJ, Bousser E, Baloukas B, Azzi M, Klemberg-Sapieha JE, Martinu L (2023) Effect of thermal ageing on the optical properties and pore structure of thermal barrier coatings. *Surf Coat Technol* 452:129080.
77. Zeng Y, Song N, Lim S, Keevers M, Wu Y, Yang Z, Pillai S, Jiang JY, Green M (2023) Comparative durability study of commercial inner-pore antireflection coatings and alternative dense coatings. *Sol Energy Mater Sol Cells* 251:112122.
78. Hoai NT, Hien PV, Vu NSH, Son DL, Man TV, Tri MD, Nam ND (2019) An improved corrosion resistance of steel in hydrochloric acid using Hibiscus sabdariffa leaf extract. *Chem Pap* 73:909–925.
79. Nam ND, Thang VQ, Hoai NT, Hien PV (2016) Yttrium 3-(4-nitrophenyl)-2-propenoate used as inhibitor against copper alloy corrosion in 0.1 M NaCl solution. *Corros Sci* 112:451–461.
80. Vu NSH, Hien PV, Mathesh M, Thu VTH, Nam ND (2019) An improved corrosion resistance of steel in ethanol fuel blend by titania nanoparticles and aganonerion polymorphum leaf extract. *ACS Omega* 4:146–158.
81. Bach LX, Dao TBN, Duong-Ngo KL, Tran TN, Le Minh T, Nguyen Trong H, Hoang Ngoc CT, Panaitescu C, Hoai NT, Nguyen Dang N (2023) Inhibitive behaviours of unripe banana peel extract for mitigating electrochemical corrosion of carbon steel in aggressively acidic solutions. *J Taibah Univ Sci* 17:2247633.
82. Vu NSH, Hien PV, Man TV, Thu VTH, Tri MD, Nam ND (2018) A study on corrosion inhibitor for mild steel in ethanol fuel blend. *Materials* 11:59.
83. Xu CS, Hayworth KJ, Lu Z, Grob P, Hassan AM, Garcia-Cerdan JG, Niyogi KK, Nogales E, Weinberg RJ, Hess HF (2017) Enhanced FIB-SEM systems for large-volume 3D imaging. *eLife* 6:e25916.
84. Yang J, Roa JJ, Odén M, Johansson-Jõesaar MP, Esteve J, Llanes L (2015) Substrate surface finish effects on scratch resistance and failure mechanisms of TiN-coated hardmetals. *Surf Coatings Technol* 265:174–184.
85. Bull SJ, Rickerby DS (2001) Characterization of hard coatings, In: *Handbook of Hard Coatings*, Elsevier.
86. Lešer V, Drobne D, Pipan Ž, Milani M, Tatti F (2009) Comparison of different preparation methods of biological samples for FIB milling and SEM investigation. *J Microsc* 233:309–319.
87. Knott G, Marchman H, Wall D, Lich B (2008) Serial section scanning electron microscopy of adult brain tissue using focused ion beam milling. *J Neurosci* 28:2959–2964.
88. Nam ND, Vaka M, Hung NT (2014) Corrosion behavior of TiN, TiAlN, TiAlSiN-coated 316L stainless steel in simulated proton exchange membrane fuel cell environment. *J Power Sources* 268:240–245.
89. Perrey CR, Carter CB, Michael JR, Kotula PG, Stach EA, Radmilovic VR (2004) Using the FIB to characterize nanoparticle materials. *J Microsc* 214:222–236.
90. Frank F, Tkadletz M, Saringer C, Czettl C, Pohler M, Burghmmer M, Todt J, Zalesak J, Keckes J, Schalk N (2023) Investigation of the microstructure of a graded ZrN/Ti0.33Al0.67N multilayer coating using cross-sectional characterization methods. *Surf Coatings Technol* 453:129126.
91. Aslannejada H, Hassanizadeha SM, Raoof A, de Wintera DAM, Tomozeiub N, van Genuchten MTh (2017) Characterizing the hydraulic properties of paper coating layer using FIBSEM tomography and 3D pore-scale modeling. *Chem Eng Sci* 160:275–280.
92. Jiang Z, Pan Y, Fe C, Li W, Wang Y, Long WJ (2023) Three-dimensional pore structure characterization of cement paste by X-ray computed tomography (XCT) and focused ion beam/scanning electron microscopy (FIB/SEM). *Constr Build Mater* 383:131379.
93. Osenberg M, Hilger A, Neumann M, Wagner A, Bohn N, Binder JR, Schmidt V, Banhart J, Manke I (2023) Classification of FIB/SEM-tomography images for highly porous multiphase materials using random forest classifiers. *J Power Sources* 570:233030.
94. Ottenwelter E, Josse C, Proietti A, Robbiola L (2022) Fire gilding investigation on early medieval copper-based jewellery by focused ion beam (FIB) on FEG-SEM. *J Archaeol Sci: Rep* 46:103602.
95. Cui T, Qian H, Lou Y, Chen X, Sun T, Zhang D, Li X (2022) Single-cell level investigation of microbiologically induced degradation of passive film of stainless steel via FIB-SEM/TEM and multi-mode AFM. *Corros Sci* 206:110543.

96. Sailer S, Mundszinger M, Martin J, Mancini M, Wohlfahrt-Mehrens M, Kaiser U (2023) Quantitative FIB/SEM tomogram analysis of closed and open porosity of spheroidized graphite anode materials for LiBs applications. *Micron* 166:103398.
97. Dumoux M, Smith JLR, Glen T, Grange M, Darrow MC, Naismith JH (2023) A protocol for cryogenic volumetric imaging using serial plasma FIB/SEM. *Methods Cell Biol* 177:327–358.
98. Vattikuti SVP, Hoang Ngoc CT, Nguyen H, Nguyen Thi HN, Shim J, Nguyen Dang N (2023) Carbon nitride coupled Co_3O_4: A pyrolysis-based approach for high-performance hybrid energy storage. *J Phys Chem Lett* 14:9412–9423.
99. Rajavaram R, Vattikuti SVP, Shim J, Lui X, Hoai NT, Nguyen Dnag N (2023) Enriched photocatalytic and photoelectrochemical activities of a 2D/0D g-C_3N_4/CeO_2 nanostructure. *Nanoscale Adv* 5:6489–6500.
100. Bonnamy S, Oberlin A (2016) Transmission electron microscopy. In: *Materials Science and Engineering of Carbon — Characterization*, Elsevier.
101. Chakraborty J, Daneu N, Rečnik A, Chakraborty M, Dasgupta S, Ghosh J, Sengupta S, Mazumdar S, Sinha MK, Basu D (2011) Stepwise formation of crystalline apatite in the biomimetic coating of surgical grade SS 316L substrate: A TEM analysis. *J Taiwan Inst Chem Eng* 42:682–687.
102. Kovalov D, Taylor CD, Heinrich H, Kelly RG (2022) Operando electrochemical TEM, ex-situ SEM and atomistic modeling studies of MnS dissolution and its role in triggering pitting corrosion in 304L stainless steel. *Corros Sci* 199:110184.
103. Liu Z, Chu Q, Qiang YJ, Zhang X, Ye YW (2023) Experimental and molecular simulation studies of N, S-doped Carbon dots as an eco-friendly corrosion inhibitor for protecting Cu in HCl environment. *Colloids Surf A* 669:131504.
104. Zhao Q, Yan K, Cui Z, Wen B, Xue F, Li J, Guo J, Xu A, Qiao K, Ye R, Long Y, Zhang D, Luo H, Taskaev S, Zhang H (2023) Data driven accelerated design of high-efficiency corrosion inhibitor for magnetic refrigeration materials. *Corros Sci* 216:111115.
105. Gåhlin R, Jacobson S (1998) A novel method to map and quantify wear on a micro-scale. *Wear* 222:93–102.
106. Kaneko R, Miyamoto T, Andoh Y, Hamada E (1996) Microwear. *Thin Solid Films* 273:105–111.
107. Miyahara K, Nagashima N, Ohmura T, Matsuoka S (1999) Evaluation of mechanical properties in nanometer scale using AFM-based nanoindentation tester. *Nanostruct Mater* 12:1049–1052.
108. Jalili N, Laxminarayana K (2004) A review of atomic force microscopy imaging systems: Application to molecular metrology and biological sciences. *Mechatronics* 14:907–945.
109. Sundararajan S, Bhushan B (2001) Development of a continuous microscratch technique in an atomic force microscope and its application to study scratch resistance of ultrathin hard amorphous carbon coatings. *J Mater Res* 16:437–445.
110. Yang XF, Tallman DE, Gelling VJ, Bierwagen GP, Kasten LS, Berg J (2001) Use of a sol–gel conversion coating for aluminum corrosion protection. *Surf Coat Technol* 140:44–50.
111. Chen J, Wu Y, Guo L, Li W, Tan B, Brahmia A (2023) Insight into the anti-corrosion mechanism of Pisum sativum L leaves extract as the degradable inhibitor for Q235 steel in sulfuric acid medium. *J Taiwan Inst Chem Eng* 143:104664.
112. Obot IB, Ul-Haq MI, Sorour AA, Alanazi NF, Al-Abeedi TM, Ali SA, Al-Muallem HA (2022) Modified-polyaspartic acid derivatives as effective corrosion inhibitor for C1018 steel in 3.5% NaCl saturated CO_2 brine solution. *J Taiwan Inst Chem Eng* 135:104393.
113. Xu C, Tan B, Zhang S, Li W (2022) Corrosion inhibition of copper in sulfuric acid by Leonurus japonicus Houtt. extract as a green corrosion inhibitor: Combination of experimental and theoretical research. *J Taiwan Inst Chem Eng* 139:104532.
114. Nam ND, Kim JG (2009) Corrosion properties of a RF-magnetron-sputtered TiN coating deposited on 316L stainless steel. *J Korean Phys Soc* 54:1104–1108.
115. Nam ND, Kim JG, Hwang WS (2009) Effect of bias voltage on the electrochemical properties of TiN coating for polymer electrolyte membrane fuel cell. *Thin Solid Films* 517:4772–4776.
116. Nam ND, Jo DS, Kim JG, Yoon DH (2011) Corrosion protection of CrN/TiN multi-coating for bipolar plate of polymer electrolyte membrane fuel cell. *Thin Solid Films* 519:6787–6791.
117. Nam ND, Kim MJ, Kim JG, Yoon DH (2013) Corrosion protection of Ti/TiN, Cr/TiN, Ti/CrN, and Cr/CrN multi-coatings in simulated proton exchange membrane fuel cell environment. *Thin Solid Films* 545:380–384.

118. Anadebe VC, Onukwuli OD, Abeng FE, Okafor NA, Ezeugo JO, Okoye CC (2020) Electrochemical-kinetics, MD-simulation and multi-input single-output (MISO) modeling using adaptive neuro-fuzzy inference system (ANFIS) prediction for dexamethasone drug as eco-friendly corrosion inhibitor for mild steel in 2 M HCl electrolyte. *J Taiwan Inst Chem Eng* 115:251–265.
119. Thanh LT, Vu NSH, Binh PMQ, Dao VA, Thu VTH, Hien PV, Panaitescu C, Nam ND (2020) Combined experimental and computational studies on corrosion inhibition of Houttuynia Cordata leaf extract for steel in HCl medium. *J Mol Liq* 315:113787.
120. Vahabi S, Salman BN, Javanmard A (2013) Atomic force microscopy application in biological research: A review study. *Iran J Med Sci* 38:76–83.
121. Hushegyi A, Bertok T, Damborsky P, Katrlik J, Tkac J (2015) An ultrasensitive impedimetric glycan biosensor with controlled glycan density for detection of lectins and influenza hemagglutinins. *Chem Commun* 51:7474–7477.
122. Hsieh YP, Hofmann M, Chang KW, Jhu JG, Li YY, Chen KY, Yang CC, Chang WS, Chen LC (2014) Complete corrosion inhibition through graphene defect passivation. *ACS Nano* 8:443–448.
123. Bunch JS, Verbridge SS, Alden JS, van der Zande AM, Parpia JM, Craighead HG, McEuen PL (2008) Impermeable atomic membranes from graphene sheets. *Nano Lett* 8:2458–2462.
124. Gu X, Nguyen T, Oudina M, Martin D, Kidah B, Jasmin J, Rezig A, Sung L, Byrd E, Martin JW, Ho DL, Jean YC (2005) Microstructure and morphology of amine-cured epoxy coatings before and after outdoor exposures—An AFM study. *J Coat Technol Res* 2:547–556.
125. Morsch S, Lui Y, Lyon SB, Gibbon SR (2016) Insights into epoxy network nanostructural heterogeneity using AFM-IR. *ACS Appl Mater Interfaces* 8:959–966.
126. Casanova L, Menegazzo M, Goto F, Pedeferri M, Duò L, Ormellese M, Bussetti G (2023) Investigating the activation of passive metals by a combined in-situ AFM and Raman spectroscopy system: A focus on titanium. *Sci Rep* 13:6117.
127. Jiao F, Cannon KS, Lin YC, Gladfelter AS, Scheuring S (2020) The hierarchical assembly of septins revealed by high-speed AFM. *Nat Commun* 11:5062.
128. Heath GR, Scheuring S (2018) High-speed AFM height spectroscopy reveals µs-dynamics of unlabeled biomolecules. *Nat Commun* 9:4983.
129. Bose K, Lech CJ, Heddi B, Phan AT (2018) High-resolution AFM structure of DNA G-wires in aqueous solution. *Nat Commun* 9:1959.
130. Paiva TO, Viljoen A, Dufrêne YF (2022) Seeing the unseen: High-resolution AFM imaging captures antibiotic action in bacterial membranes. *Nat Commun* 13:6196.
131. Cross SE, Jin YS, Rao J, Gimzewski JK (2009) Applicability of AFM in cancer detection. *Nat Nanotechnol* 4:72–73.
132. Shin C, Kim K, Kim J, Ko W, Yang Y, Lee S, Jun CS, Kim YS (2013) Fast, exact and non-destructive diagnoses of contact failures in nano-scale semiconductor device using conductive AFM. *Sci Rep* 3:2088.
133. Li X, Ramazani A, Prahl U, Bleck W (2018) Quantification of complex-phase steel microstructure by using combined EBSD and EPMA measurements. *Mater Charact* 142:179–186.
134. Oh SJ, Cook DC, Kwon SJ, Townsend HE (2002) Studying the atmospheric corrosion behavior of weathering steels at a mild marine environment. In *Proceedings of the LACME'98 Sixth Latin American Conference on Applications of the Mössbauer Effect*. Springer Science & Business Media.
135. McGee JJ, Keil K (2001) Application of electron probe microanalysis to the study of geological and planetary materials. *Microsc Microanal* 7:200–210.
136. Tu J, Chen T, Zhang C, Shi Z, Wu C (2002) GaAs thin films grown on conducting glass by hot wall epitaxy for solar cell. *J Korean Vac Sci Technol* 6:71–75.
137. Sababi M, Zhang F, Krivosheeva O, Forslund M, Pan J, Claesson PM, Dedinaite A (2012) Thin composite films of mussel adhesive proteins and ceria nanoparticles on carbon steel for corrosion protection. *J Electrochem Soc* 159:C364–C371.
138. Bart JJ (1969) The analysis of chemical and metallurgical changes in microcircuit metalization systems. *IEEE Trans on Electron Devices* 16:351–356.
139. Choi J, Choi S, Choi J, Park Y, Park, HM, Lee HW, Woo BC, Cho S (2004) Magnetic properties of Mn-doped Bi_2Te_3 and Sb_2Te_3. *Phys Status Solidi B* 241:1541–1544.
140. Nam ND (2016) Role of zinc in enhancing the corrosion resistance of Mg-5Ca alloys. *J Electrochem Soc* 163:C76–C84.

141. Hao S, Zhao L, He D (2013). Surface microstructure and high temperature corrosion resistance of arc-sprayed FeCrAl coating irradiated by high current pulsed electron beam. *Nucl Inst Methods Phys Res Sec B: Beam Interactions Mater Atoms* 312:97–103.
142. Schiffmann KI, Steinberg C (2019) EPMA-analyse dünner PVD- und CVD-Schichten. *Vak Forsch Prax* 31:26–36.
143. Lee K, Park SW, Ko MJ, Kim K, Park NG (2009) Selective positioning of organic dyes in a mesoporous inorganic oxide film. *Nature Mater* 8:665–671.
144. Yu C, et al (2023) Superior zero thermal expansion dual-phase alloy via boron-migration mediated solid-state reaction. *Nat Commun* 14:3135.
145. Yamashita T, Tanaka Y, Nagoshi M, Ishida K (2016) Novel technique to suppress hydrocarbon contamination for high accuracy determination of carbon content in steel by FE-EPMA. *Sci Rep* 6:29825.
146. Tang Y, Hanus R, Chen SW, Snyder GJ (2015) Solubility design leading to high figure of merit in low-cost Ce-CoSb3 skutterudites. *Nat Commun* 6:7584.
147. Kumar A, Pandey C (2023) Structural integrity assessment of Inconel 617/P92 steel dissimilar welds for different groove geometry. *Sci Rep* 13:8061.
148. Bénard A, Klimm K, et al. (2018) Oxidising agents in sub-arc mantle melts link slab devolatilisation and arc magmas. *Nat Commun* 9:3500.
149. Nakashima D, Nakamura T, Zhang M, et al. (2023) Chondrule-like objects and Ca-Al-rich inclusions in Ryugu may potentially be the oldest solar system materials. *Nat Commun* 14:532.
150. Murray SP, Pusch KM, et al. (2020) A defect-resistant Co-Ni superalloy for 3D printing. *Nat Commun* 11:4975.
151. Titus D, Samuel EJJ, Roopan SM (2019) Nanoparticle characterization techniques. In: *Green Synthesis, Characterization and Applications of Nanoparticles, Micro and Nano Technologies*, Elsevier.
152. Goldstein JI, Newbury DE, Michael JR, Ritchie NWM, Scott JHJ, Joy DC (2018) *Scanning Electron Microscopy and X-Ray Microanalysis*. Springer Science + Business Media LLC.
153. Imashuku S, Kawakami T, Ze L, Kawai J (2012) Possibility of scanning electron microscope observation and energy dispersive X-ray analysis in microscale region of insulating samples using diluted ionic liquid. *Microsc Microanal* 18:365–370.
154. Houk CS, Burgoine GA, Page CJ (1995) Application of energy-dispersive X-ray elemental mapping to probe the homogeneity of sol-gel derived $YBa_2Cu_3O_7$-delta. and related phases. *Chem Mater* 7:649–656.
155. Appapillai AT, Mansour AN, Cho J, Shao-Horn Y (2007) Microstructure of $LiCoO_2$ with and without "$AlPO_4$" nanoparticle coating: Combined STEM and XPS studies. *Chem Mater* 19:5748–5757.
156. John JVS, Coffer JL, Rho YG, Diehl P, Pinizzotto RF, Culp TD, Bray KL (1997) Erbium doped SiO_2 layers formed on the surface of silicon by spark processing. *Chem Mater* 9:3176–3180.
157. Lu P, Zhou L, Kramer MJ, Smith DJ (2014) Atomic-scale chemical imaging and quantification of metallic alloy structures by energy-dispersive x-ray spectroscopy. *Sci Rep* 4:3945.
158. Feng Y, Li Y, Lin J, Wu H, Zhu L, Zhang X, Zhang L, Sun CF, Wu M, Wang Y (2023) Production of high-energy 6-Ah-level Li | |LiNi0.83$Co_{0.11}Mn_{0.06}O_2$ multi-layer pouch cells via negative electrode protective layer coating strategy. *Nat Commun* 14:3639.
159. Hagemann N, Joseph S, Schmidt HP, et al. (2017) Organic coating on biochar explains its nutrient retention and stimulation of soil fertility. *Nat Commun* 8:1089.
160. Suguro T, Kishimoto F, Kariya N, Fukui T, Nakabayashi M, Shibata N, Takata T, Domen K, Takanabe K (2022) A hygroscopic nano-membrane coating achieves efficient vapor-fed photocatalytic water splitting. *Nat Commun* 13:5698.
161. Liang J, Zhu Y, Li X, Luo J, Deng S, Zhao Y, Sun Y, Wu D, Hu Y, Shan TK, Gu M, Sun X (2023) A gradient oxy-thiophosphate-coated Ni-rich layered oxide cathode for stable all-solid-state Li-ion batteries. *Nat Commun* 14:146.
162. Yu H, Cao Y, Chen L, Hu Y, Duan X, Dai S, Li C, Jiang H (2021) Surface enrichment and diffusion enabling gradient-doping and coating of Ni-rich cathode toward Li-ion batteries. *Nat Commun* 12:4564.
163. Amini M, Nederi R, Mahdavian M, Badiei A (2020) Effect of piperazine functionalization of mesoporous silica type sba-15 on the loading efficiency of 2-mercaptobenzothiazole corrosion inhibitor. *Ind Eng Chem Res* 59:3394–3404.

164. Ress J, Martin U, Bosch J, Bastidas DM (2020) pH-triggered release of $NaNO_2$ corrosion inhibitors from novel colophony microcapsules in simulated concrete pore solution. *ACS Appl Mater Interfaces* 12:46686–46700.
165. Tang G, Chen C, Wu D, Yan Z, Ren T, Hou X, Zhao M, Fu S, Fan L (2023) Halloysite@polyaniline nanoparticles loaded with the praseodymium (III) cation for improving active corrosion protection of waterborne epoxy coating. *ACS Appl Mater Interfaces*. https://doi.org/10.1021/acsami.3c03461.
166. Farhadian A, Rahimi A, Safaei N, Shaabani A, Sadeh E, Abdouss M, Alavi A (2021) Exploration of sunflower oil as a renewable biomass source to develop scalable and highly effective corrosion inhibitors in a 15% HCl medium at high temperatures. *ACS Appl Mater Interfaces* 13:3119–3138.
167. Nam ND, Hien PV, Hoai NT, Thu VTH (2018) A study on the mixed corrosion inhibitor with a dominant cathodic inhibitor for mild steel in aqueous chloride solution. *J Taiwan Inst Chem Eng* 91:556–569.
168. Koningstein JA (1972) *Introduction to the Theory of the Raman Effect*. D. Reidel Publishing Company, Dordrecht.
169. Nie S, Emory SR (1997) Probing single molecules and single nanoparticles by surface-enhanced Raman scattering. *Science* 275:1102–1106.
170. Kneipp J, Kneippa H, Kneippac K (2008) SERS—A single-molecule and nanoscale tool for bioanalytics. *Chem Soc Rev* 37:1052–1060.
171. Ledinský M, Paviet-Salomon B, Vetushka A, Geissbuhler J, Tomasi A, Despeisse M, Wolf SD, Ballif C, Fejfar A (2016) Profilometry of thin films on rough substrates by Raman spectroscopy. *Sci Rep* 6:37859.
172. Cong C, Li K, Zhang X, Yu T (2013) Visualization of arrangements of carbon atoms in graphene layers by Raman mapping and atomic-resolution TEM. *Sci Rep* 3:1195.
173. Zhang Q, Li X, Ma Q, Zhang Q, Bai H, Yi W, Liu J, Han J, Xi G (2017) A metallic molybdenum dioxide with high stability for surface enhanced Raman spectroscopy. *Nat Commun* 8:14903.
174. Chen N, Xiao TH, Luo Z, Kitahama Y, Hiramatsu K, Kishimoto N, Itoh T, Cheng Z, Goda K (2020) Porous carbon nanowire array for surface-enhanced Raman spectroscopy. *Nat Commun* 11:4772.
175. Finnie P, Quyang J, Lefebvre J (2020) Full spectrum Raman excitation mapping spectroscopy. *Sci Rep* 10:9172.
176. Tsukada S, Fujii Y, Kanagawa A, Akishige Y, Ohwada K (2023) Polarization behavior in a compositionally graded relaxor–ferroelectric crystal visualized by angle-resolved polarized Raman mapping. *Commun Phys* 6:107.
177. Yang G, Li X, Cheng Y, Wang M, Ma D, Sokolov AP, Kalinin SV, Veith GM, Nanda J (2021) Distilling nanoscale heterogeneity of amorphous silicon using tip-enhanced Raman spectroscopy (TERS) via multiresolution manifold learning. *Nat Commun* 12:578.
178. Choobbari ML, Ferguson J, den Brande NV, Smith T, Chalyan T, Meulebroeck W, Ottevaere H (2023) Studying the concentration of polymers in blended microplastics using 2D and 3D Raman mapping. *Sci Rep* 13:7771.
179. Ma L, Wang J, Ren C, Ju P, Huang Y, Zhang F, Zhao F, Zhang Z, Zhang D (2020) Detection of corrosion inhibitor adsorption via a surface-enhanced Raman spectroscopy (SERS) silver nanorods tape sensor. *Sens Actuators B* 321:128617.
180. Kharitonov DS, Sommertune J, Örnek C, Ryl J, Kurilo II, Claesson PM, Pan J (2019) Corrosion inhibition of aluminium alloy AA6063-T5 by vanadates: Local surface chemical events elucidated by confocal Raman micro-spectroscopy. *Corros Sci* 148:237–250.
181. Shipp DW, Sinjab F, Notingher I (2017) Raman spectroscopy: Techniques and applications in the life sciences. *Adv Opt Photonics* 9:315–428.
182. Koch S, Ziegler G, Hutter H (2013) ToF-SIMS measurements with topographic information in combined images. *Anal Bioanal Chem* 405:7161–7167.
183. Singh AV (2020) ToF-SIMS 3D imaging unveils important insights on the cellular microenvironment during biomineralization of gold nanostructures. *Sci Rep* 10:261.
184. Vickerman JC (2001) ToF-SIMS—An overview. In: *ToF-SIMS: Surface Analysis by Mass Spectrometry*. UMIST, Manchester, UK.
185. Visser P, Marcoen K, Trindade GF, Abel ML, Watts JF, Hauffman T, Mol JMC, Terryn H (2019) The chemical throwing power of lithium-based inhibitors from organic coatings on AA2024-T3. *Corros Sci* 150:194–206.

186. Marcoen K, Visser P, Trindade GF, Abel ML, Watts JF, Mol JMC, Terryn H, Hauffman T (2018) Compositional study of a corrosion protective layer formed by leachable lithium salts in a coating defect on AA2024-T3 aluminium alloys. *Prog Org Coat* 119:65–75.
187. Trindade GF, Abel ML, Lowe C, Tshulu R, Watts JF (2018) A time-of-flight secondary ion mass spectrometry/multivariate analysis (ToF-SIMS/MVA) approach to identify phase segregation in blends of incompatible but extremely similar resins. *Anal Chem* 90:3936–3941.
188. Bamford SE, Yalcin D, Gardner W, Winkler DA, Kohl TM, Muir BW, Howard S, Bruton EA, Pigram PJ (2023) Multi-dimensional machine learning analysis of polyaniline films using stitched hyperspectral ToF-SIMS data. *Anal Chem* 95:7968–7976.
189. Gardner W, Winkler DA, Muir BW, Pigram PJ (2022) Applications of multivariate analysis and unsupervised machine learning to ToF-SIMS images of organic, bioorganic, and biological systems. *Biointerphases* 17:020802.
190. Minhas B, Dino S, Huang L and Wu D (2022) Active corrosion protection by epoxy coating on Li_2CO_3-pretreated anodized aluminum alloy 2024-T3. *Front Mater* 8:804328.
191. Mei H, Laws TS, Terlier T, Verduzco R, Stein GE (2021) Characterization of polymeric surfaces and interfaces using time-of-flight secondary ion mass spectrometry. *J Polym Sci* 60:1174–1198.
192. He Y, Abdi M, Trindade GF, Begines B, et al. (2021) Exploiting generative design for 3D printing of bacterial biofilm resistant composite devices. *Adv Sci* 8:2100249.
193. Bañuls-Ciscar J, Trindade GF, Abel ML, Phanopoulos C, Pans G, Pratelli D, Marcoen K, Hauffman T, Watts JF (2020) A study of the interfacial chemistry between polymeric methylene diphenyl diisocyanate and a Fe-Cr alloy. *Surf Interface Anal* 53:340–349.
194. Abbasi K, Smith H, Hoffman M, Farghadany E, Bruckman LS, Sehirlioglu A (2020) Dimensional stacking for machine learning in ToF-SIMS analysis of heterostructures. *Adv Mater Interfaces* 8:2001648.
195. Bañuls-Ciscar J, Fumagalli F, Ruiz-Moreno A, Rossi F, Suraci SV, Fabiani D, Ceccone G (2020) A methodology to investigate heterogeneous oxidation of thermally aged cross-linked polyethylene by ToF-SIMS. *Surf Interface Anal* 52:1178–1184.
196. Shimizu R (1983). Quantitative analysis by Auger electron spectroscopy. *Jpn J Appl Phys* 22:1631–1642.
197. Prasittichai C, Zhou H, Bent SF (2013) area selective molecular layer deposition of polyurea films. *ACS Appl Mater Interfaces* 5:13391–13396.
198. Dong WJ, Lim JW, Hong DM, Park JY, Cho WS, Baek S, Yoo CJ, Kim W, Lee JL (2020) Evidence of local corrosion of bimetallic Cu–Sn catalysts and its effects on the selectivity of electrochemical CO_2 reduction. *ACS Appl Energy Mater* 3:10568–10577.
199. Gao J, Li B, Tan J, Chow P, Lu TM, Koratkar N (2016) Aging of transition metal dichalcogenide monolayers. *ACS Nano* 10:2628–2635.
200. Lee S, Kim HM, Baek GH, Park JS (2021) Dry-etchable molecular layer-deposited inhibitor using annealed indicone film for nanoscale area-selective deposition. *ACS Appl Mater Interfaces* 13:60144–60153.
201. Quesne-Turin A, Flahaut D, Croguennec L, Vallverdu G, Allouche J, Charles-Blin Y, Chotard JN, Ménétrier M, Baraille I (2017) Surface reactivity of Li_2MnO_3: First-principles and experimental study. *ACS Appl Mater Interfaces* 9:44222–44230.
202. Lin WC, Lo WC, Li JX, Huang PC, Wang MY (2021) Auger electron spectroscopy analysis of the thermally induced degradation of $MAPbI_3$ perovskite films. *ACS Omega* 6:34606–34614.
203. Danty PMP, Mazel A, Cormary B, De Marco ML, Allouche J, Flahaut D, Jimenez-Lamana J, Lacomme S, Delville MH, Drisko GL (2020) Microwave-assisted and metal-induced crystallization: A rapid and low temperature combination. *Inorg Chem* 59:6232–6241.
204. Wood KN, Xerxes Steirer K, Hafner SE, Ban C, Santhanagopalan S, Lee SH, Teeter G (2018) Operando X-ray photoelectron spectroscopy of solid electrolyte interphase formation and evolution in Li2S-P2S5 solid-state electrolytes. *Nate Commun* 9:2490.
205. Arioka K, Staehle RW, Yamada T, Miyamoto T, Terachi T (2016) Degradation of alloy 690 after relatively short times. *Corrosion* 72:1252–1268.
206. Zhao H, Chakraborty P, Ponge D, Hickel T, Sun B, Wu CH, Gault B, Raabe D (2022) Hydrogen trapping and embrittlement in high-strength Al alloys. *Nature* 602:437–441.

207. Seuss F, Gaag N, Virtanen S (2017) Corrosion mechanism of $CuZn_{21}Si_3P$ in aggressive tap water. *Mater Corros* 68:42–49.
208. Uhart A, Ledeuil JB, Gonbeau D, Dupin JC, Bonino JP, Ansart F, Esteban J (2016) An Auger and XPS survey of cerium active corrosion protection for AA2024-T3 aluminum alloy. *Appl Surf Sci* 390:751–759.
209. Montemor MF, Simões AM, Ferreira MGS, Williams B, Edwards H (2000) The corrosion performance of organosilane based pre-treatments for coatings on galvanised steel. *Prog Org Coat* 38:17–26.
210. Duchoslav J, Arndt M, Steinberger R, Keppert T, Luckeneder G, Stellnberger KH, Hagler J, Riener CK, Angeli G, Stifter D (2014) Nanoscopic view on the initial stages of corrosion of hot dip galvanized Zn-Mg-Al coatings. *Corros Sci* 83:327–334.
211. Powell CJ, Seah MP (1990) Precision, accuracy, and uncertainty in quantitative surface analyses by Auger-electron spectroscopy and x-ray photoelectron spectroscopy. *J Vac Sci Technol A* 8:735–763.

14 Coating Characterization with Computational Techniques

Nkechinyere Amaka Chikaodili, Patrick Chukwudi Nnaji, Joseph Okechukwu Ezeugo, Abhinay Thakur, Valentine Chikaodili Anadebe, and Okechukwu Dominic Onukwuli

14.1 INTRODUCTION

Every metallic piece of process equipment including those used in the automotive, pipeline, and construction industries is susceptible to corrosion [1–2]. Corrosion has been regarded as the major nightmare facing the chemical industries at large; this is due to severe acidic conditions constantly experienced in the process [3]. The viewpoint of every industry is to reduce the cost of production and repair, and at the same time meet the production output without disrupting industrial activities. Therefore, smart healable materials with sensing and self-healing capabilities are essential for extending the lifespan of metals to avoid future catastrophic failures [4–5]. Numerous early damage detection methods are based on specific indicators that can noticeably change color in response to corrosion [6]. The tier and wear of metals are mainly attributed to localized physical and chemical changes within a specific system. Thus, the application of microcapsules or encapsulation of inhibitors via coating is a flexible and straightforward method to ameliorate mechanical damage or corrosion [7]. In view of the above, efforts have been geared toward achieving durable and sustainable coatings. According to Çömlekçi et al. [8], the self-healing ability of an epoxy coating via microcapsules having linseed oil was studied. The microcapsules embedded with linseed oil were incorporated into an epoxy framework to formulate the self-healing coating. The authors further observed that the in-built capsule does not alter the film layer growth and the water vapor permeability of the epoxy film was enhanced up to 80% and later declined by 43% by incorporation of line seed oil and its composite form LO-Alkyd MCs. Their electrochemical findings prove the systematic increase in the film with the addition of the capsules into the epoxy. Similarly, a steel plate coated with epoxy-based coatings modified with microcapsules loaded with isophorone diisocyanate was investigated as an inhibitor [9]. The healing effect was scrutinized via different localized impedance studies. The impedance techniques evidenced multilevel protection by combining both additives, which shows a more stable coating layer with the absence of interfacial corrosion mechanism. Also, a novel cerium organic network modified graphene oxide (GO)-prepared waterborne epoxy coating was studied [10]. A one-step procedure was used to in situ design the cerium organic network (GO-CN/CONU) coated with ceria nanoparticles (NPs) on GO. The authors concluded that due to the interaction between composite materials and the epoxy resin, the coating surprisingly demonstrates remarkable mechanical qualities and hydrophobicity, which was attributed to the optimized crosslinking degree, compactness, and toughness. Further test reveals that even after 60 days of immersion in the corrosive medium, its anticorrosion performance (about $10^9\ \Omega.\ cm^2$) is three orders of magnitude greater than that of pure epoxy coating

DOI: 10.1201/9781032677255-14

(less than 10^6 Ω. cm^2). In another study, a durable smart coating based on a pH-controlled release polyaniline encapsulating inhibitor was investigated [11]. The pH-controlled release polyaniline encapsulating inhibitor provided an adequate release of 2-mercapto-benzothiazole (MBT) for a long period when incorporated into an epoxy matrix. The surface analysis evidenced that the self-healing effect was due to the release of MBT. Therefore, the current chapter is focused on the development of sustainable self-healing coatings, advanced characterizations, and computational techniques for metal protection.

14.2 COATINGS

Coatings or self-healing technology have been used for surface treatment to avert corrosion of metals in varying corrosive environments. One of the major features of these coatings is their tendency to self-heal. When a self-healing protective coating experiences damage and the corrosive moieties permeate the metal surface, a gradual dissolution process will take place; then, the damaged metal surface is automatically self-healed by a chemical composition of the coating which has been introduced into the system. Prior to coating, there are many operating conditions to be considered, such as the material of construction involved, the nature of the corroding moieties or environment, coating compatibility with the system, etc. Currently, protective coatings have advanced toward corrosion sensing with respect to the above-mentioned operating conditions. When these conditions are known, achieving durable and efficient barrier coatings is possible. The main aim of barrier coatings is primarily to offer physical barriers to obstruct corrosion [12].

14.3 ENCAPSULATED CORROSION INHIBITORS FOR SMART COATINGS

A feasible option to enhance the coating's anticorrosive effect is to incorporate inhibitive pigments into a coating matrix or formulation [13]. The inhibitive pigments are partially dissolved when water permeates the coating film, and the inhibitive ions can then move to the coating/metal interface and prevent corrosion by forming a barrier film on the Fe substrate [14]. Also, a quick and sufficient supply of inhibitive ions is required during the early exposure stages to accelerate the inhibitive film in order to form a sufficient inhibitory activity. On the contrary, in conventional coating incorporated with inhibitive ions, prolonged exposure to moist conditions can affect the release and movement of inhibitive species. This could quickly deplete the coating of the inhibitive pigment and compromise the coating film's barrier properties [15–16]. Hence, a higher pigment concentration is required in the coating formulation to guarantee the durability and sustainability of the coating [17]. The regulated release of the inhibitive agents from a smart carrier is another desirable development in smart coatings technology to guarantee that there are enough inhibitive agents and to avoid unwanted loss [18]. Smart carriers such as capsules [19] and mesoporous NPs [20] are nanocontainers for inhibitive agents [21]. The regulated pigment release can increase the coating's longevity by preventing spontaneous leaching of the inhibitive agents from the coating layer, thus improving their barrier film properties [22]. Also, cerium containing pH-responsive microcapsule for smart coating was synthesized as shown in Figure 14.1 [23]. The experiments showed that adding some portion of the microcapsules into the epoxy coating significantly improved the aluminum substrate's anticorrosion strength compared to the neat epoxy coating and provided active corrosion protection of the substrate through the self-healing action of cerium ions.

14.4 ELECTROCHEMICAL ANALYSIS OF SMART COATINGS

Electrochemical studies are regarded as an advanced technique to measure and observe the reactions occurring at the metal solution interface. This study comprises two sections: (1) polarization measurement which is destructive in nature and (2) electrochemical impedance spectroscopy

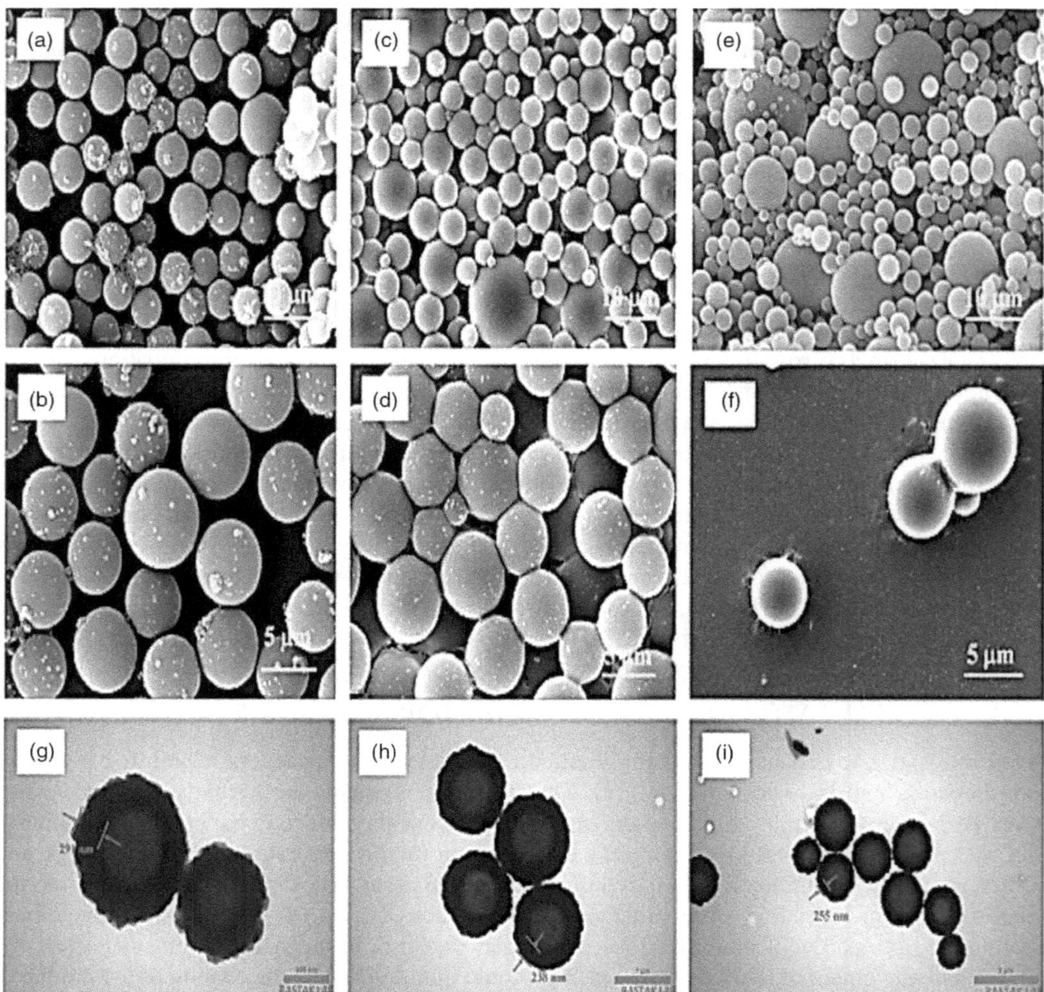

FIGURE 14.1 SEM images of the melamine-formaldehyde microcapsule containing different concentrations of cerium nitrate: (a and b) 500 ppm, (c and d) 1000 ppm, (e and f) 5000 ppm, and (g, h, and i) TEM images of the microcapsules containing ethylene glycol. (Reproduced with permission from Ghahremani et al. [23]. © 2020 Elsevier B.V.)

(EIS) (nondestructive). However, in order to identify the type of coating existing on a metal with respect to anodic, cathodic, and mixed nature, polarization measurement will be useful [24–25]. On the contrary, for proper evaluation of coating film adherence on the Fe surface, EIS is always recommended due to its accuracy with respect to charge transfer resistance (Rct) [26]. Many studies on coatings have been successfully evaluated for long-term stability using EIS analysis. A novel hybrid plasma electrolytic oxidation (PEO)/sol-gel coating incorporated with Ce was tested for AA2024T-3 [27]. The long-term stability of the coating over 21 days in the chloride system was evaluated via the electrochemical impedance method. On the Flash PEO coating, hybrid sol-gel (HSG) and HSG-Ce coatings were uniformly distributed. Also, within 21 days of immersion in 3.5 wt.% NaCl, HSG-Ce demonstrated significant scratch protection with regard to corrosion resistance. Also, linseed oil was encapsulated in a GO shell and further used as a self-healing composite inhibitor [28]. The therapeutic component, linseed oil, was encapsulated in the microcapsules which serve as the healing agent, while the GO acts as a stabilizer. From the impedance studies, the unmodified (pure) PU coatings were in the order of 10^4 Ω, whereas that of the composite coatings containing microcapsules was in the scale of 10^8 Ω. The observed coating trend denotes the capacity of the coating to prevent current from flowing between anodic and cathodic regions. The impedance modulus of the composite coatings increased by four orders of magnitude, demonstrating the healing impact of the microcapsules. In the same context, the EIS and electrochemical noise (EN) methods were adopted to evaluate the self-healing capability of coatings based on microcapsules loaded with linseed oil and CeO_2 NPs [29]. The scratch sealing efficiency (SSE) was measured via EIS. It was observed that due to the small amount of released linseed oil in the coated samples with 5% microcapsule concentration, the artificial scratch could not be adequately repaired. Contrarily, the amount of released linseed oil was sufficient to seal the scratch when the microcapsule concentration was increased above 10%. However, although having a lower microcapsule concentration, the coating sample with 15% NP-loaded microcapsules was the best self-healing coating because it exhibited SSE values that were comparable to those of samples with 20% microcapsule dose. Additionally, an epoxy-based polyphenalkaline (Poly-PhAA) coating derived from biomass was fabricated and tested electrochemically [30]. The authors reported the polymerization and cross-linking of a virgin phenalkaline (PhAA) resin and its modification as a corrosion inhibitor for mild steel. It was observed that within the 40 days of immersion time, the electrochemical impedance test and open-circuit potential measurements showed the formation of a stable, well-adhered coating with a uniform impedance value in the 10^6 Hz range. The Poly-PhAA coatings exhibited very high hydrophobic behavior, mechanical toughness, and thermal stability, which also conform to the electrochemical findings. Furthermore, a smart coating with a dual purpose of self-reporting and self-healing was investigated [31].

The smart composite coating was fabricated by simply incorporating an ionic liquid grafted carboxylated carbon nanotube (CNT) and phenolphthalein into a micro-arc oxidation coating. The electrochemical approaches evidenced that the smart composite coating has good corrosion protection for AZ31B Mg alloy in 5 wt.% NaCl. In addition, the smart composite coating presents a unique self-healing mechanism by coordination of CCNT-IL and Mg ions from the based metal to form a complex to heal the damaged area. The observed coating offers a promising method for safeguarding Mg alloys in an effective and economical way. Hence, the electrochemical study on protective coatings provides corrosion engineers with important information regarding coating formulations and reactions occurring at the metal solution interface.

14.5 CHARACTERIZATION OF COATINGS VIA SEM, OPTICAL, AND SCRATCH TEST

Surface characterizations like scanning electron microscope (SEM), scratch test, atomic force microscope, adhesion test, optical test, and color mapping are important post-characterization techniques usually adopted to evaluate a protective coating on a metal surface. Each of the above methods has its unique contribution to the field of corrosion. Many corrosion engineers have consistently worked with each of these methods to justify the presence and long-term stability of the protective coatings. Detailed fracture conditions and material response of TiN film/stainless steel substrate were critically analyzed using a cross-sectional scratch test approach. The scratch tracks were examined via cross-sectional imaging. Some parameters like pile-up in front of the stylus, substrate plastic deformation, fragmentation of TiN film, and synergistic bulge deformation at the scratch track edge were the material response. The first spallation at the edge of the track, which is typically classified as critical load (Lc) and has an adhesive failure was discovered to be a cohesive failure [32]. The findings suggest that greater caution should be exercised when interpreting other film/substrate systems' adhesion qualities using Lc. Also, an epoxy coating with dual-function-based 2-diethyl amino ethyl methacrylate (DEAEMA) modified on mesoporous silica NPs was studied [33]. The study included detailed electrochemical, surface analysis, and color mapping to prove the coating adherence to the metal surface. The authors observed that the thickness of all coatings is about 310 ± 10 μm. The observed phenomenon indicated that excessive dosage of MSN-BTA@PDEAEMA affects the epoxy coating matrix negatively. Based on this, using 7% MSN-BTA@PDEAEMA in a coating formulation caused the coating's corrosion resistance to decline. The major constituents of the epoxy coating devoid of MSN-BTA@PDEAEMA include nitrogen, carbon, oxygen, and titanium, while silicon was identified in the coatings with MSN-BTA@PDEAEMA. Similarly, the role of cobalt additive in coating formulation for AA2024 alloy in alkali-silicate electrolyte was studied [34]. For a short-term study (24 h), the thin coating (~17 μm) exhibits minimum current densities and the highest value of impedance modulus. Also, with an increase in time (168 h) the coating thickness expedites (~82 μm) with the best long-term study. Also, the SEM analysis further reveals that the systematic enhancement in the micro discharge power rate and the drop in their quantity can only result in the presence of crater-shaped defects on the coatings. From the color mapping, Co is dispersed across the entire coating surface, with average concentration with the rise in coating thickness. In another study [35], linseed oil-urea formaldehyde micro/nanocapsules and their self-healing property in an epoxy coating were evaluated. The study was based on SEM imaging and optical tests on the substrate. As shown in Figure 14.2, with increasing epoxy capsule content in all the samples, the capsules tend to agglomerate. This problem is most noticeable in capsules with a greater content (9 wt.%). Since there are no suitable contacts between epoxy and urea-formaldehyde capsules, the agglomeration potential rises as the interparticle distance declines gradually.

In the same vein, a hybrid self-healing coating containing linseed oil loaded with nanocapsules on the Cu surface was examined [36]. The crack healing time was studied using the optical test method. As presented in Figure 14.3, almost all the coatings that contain nanocapsules underwent a self-healing process. Additionally, a comparison of images (c) and (d) (from the 1st and 7th days) revealed that the scratch had filled in more than it had on the first day, indicating that the self-healing process continued even after being submerged in the corrosive solution. Also, the effect of temperature on water desorption and oxide formation of MoS_2 and its tribological performance was studied [37].

In order to determine how temperature affects wear mechanisms in air, MoS_2 coatings were applied on AISI 440C steel. At temperatures of about 75 °C, the surface oxides had no effect on the tribological characteristics of MoS_2. Figure 14.4(a) shows large debris particles in a wear scar caused by room temperature (RT), while Figure 14.4(b) shows very minute particles at 75 °C. At higher temperatures, the re-distributed debris forms a third body film, as evidenced by the difference between worn-out MoS_2 at RT (c) and 75 °C (d).

14.6 COMPUTATIONAL TECHNIQUES FOR SMART COATINGS

The advent of the computer is a desirable development with respect to artificial intelligence models. Computational studies via density functional theory (DFT) and molecular dynamics (MD)/monte carlo (MC) simulations have been useful in the area of corrosion inhibitor formulations. Considering the complex nature of most corrosion inhibitors, computational techniques can be used to optimize the structure and locate the active sites of the compound (highest occupied molecular orbital (HOMO) and lowest unoccupied molecular orbital (LUMO) regions). The advantage of computational studies has necessitated very high theoretical research not only in corrosion studies but also in every aspect of electrochemistry and engineering [38–39]. Recently, computational techniques have been incorporated into smart coating technology to probe the adherence of the barrier film onto the substrate. A GO-based 2D modified via phytic acid decorated by imidazole ZIF-9 MOFs for multifunctional smart coating was investigated

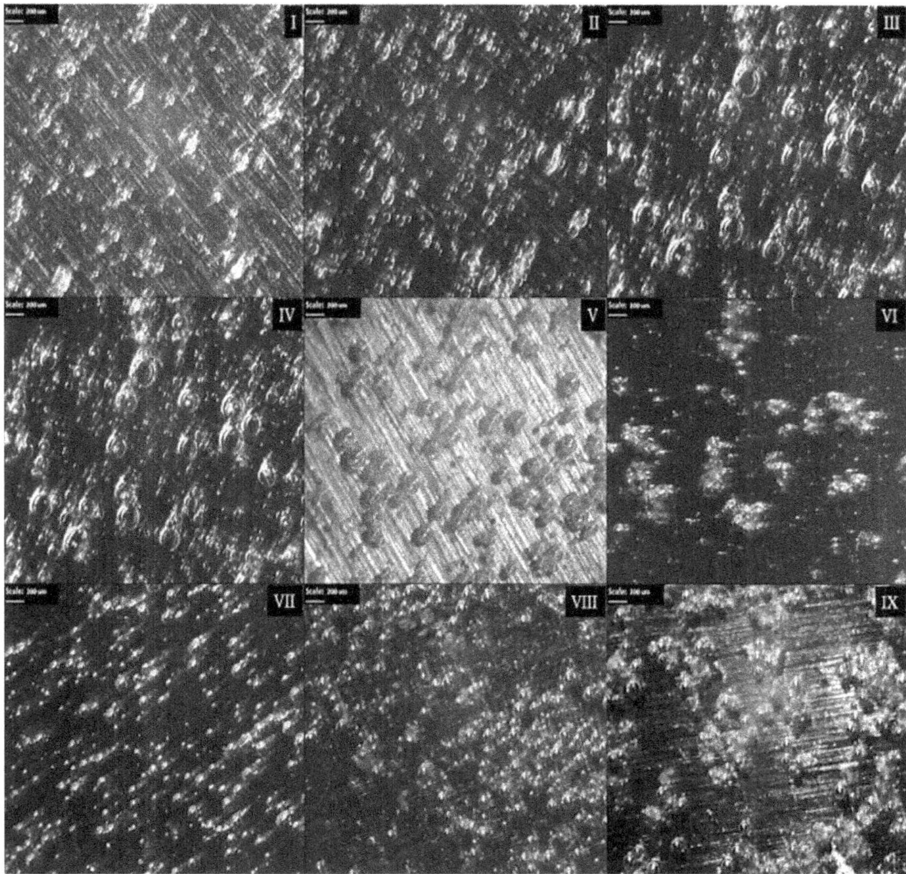

FIGURE 14.2 Optical microscopy images from embedded capsules in an epoxy coating containing (a) 3 wt. % of type A, (b) 6 wt. % of type A, (c) 9 wt. % of type A, (d) 3 wt. % of type B, (e) 6 wt. % of type B, (f) 9 wt. % of type B, (g) 3 wt. % of type C, (h) 6 wt. % of type C, and (i) 9 wt. % of type C capsules. (Reproduced with permission from Abdipour et al. [35]. © 2018 Elsevier B.V.)

using DFT-D computation and empirical approach [40]. According to computer modeling studies, binding energies of 42.88, 51.15, and 48.08 kcal/mol were associated with the stabilization of phytic acid over the GO(O), GO(COOH), and GO(OH) adsorbent models, respectively. Similarly, reduced GO nanosheets decorated with L-histidine loaded-β-cyclodextrin for smart epoxy coating were tested using detailed DFT-D modeling and experimental route [41]. The geometry of optimized hybrid histidine/β-CD clusters was done in gas and aqueous phases, respectively. From the results, the side/top views of the clusters show that both in the absence and presence of solvent molecules, the histidine substance may be engulfed by β-CD, facilitating the formation of the combination of histidine and β-CD complex. Additionally, the development of the aqueous-phase inclusion complex occurred at energies of 24.36. It was observed that the weakened inclusion complex formation in the presence of solvent molecules is shown by the much lower binding energies in aqueous-phase conditions as compared to gas-phase environments. Also, the smart coating effect of benzimidazole-loaded b-cyclodextrin on mild steel in marine environments was evaluated using combined experimental and computational assessment [42]. The interactive effects of bendamustine (BM) and its two β-CD-BM complexes on the Fe

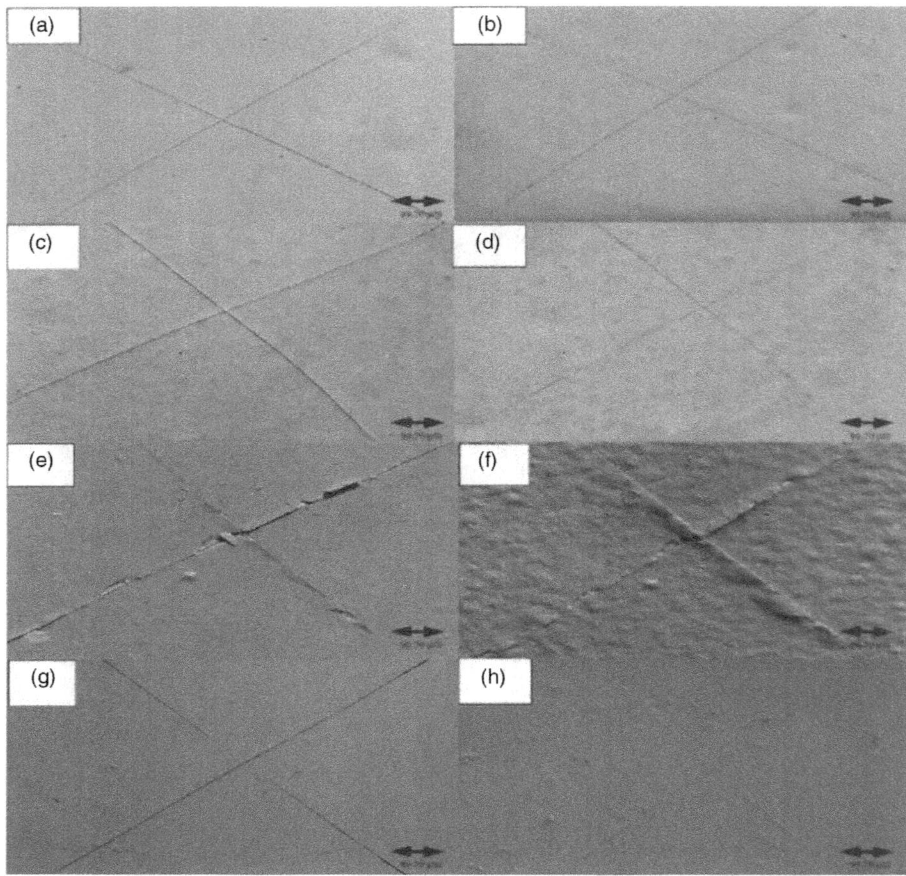

FIGURE 14.3 Optical microscopy images. (a and b) 1st and 7th days of Cu1, respectively; (c and d) 1st and 7th days of Cu5, respectively; (e and f) 1st and 7th days of Cu7, respectively; and (g and h) 1st and 7th days of Cu13, respectively. (Reproduced with permission from Mahmoudian et al. [36]. © 2018 Elsevier B.V.)

surface were simulated. Analysis suggests that the ideal mode of BM molecule adsorption in the equilibrated cells is planar, which gave rise to total coverage of the surface by BM molecules. In summary of their findings, the BM and β-CD-BM complexes are capable of coating onto the Fe (110) surface. Also, the anticorrosion capacity of a silane film with β-cyclodextrin-based nanocontainer synergized with L-histidine was theoretically analyzed [43]. The computational study was done via MC/MD simulations in aqueous and gas environments. As shown in Figure 14.5, both perspectives (side and top) evidenced that the β-CD/histidine inclusion complexes as well as histidine were able to localize over the topmost layer of iron (110). From the theoretical point of view, the energies were discovered to be negative, further describing the interfacial adsorption of histidine molecules in both pure and inclusion complex forms.

The strong adsorption of β-CD/histidine complexes was further supported by the inclusion of complexes compared to the pure histidine. Also, the electronic DFT-D modeling of P, Zn-codoped reduced-graphene oxide (rGO) embedded in an epoxy coating was investigated [44]. As presented in Figure 14.6, the pure AMP and $[Zinc(AMP)n]^{+2}$ complexes were arranged quite planarly over the surface of GO. Then, DFT-D equilibration was performed on the prepared GO/AMP and GO/

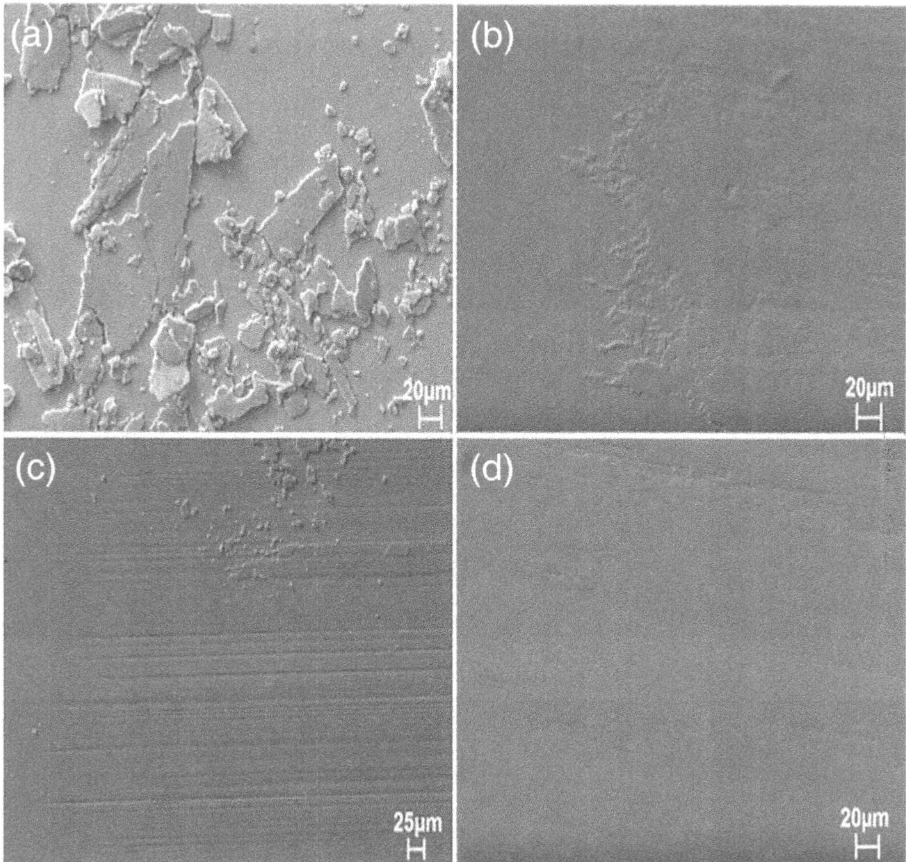

FIGURE 14.4 SEM image of (a) debris at RT, (b) wear scar tip at 75 °C, (c) wear scar tip at RT, and (d) wear scar at 75 °C. (Reproduced with permission from Dreva et al. [37]. © 2022 Elsevier B.V.)

$[Zinc(AMP)n]^{+2}$ complex geometries. Furthermore, an ultrastable porous covalent organic framework for smart coatings was studied theoretically [45].

The interaction between covalent organic frameworks (COF) and the CNT NPs was assessed via the DFT-D model. It is clear that the SNW-1 model B can aid intermolecular hydrogen bonding interactions with COOH group-containing nanotube surfaces. From the computational data obtained, both SNW-1 models have negative binding energies for all the oxidized CNT structures. The obtained energy values quantitatively demonstrate the COF structure propensity to adsorb onto the oxidized CNT-based nanostructures.

14.7 CONCLUSION AND FUTURE PERSPECTIVES

The emerging trend of low-cost, renewable, and eco-friendly self-healing materials is a desirable development in the area of corrosion and materials protection. The design of multifunctional coatings with corrosion-sensing capability will play an important role in corrosion mitigation. Hence, surface barrier coatings are highly recommended for process equipment that experiences severe acidic conditions. Nevertheless, there are still challenges faced by the smart coating

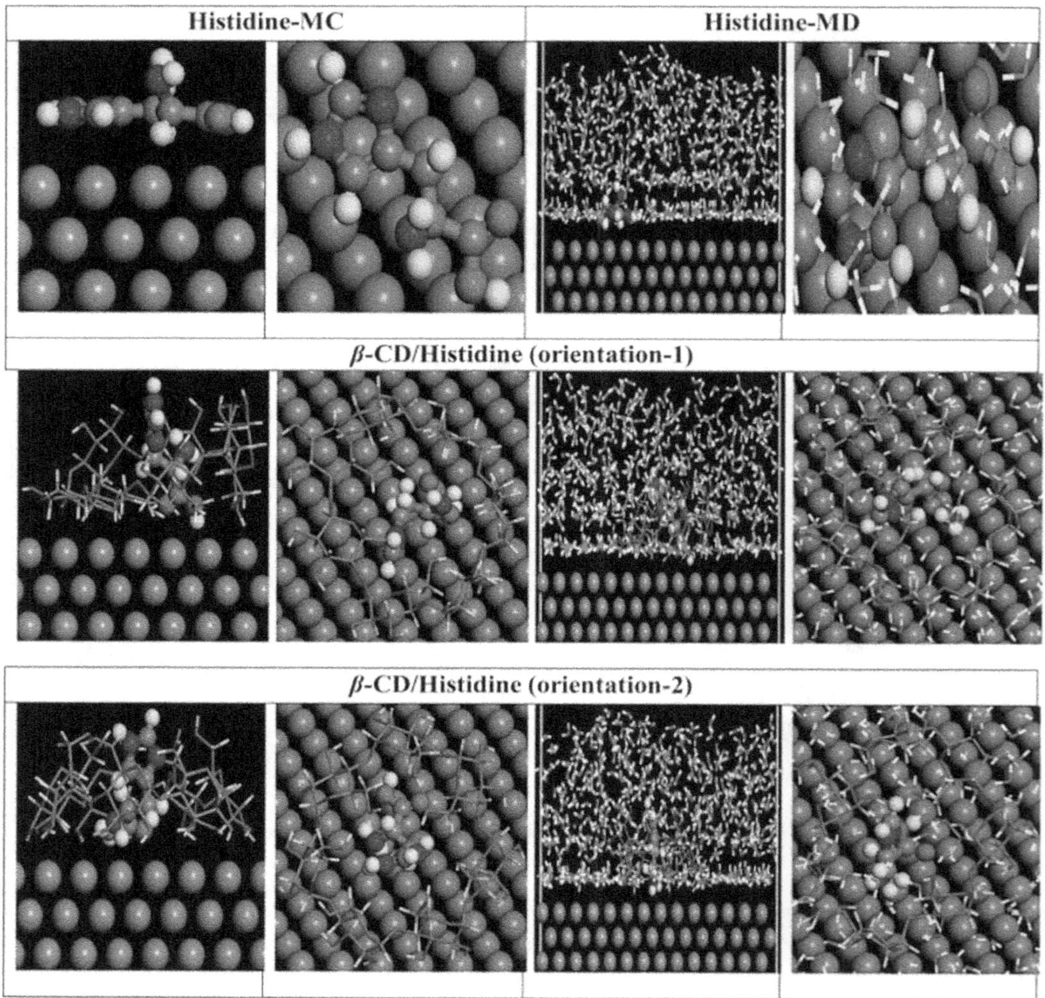

FIGURE 14.5 The final snapshots of histidine, and β-CD/histidine inclusion complexes with two different orientations above Fe (110) surface obtained from molecular simulations (MC and MD). The histidine molecule, β-CD and Fe (110) surface are shown in Ball & Stick, Stick and CPK styles, respectively. (Reproduced with permission from Dehghani et al. [43]. © 2021 Elsevier B.V.)

technology. Hence, the following points should be taken into account prior to coating formulation: (1) the compatibility of the proposed coating, corrosive environment, and the metal should be known; (2) multifunctional smart coatings are needed; (3) the high cost of smart coating is one of the main drawbacks, so it is crucial to reduce the cost of smart coatings so that they can be applied more frequently; and (4) a new synthesis method should be developed considering the low cost.

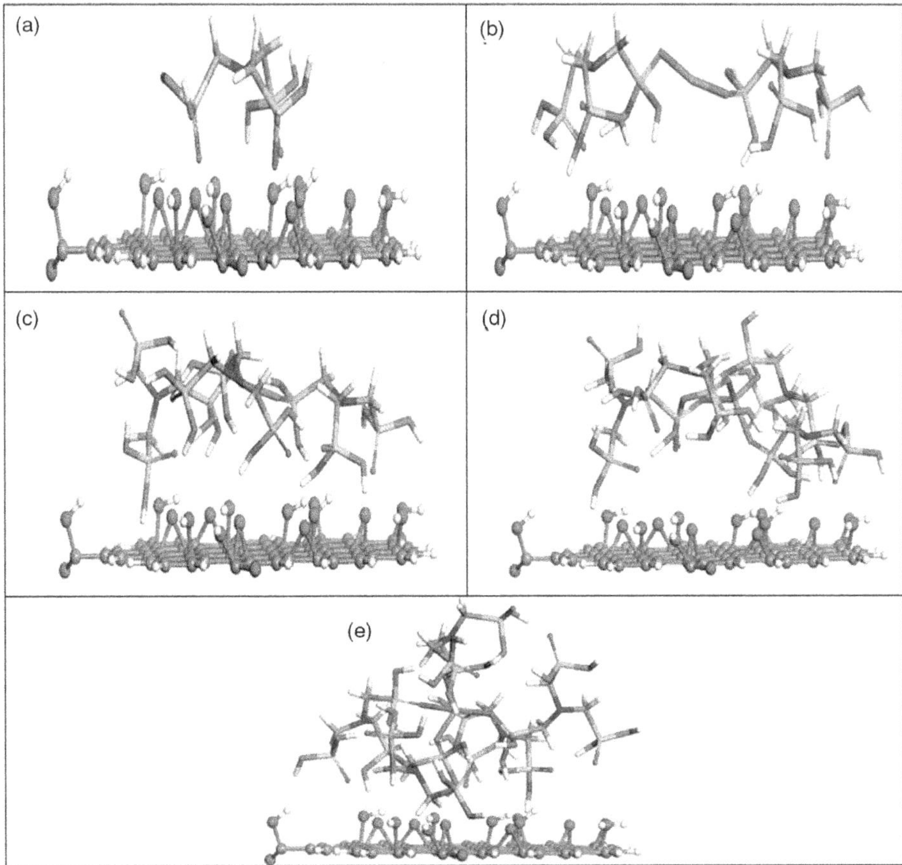

FIGURE 14.6 The initial geometries of (a) AMP-GO, (b) [Zinc (AMP)2] +2-GO, (c) [Zinc(AMP)3] +2-GO, (d) [Zn(AMP)4] +2-GO, and (e) [Zinc(AMP)5] +2-GO clusters prepared for DFT-D computations. The AMP and Zn+2/AMP complexes are shown in Stick style, and the GO surface is displayed in Ball & Stick model. (Reproduced with permission from Keshmiri et al. [44]. © 2021 Elsevier B.V.)

REFERENCES

[1] Anadebe, VC, Chukwuike, VI, Selvaraj, V, Pandikumar, A, Barik, RC (2022) Sulfur-doped graphitic carbon nitride (Sg-C_3N_4) as an efficient corrosion inhibitor for X65 pipeline steel in CO_2-saturated 3.5 wt. % NaCl solution: Electrochemical, XPS and nanoindentation studies. *Process Safety and Environmental Protection*, 164, 715–728.

[2] Anadebe, VC, Chukwuike, VI, Ramanathan, S, Barik, RC (2022) Cerium-based metal organic framework (Ce-MOF) as corrosion inhibitor for API 5L X65 steel in CO_2-saturated brine solution: XPS, DFT/MD-simulation, and machine learning model prediction. *Process Safety and Environmental Protection*, 168, 499–512.

[3] Anadebe, VC, Chukwuike, VI, Chidiebere, MA, Barik, RC (2023) Synthesis, characterization, and evaluation of Co-MOF based ZIF-67 for CO_2 corrosion inhibition of X65 steel: Insights from electrochemical studies and a machine learning algorithm. *The Journal of Physical Chemistry C*, 123(20), 9871–9886.

[4] Kling, S, Czigány, T (2014) Damage detection and self-repair in hollow glass fiber fabric-reinforced epoxy composites *via* fiber filling. *Composites Science and Technology*, 99, 82–88.

[5] Wang, JP, Wang, JK, Zhou, Q, Li, Z, Han, Y, Song, Y, Yang, S, Song, X, Qi, T, Möhwald, H, Shchukin, D (2018) Adaptive polymeric coatings with self-reporting and self-healing dual functions from porous core–shell nanostructures. *Macromolecular Materials and Engineering*, 303(4), 1700616.

[6] Liu, C, Wu, H, Qiang, Y, Zhao, H, Wang, L (2021) Design of smart protective coatings with autonomous self-healing and early corrosion reporting properties. *Corrosion Science*, 184, 109355.

[7] Calvino, C, Weder, C (2018) Microcapsule-containing self-reporting polymers. *Small*, 14(46), 1802489.

[8] Çömlekçi, GK, Ulutan S (2019) Acquired self-healing ability of an epoxy coating through microcapsules having linseed oil and its alkyd. *Progress in Organic Coatings*, 129, 292–299.

[9] Attaei, M, Calado, LM, Morozov, Y, Taryba, MG, Shakoor, RA, Kahraman, R, Marques, AC, Montemor, MF (2020) Smart epoxy coating modified with isophorone diisocyanate microcapsules and cerium organophosphate for multilevel corrosion protection of carbon steel. *Progress in Organic Coatings*, 147, 105864.

[10] Li, H, Zhang, QH, Meng, XZ, Liu, P, Wu, LK, Cao, FH (2023) A novel cerium organic network modified graphene oxide prepared multifunctional waterborne epoxy-based coating with excellent mechanical and passive/active anti-corrosion properties. *Chemical Engineering Journal*, 465, 142997.

[11] Cao, Y, Wu, H, Wang, X, Wang, G, Yang, H (2022) Novel long-acting smart anticorrosion coating based on pH-controlled release polyaniline hollow microspheres encapsulating inhibitor. *Journal of Molecular Liquids*, 359, 119341.

[12] Ulaeto, SB, Ravi, RP, Udoh, II, Mathew, GM, Rajan, TPD (2023) Polymer-based coating for steel protection, highlighting metal–organic framework as functional actives: A Review. *Corrosion and Materials Degradation*, 4(2), 284–316.

[13] Lamprakou, Z, Bi, H, Weinell, CE, Dam-Johansen, K (2023) Encapsulated corrosion inhibitive pigment for smart epoxy coating development: An investigation of leaching behaviour of inhibitive ions. *ACS Omega*, 8(16), 14420–14429.

[14] Rammelt, U, Reinhard, G (1994) Characterization of active pigments in damage of organic coatings on steel by means of electrochemical impedance spectroscopy. *Progress in Organic Coatings*, 24(1–4), 309–322.

[15] Hao, Y, Liu, F, Han, EH, Anjum, S, Xu, G (2013) The mechanism of inhibition by zinc phosphate in an epoxy coating. *Corrosion Science*, 69, 77–86.

[16] Ashrafi-Shahri, SM, Ravari, F, Seifzadeh, D (2019) Smart organic/inorganic sol-gel nanocomposite containing functionalized mesoporous silica for corrosion protection. *Progress in Organic Coatings*, 133, 44–54.

[17] Habib, S, Shakoor, RA, Kahraman, R (2021) A focused review on smart carriers tailored for corrosion protection: Developments, applications, and challenges. *Progress in Organic Coatings*, 154, 106218.

[18] Ding, C, Fu, J (2020) Smart anticorrosion coatings based on nanocontainers. In *Smart Nanocontainers*, 413–429. Elsevier.

[19] Wei, H, Wang, Y, Guo, J, Shen, NZ, Jiang, D, Zhang, X, Yan, X, Zhu, J, Wang, Q, Shao, L, Lin, H (2015) Advanced micro/nanocapsules for self-healing smart anticorrosion coatings. *Journal of Materials Chemistry A*, 3(2), 469–480.

[20] Yeganeh, M, Asadi, N, Omidi, M, Mahdavian, M (2019) An investigation on the corrosion behavior of the epoxy coating embedded with mesoporous silica nanocontainer loaded by sulfamethazine inhibitor. *Progress in Organic Coatings*, 128, 75–81.

[21] Ma, IW, Ammar, S, Bashir, S, Selvaraj, M, Assiri, MA, Ramesh, K, Ramesh, S (2020) Preparation of hybrid chitosan/silica composites via ionotropic gelation and its electrochemical impedance studies. *Progress in Organic Coatings*, 145, 105679.

[22] Falcón, JM, Otubo, LM, Aoki, IV (2016) Highly ordered mesoporous silica loaded with dodecylamine for smart anticorrosion coatings. *Surface and Coatings Technology*, 303, 319–329.

[23] Ghahremani, P, Sarabi, AA, Roshan, S (2021) Cerium containing pH-responsive microcapsule for smart coating application: Characterization and corrosion study. *Surface and Coatings Technology*, 427, 127820.

[24] Udunwa, DI, Onukwuli, OD, Anadebe, VC (2022) Synthesis and evaluation of 1-butyl-3-methylimidazolium chloride based ionic liquid for acid corrosion inhibition of aluminium

alloy: Empirical, DFT/MD-simulation and RSM modeling. *Journal of Molecular Liquids*, 364, 120019.

[25] Anadebe, VC, Onukwuli, OD, Omotioma, M, Okafor, NA (2019) Experimental, theoretical modeling and optimization of inhibition efficiency of pigeon pea leaf extract as anti-corrosion agent of mild steel in acid environment. *Materials Chemistry and Physics*, 233, 120–132.

[26] Anadebe, VC, Nnaji, PC, Onukwuli, OD, Okafor, NA, Abeng, FE, Chukwuike, VI, Okoye, CC, Udoh, II, Chidiebere, MA, Guo, L, Barik, RC (2022) Multidimensional insight into the corrosion inhibition of salbutamol drug molecule on mild steel in oilfield acidizing fluid: Experimental and computer aided modeling approach. *Journal of Molecular Liquids*, 349, 118482.

[27] del Olmo, R, López, E, Matykina, E, Tiringer, U, Mol, JMC, Mohedano, M, Arrabal, R (2023) Hybrid PEO/sol-gel coatings loaded with Ce for corrosion protection of AA2024-T3. *Progress in Organic Coatings*, 182, 107667.

[28] Li, J, Li, Z, Feng, Q, Qiu, H, Yang, G, Zheng, S, Yang, J (2019) Encapsulation of linseed oil in graphene oxide shells for preparation of self-healing composite coatings. *Progress in Organic Coatings*, 129, 285–291.

[29] Hasanzadeh, M, Shahidi, M, Kazemipour, M (2015) Application of EIS and EN techniques to investigate the self-healing ability of coatings based on microcapsules filled with linseed oil and CeO_2 nanoparticles. *Progress in Organic Coatings*, 80, 106–119.

[30] Masood, S, Ghosal, A, Zafar, F, Alam, M, Nishat, N (2023) Epoxy-free polyphenalkamine one-pack sustainable coating from technical cashew nut shell liquid: Fabrication and electrochemical studies for anti-corrosive applications. *Materials Today Sustainability*, 22, 100362.

[31] Huangfu, H, Guo, X, Li, N, Xiong, Y, Huang, Y, Zhang, J, Wang, L (2023) A smart composite coating with self-reporting and self-healing functions to enhance corrosion protection for magnesium alloys. *Progress in Organic Coatings*, 181, 107598.

[32] Wu, G, Li, Y, Brittain, R, Lu, Z, Yang, L (2022) Understanding of fracture conditions and material response in a model TiN film/stainless steel substrate system–A cross-sectional scratch test study. *Surface and Coatings Technology*, 442, 128340.

[33] Wang, J, Yang, H, Meng, Z, Xie, B, Yu, X, Su, G, Wang, L (2022) Epoxy coating with excellent anticorrosion and pH-responsive performances based on DEAEMA modified mesoporous silica nanomaterials. *Colloids and Surfaces A: Physicochemical and Engineering Aspects*, 634, 127951.

[34] Rakoch, AG, Van Tuan, T, Khabibullina, ZV, Blawert, C, Serdechnova, M, Scharnagl, N, Zheludkevich, ML, Gladkova, AA (2022) Role of cobalt additive on formation and anticorrosion properties of PEO coatings on AA2024 alloy in alkali-silicate electrolyte. *Surface and Coatings Technology*, 433, 128075.

[35] Abdipour, H, Rezaei, M, Abbasi, F (2018) Synthesis and characterization of high durable linseed oil-urea formaldehyde micro/nanocapsules and their self-healing behaviour in epoxy coating. *Progress in Organic Coatings*, 124, 200–212.

[36] Mahmoudian, M, Nozad, E, Kochameshki, MG, Enayati, M (2018) Preparation and investigation of hybrid self-healing coatings containing linseed oil loaded nanocapsules, potassium ethyl xanthate and benzotriazole on copper surface. *Progress in Organic Coatings*, 120, 167–178.

[37] Dreva, K, Morina, A, Yang, L, Neville, A (2022) The effect of temperature on water desorption and oxide formation in MoS2 coatings and its impact on tribological properties. *Surface and Coatings Technology*, 433, 128077.

[38] Dhandapani, HN, Mahendiran, D, Karmakar, A, Devi, P, Nagappan, S, Madhu, R, Bera, K, Murugan, P, Babu, BR, Kundu, S (2022) Boosting of overall water splitting activity by regulating the electron distribution over the active sites of Ce doped NiCo-LDH and atomic level understanding of the catalyst by DFT study. *Journal of Materials Chemistry A*, 10(34), 17488–17500.

[39] Montejo-Valencia, BD, Pagán-Torres, YJ, Martínez-Iñesta, MM, Curet-Arana, MC (2017) Density functional theory (DFT) study to unravel the catalytic properties of M-exchanged MFI (M= Be, Co, Cu, Mg, Mn, Zn) for the conversion of methane and carbon dioxide to acetic acid. *ACS Catalysis*, 7(10), 6719–6728.

[40] Rahmani, MH, Dehghani, A, Bahlakeh, G, Ramezanzadeh, B (2022) Introducing GO-based 2D-platform modified via phytic acid molecules decorated by zeolite imidazole ZIF-9 MOFs for designing multi-functional polymeric anticorrosive system; DFT-D computations and experimental studies. *Journal of Molecular Liquids*, 364, 119945.

[41] Dehghani, A, Bahlakeh, G, Ramezanzadeh, B, Mofidabadi, AHJ (2021) 2D reduced-graphene oxide (rGO) nanosheets decorated with l-histidine loaded-β-cyclodextrin for efficient epoxy nanocomposite anti-corrosion properties; DFT-D modeling/experimental assessments. *Flat Chemistry* 30, 100309.

[42] Dehghani, A, Bahlakeh, G, Ramezanzadeh, B, Mofidabadi, AHJ, Mostafatabar, AH (2021) Benzimidazole loaded β-cyclodextrin as a novel anti-corrosion system; Coupled experimental/computational assessments. *Journal of Colloid and Interface Science*, 603, 716–727.

[43] Dehghani, A, Bahlakeh, G, Ramezanzadeh, B, Mofidabadi, AHJ (2021) Improvement of the anti-corrosion ability of a silane film with β-cyclodextrin-based nanocontainer loaded with L-histidine: Coupled experimental and simulations studies. *Progress in Organic Coatings*, 157, 106288.

[44] Keshmiri, N, Najmi, P, Ramezanzadeh, B, Ramezanzadeh, M, Bahlakeh, G (2021) Nano-scale P, Zn-codoped reduced-graphene oxide incorporated epoxy composite; Synthesis, electronic-level DFT-D modeling, and anti-corrosion properties. *Progress in Organic Coatings*, 159, 106416.

[45] Keshmiri, N, Najmi, P, Ramezanzadeh, M, Ramezanzadeh, B, Bahlakeh, G (2022) Ultrastable porous covalent organic framework assembled carbon nanotube as a novel nanocontainer for anti-corrosion coatings: Experimental and computational studies. *ACS Applied Materials & Interfaces*, 14(17), 19958–19974.

15 Analytical Imaging Techniques in Corrosion-Resistant Coating Studies

Purnima, Ashish Kumar Tiwari, and Shweta Goyal

15.1 INTRODUCTION

In this chapter, the surface morphological and topographical techniques that are used to ascertain the coating characteristics are discussed in detail. The important coating characteristics include uniformity of the coating compound, thickness of the coating, surface roughness and hydrophobicity. These properties emerge from the structure of coating and the process of development of the coating on a given metal. The information about these properties is measured in terms of surface morphology and topography, and is essential for the future application of coatings in various industrial sectors.

The goal of characterizing the surface of coating can be addressed by a wide range of techniques. Each applied technique provides information about a particular property of the coating.

This chapter includes the description of various surface morphological and topographical techniques generally applied for the surface analysis of the anticorrosive coatings. The techniques are based on scanning probe and optical/electron microscopy. The implementation of the discussed techniques (individually and in combination) is necessary for multidisciplinary and complete study of the coatings.

15.2 CHARACTERIZATION TECHNIQUES

The morphology and/or topography of the metal surface after the development of an anticorrosive coating can be directly analyzed by different microscopic techniques. Some of the most commonly used techniques involve optical microscopy (OM), scanning electron microscopy (SEM), field emission scanning electron microscopy (FESEM), tunneling electron microscopy (TEM), scanning tunneling microscopy (STM), atomic field microscopy (AFM) and others. The use of a particular technique is related to the typical characteristic of the coatings. For instance, TEM and STM are employed for the analysis of ultrafine formations/structures on the coating, while SEM and OM can be employed for examination of any macro formations. Further, in order to develop an understanding about the surface roughness/smoothness, topography can be analyzed by AFM.

Table 15.1 presents the overview of morphological techniques (analytical imaging techniques) used by various researchers to characterize the coatings and the microcapsules that the coatings are doped with.

DOI: 10.1201/9781032677255-15

TABLE 15.1
Review of papers in context with the various morphological techniques used for characterization

Ref.	Corrosion Inhibitor	Metal	Medium	Application Form	OM	SEM	TEM	EDS	AFM	WCA
[1]	Modified eco-friendly chitosan	N80 Steel	CO_2 NaCl	Encapsulated	-	✓	-	✓	-	-
[2]	Twi synthesized dithiane molecules	Cu	0.5M H_2SO_4	Monolayer development in ethanol	-	✓	-	-	-	✓
[3]	Smart micro-arc oxidation (MAO)/epoxy resin (EP) composite coatings (EPM1 and EPM2)	AZ31 Mg alloy	3.5% NaCl	Composite coating in vacuum	-	✓	-	✓	-	-
[4]	Hybrid zinc coatings with chitosan/alginate Encapsulated CuO-nanoparticles	MS	3.5% NaCl and artificial sea water	Electro-deposition	-	✓	-	✓	-	✓
[5]	Vinyl trimethyl silane and acrylic acid-modified silica nanoparticles (Si-PAA-VTMS)	316L SS	3.5% NaCl and seawater	Admixed	-	✓	-	✓	-	✓
[3]	MAO coatings incorporated with organic–inorganic hybrid materials (with and without PVA)	AZ31 Mg alloy	NaCl	MAO	-	✓	-	✓	✓	✓
[6]	Dimethyl sulfide (DMS)	MS	0.5N HCl	Admixed	-	✓	-	✓	✓	-
[7]	Dibutyl pthalate nanocapsules (DBT as core with urea and formaldehyde-based shell)			Sonochemical method to form nanocapsules	-	✓	✓	-	-	-
[8]	GO-incorporated silane nanocomposite coating	AZ91 Mg alloy		Electroless co-deposition		✓	-	✓	-	✓
[9]	Multifunctional cobalt coating	Low carbon steel		Electrodeposition		✓	-	✓	✓	-
[10]	Aniline derivatives and their molybdate doped coatings	Cu	0.5M H_2SO_4	Electro- polymerization via cyclic voltammetry		✓	-	-	✓	-
[11]	Superhydrophobic corrosion inhibitor double hydroxide coating (Mg-Al LDH)	AZ31 Mg alloy	3.5% NaCl	Anodization and hydrothermal treatment		✓	-	✓	-	✓
[12]	N-doped graphene quantum dots (N-GQDs) and polymethyltrimethoxysilane (PMTMS) coating	Magnesium alloy (AZ91D)	3.5% NaCl	Electrodeposition and subsequent silane treatment	-	✓	✓	✓	✓	-
[13]	Zirconium (Zr)-based conversion coating using hexafluorozirconic acid with Cu additive	LCS	-	-		✓	-	✓	-	-

[14]	Ti-Si-B-C nanocomposite coating	SS 304	3.5% NaCl	Magnetron sputtered		✓	-	✓	-	✓
[15]	Cerium and quinoline-based microcapsules' epoxy-based coating	MS	3.5% NaCl	Polymerization for micro-capsulation and epoxy-based coating of micro-capsules was manually applied on MS substrate	✓	✓	-	-	-	-
[16]	Soyabean oil	Glass	-	Microencapsulation: in urea-formaldehyde microcapsules	✓	✓	-	-	-	-
[17]	Electrospun fiber coating	Hot rolled carbon sheet	5% NaCl	Electrospinning	✓	✓	-	-	-	-
[18]	Quinoline	MS	5% HCl	Micro-encapsulation	✓	✓	-	-	-	-
[19]	PAMAM dendrimer synthesized using ethylenediamine and methacrylate	MS	5% NaCl	Micro-encapsulation	✓	✓	-	-	-	-
[20]	Tetraethyl orthosilicate and methyltriethoxysilane	MS	3.5% NaCl	Hydrolysis and polycondensation	✓	✓	-	✓	✓	-

Note: Cu: copper, Mg: magnesium, SS: stainless steel. MS: mild steel, LCS: low carbon steel, MAO: micro arc oxidation.

15.3 OPTICAL MICROSCOPY

OM exhibits remarkable versatility as an imaging technique that is commonly used for analyzing the coating homogeneity (morphology/topography), defects in the coatings after coating deposition and also after subjecting to corrosive environment. The fundamental principle underlying OM involves illuminating an object and collecting the light that is either scattered or transmitted through a lens system to create a microscopic image. The collection of scattered and transmitted lights is used for analyzing the opaque objects and to research the transparent materials, respectively. The images produced by OM provide information at a magnification range of 2X to 2000X and at a resolution of ~0.5 μm [21]. Information about the visible features like shape, size and arrangement can be obtained through OM as it uses visible light, i.e., wavelength of radiation between 380 and 760 nm (depending on the color of the light or specimen nature).

Various types of microscopes such as simple microscope, compound microscope, polarized light microscope, confocal scanning laser microscope and others, are employed for coating characterization. The application of a particular type of microscope depends on the information required. For example, conventional microscopes produce poor images of rough surfaces, while confocal microscopes have no limitation in terms of surface roughness. In addition, these types of microscopes have better lateral resolution also, and hence can be used for analyzing the surfaces that are rough and need to be analyzed spatially. Additionally, OM can be used in combination with interferometer in order to study the topographical features of the coated films, which is termed as optical profilometry. High vertical (up to 2 nm) and lateral resolutions (up to 150 nm) are possible because of its high aperture (0.95) in air. Employing such types of OMs makes it possible to measure the thickness of the coating materials in the range of 10 nm to 20 μm (approximately). Further, for analyzing the morphology of the microcapsules loaded with inhibitors, the microcapsules are placed over a glass slide and observed under the microscope; for the characterization of the metals coated with the capsule doped coatings, the metal samples are cut into a coupon size compatible with the microscope machine. Once the images are captured under the OM, the size of the capsules or any particular feature can be determined using particle size analyzer softwares.

15.3.1 Application of OM for Coating Characterization

The influence of temperature, material of coating, duration, etc., on the coating morphology can be analyzed and optimized through OM. Phanasgaonkar et al. [20] employed OM to study the coating surface microscopically in order to examine the effect of temperature and three different coating materials. The authors reported that with the rise in temperature with type 1 coating, cracks were seen in the optical micrographs; while for types 2 and 3 coatings, the micrographs were crack-free, hence suggesting the better performance of types 2 and 3 coatings. The study involved the quick analysis of the surface morphology via optical analysis which was further strengthened with the help of other morphological techniques (SEM and AFM).

Figure 15.1 clearly shows the results of the surface morphology of type 1 coating with increasing temperature. Cracks are evident at higher temperatures owing to the thinner coating thickness [20]. In another interesting study by Gite et al. [18], OM was employed for preliminary confirmation of the shape and size of the prepared quinoline-based microcapsules incorporated in polyurethane (PU) coatings for future encapsulation as corrosion inhibitors for mild steel specimens in 5% HCl solution (by weight). As shown in Figure 15.2, well-defined spherical shaped capsules with intact morphology and shell membrane were evinced [18]. Shisode et al. also employed OM to analyze the prepared microcapsules containing soybean oil at 40x and 100x [16]. In a study conducted by Tatiya et al. [19], the newly developed polyurea microcapsules were firstly assessed using OM for their size and size distribution variation with varying agitation rates (from 3,000 to 8,000 rpm) at 40x to 100x resolution. The authors reported nearly spherical shaped microcapsules for all the agitation rates. In addition, the employment of OM revealed that with the rise in the rate of agitation,

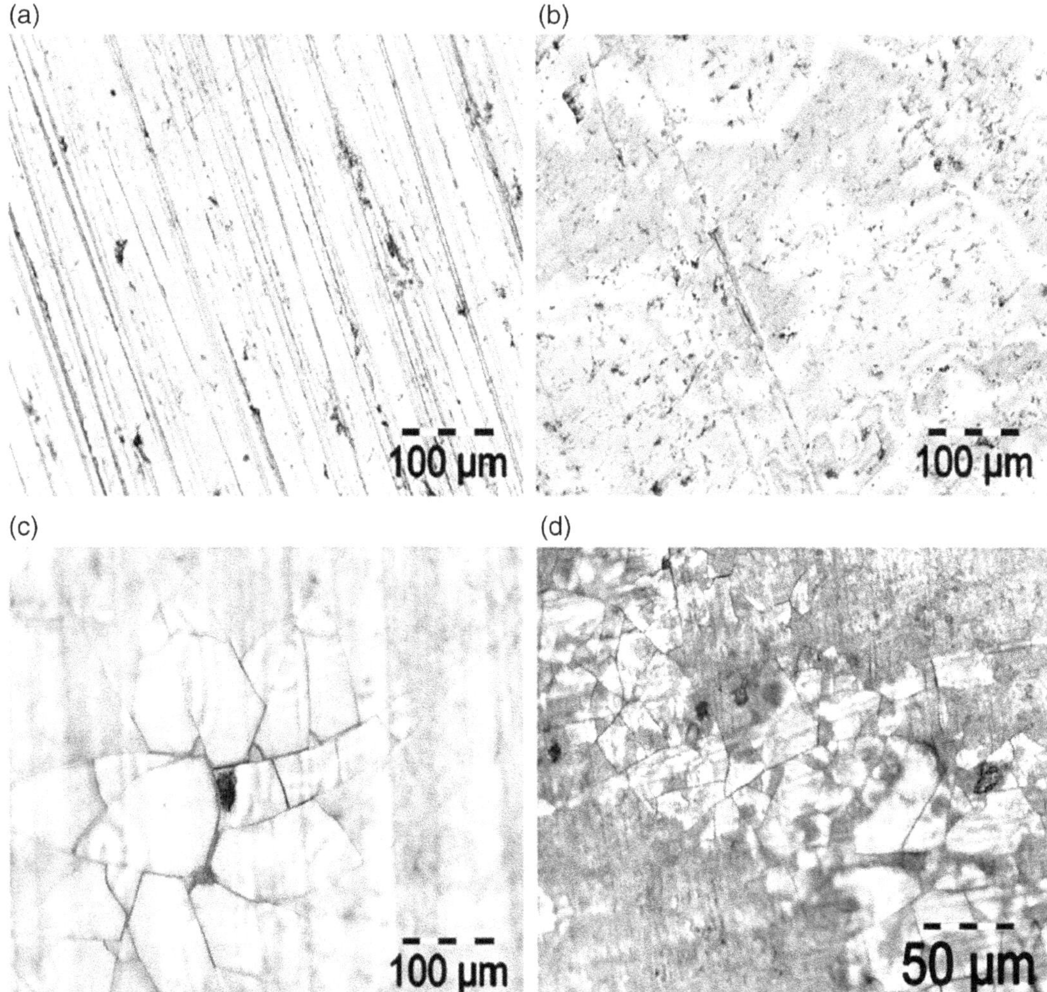

FIGURE 15.1 Optical micrograph of MS substrate coated with 1 hour aging without curing (a) PHC, (b) PHC, 200°C, (c) PHC, 300°C, and (d) PHC, 400°C. (Reproduced from [20] by permission of Elsevier Science Ltd.)

the size of the microcapsule decreased. When the agitation rate was high, larger droplets were subjected to strong shear forces, causing them to break into smaller droplets. Conversely, at a low agitation rate, the dominant factor was the interfacial tension, which led to the formation of larger-sized microcapsules [19].

15.3.2 Application of OM for Evaluating Coating Resistance against Corrosion

The use of OM has been extensively employed to examine the anticorrosion properties and self-healing capability of diverse encapsulated corrosion inhibitors. One such interesting study was conducted by Fan et al. [22]. Figure 15.3 shows the optical microscopic (a–d) and thermal (e–h) images of scratch at different durations (up to 1.5 hours) and cross-sectional image (i) of the healed defect from their study. Alodine 5200 inhibitor and a healing agent were encapsulated in a PU coating. When the scratch was created, the released inhibitor from the capsule healed the surface,

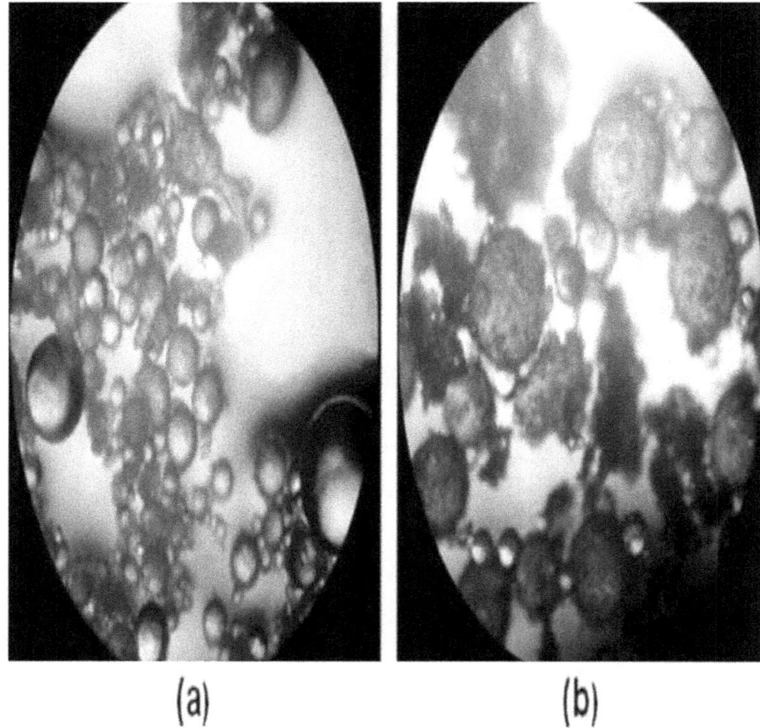

FIGURE 15.2 Optical microscopic images of PU microcapsules containing quinoline with solvent (a) xylene and (b) ethyl acetate (magnification=10X). (Reproduced from [18] by permission of Elsevier Science Ltd.)

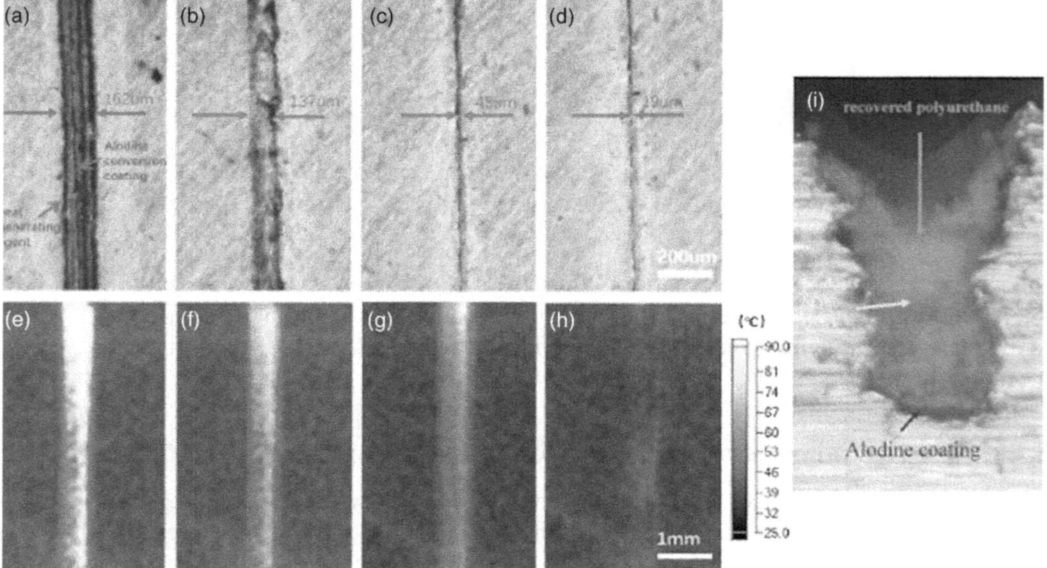

FIGURE 15.3 Optical microscopic and thermal images of scratch at different durations after scratch formation: (a and e) 2 min, (b and f) 0.5 hr, (c and g) 1 hr, and (d and h) 1.5 hrs; (i) cross-sectional view of healed coating. (Reproduced from [22] by permission of Elsevier Science Ltd.)

which could be clearly seen from the microscopic images taken at various times in ascending order. Furthermore, cross-sectional OM also displayed a clear evidence of the healing effect of the inhibitor [22].

Doan et al. [17] characterized the self-healing performance of electrospun fiber coating via OM after the conductance of electrochemical tests on metal substrates exposed to corrosive aqueous salt solution for four months. OM was used to demonstrate that the coating integrated with fiberspun exhibited a greater healing effect compared to its counterpart without fibers [17].

15.4 SCANNING ELECTRON MICROSCOPY

Micro and nanostructural details of the coating surface can be inspected through SEM technique. In this technique, an electron beam is focused into fine probe and scanned over a confined area. Various signals are created as the beam interacts with the sample in the form of secondary electrons, photon emission, etc. All these signals can be captured easily. By utilizing these signals to modulate the luminosity of the cathode ray tube, which undergoes raster scanning synchronized with the electron beam, an image is generated on the display surface. In addition to this, SEM has also the potential to extract microchemical information from the structural details. FESEM, i.e., SEM equipped with field emission gun, can be employed to differentiate between surface features that are only 1 nm apart from each other (spatial resolution of 1 nm). Three-dimensional (3D) characteristics are yielded through SEM images while keeping large areas of the surface in focus as SEM has the excellent ability to depict large depths of field (~100% of horizontal field depth). Topographic contrasts can be obtained via secondary electron imaging, while topographic and compositional contrast can be captured via backscattered electron imaging. In most research applications, energy dispersive X-ray spectroscopy (EDS) is utilized in combination with SEM to obtain the microchemical information, both qualitatively and quantitatively. Such combination is helpful for analyzing the coating morphology in a holistic manner and to gain complete information. Typically, bulk sample analysis (i.e., morphology, inclusions, precipitates, grain boundaries, etc.) and localized examinations (i.e., scratch thickness and depth, corrosion product analysis, coating material analysis, etc.) are the two widespread applications for coating characterization through SEM.

The sample preparation requirements are minimal and occasionally require to be coated with conducting film (usually gold). An important factor to be taken care of is that the sample must be vacuum compatible, that is, it has to be analyzed under SEM.

15.4.1 APPLICATION OF SEM/FESEM FOR COATING CHARACTERIZATION

The spatial resolution and sampling depth of SEM is generally 1 nm and submicron to several microns, respectively. Alongside, it has an elemental detection limit from 0.2% to 0.5%. Figure 15.4 shows the surface morphologies of passivated copper surface coated with nanostructured polyaniline (PANi), poly (N-ethylaniline) (PNEA) and their molybdate-doped coatings (PANi-Mo and PNEA-Mo) via voltammetric cycles in oxalic acid through electropolymerization method. Figure 15.4(a) presents a spindle granular morphology, while nano-flat sheet morphology is reported in Figure 15.4(b). Although the morphologies in Figure 15.4(c) and (d) are similar to that of Figure 15.4(a) and (b), a clear variation with respect to their denser packaging and smaller aggregate particles can be realized through SEM analysis. All the imaging was reported to be conducted at 5 μm. Furthermore, SEM can also be utilized to examine and determine the coating thickness for different coating types and coating methods. Figure 15.5 shows the corresponding cross-sectional images for Figure 15.4. The average thicknesses of PANi, PNEA, PANi-Mo and PNEA-Mo coatings are 27.20, 34.81, 56.92 and 61.08 μm, respectively. The study, hence, used surface SEM and cross-sectional SEM imaging to prove that molybdate-doped coatings exhibit a discernible augmentation in thickness relative to their undoped counterparts [10].

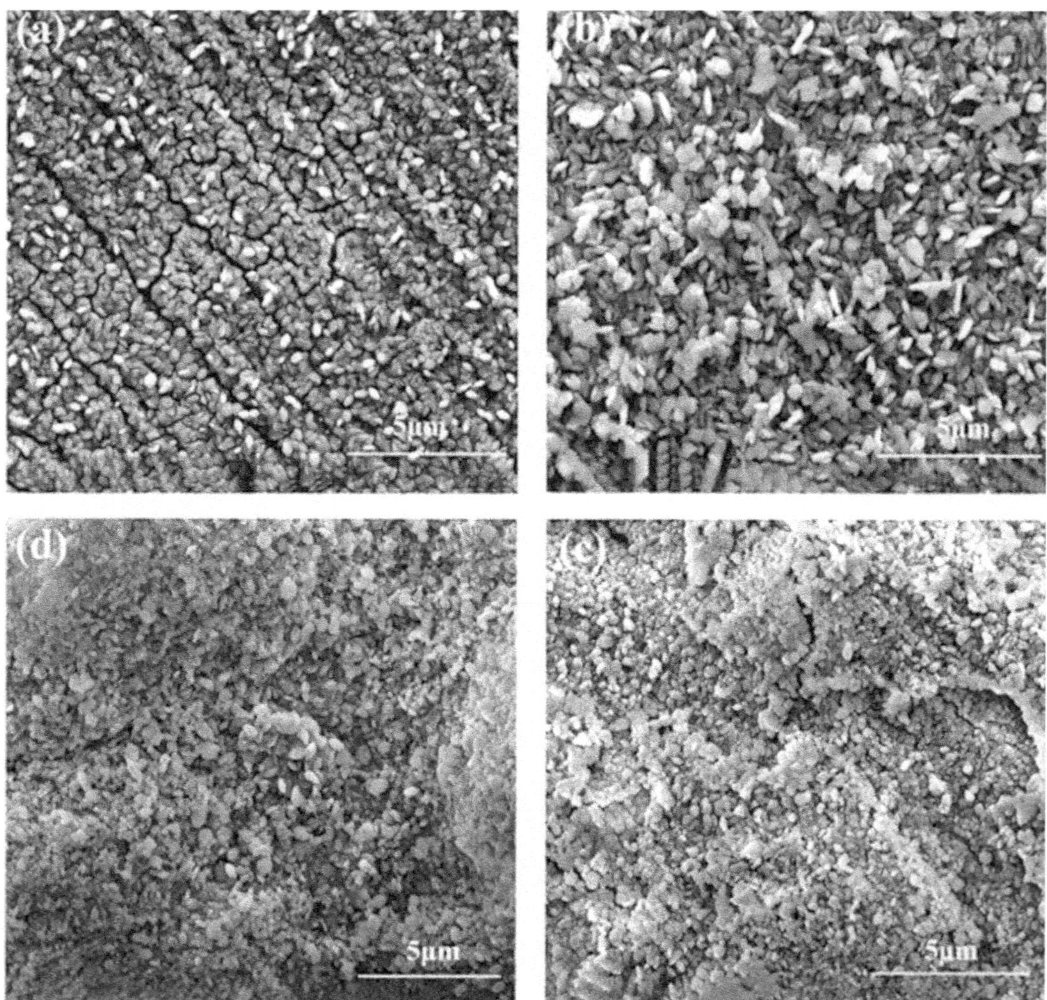

FIGURE 15.4 Surface morphologies of (a) PANi, (b) PNEA, (c) PANi-Mo and (d) PNEA-Mo coatings on copper surface. (Reproduced from [10] by permission of Elsevier Science Ltd.)

Zhang et al. [23] analyzed smart micro arc oxidation epoxy resin (EP) composite coatings on AZ31 magnesium (Mg) alloy for inhibition against 3.5% NaCl corrosive solution via SEM equipped with EDS. Cross-sectional images of composite coatings displayed denser morphology with negligible pores as compared to pure EP coating. Noticeable signals of silica (Si) and sodium (Na) which were homogeneously distributed throughout the depth in the EDS analysis explained the compactness of the composite layer. The surface analysis under SEM also unveiled a smoother surface for composite coating instead of rough surface with white spots for pure EP coatings [23]. Further, SEM analysis can also be adopted to optimize the coating characteristics varied by dopant concentration, current density employed for coating, pH, duration and others [3, 8].

Figure 15.6 (b, c and d) shows FESEM images of cobalt coatings (analyzed at 5 μm) electrodeposited on low carbon steel surface. As the current density rose, the morphology transitioned from smaller and flatter morphology to larger and taller morphology (of size 2–4 μm), which increased the roughness and hydrophobicity of the coating. Larger pyramidal morphology

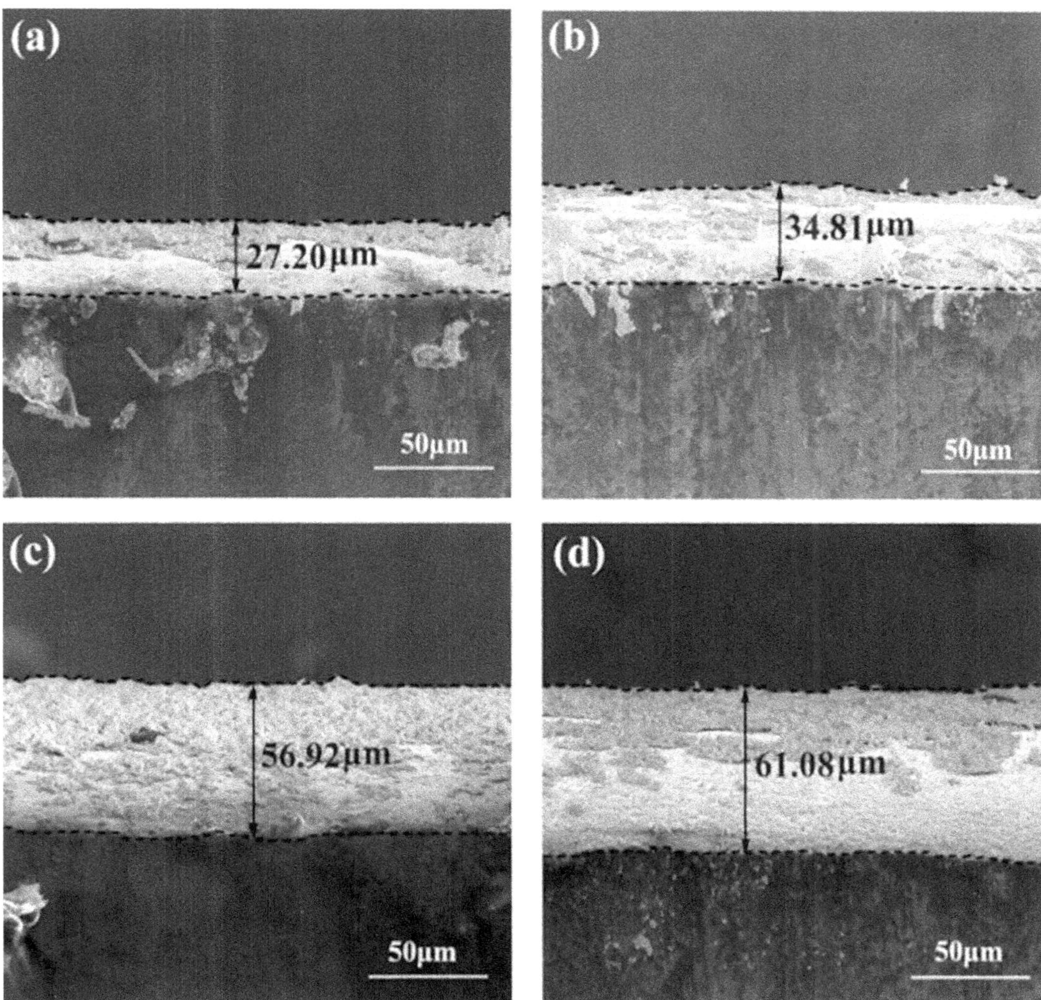

FIGURE 15.5 Cross-sectional morphologies of (a) PANi, (b) PNEA, (c) PANi-Mo and (d) PNEA-Mo coatings on copper surface. (Reproduced from [10] by permission of Elsevier Science Ltd.)

can trap more air and raise the water contact angle, which in turn can isolate the metal from the attack of corrosive media [9]. In another study, hyperbranched chitosan-based microcapsules (prepared to perform as corrosion inhibitors) were coated with an acrylamide and a cross-linking agent (bentonite). The concentration of acrylamide was varied from 10% to 30% in order to optimize for higher resistance against temperature and pressure. For this purpose, SEM imaging (shown in Figure 15.7) was conducted, which revealed that with a rise in concentration the 3D mesh structure became denser, ensuring better performance [1]. Wu et al. [24] also employed SEM (surface morphology) in order to analyze the plasma electrolytic oxidation coatings on long period stacking ordered Mg-Gd-Y-Zn alloy. Average pore size (in μm) and open porosity (in percentage) were also measured from the observed crater morphology structure. Pores upto the nanoscale level were affirmed with the help of SEM analysis [24].

In a study conducted by Rahimi et al. [25], cyclodextrine cores were synthesized for the encapsulation of two inhibitors [2-mercaptobenzothiazole and 2-mercaptobenzimidazole (MBI)] in the

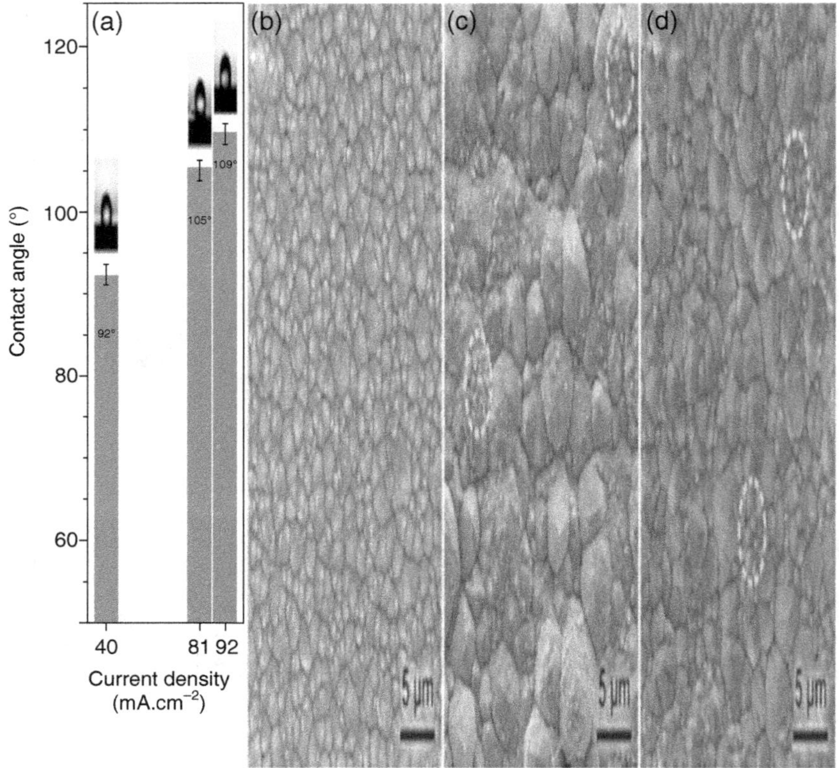

FIGURE 15.6 (a) Effect of hydrophobicity on cobalt coating, FESEM analysis with varying current densities: (b) 40 mA/cm^2, (c) 81 mA/cm^2 and (d) 90 mA/cm^2 (deposited for 20 min at pH = 3.25). (Reproduced from [9] by permission of Elsevier Science Ltd.)

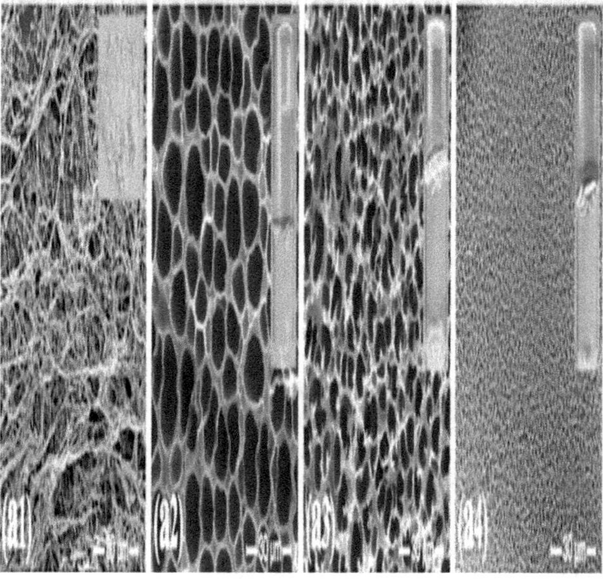

FIGURE 15.7 SEM images of (a1) acrylamide, (a2) cross-linked 10% acrylamide, (a3) cross-linked 20% acrylamide and (a4) cross-linked 30% acrylamide. (Reproduced from [1] by permission of Elsevier Science Ltd.)

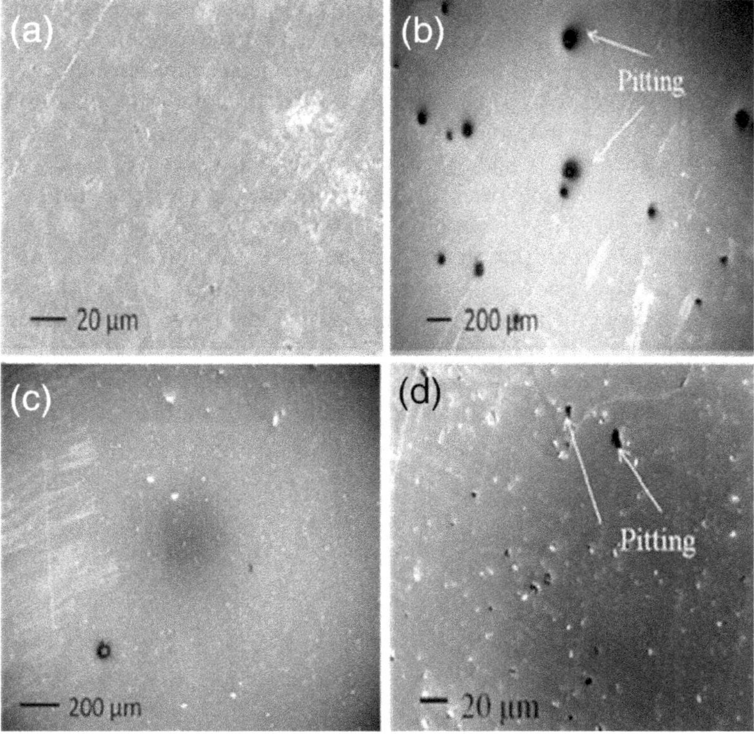

FIGURE 15.8 The FESEM images of (a, b) uncoated SS 304 and (c, d) Ti-Si-B-C coated surface for 1 day and 70 days exposure in 3.5 wt% NaCl, respectively. (Reproduced from [14] by permission of Elsevier Science Ltd.)

presence of β -cyclodextrin and utilized as hybrid coatings for metallic substrates. The authors analyzed the SEM results in order to study the influence of temperature at the time of synthesis of the core and found that smaller sized nanocontainers (for encapsulation of corrosion inhibitors) were formed at room temperature rather than sonic energy. Furthermore, the extent of orderliness of the coating (thickness ~100 nm) on the metallic substrates was also dependent on the dispersion of inhibitors, which was in turn dependent on the mixing conditions [25]. Similarly, Feng et al. [26] also studied the effect of doped concentration of benzotriazole-encapsulated nanocontainers on its dispersity on the metal substrate and found that the highest concentration of the doped nanocontainers resulted in uniform dispersion of coating. EDS elemental mapping of the coating surface along with SEM was employed for this purpose [26]. Marathe et al. [27] also utilized SEM for microcapsules' morphological studies. The authors visualized the presence of spherical polyurea microcapsules encapsulated with corrosion inhibitors. The morphology appeared rough owing to the clearly visible certain nanoparticles on the capsules [27].

15.4.2 Application of SEM/FESEM for Evaluating Coating Resistance against Corrosion

The application of SEM is widespread and can be successfully employed to examine the anticorrosive properties of various types of coatings. This can be done by analyzing the SEM images before and after the attack of corrosive species or with and without the coating on metal surface. Figure 15.8(a) and (b) display the FESEM image of uncoated stainless steel surface, while Figure 15.8(c) and (d) display the images of Ti-Si-B-C nanocomposite coating in 3.5 wt% NaCl at 20 and 200 µm, respectively. The coated surface (10 μm thick, confirmed via cross-sectional SEM analysis) was seen to remain intact, with no significant signs of corrosion even after 70 days of immersion, whereas the uncoated surface showed large pits distributed all over the metal substrate [14].

Jamshidnejad et al. [28] developed a self-healing smart coating (epoxy and PU-based) with a green corrosion inhibitor extracted from olive leaves encapsulated within it. The self-healing capacity of the coating doped with 3, 4 and 5% of inhibitor capsules was identified through analytical imaging (SEM). It was realized that after one day immersion of the scratched coated samples in 0.6 M NaCl solution, the healing agent from the submicron capsules got released, which effectively healed the crack. In high-resolution image of the metal substrate at the location of the artificial crack, no evidence of corrosion or corrosion product was found. The capsules broke open and chemically reacted to recover the damage caused to the coating. Further, out of 3% and 5% concentration, best inhibition results were displayed by the higher concentration of capsules in the coating (see Figure 15.9) [28]. Another research study reported by Ji et al. [29] also worked on similar lines and reported the self-healing ability of synthesized core-shell nanofibers encapsulated with oleic

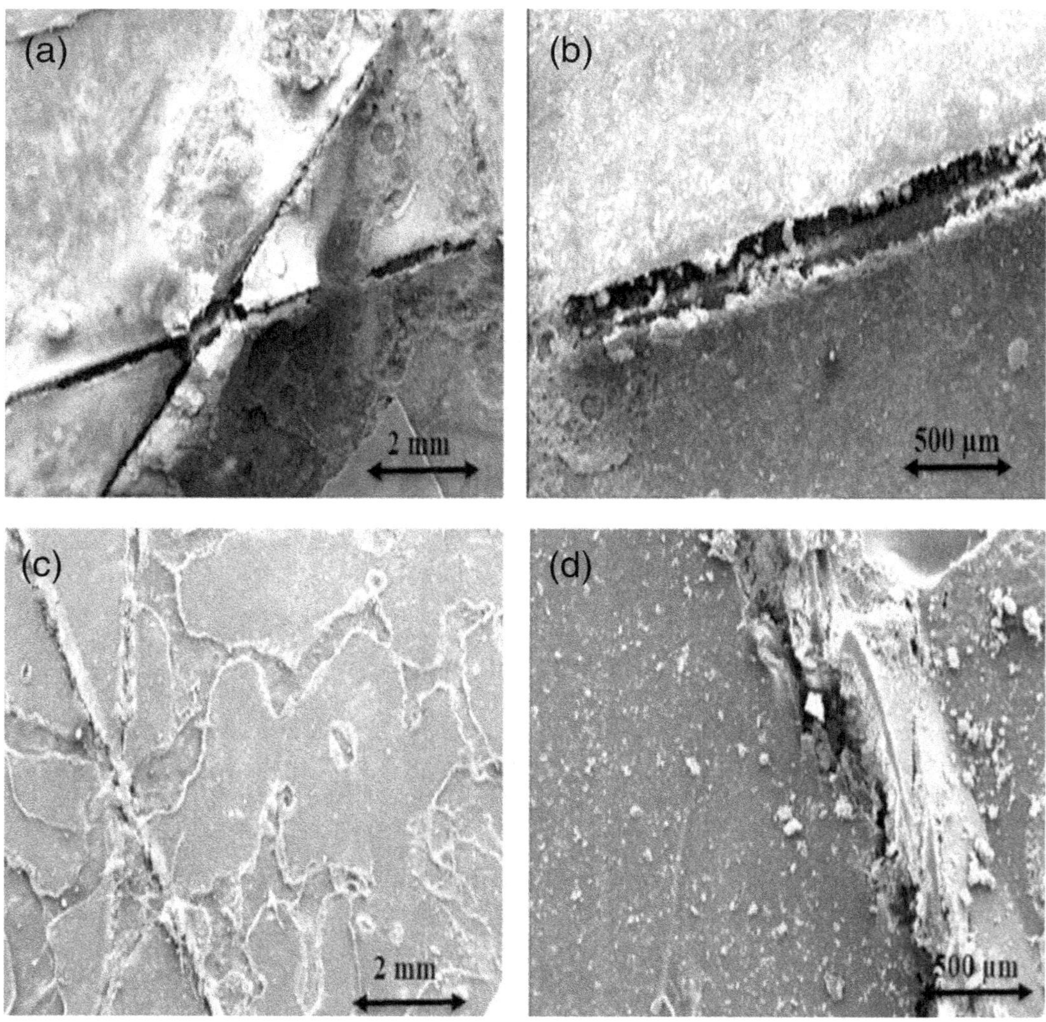

FIGURE 15.9 SEM images of scratch on epoxy coating and its healing: (a, b) 3 wt% and (c, d) 5 wt% submicron capsules. (Reproduced from [29] by permission of Elsevier Science Ltd.)

acid and 2-mercaptobenzimidazolefor Q235 steel coupons against saline corrosive media via SEM and electrochemical tests [29].

Hamdy et al.[30] employed an environmentally friendly chemically convertible coating based on vanadia for utilization to enhance the corrosion resistance of a recently developed magnesium AZ31 HP-O alloy. The effect of pH and concentration of vanadia was studied via OM, SEM and electrochemical techniques before and after the immersion of coated Mg alloy substrates in NaCl corrosive media. The results clearly revealed that micro cracks develop with increasing pH (from 7 to 9) and adversely affect the corrosion resistance performance of the coating. Furthermore, the vanadia-based coating reflected self-healing properties, which was evinced by the optical micrographs taken at various times [30]. Several other researchers had adopted a similar methodology to confirm the anticorrosive properties of coatings revealing a smoother and less damaged surface [1, 2, 5, 6, 23, 26]. Decreased thickness of the corrosion scales was also confirmed by cross-sectional analysis of coated metallic substrates via SEM/ FESEM.

15.5 TRANSMISSION ELECTRON MICROSCOPY (TEM)

In TEM, a vacuum environment is used to bombard a thin solid specimen, typically less than 200 nm in thickness, with a highly focused and monoenergetic electron beam. The energy of the beam is sufficient to ensure its propagation through the specimen. Subsequently, a succession of electromagnetic lenses is employed to amplify the transmitted electron signal. The diffracted electrons manifest as a diffraction pattern beneath the specimen, which provides valuable insights into the material's atomic structure. The transmitted electron images, acquired from localized regions of the sample exhibiting contrast resulting from diverse scattering mechanisms originating from electron interactions with the atomic constituents, enable analysis of both the atomic structure and the presence of defects within the material.

It does not identify the elemental composition and the testing can be sometimes slightly destructive in nature. The requirements of sample preparation are similar to SEM (conducting metal or coating insulator). Alongside, most commonly used sample specification is 3 mm in diameter and <200 mm thickness in the center in particular.

15.5.1 Application of TEM for Coating Characterization

TEM is very useful in studying the layered characteristics of the coating material. Multilayering can be distinguished via darker and lighter shades of black and grey in the resultant TEM images. A few significant researches that employed TEM for coating characterization are discussed in this section. Bagale et al. [7] prepared nanocapsules with dibutyl phthalate (DBT) as corrosion inhibitor encapsulated in urea formaldehyde shell. The stability, size and aggregation of capsule encapsulation within an epoxy matrix, as well as the thickness of the shell material's morphology, were thoroughly investigated utilizing FESEM and TEM imaging techniques. Figure 15.10 shows the SEM and TEM images taken by the authors. The images display spherical shaped thick polymer shells with corrosion inhibitor enclosed. The TEM images clearly revealed that the capsule consisted of two layers: an outer shell and an inner smooth layer that safeguarded the core, which was DBT. The average thicknesses of the outer and inner walls were 250–270 nm and 80–100 nm, respectively. The average capsule size was measured as 330–350 nm. The concentration of the nanocapsules encapsulated in epoxy-amine-based coating was optimized and confirmed via electrochemical test methods. The study stated that using nanocapsules with a polymeric shell to prevent corrosion was more effective than using regular polymeric coatings alone [7].

Yang et al.[31], on the other hand, employed high velocity oxygen fuel technique to deposit Ni-Mo coating on cast iron substrate to increase its corrosion resistance to HCl. Thereafter, the coating was annealed to change its microstructure and enhance the corrosion resistance. The microstructural

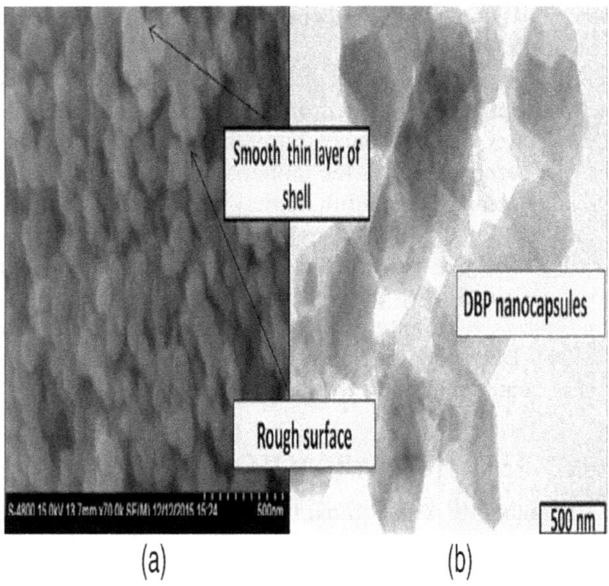

FIGURE 15.10 (a) FESEM and (b) TEM images of the DBT-based nanocapsules. (Reproduced from [7] by permission of Elsevier Science Ltd.)

evolution of the coating with and without the annealing process was investigated using OM, SEM and TEM equipped with EDS. In the study, TEM was used extensively: (i) for analyzing the surface of as-sprayed Ni-Mo alloy coating, (ii) to analyze the 'near surface' features of the Ni-Mo alloy coating annealed at 850°C and (iii) to analyze the 'interfacial features' (between cat iron substrate and the coating) of the Ni-Mo alloy coating annealed at 850°C. Signs of significant plastic deformation can be observed in the deposited particles, characterized by the presence of Zone 1: where dislocation density is high, and Zone 2: where grains are elongated (shown in Figure 15.11(a)). The post-spray annealing treatment notably influenced the microstructural evolution of Ni-Mo coating, such as interfacial bonding between deposited particles, coalescence of micro-voids, phase transformation within the coating, as well as the element diffusion reactions from the substrate into the coating, resulting in the highest level of corrosion resistance [31]. In another study, Jiang et al. [12] attempted to enhance the corrosion resistance capacity of a magnesium alloy (Mg-AZ91) by electrodeposition of graphene coating followed by polymethyltrimethoxysilane coating. The graphene coating was observed carefully though analytical imaging, FESEM, AFM and TEM. Randomly oriented graphene nanoparticles were observed under TEM; alongside, the HRTEM images confirmed that the crystallographic structure of the coating was comprised of hexagonal graphene arrangement [12]. The composite coating of graphene and polymethyltrimethoxysilane showed better corrosion resistance than the sole graphene coating. Furthermore, TEM is also applied to observe the shell and core morphology for the core–shell inhibitors. Ji et al. [29] employed SEM, TEM and confocal fluorescence microscopy to characterize the morphology of the electrospun fibers. Self-healing and pH-responsive coatings for a steel substrate were created by incorporating core–shell electrospun fibers into poly(dimethyl siloxane). These fibers consisted of a cellulose acetate shell and a core composed of oleic acid and alkyd varnish resin, which were synthesized specifically for this purpose. The TEM image very clearly depicted the structure of the fibers with shell and core sizes as 75 and 300 nm (±10 nm), respectively [29].

Izadi et al. [32] prepared Fe_3O_4 nanocontainers which were loaded with a green corrosion inhibitor and the morphology was analyzed via FESEM and TEM imaging. FESEM confirmed

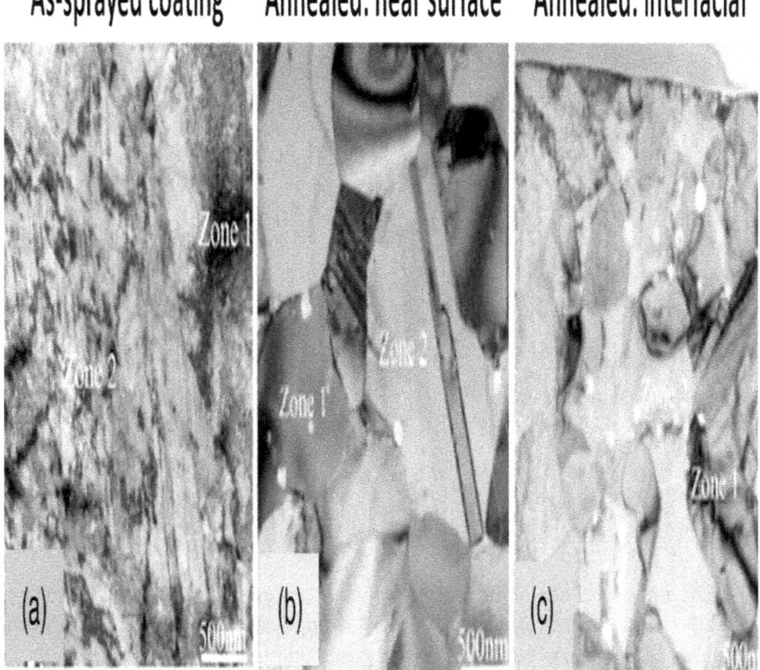

FIGURE 15.11 TEM images of (a) as-sprayed, (b) annealed at 850°C: near surface and (c) annealed at 850°C: interfacial Ni-Mo alloy coating. (Reproduced from [31] by permission of Elsevier Science Ltd.)

the spherical size of the nanocontainers along with its monodispersed form. TEM images, on the other hand, helped to realize that the Fe_3O_4 nanocontainers were surrounded by polyelectrolyte layer. The images also showed a layered structure with bright intensity surrounding the dark Fe_3O_4 nanoparticles, indicating the formation of a shell around the solid core which increased the average size of the nanocontainers [32]. Based on the SEM images, the hollow mesoporous silica particles exhibit a spherical shape. Also, they were observed to be semitransparent, validating their hierarchical hollow structure. Shun et al. [33] prepared hollow mesoporous silica spheres (HMSS) that were redox-responses and were encapsulated with corrosion inhibitors. The morphology was characterized by SEM, while the microstructure was realized via TEM. The SEM image further reveals that the silica spheres possess a uniform morphology, excellent mono-dispersity and consistent particle size distribution. The TEM images, however, particularly highlighted the hollow cavity and mesoporous shell. Additionally, an average diameter of 260 nm for the HMSSs with a mesoporous shell thickness of approximately 50 nm was evaluated from the TEM image [33]. Haddadi et al. [34] synthesized carbon hollow spheres (CHSs) using the silica templating method, enclosed the MBI inhibitor within the CHSs and assessed their effectiveness in inhibiting corrosion against corrosive saline media. The structural morphology of the synthesized silica core/carbon shell was examined using TEM both before and after the removal of the silica core. Figure 15.12(a) shows that the synthesized silica core/carbon shell structures have a spherical shape, which matches the FE-SEM results. The dark regions at the center of the particles are caused by the silica core. In Figure 15.12(b), transparent regions appear at the center, indicating that the silica cores were removed after treatment with an aqueous hydrofluoric acid (HF) solution. From the comparison of Figures 15.12(a) and 15.12(b), we can see that removing the silica cores did not significantly change the original spherical shape of the particles [34].

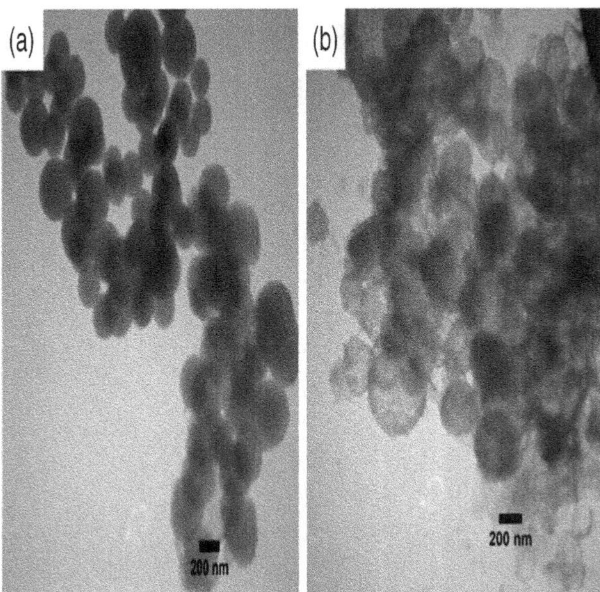

FIGURE 15.12 TEM images of the shell morphology (a) before and (b) after the removal. (Reproduced from [34] by permission of Elsevier Science Ltd.)*Note*: The application of TEM for confirming the corrosion-resistant characteristics of the coatings is not extensively used and hence not presented here.

15.6 ATOMIC FORCE MICROSCOPY

AFM, unlike STM, does not require oxide-free, electrically conductive surfaces for measurement (repulsive as well as attractive). AFM is a type of scanning probe microscopy that uses a micromachined cantilever with a sharp tip to perceive the deflections caused by interatomic forces on the measured surface and is sometimes called as scanning force microscopy. AFM images surfaces by measuring the force between the tip and the sample. The interatomic forces between the sample and the microfabricated cantilever cause its deflection. This deflection, in turn, can be quantified either by measuring the magnitude of the tunneling current or by monitoring the displacement of the cantilever. The sensitivity of these measurements to the separation between the surface and the atoms at the tip allows for atomic-scale topographical mapping in all three dimensions. This becomes the principle of AFM imaging. Different modes of operation, such as contact, tapping, noncontact and peak-force tapping, vary in tip motion and tip–sample interaction, with interactions being either attractive or repulsive. The maximum force, frequency and amplitude at the point of contact influences the feedback that is presented.

Further, the dynamics of a vibrating tip near a surface and the tip–surface interactions are generally independent of the specific dynamic method used. However, different experimental quantities, such as amplitude and phase shift in amplitude modulation, and resonance frequency and amplitude in frequency modulation, indicate distinct experimental setups. As a result, the comprehension of tip motion has developed autonomously, leading to a perception that amplitude and frequency modulation methods have only little in common among two distinct scientific groups. The sample to be tested under AFM does not require any typical preparations as solid conductors, semiconductors and insulators are all compatible.

Table 15.2 presents the comparison between different modes of AFM with complete information including sample surface preparation and application.

TABLE 15.2
Comparison between different modes of AFM with complete information including sample surface preparation and application

Mode	Contact Mode AFM	Lateral Force Microscopy (LFM)	Noncontact Mode
Description	Primary and original mode of operation	Measures frictional forces on a surface	Oscillating cantilever in attractive regime, not in direct contact
Tip-surface interaction	Tip scans over the surface topography, lever is deflected	Tip is in contact with the sample surface, lateral twist of the cantilever	Tip is close but not in direct contact with the sample
Imaging techniques	Constant force or constant height	Measures lateral twist of the cantilever	Changes in resonant frequency or amplitude of the cantilever
Feedback control	Tip adjustment to maintain specified deflection (constant force mode)	Lateral deflection signal from the cantilever's twisting motion	Dampening of cantilever oscillation due to tip–sample interaction
Sample surface requirement	Relatively flat surface for feedback control (constant height mode)	-	-
Application	Small, high-speed atomic resolution scans	Imaging areas of higher and lower friction	Imaging with minimal contact forces and high sensitivity
Tip-surface contact	Tip in hard contact with the surface	Tip in contact with the surface	Tip close but not in direct contact with the sample
Lever stiffness	Lever stiffness <1 N/m	-	-

15.6.1 Application of AFM for Coating Characterization

AFM analysis offers valuable data related to the surface morphology and roughness of the coating. The surface roughness obtained from AFM measurements can be effectively linked to the grain size identified through SEM analysis. Karnik et al. [35] employed AFM and SEM analysis on different types of filtration membranes using SEM and AFM. The surface morphology and roughness were evaluated, and their correlations with surface porosity and filtration performance were studied to gain insights into membrane structure. AFM images were examined carefully to study the transitions in surface features and height as membranes underwent sintering and coating processes. The height of features reduced with coating and sintering. The number of features, however, increased with the number of coats; their height remained consistent. Roughness analysis revealed a significant decrease in roughness upon sintering, and the roughness of coated membranes was lower compared to the uncoated membrane. The increase in roughness was attributed to interparticle sintering between iron oxide nanoparticles, indicating the integrity of the iron oxide nanocoating [35]. In a recent study by Uc-Fernandez et al.[36], the roughness analysis of two different coatings (MTES: triethoxymethylsilane-based coating and MTES-HDTMS: MTES functionalized with hexadecyltrimethoxysilane coating) on the surfaces of copper (Cu) and iron (Fe) was conducted via AFM in tapping mode (scanning frequency 0.3 Hz).

Figure 15.13 displays AFM microscopies of coated samples from their study. The MTES-HDTMS copper-coated samples exhibit higher roughness compared to other samples. Roughness values for Cu-MTES and Cu-MTES-HDTMS were 222.5 ± 40.3 nm and 282.8 ± 59.4 nm, respectively, indicating increased roughness in the nanoparticle-containing coating. For Fe coatings, roughness values for Fe-MTES and Fe-MTESHDTMS were 298.4 ± 54.4 nm and 421.8 ± 76.2 nm,

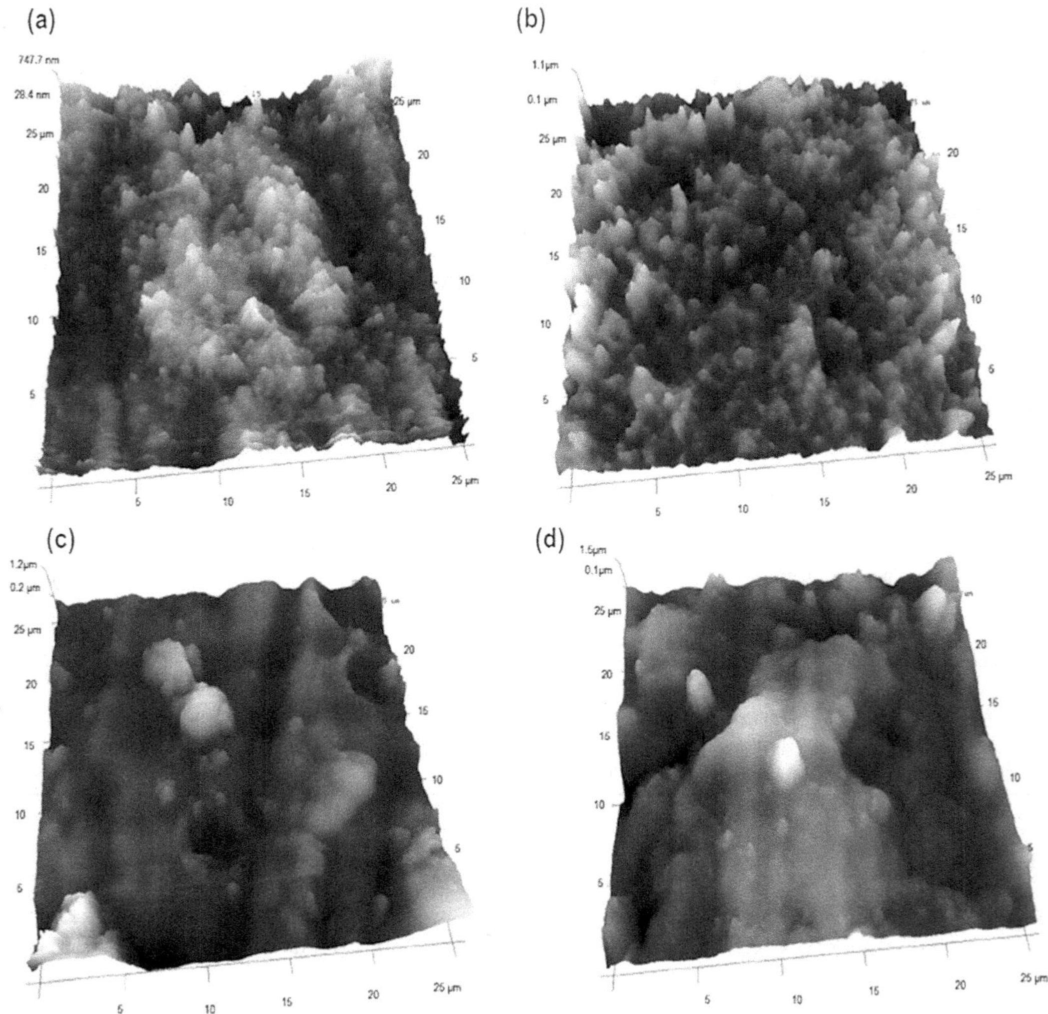

FIGURE 15.13 3D AFM micrographs of (a) Cu-MTES, (b) Cu-MTES-HDTMS, (c) Fe-MTES and (d) Fe-MTESHDTMS surfaces. (Reproduced from [36] by permission of Elsevier Science Ltd.)

respectively. In addition, the images also show localized areas with higher particle conglomerates. In summary, the authors utilized AFM analysis results to relate the double-scale roughness to the superhydrophobic nature of the Cu-MTES-HDTMS sample [36]. Jia et al. also employed AFM and found a two-dimensional nanosheet morphology with lateral size of 1 to 2 μm and coating thickness of nearly 1 nm [37]. A similar study was conducted by Izadi et al. [38], where the authors had employed AFM (3D) to find the roughness measurements and to visualize the topography of the of the neat alkyd coating and coating incorporated with alkyl coating (AFM 3D image obtained in the study is shown in Figure 15.14).

15.6.2 Application of AFM for Evaluating the Coating Resistance against Corrosion

The application of AFM like other techniques can also be employed to study the anticorrosive potential of various types of coatings against saline, acidic or alkaline media.

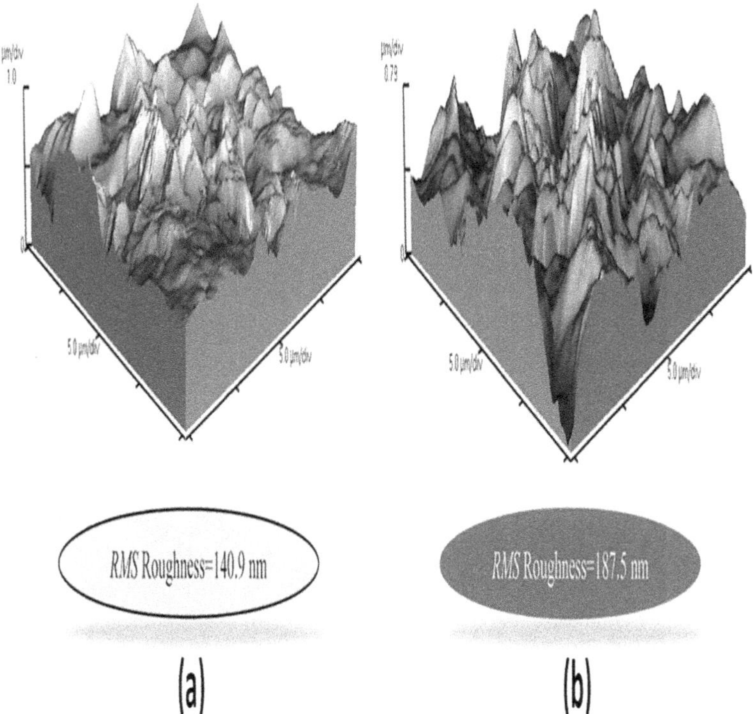

FIGURE 15.14 3D AFM micrographs of (a) neat alkyd coating and (b) coating incorporated with alkyl coating (1 wt %) samples and their corresponding roughness value. (Reproduced from [38] by permission of Elsevier Science Ltd.)

One such example is the study reported by Ma et al. [10] where four different types of coatings developed on Cu surface were immersed in H_2SO_4 in order to prove the coatings' inhibition efficiency through AFM. The samples with no coating displayed an average surface roughness of 280 nm before and after the immersion, while the coated surfaces showed only a change of 4–24 nm, signifying the protective nature of coatings against acidic attack (see Figure 15.15) [10]. In another study reported by Mahato et al., the uniform distribution of coating elements was seen under AFM analysis and no relevant changes in the coating roughness were recorded with the increasing immersion time of coated metal substrate in corrosive 3.5% NaCl solution, signifying the excellent adhesion of coating on metal substrate [14].

15.7 CONTACT ANGLE MEASUREMENT

The penetration of water through the protective polymer coatings is recognized as a significant factor causing adhesion loss, delamination and corrosion beneath the coating. This, in turn, leads to frequent coating replacement and incurs substantial maintenance expenses [39]. The idea of using hydrophobic/superhydrophobic coatings to safeguard steel against corrosion appears to be an attractive solution. These coatings are highly resistant to wetting, and the concept of the self-cleaning lotus leaf effect, which has been observed in Asia for thousands of years, has gained attention relatively recently. It was not until the early 1970s, with the advent of electronic microscopy, that botanist Wilhelm Barthlott began to explore the lotus leaf phenomenon in detail. Since the 1990s, researchers have gained interest in developing new hydrophobic/superhydrophobic

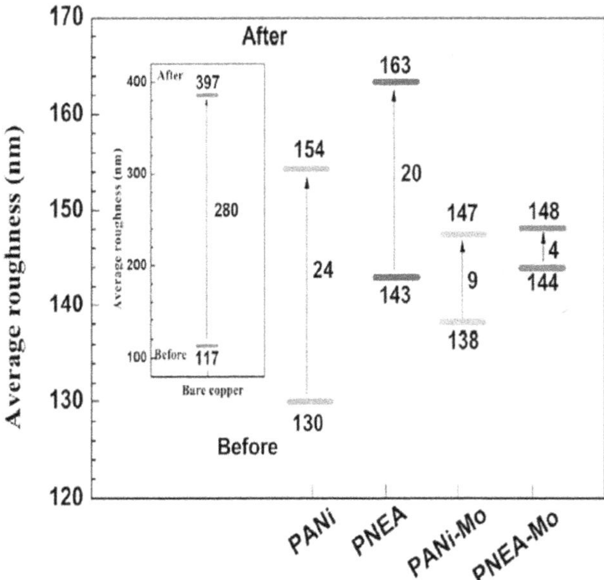

FIGURE 15.15 Average roughness of bare copper and the four different coatings before and after 24 hours of immersion in 0.5 M H_2SO_4 solution at 25 °C. (Reproduced from [10] by permission of Elsevier Science Ltd.)

surfaces due its large-scale industrial applications. There are various processes through which such films can be developed, including the newly emerging nanotechnology. For the superhydrophobic surfaces, two properties must exist concurrently: physical creation of roughness and chemical creation of low energy of the coating material. The combination of these two develops an air layer over the coating surface, further resulting in an air–water interface. This interface development produced a large contact angle (CA), typically >150°. The change in the boundary conditions transforms the partial slip condition to no-slip condition [40]. For the analysis of such types of coatings, CA measurement technique can be indirectly used to determine the hydrophobic/hydrophilic nature of the coated surface. The angle that is made between the edge of any droplet and the coated surface is called as the CA. The CA that will be made is dependent upon the properties of both, the droplet and the coating. According to Mishra et al. [41], the tension between the formed interfaces (γ_{sa}: solid–air, γ_{sl}: solid–liquid and γ_{la}: liquid–air) can be used to determine the CA (in terms of cosθ) of an ideally smooth and homogeneous surface as per equation 15.1.

$$\cos\theta = \frac{\gamma_{sa} - \gamma_{sl}}{\gamma_{la}} \tag{15.1}$$

In reality, no surface can be ideally smooth, rather it possesses some degree of deformations, defects and irregularities that can be said as roughness of the surface. These surfaces are referred to as chemically heterogeneous surfaces. According to Manoj et al. [42], the droplet of water will percolate through the irregularities of the surface and it also introduced a roughness factor (r). Hence, equation 15.1 was modified with θ as the ideal CA and θ_w as the actual CA (see equation 15.2). The value of r is always ≥1. Equations 15.1 and 15.2 coincide only when $r = 0$, i.e., for an ideally smooth surface.

$$\cos\theta_w = r.\frac{\gamma_{sa} - \gamma_{sl}}{\gamma_{la}} = r.Cos\theta \tag{15.2}$$

TABLE 15.3
Summary of the methodology and findings conducted on contact angle measurement by two prominent researchers

Researcher	Methodology	Findings
Bico et al.	Modeled regularly textured superhydrophobic surfaces with spikes, holes, and stripes	All surfaces exhibited contact angles in accordance with the Cassie-Baxter prediction, regardless of the roughness factor. The main parameter determining the contact angle was the solid or air fraction. Roughness alone did not significantly affect the contact angle.
Patankar	Investigated modelling of hydrophobic contact angles on rough surfaces with square posts	Found that minimizing energy during the drop forming process resulted in excellent superhydrophobic surfaces with high pillars. Contrary to Bico's findings, Patankar suggested that using high pillars at a given pillar size was preferable. Further periodic spacing between the pillars made the substrate insensitive to wetted or composite contact formation.

The surface is hydrophobic if $\theta > 90°$ and it is hydrophilic if $\theta < 90°$. Under certain roughness conditions, when $\theta > 90°$, the water droplet is considered to be situated upon a composite surface. In this model, specifically in the heterogeneous regime, the liquid is held above the surface bumps and doesn't seep into the recesses of the surface features. This is illustrated in Figure 15.13(c); the interface between the liquid and the surface is made up of two different materials, labeled as 1 and 2, with surface fractions represented by $f1$ and $f2$, respectively. The CA can be determined by equation 15.3 (two CA for two surfaces as θ_1 and θ_2).

$$\cos\theta_b = f_1.\cos\theta_1 + f_2.\cos\theta_2 \tag{15.3}$$

where f value ranges from 0 to 1 ($f = 1$ being the perfectly flat surface).

Several other researchers also attempted to analyze which formula/modified formula could be applied in different conditions for assessing the nature of the coated surface. Two important works are discussed hereunder in Table 15.3.

CAs are commonly determined using optical or force tensiometers. Optical methods allow for the measurement of static, dynamic and roughness-corrected CAs, while force tensiometry is limited to dynamic measurements. Optical methods encompass the sessile drop (static), needle method, tilting (dynamic) and meniscus methods. The force-based measurement is known as the Wilhelmy method. Table 15.4 provides a brief description of each method. The optical tensiometer is the main method for CA measurement since it allows for both static and dynamic measurements. It also provides the ability to assess sample homogeneity by measuring contact angles at multiple locations. In contrast, the Wilhelmy method calculates the average contact angle over the whole immersed area and requires homogeneity on both sides of the sample.

15.7.1 Application of CA for Coating Characterization

In a very recent study, Cai et al. [3] coated the surface of a Mg alloy via micro arc oxidation process (MAO) followed by immersion of the Mg substrates in $Nd(NO_3)_3$ and PVA + $Nd(NO_3)_3$ solutions at 50°C to seal the MAO coating. The former was referred to as NMAO and the latter as PNMAO. For the coating morphology characterization, macroscopic visual examination using a digital camera and microscopic analysis using FESEM technique were carried out. Furthermore, the hydrophobicity of different layers was analyzed via CA measurement technique as shown in Figure 15.16. Out of the three coatings analyzed, the order of CA is PN-MAO>N-MAO>MAO; it signifies that PN-MAO coating has

TABLE 15.4
Brief description of methods applied for the determination of contact angle

S.No.	Method	Principle	Advantages	Limitations
1.	Sessile drop	Droplet placed on the surface; image recorded	Most commonly used, allows roughness correction	Limited to static measurements
2.	Needle method	Droplet volume increased/ decreased while measuring contact angle	Measures both advancing and receding angles	Requires precise control of droplet volume
3.	Tilting method	Substrate gradually tilted to measure advancing and receding angles	Can determine roll-off angle and hysteresis	Requires precise tilting control and equipment
4.	Meniscus method	Sample immersed in liquid, meniscus formed and contact angle measured	Suitable for thin objects like rods and fibers	Wetting conditions differ from static/dynamic measurements, limited to angles below 90°
5.	Wilhelmy method	Force tensiometry measures mass affecting balance, contact angle calculated	Measures dynamic contact angles	Sample must be homogeneous on both sides

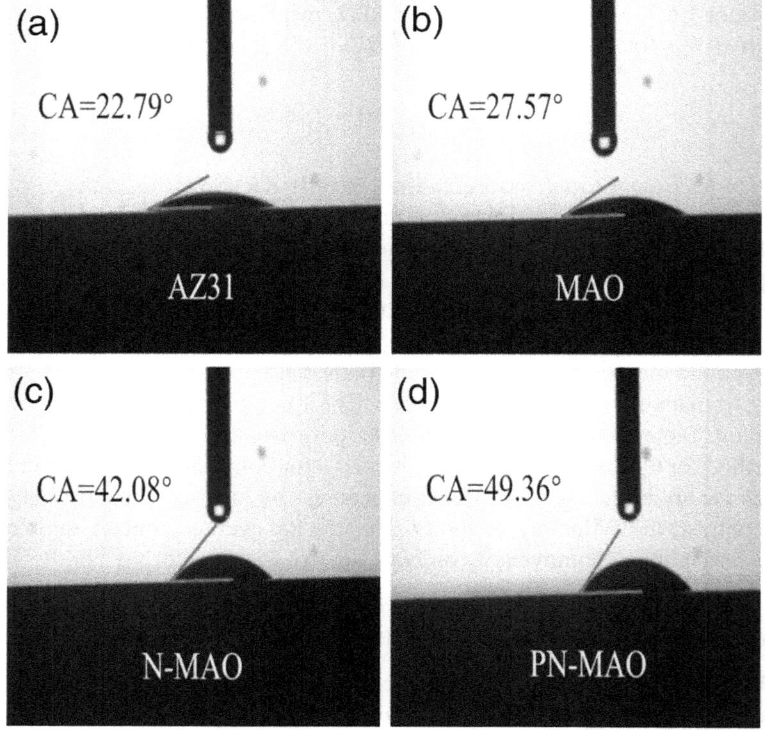

FIGURE 15.16 Contact angle of (a) AZ31-Mg matrix, (b) MAO, (c) N-MAO and (d) PN-MAO coatings with water. (Reproduced from [3] by permission of Elsevier Science Ltd.)

the highest contact angle; hence, the water repelling characteristics developed by the hybrid coating are the highest. The hybrid coating was observed to be more adherent, compact and thick than any other counterparts. Even from the salt spray and polarization test results, it was concluded that PN-MAO showed very high corrosion-resistant abilities (very low corrosion density and high R values were recorded after 360 hours). The PN-MAO coating additionally displayed self-healing characteristics. The complexes of sealant and Mg developed certain carbonate compounds (water insoluble Ng-carbonates) that healed the scratch marked on the Mg substrate and provided prolonged corrosion protection to the alloy [3]. Malik et al. [8] used the electroless co-deposition technique on AZ91 Mg alloy for developing a coating composed of (3-glycidyloxypropyl) trimethoxysilane (GPTMS) and graphene oxide (GO). In this process, the GO sheets were modified with silanol groups and then grafted onto the surface. The application of the GPTMS-GO coating on the Mg alloy involved immersing the sample in a solution containing GPTMS silane with grafted GO, followed by curing in an oven at 120°C. For analyzing the coating characterizations, water CA (WCA) measurements, SEM-EDS analysis was carried out. WCA of 39°, 97° and 108° was measured for bare Mg alloy, GPTMS coating and GPTMS-GO coating, respectively. The GPTMS-GO coating was found to be hydrophobic in nature. Furthermore, as there was an increase in water WCA results into low water penetration ability of water into the coatings, further resulting into highly corrosion-resistant nature of the coated alloys, which was evinced by the authors through electrochemical tests [8]. Superhydrophobic corrosion inhibitor intercalated Mg–Al layered double hydroxide coating (Mg-Al LDH) was prepared via anodization and hydrothermal treatment for AZ31 Mg Alloy against 3.5% NaCl corrosive media by Wang et al.[43]. Lamellar aggregation, nest-like formations and thin and abundant sheet coverings are the different morphologies observed with LDH-coated metal substrate modified with different chemicals to make the coating more hydrophobic. The coatings covered the metal's surface homogeneously as affirmed by EDS analysis [43]. The WCA were in the range of 139°–145°, signifying the highly hydrophobic nature of the coatings. Further, a review by Fihri et al. [44] on the progress of superhydrophobic coatings for steel substrates emphasized on the application of CA measurement to analyze the coating characteristics along with other morphological and compositional techniques.

15.7.2 Application of CA for Corrosion-Resistant Characteristics of Coatings

Feng et al. [2] synthesized dithiane molecules for Cu substrates against acidic corrosive environment (0.5M H_2SO_4) by monolayer development method. The coating was characterized morphologically via SEM and CA. A highly rough surface with evident corrosion pits was observed on the samples with no coating. The CA was 57.7°. However, when the coated surface was analyzed, a smoother surface with an increased CA of up to 103.2° was noted, indicating the increased hydrophobic characters [2]. Figure 15.17 shows the surface of an aluminum substrate after coating; the cross-sectional image and lateral image are shown in the figure. The thickness of the coating was measured as nearly 55 μm. After 30 minutes immersion of the coated aluminum in NaCl media, a WCA of 153° was recorded, implying that the surface behaved to be highly hydrophobic and had the potential to repel water molecules from the metal substrate's surface, preventing the probability of occurrence of corrosion. The study was carried out by Liang et al. [45]. Although, with the increase in immersion time, the WCA kept on decreasing, indicating some loss of hydrophobic nature with time, but even after 1,440 minutes, the angle recorded was 138.1°, which was still in the hydrophobic range (i.e., >90°). The highly anticorrosive nature of the film coating was evinced by electrochemical tests also [45].

Ammar et al. [46] carried out a research focused on incorporating ZnO nanoparticles at different concentrations (ranging from 1 to 8% by weight) into a hybrid acrylic-silicone polymer-based matrix. The aim was to create nanocomposite coatings with outstanding anticorrosion and hydrophobic properties. FESEM and WCA were utilized for the morphological and wettability characteristics of the developed coating. The coating system incorporated with 3% ZnO nanoparticles exhibited

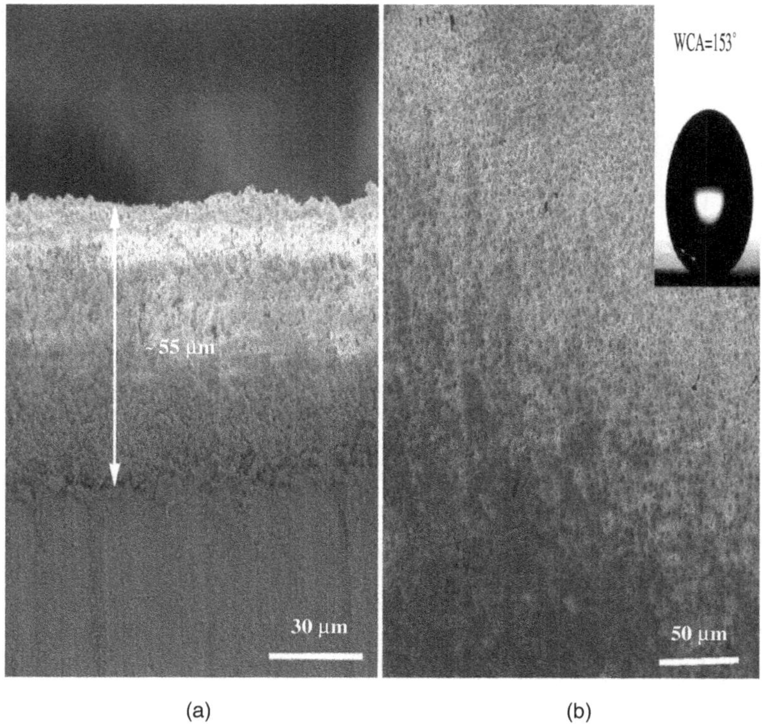

FIGURE 15.17 (a) Aluminum substrate surface with coating and (b) structure after immersion in NaCl for 30 minutes. (Reproduced from [45] by permission of Elsevier Science Ltd.)

the strongest anticorrosion properties (which were proved by the electrochemical test results) and achieved a high CA value of 95.6° [46].

15.8 CONCLUSIONS

To summarize, the techniques mentioned in this chapter have been used over the last one and half decades to characterize the anticorrosive coatings on metals and their alloys. More importantly, it is critical to analyze the synthesis process of the encapsulation of corrosion inhibitors into different coatings, as this is the ruling parameter that will determine the anticorrosive characteristics of the coating once coated on the metal/alloy. Another point that is worth mentioning is that, in order to fully characterize the inhibitor-loaded capsules/coating materials, a systematic and step-wise employment of multiple techniques is recommended. Each technique provides us with a unique characteristic of the coating to be characterized, and hence their employment.

REFERENCES

[1] A. Sun, G. Cui, Q. Liu, Capsule corrosion inhibitor loaded with hyperbranched chitosan: Carbon dioxide corrosion protection for downhole pipelines in oil fields, Colloids Surf. A Physicochem. *Eng. Asp.* 664 (2023) 131106. https://doi.org/10.1016/j.colsurfa.2023.131106

[2] L. Feng, S. Zheng, H. Zhu, X. Ma, Z. Hu, Detection of corrosion inhibition by dithiane self-assembled monolayers (SAMs) on copper, *J. Taiwan Inst. Chem. Eng.* 142 (2023) 104610. https://doi.org/10.1016/j.jtice.2022.104610

[3] L. Cai, X. Song, C.B. Liu, L.Y. Cui, S.Q. Li, F. Zhang, M. Bobby Kannan, D.C. Chen, R.C. Zeng, Corrosion resistance and mechanisms of Nd(NO3)3 and polyvinyl alcohol organic-inorganic hybrid material incorporated MAO coatings on AZ31 Mg alloy, *J. Colloid Interface Sci.* 630 (2023) 833–845. https://doi.org/10.1016/j.jcis.2022.10.087

[4] N. Boshkova, K. Kamburova, T. Radeva, S. Simeonova, N. Grozev, N. Boshkov, Hybrid zinc coatings with chitosan/alginate encapsulated CuO-nanoparticles for anticorrosion and antifouling protection of mild steel, *Coatings.* 13 (2023) 895. https://doi.org/10.3390/coatings13050895

[5] T.A. Saleh, M. Satria, M.M. Nur, N. Aljeaban, B. Alharbi, Synthesis of vinyl trimethyl silane and acrylic acid modified silica nanoparticles as corrosion inhibition protocols in saline medium, *Fuel.* 339 (2023). https://doi.org/10.1016/j.fuel.2022.127277

[6] M.S. Morad, A.K. El-Dean, , 2006. 2, 2′-Dithiobis (3-cyano-4, 6-dimethylpyridine): A new class of acid corrosion inhibitors for mild steel. *Corr Sci.* 48(11) (2006) 3398–3412.

[7] U. Bagale, A. Kadi, I. Potoroko, V. Rangari, M. Mahale, Ultrasound-assisted dibutyl phthalate nanocapsules preparation and its application as corrosion inhibition coatings, *Karbala Int. J. Mod. Sci.* 8 (2022) 154–168. https://doi.org/10.33640/2405-609X.3218

[8] M.U. Malik, M. Tabish, G. Yasin, M.J. Anjum, S. Jameel, Y. Tang, X. Zhang, S. Manzoor, S. Ibraheem, W.Q. Khan, Electroless codeposition of GO incorporated silane nanocomposite coating onto AZ91 Mg alloy: Effect of GO content on its morphology, mechanical and corrosion protection properties, *J. Alloys Compd.* 883 (2021) 160790. https://doi.org/10.1016/j.jallcom.2021.160790

[9] M. Alinezhadfar, S. Nasiri Khalil Abad, M. Mozammel, Multifunctional cobalt coating with exceptional amphiphobic properties: Self-cleaning and corrosion inhibition, *Surf. Interfaces.* 21 (2020). https://doi.org/10.1016/j.surfin.2020.100744

[10] Y. Ma, B. Fan, H. Liu, G. Fan, H. Hao, B. Yang, Enhanced corrosion inhibition of aniline derivatives electropolymerized coatings on copper: Preparation, characterization and mechanism modeling, *Appl. Surf. Sci.* 514 (2020). https://doi.org/10.1016/j.apsusc.2020.146086

[11] X. Wang, C. Jing, Y. Chen, X. Wang, G. Zhao, X. Zhang, L. Wu, X. Liu, B. Dong, Y. Zhang, Active corrosion protection of super-hydrophobic corrosion inhibitor intercalated Mg–Al layered double hydroxide coating on AZ31 magnesium alloy, *J. Magnes. Alloy.* 8 (2020) 291–300. https://doi.org/10.1016/j.jma.2019.11.011

[12] B.K. Jiang, A.Y. Chen, J.F. Gu, J.T. Fan, Y. Liu, P. Wang, H.J. Li, H. Sun, J.H. Yang, X.Y. Wang, Corrosion resistance enhancement of magnesium alloy by N-doped graphene quantum dots and polymethyltrimethoxysilane composite coating, *Carbon N. Y.* 157 (2020) 537–548. https://doi.org/10.1016/j.carbon.2019.09.013

[13] X. Liu, D. Vonk, H. Jiang, K. Kisslinger, X. Tong, M. Ge, E. Nazaretski, B. Ravel, K. Foster, S. Petrash, Y.C.K. Chen-Wiegart, Environmentally friendly Zr-based conversion nanocoatings for corrosion inhibition of metal surfaces evaluated by multimodal X-ray analysis, *ACS Appl. Nano Mater.* 2 (2019) 1920–1929. https://doi.org/10.1021/acsanm.8b02309

[14] P. Mahato, S.K. Mishra, M. Murmu, N.C. Murmu, H. Hirani, P. Banerjee, A prolonged exposure of Ti-Si-B-C nanocomposite coating in 3.5 wt% NaCl solution: Electrochemical and morphological analysis, *Surf. Coatings Technol.* 375 (2019) 477–488. https://doi.org/10.1016/j.surfcoat.2019.07.039

[15] H. Eivaz Mohammadloo, S.M. Mirabedini, H. Pezeshk-Fallah, Microencapsulation of quinoline and cerium based inhibitors for smart coating application: Anti-corrosion, morphology and adhesion study, *Prog. Org. Coatings.* 137 (2019). https://doi.org/10.1016/j.porgcoat.2019.105339

[16] P.S. Shisode, C.B. Patil, P.P. Mahulikar, Preparation and characterization of microcapsules containing soybean oil and their application in self-healing anticorrosive coatings, *Polym. – Plast. Technol. Eng.* 57 (2018) 1334–1343. https://doi.org/10.1080/03602559.2017.1381248

[17] T.Q. Doan, L.S. Leslie, S.Y. Kim, R. Bhargava, S.R. White, N.R. Sottos, Characterization of core-shell microstructure and self-healing performance of electrospun fiber coatings, *Polymer (Guildf).* 107 (2016) 263–272. https://doi.org/10.1016/j.polymer.2016.10.062

[18] V. V. Gite, P.D. Tatiya, R.J. Marathe, P.P. Mahulikar, D.G. Hundiwale, Microencapsulation of quinoline as a corrosion inhibitor in polyurea microcapsules for application in anticorrosive PU coatings, *Prog. Org. Coatings.* 83 (2015) 11–18. https://doi.org/10.1016/j.porgcoat.2015.01.021

[19] P.D. Tatiya, R.K. Hedaoo, P.P. Mahulikar, V. V. Gite, Novel polyurea microcapsules using dendritic functional monomer: Synthesis, characterization, and its use in self-healing and anticorrosive polyurethane coatings, *Ind. Eng. Chem. Res.* 52 (2012) 1562–1570. https://doi.org/10.1021/ie301813a

[20] A. Phanasgaonkar, V.S. Raja, Influence of curing temperature, silica nanoparticles- and cerium on surface morphology and corrosion behaviour of hybrid silane coatings on mild steel, *Surf. Coatings Technol.* 203 (2009) 2260–2271. https://doi.org/10.1016/j.surfcoat.2009.02.020

[21] C.K. Schoff, Optical microscopy, *Coatings Technol.* 3 (2006) 36–43. https://doi.org/10.4324/9780429399916-21

[22] W. Fan, Y. Zhang, W. Li, W. Wang, X. Zhao, L. Song, Multi-level self-healing ability of shape memory polyurethane coating with microcapsules by induction heating, *Chem. Eng. J.* 368 (2019) 1033–1044. https://doi.org/10.1016/j.cej.2019.03.027

[23] A. meng Zhang, C. Liu, P. sheng Sui, C. Sun, L. yue Cui, M.B. Kannan, R.C. Zeng, Corrosion resistance and mechanisms of smart micro-arc oxidation/epoxy resin coatings on AZ31 Mg alloy: Strategic positioning of nanocontainers, *J. Magnes. Alloy.* 11 (2023) 4562–4574. https://doi.org/10.1016/j.jma.2022.12.013

[24] J. Wu, L. Wu, W. Yao, Y. Chen, Y. Chen, Y. Yuan, J. Wang, A. Atrens, F. Pan, Effect of electrolyte systems on plasma electrolytic oxidation coatings characteristics on LPSO Mg-Gd-Y-Zn alloy, *Surf. Coatings Technol.* 454 (2023) 129192. https://doi.org/10.1016/j.surfcoat.2022.129192

[25] A. Rahimi, S. Amiri, Anticorrosion hybrid nanocomposite coatings with encapsulated organic corrosion inhibitors, *J. Coatings Technol. Res.* 12 (2015) 587–593. https://doi.org/10.1007/s11998-015-9657-4

[26] Y. Feng, Y.F. Cheng, An intelligent coating doped with inhibitor-encapsulated nanocontainers for corrosion protection of pipeline steel, *Chem. Eng. J.* 315 (2017) 537–551. https://doi.org/10.1016/j.cej.2017.01.064

[27] R. Marathe, P. Tatiya, A. Chaudhari, J. Lee, P. Mahulikar, D. Sohn, V. Gite, Neem acetylated polyester polyol-renewable source based smart PU coatings containing quinoline (corrosion inhibitor) encapsulated polyurea microcapsules for enhance anticorrosive property, *Ind. Crops Prod.* 77 (2015) 239–250. https://doi.org/10.1016/j.indcrop.2015.08.054

[28] Z. Jamshidnejad, A. Afshar, M.A. RazmjooKhollari, Synthesis of self-healing smart epoxy and polyurethane coating by encapsulation of olive leaf extract as corrosion inhibitor, *Int. J. Electrochem. Sci.* 13 (2018) 12278–12293. https://doi.org/10.20964/2018.12.83

[29] X. Ji, W. Wang, J. Duan, X. Zhao, L. Wang, Y. Wang, Z. Zhou, W. Li, B. Hou, Developing wide pH-responsive, self-healing, and anti-corrosion epoxy composite coatings based on encapsulating oleic acid/2-mercaptobenzimidazole corrosion inhibitors in chitosan/poly(vinyl alcohol) core-shell nanofibers, *Prog. Org. Coatings.* 161 (2021) 106454. https://doi.org/10.1016/j.porgcoat.2021.106454

[30] A.S. Hamdy, I. Doench, H. Möhwald, Smart self-healing anti-corrosion vanadia coating for magnesium alloys, *Prog. Org. Coatings.* 72 (2011) 387–393. https://doi.org/10.1016/j.porgcoat.2011.05.011

[31] X. Yang, S. Fu, Q. Wang, Q. Sun, J. Zhang, Y. Peng, Z. Liang, J. Li, Protective behaviour of naphthylamine derivatives for steel reinforcement in the simulated concrete pore solutions: Detailed experimental and computational explorations, *J. Mol. Struct.* 1270 (2022) 133898. https://doi.org/10.1016/j.molstruc.2022.133898

[32] M. Izadi, T. Shahrabi, B. Ramezanzadeh, Electrochemical investigations of the corrosion resistance of a hybrid sol–gel film containing green corrosion inhibitor-encapsulated nanocontainers, *J. Taiwan Inst. Chem. Eng.* 81 (2017) 356–372. https://doi.org/10.1016/j.jtice.2017.10.039

[33] S. Sun, X. Zhao, M. Cheng, Y. Wang, C. Li, S. Hu, Facile preparation of redox-responsive hollow mesoporous silica spheres for the encapsulation and controlled release of corrosion inhibitors, *Prog. Org. Coatings.* 136 (2019). https://doi.org/10.1016/j.porgcoat.2019.105302

[34] S.A. Haddadi, S.A.A. Ramazani, M. Mahdavian, P. Taheri, J.M.C. Mol, Fabrication and characterization of graphene-based carbon hollow spheres for encapsulation of organic corrosion inhibitors, *Chem. Eng. J.* 352 (2018) 909–922. https://doi.org/10.1016/j.cej.2018.06.063

[35] B.S. Karnik, M.J. Baumann, S.J. Masten, S.H. Davies, AFM and SEM characterization of iron oxide coated ceramic membranes, *J. Mater. Sci.* 41 (2006) 6861–6870. https://doi.org/10.1007/s10853-006-0943-5

[36] E. Uc-Fernández, J. González-Sánchez, A. Ávila-Ortega, Y. Pérez-Padilla, J.M. Cervantes-Uc, J. Reyes-Trujeque, W.A. Talavera-Pech, Anticorrosive properties of a superhydrophobic coating based on an ORMOSIL enhanced with MCM-41-HDTMS nanoparticles for metals protection, *J. Coatings Technol. Res.* 20 (2023) 347–357. https://doi.org/10.1007/s11998-022-00675-1

[37] W. Jia, H. Cao, T. Wang, Y. Min, Q. Xu, Electrochemical behavior and anti-corrosion property of Ti3C2Tx MXene/LDH heterostructured coating on aluminum alloy, *Surf. Coatings Technol.* 463 (2023) 129551. https://doi.org/10.1016/j.surfcoat.2023.129551

[38] M. Izadi, T. Shahrabi, I. Mohammadi, B. Ramezanzadeh, A. Fateh, The electrochemical behavior of nanocomposite organic coating based on clay nanotubes filled with green corrosion inhibitor through a vacuum-assisted procedure, *Compos. Part B Eng.* 171 (2019) 96–110. https://doi.org/10.1016/j.compositesb.2019.04.019

[39] R.E. Melchers, R. Jeffrey, Corrosion science section modeling of long- term corrosion loss and pitting for chromium-bearing and stainless steels in sea water. *Corrosion* 64(2) (2008) 143–154.

[40] J.H. Kim, P. Kavehpour, J.P. Rothstein, Dynamic contact angle measurements on superhydrophobic surfaces, *Phys. Fluids.* 27 (2015). https://doi.org/10.1063/1.4915112

[41] V.K. Mishra, R. Saini, N. Kumar, A review on superhydrophobic materials and coating techniques, *IOP Conf. Ser. Mater. Sci. Eng.* 1168 (2021) 012026. https://doi.org/10.1088/1757-899x/1168/1/012026

[42] A. Manoj, R. Ramachandran, P.L. Menezes, Self-healing and superhydrophobic coatings for corrosion inhibition and protection, *Int. J. Adv. Manuf. Technol.* 106 (2020) 2119–2131. https://doi.org/10.1007/s00170-019-04758-z

[43] P. Wang, L. Xiong, Z. He, X. Xu, J. Hu, Q. Chen, R. Zhang, J. Pu, L. Guo, Synergistic effect of imidazoline derivative and benzimidazole as corrosion inhibitors for Q235 steel: An electrochemical, XPS, FT-IR and MD study, *Arab. J. Sci. Eng.* 47 (2022) 7123–7134. https://doi.org/10.1007/s13369-021-06540-4

[44] A. Fihri, E. Bovero, A. Al-Shahrani, A. Al-Ghamdi, G. Alabedi, Recent progress in superhydrophobic coatings used for steel protection: A review, *Colloids Surf. A Physicochem. Eng. Asp.* 520 (2017) 378–390. https://doi.org/10.1016/j.colsurfa.2016.12.057

[45] J. Liang, Y. Hu, Y. Wu, H. Chen, Facile formation of superhydrophobic silica-based surface on aluminum substrate with tetraethylorthosilicate and vinyltriethoxysilane as co-precursor and its corrosion resistant performance in corrosive NaCl aqueous solution, *Surf. Coatings Technol.* 240 (2014) 145–153. https://doi.org/10.1016/j.surfcoat.2013.12.028

[46] S. Ammar, K. Ramesh, B. Vengadaesvaran, S. Ramesh, A.K. Arof, Formulation and characterization of hybrid polymeric/ZnO nanocomposite coatings with remarkable anti-corrosion and hydrophobic characteristics, *J. Coatings Technol. Res.* 13 (2016) 921–930. https://doi.org/10.1007/s11998-016-9799-z

16 The Role of Nanomaterials in Developing Corrosion-Resistant Coatings

Bening Nurul Hidayah Kambuna, Cindy Putri Pancaningtias, Fidya Ayuningtyas, and Siska Prifiharni

16.1 INTRODUCTION

The adverse effect of a chemical interaction between a metal or metal alloy and its surrounding environment is corrosion. Metal atoms can be found in chemical compounds or minerals in nature. When corrosion occurs, the same proportion of energy is released as when metals are extracted from minerals. The metal is brought back to its mixed condition via corrosion. Minerals from which the metals were extracted had chemical compounds that were identical to or comparable to those minerals. Corrosion is thus the reverse of extractive metallurgy [1–5].

Corrosion is causing a lot of losses, especially in industrial applications. It is obvious that prevention is the greatest defense against this. Historically, inhibitors have been widely accepted in the industry because of their excellent anticorrosive properties. However, many have emerged as unintended consequences that harm the environment. As a result, the scientific community has started looking for environmentally friendly inhibitors, such as organic inhibitors. Among the various types of inhibitors, organic inhibitors are one [6].

16.2 CORROSION MECHANISM

Corrosion is the term for the attack of corrosion on the metal brought on by a chemical or electrochemical interaction with its environment. Steel buildings can be protected in corrosive situations by organic coatings. Lead or hexavalent chromium compounds were traditionally used as the active pigment in anticorrosive paint formulations [7–9]. The breakdown of materials, typically metals and alloys, as a result of chemical or electrochemical interactions with their surroundings is known as corrosion. The corrosion process usually entails a number of phases, and the precise mechanism can change based on the type of corrosion and the particular materials involved, as shown in Figure 16.1. Red lead and zinc chromates are poisonous substances; as a consequence of growing environmental consciousness and strict national and international legislation, their usage is forbidden [4–8]. These pigments have an impact on the environment and human health during the manufacturing and removal of paint. Combining several elements, including molybdenum, phosphate, calcium, aluminum, cerium, zinc, etc., will help reduce the use of harmful organic pigments. Given that zinc is cationic, another option is to combine zinc with calcium, molybdenum, phosphate, or aluminum. There are differences between the modified anticorrosive pigments in relation to how easily they dissolve in water, inhibit aqueous extracts, and prevent corrosion in organic coatings. Significant anticorrosive capabilities are present in the nano form of zinc molybdate [9]. In contrast to these bulk forms, nanomaterials offer exceptional physical and chemical characteristics [10, 11]. There are three efficient ways to avoid corrosion on metal surfaces cathodic protection, anodic protection, and barrier mechanisms [12, 13]. Since it separates the metal surface from the corrosive

DOI: 10.1201/9781032677255-16

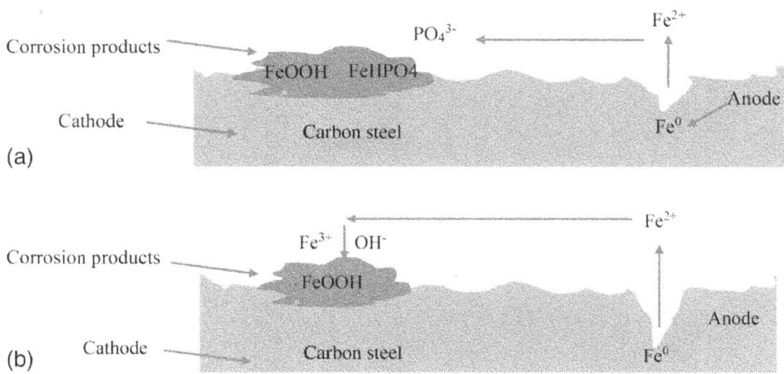

FIGURE 16.1 Depiction of corrosion mechanism.

environment by adding an additional coating of anticorrosive substance, the barrier technique is chosen as a direct corrosion prevention method [14, 15]. This reduces the absorption of corrosive substances chemicals on the metal exterior. A chemical known as a corrosion inhibitor coats the metal surface to form a barrier coating that protects it against corrosion.

16.3 NANOPARTICLES IN CORROSION PROTECTION

It is commonly known that adding nanoparticles can enhance mechanical qualities like scratch or abrasion resistance, as was already noted. Nanoparticle presence needs to be managed in terms of quantity and dispersion. The coating's barrier qualities may be improved by the nanoparticles [16].

Al_2O_3 nanoparticles are chosen to improve conductivity, calcium carbonate ($CaCO_3$) particles are chosen for their affordability, and silicon carbide nanoparticles are used for their hardness, corrosion resistance, and tensile strength. The choice of nanoparticles is dependent on the desired properties.

By adding corrosion inhibitors to inorganic oxide nanoparticles, where their porous sol-gel network architectures were exploited as reservoirs for storage and gradual release of inhibitors, a comparable result was attained. Compared to typical micro-particles, nanoparticles perform better in terms of compactness and mechanical, physical, and chemical resistance due to their small size, unusually high surface energy, and increased number of surface atoms [17–20]. An alternate approach relies on corrosion-inhibiting coatings that are created by either directly integrating corrosion inhibitors into the coatings themselves or adding them to nanocarriers distributed in coatings made of nanocomposite materials [21–38]. Polymeric nanoparticles are made up of structures called nanocapsules, which are made of a polymer shell enclosing a greasy core, or nanospheres, which are made of system with a polymer matrix [39–45]. Coating is the method used most frequently to stop metals from corroding. Although the polymer coating provides long-term resilience to the corrosion, coatings have recently incorporated nanoparticles to enhance their mechanical, chemical, and optical characteristics.

16.3.1 NANOCOATINGS

Nanocoatings consist of tiny components or layers with less than 100 nm thickness. Due to a number of benefits, including interface hardness and adhesive quality, nanocoatings are effectively used to reduce the impact of a corrosive environment due to their long-term and additional high-temperature corrosion resistance, enhancement of tribological characteristics, etc. Nanocoatings can also be applied in thinner, smoother forms [46]. Given that the surface zone for a given mass

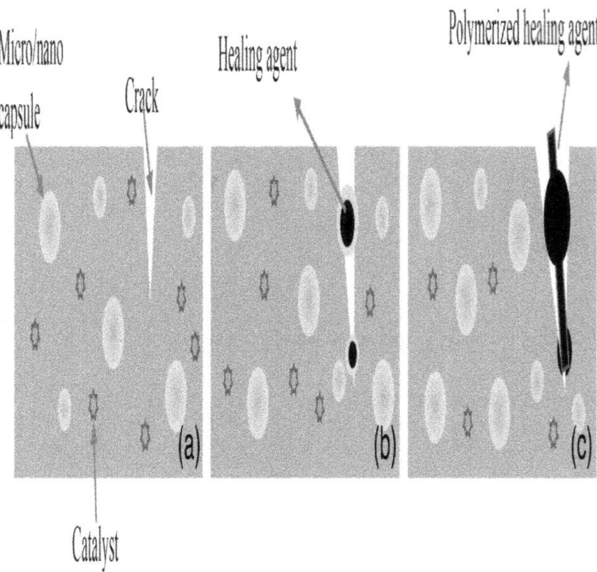

FIGURE 16.2 Depiction of nanoparticle inhibitor.

and the number of active centers rise as the molecule's size decreases, nanoparticles may be stronger corrosion inhibitors. Currently, inhibitors of corrosion and coatings based on nanotechnology perform multiple functions in addition to acting as protective layers against mechanical threats or decorative accents. By incorporating nanoparticles into coatings, a denser, more uniform, and porous covering is produced in comparison to conventional coatings. Nanocomposite coatings have superior corrosion resistance compared to conventional coatings because they are free of pores and defects and function as a barrier against corrosive electrolyte entry [46]. Metal nanoparticles work well as an inhibitor of corrosion (see Figure 16.2). The literature has a large number of publications on the use of metal nanoparticles as corrosion inhibitors [34–46].

When nanoparticles are used to prevent corrosion instead of regular organic inhibitors, a uniform thin film is created on the metal's surface to cover all surfaces flawlessly, offering advantages over regular organic inhibitors. Nano-enhanced coatings have the widest range of commercial applications in any area of nanotechnology. Nanocoatings are perfect for producing scratch-resistant, clear, cost-effective coatings with strong ultraviolet radiation absorption because they have increased durability, strength, optical characteristics, high hardness, low refractive index, and reasonable price. Corrosion is effectively inhibited by the nanomaterials' physical and chemical physisorption/chemisorption on the metal's corroded surface. Active corrosion protection will be improved by using nanoparticles as coating corrosion inhibitors, giving the metal a longer lifespan of corrosion resistance [30]. Coating performance in such environments can be improved using nanoparticles as reinforcements, such as TiO_2 [1–9, 46, 47].

16.3.2 Nanocontainers

For the controlled release of corrosion of mild steel, polyelectrolyte-encapsulated hematite nanoparticles have been used as nanocontainers [18–23, 45]. In contrast, the encapsulated corrosion inhibition technique was used to develop zinc phosphate nanoparticles as a core and modified with polyelectrolyte (polyaniline and polyacrylic acid) for corrosion inhibition applications. One of the most significant achievements of nanotechnology is using nanoparticles to manufacture anticorrosion coatings.

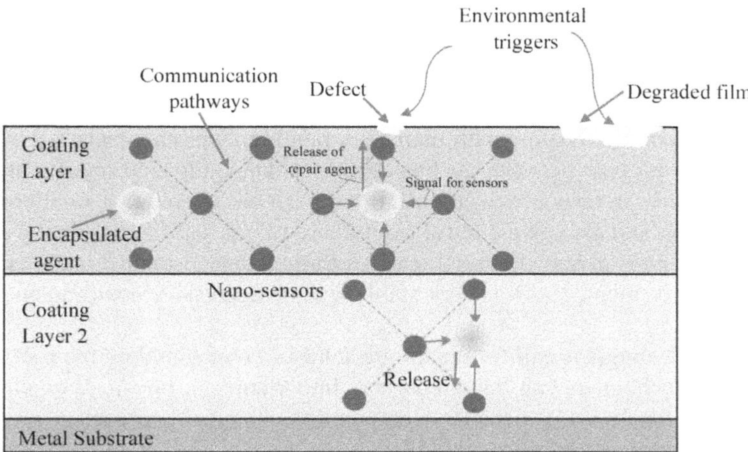

FIGURE 16.3 Illustration of the encapsulated process.

It is commonly known that adding nanoparticles can enhance mechanical qualities like scratch or abrasion resistance, as was already noted.

However, due to the creation of electrolyte channels at the contact point between the nanoparticles and matrix, this placement can also reduce corrosion resistance. As a result, it is crucial to assess how the film's corrosion properties have been altered by the production of silica nanoparticles [45].

16.4 ENCAPSULATION PROCESS

One method for addressing issues brought on by the high solubility of some corrosion inhibitors is encapsulation. The processes of distillation precipitation polymerization, emulsion polymerization, and sol-gel were used to create various kinds of core-shell structured nanocontainers. The following polymerization techniques are frequently employed to create microcapsules or microparticles: interfacial polymerization, in situ polymerization, emulsion polymerization, and solvent evaporation [1, 4–5].

Comparing an inhibited system with and without encapsulation of the inhibitor, encapsulation caused a shift in the kinetics of the inhibitor's release.

Additionally, by delaying water entry, the encapsulated corrosion inhibitors prevent coating delamination during the early stages of corrosion and are not detrimental to the longevity of the coating layer in the electrolyte [14]. The organic coating sheet shields a metal substrate by preventing the movement of corrosive ions and water. As a result, it's crucial that the coating layer and metal have good interfacial adhesion. The barrier protection the polymer film provides is reduced when nanoparticles are introduced to the coating matrix. A concentration of too high nanocapsules might cause the protective performance to suffer. So, the proper concentration range of nanocapsules must be added to the coating layer [26]. The use of (nano)containers to encapsulate corrosion inhibitors is a successful controlled release technique. Although there has increased interest in the use of encapsulated corrosion inhibitors in coating development, there have been few studies on the behavior of inhibitive pigments that leach from organic coatings and the impact of encapsulation. The formulation of the coating with controlled release technology may be improved with the knowledge of how the inhibitive pigments leach from the coating to provide corrosion protection that is more effective and lasts longer [30–31].

For the controlled release of corrosion of mild steel, polyelectrolyte-encapsulated hematite nanoparticles have been used as nanocontainers [1, 50]. In contrast, zinc phosphate nanoparticles were synthesized with ultrasound as a core and functionalized with polyelectrolyte (polyaniline and polyacrylic acid) to develop a nanocontainer for corrosion inhibition applications.

According to the literature, the most effective way for self-healing coatings is the encapsulation method using micro- or nanocapsules and containers. In the encapsulation process shown in Figure 16.3, particles or self-healers with diameters ranging from nanometers to millimeters are embedded. These capsules are between the healer and the damaging environment and act as a barrier, releasing the healers in response to stimuli. It has been discovered that smaller-sized capsules improve the dispersion and also give control over where to put loaded capsules over the polymer matrix [52]. The coating is given additional features through encapsulation, such as longer effectiveness and ease of treatment, and it is kept stable against aggressive environmental degradation [53–55].

All encapsulation techniques call for creating a solid layer surrounding the useful components [56]. Encapsulation techniques can be categorized into chemical, physical, or physicochemical procedures. Each approach has its benefits. Compared to chemical approaches, physical ones are typically quicker and more straightforward [57].

Chemical methods offer the benefit of a finely adjusted capsule structure that is particularly useful for slow-release systems. Physical methods often rely on separate phases of the material used for the shell to ensure that the substance's chemical composition does not change throughout the process. Using chemical procedures, the created shell is a new substance of former reactants [58–60].

16.5 CORROSION INHIBITOR

An inhibitor is a chemical substance that can stop or slow down a chemical reaction. On the other hand, a corrosion inhibitor is a chemical that, when introduced to an environment, can slow down the rate at which that environment attacks a metal with corrosion [58–59]. There can be more than one type of inhibitory mechanism. Several inhibitors prevent corrosion through adsorption, forming an invisible thin layer with a thickness of only a few molecules. Due to environmental influences, some form visible precipitates, defend the metal from attacks that cause corrosion, and produce products that form a passive layer.

16.5.1 Types of Films in Corrosion Inhibitor

Regardless of the reaction that the corrosion inhibitor slows down, it will interact with the metal–solution interface by generating one of three types of films:

a. Passivating film,
b. Precipitation film, or
c. Adsorption film.

Most adsorption protective film inhibitors are organic compounds. They frequently feature a surfactant-like molecular structure, with a group of hydrophilic molecules that can connect with the surface of the metal and a hydrophobic portion of the molecule that protrudes toward the bulk of the solution. Adsorbed inhibitor molecules restrict oxygen transport and access to water on the metal surface, slowing the rate of corrosion [57–60].

The absence of chemical inhibitors like chromates, which are very carcinogenic, is one of the coating's most notable benefits. The excessive consumption of chromates in conventional coatings can result in overconsumption and many environmental risks. High side region, low cost, and the formation of an available surface for proper dispersion should all be considered when choosing a nanoparticle type [61].

The capsules' self-healing capabilities make it simple to embed corrosion inhibitors into them using a variety of approaches. Physical and chemical approaches are the two main subcategories of the self-healing process [4, 13, 62]. Due to its applicability and internal adherence to the corrosion sites, the chemical approach is considered a superior source for corrosion inhibition. Organic polymerization is the chemical approach that is the quickest and most efficient. This technique uses inhibitor-containing trapped microcapsules that are dispersed in a matrix that includes a catalyst that can polymerize the healing agent and release low-viscosity self-healing substances to the damaged area for curing and completing small holes on the metal surfaces.

16.5.2 Mechanism Inhibitor

The following differences exist in the mechanism of action:

1. The inhibitor produces a thin layer on the surface of the metal made up of some inhibitor molecules. Although it is invisible to the naked eye, this layer can prevent the metal from being harmed by the environment.
2. The inhibitor precipitates and then becomes absorbed on the metal surface, preventing corrosion due to environmental factors (such as pH). The amount of precipitation that happens is sufficient for the layers to be visible to the naked eye.
3. The inhibition first causes the metal to erode, producing a chemical compound that creates a passive coating on the metal surface when the corrosion product is absorbed.
4. Inhibitors reduce away. Inhibitors eliminate aggressive elements from their surroundings.

The microcapsules and particles were created to detect the pH changes linked to the beginning of corrosion and respond automatically to release indications and inhibitors for early recognition of corrosion and prevention. Overprotective coatings, trigger mechanisms can be induced autonomously (without human intervention) or non-autonomously (with an external trigger or manual intervention). Most studies focus on intrinsic/automatic responses, which can react without any outside trigger. This inherent response is already attained when utilizing protective coatings that are coated with chromate, which dissolves at low pH. The nanoparticles of the current, as a consequence of pH, do not alter noticeably, and the E_{corr} (corrosion density) only slightly shifts to more negative potentials [25–31].

Inhibitors can change the anodic and cathodic polarization of current depending on the electrochemical characteristics of metal corrosion. Corrosion inhibitors as shown in Figure 16.4 offer the chance to increase the anodic polarization, cathodic polarization, or electrical resistance of the circuit by formation of thin deposits on the metal surface if a corrosion cell can be thought of as consisting of four components, namely, anode, cathode, electrolyte, and electronic conductors. An experimentally obtained polarization curve can be used to observe this mechanism [24].

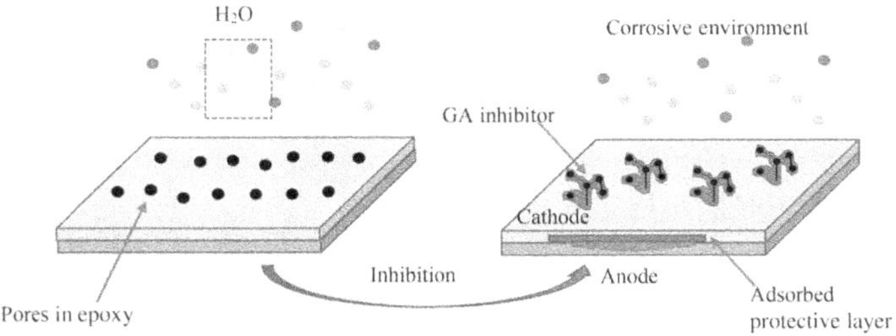

FIGURE 16.4 Schematic depiction of the inhibitor forming defensive layer.

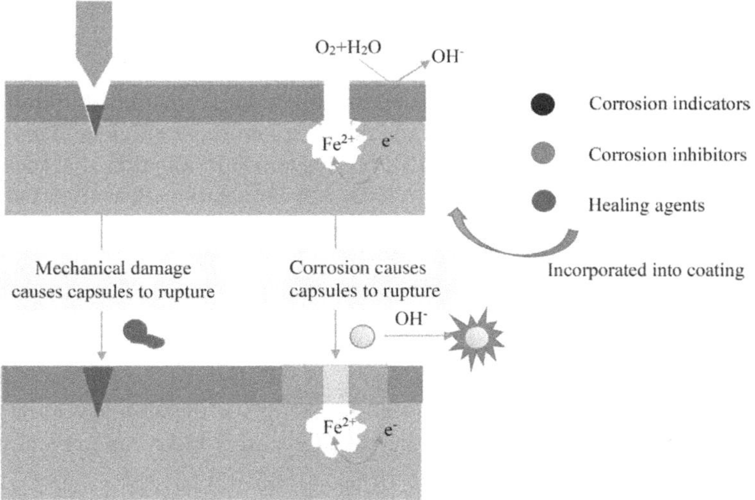

FIGURE 16.5 Illustration of the smart coating.

In addition to their barrier qualities, coatings can also be made more anticorrosive by using corrosion inhibitors. Most chemical inhibitors slow corrosion by creating a layer of passive adsorption on the metal surface [21]. For various metals and alloys, chromate-based inhibitors used in coatings are the most effective anticorrosive technique, suppressing both the anodic and cathodic processes that cause corrosion and metal loss [22]. However, hexavalent chromium has been outlawed due to its high toxicity; as a result, substitute, non-toxic chemical inhibitors are required to replace these extremely effective chromate-based compounds. New chemical and inorganic inhibitors have been created in recent years. Corrosion inhibitors have a cathodic or anodic inhibitory mechanism [3–4]. Ion-exchange pigments (such as cation exchange or anion exchange [52, 58]) and nanoparticles (such as cerium oxide (CeO_2) [12, 57], silica (SiO_2) [33], and zinc oxide (ZnO) [14, 58]) have demonstrated encouraging results as corrosion inhibitors for inorganic materials. Given that several researchers have discussed the inhibiting effect of cerium salts, it is debatable whether nanoparticles have an inhibitory effect on cerium compounds [37–40]. In any case, it would seem that the uniformity with which the nanoparticles are distributed across the polymeric layer affects the anticorrosion performance [35, 40, 43, 65].

The following equation can be used to calculate the inhibitor efficiency:

$$E_f = \frac{R_{i-}R_O}{R_0} \times 100$$

where R_i is the metal's corrosion rate with an inhibitor [49, 49, 51], R_o is the metal's corrosion rate without an inhibitor [63, 64], and E_f is the inhibitor efficiency (percentage) [65].

16.6 SMART COATING

According to one definition, a smart coating has characteristics that react to outside influences, including pressure, wetness, and electricity. A significant difficulty is figuring out how long coatings will last on industrial equipment [55]. Systems that can deliver information regarding a coating's thickness and flaws are referred to as smart coatings. In most cases, they can react to

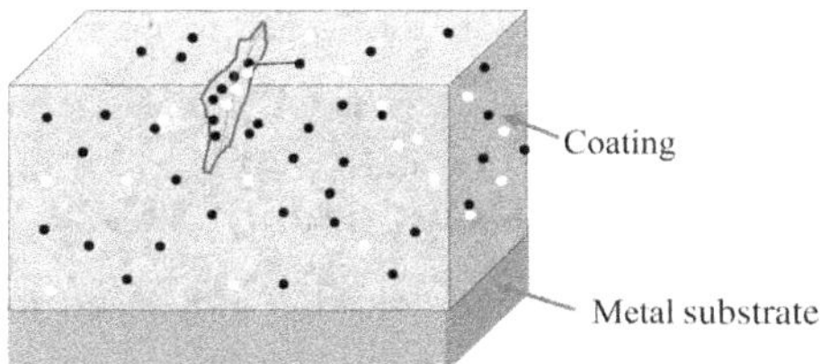

FIGURE 16.6 Nanoparticle migration at a crack in materials and coatings to aid in self-healing.

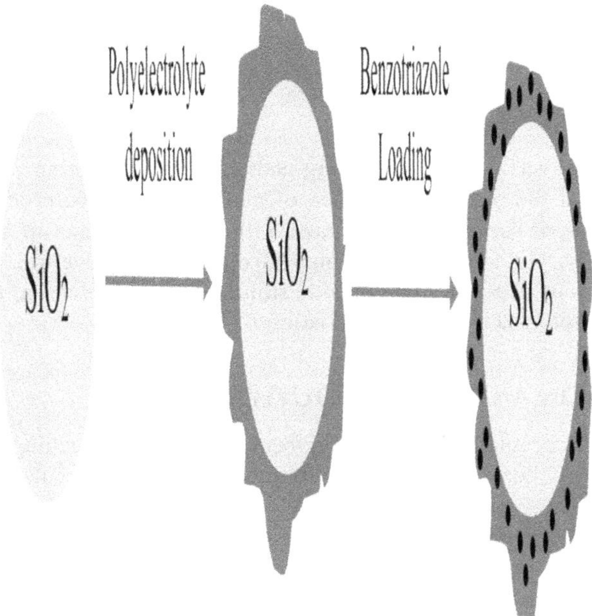

FIGURE 16.7 Mechanism of the smart coating.

external stimuli like heat, stress, strain, or corrosive environmental conditions and heal themselves when hurt.

To address the shortcomings of organic coating layers, smart coating systems like self-healing protection have been created [66]. Over the past decade, there has been significant progress in developing smart coatings (see Figure 16.5). Such coatings can both dramatically extend the durability of structures and provide coated systems with additional capabilities.

The most popular method of preventing corrosion in materials is painting and coating. Due to its exceptional self-healing qualities, which can immediately repair the damage caused by chemical or mechanical stimuli, hexavalent chromate is the most commonly employed ingredient by industry in the manufacture of painting and pigment application, as well as other surface coating processes. Hexavalent chromate is one of the 186 most hazardous compounds, and in addition to its unfavorable effects on the environment, it is also known to cause human cancer when inhaled.

As previously indicated, self-healing smart coatings shown in Figure 16.6 can repair cracks if the immediate environment varies or an outside force damages the active surface. Micro/nanocontainers are frequently filled with polymerizable reagents or inhibitors and then dispersed in a coating matrix to accomplish this capability. It is crucial to monitor and measure the oxidation phenomenon of coatings using materials that only react when there is damage (thin cracks or scratches).

16.6.1 Methods for Evaluation of Smart Coating

The study of smart coatings shows that these coatings may need many clusters and cannot be classified under a single column [17, 18]. The classification of smart coatings was attempted based on a variety of materials, applications, sensitivity, functionality, and complexity level [63, 64]. However, no specific coatings could be placed under these classifications [55, 67].

The key to using nanoparticles for smart coatings is choosing a kind that will temporarily connect with the inhibitors in order to transfer inhibitors into the homogeneous medium and prevent the release of corrosion by-products when the bonds are broken (Figure 16.7).

16.6.2 Mechanism of Smart Coating

Even if nanotechnology was crucial in creating and using smart coatings, there is still a need to make even more use of the distinct properties of nanostructured materials. Smart coatings can recognize changes in their surroundings, engage with them, and react to them while preserving their structural integrity. The intelligent coatings have auto-responsive traits and can dynamically change their properties in response to an outside stimulus. In the corrosion prevention world, this coating type has been referred to as healing coatings [68-74].

16.7 CONCLUSION AND FUTURE OUTLOOK

The coating with the encapsulated inhibitive pigment has a greater leaching rate of inhibitive ions, which results in more effective corrosion inhibition. Pigment encapsulation helped to create a more homogeneous coating microstructure, which provided the coated film with capable barrier qualities during exposure. Distributing sufficient healers throughout the damaged area or fracture over-coating is a difficult problem in smart coatings. This was accomplished successfully using pH-responsive containers and capsules [35, 75].

One efficient/successful development in smart coatings involves the encapsulation of corrosion inhibitors and healers over nanocapsules or containers embedded in conducting polymer. The key to employing nanoparticles for smart coatings is choosing a type that will form brief connections with the inhibitors, allowing for the delivery of inhibitors into the homogeneous medium while preventing the release of corrosion by-products when the bonds are broken. The anti-corrosion impact of the nanofillers in the coating can be studied further using the potentiodynamic-polarization curve [76]

The active chemicals are transferred to the damaged areas when the coating breaks down after being discharged from the container [77]. The repair method of the coating is determined by the properties of these active ingredients. It is difficult to synthesize the micro/nanocontainers. In order to avoid the leaking of the active substance under normal circumstances, they must meet certain parameters, including strong compliance with the matrix of organic matter, large storage ability, and promptly responding following a sense of the commencement of the corrosion [78].

REFERENCES

[1] Ashish Kumar and Abhinay Thakur, "Encapsulated nanoparticles in organic polymers for corrosion inhibition," *Micro and Nano Technologies, Chapter 18*, pp. 345–362, 2020. doi: 10.1016/B978-0-12-819359-4.00018-0

[2] Castaldo, R., de Luna, M. S., Siviello, C., Gentile, G., Lavorgna, M., Amendola, E., and Cocca, M., "On the acid-responsive release of benzotriazole from engineered mesoporous silica nanoparticles

for corrosion protection of metal surfaces," *Journal of Cultural Heritage*, vol. 44, pp. 317–324, 2020. doi: 10.1016/j.culher.2020.01.016

[3] Edurne González, Robin Stuhr, Jesús Manuel Vega, Eva García-Lecina, Hans-Jürgen Grande, Jose Ramon Leiza, and María Paulis, "Assessing the effect of CeO2 nanoparticles as corrosion inhibitor in hybrid biobased waterborne acrylic direct to metal coating binders," *Polymers*, vol. 13, pp. 1–13, March 2021. doi: 10.3390/polym13060848

[4] Hana Choi, Kyoo Young Kim, and Jong Myung Park, "Encapsulation of aliphatic amines into nanoparticles for self-healing corrosion protection of steel sheets," *Progress in Organic Coatings*, vol. 76, pp. 1316–1324, October 2013. doi: 10.1016/j.porgcoat.2013.04.005

[5] Falcon J.M., Batista F.F., and Aoki I.V., "Encapsulation of dodecylamine corrosion inhibitor on silica nanoparticles," *Electrochimica Acta*, vol. 124, pp. 109–118, June 2013. doi: 10.1016/j.electacta.2013.06.114

[6] Judit, T., Abdul, S., and Gyöngyi, V., "Micro/nanocapsules for anticorrosion coatings," In *Fundamentals of Nanoparticles*, pp. 521–551. Elsevier, 2018. doi: 10.1016/B978-0-323-51255-8.00017-3

[7] Jana, N. R., Earhart, C., and Ying, J. Y., "Synthesis of water-soluble and functionalized nanoparticles by silica coating," *Chemistry of Materials*, vol. 19, No. 21, pp. 5074–5082, 2007. doi: 10.1021/cm071368z

[8] JiaJun Fu, Tao Chen, MingDong Wang, NianWang Yang, SuNing Li, Ying Wang, and XiaoDong Liu, "Acid and alkaline dual stimuli-responsive mechanized hollow mesoporous silica nanoparticles as smart nanocontainers for intelligent anticorrosion coatings," *American Chemical Society*, vol. 7, pp. 11397–11408, November 2013. doi: 10.1021/nn4053233

[9] Olivieri, F., Castaldo, R., Cocca, M., Gentile, G., and Lavorgna, M., "Mesoporous silica nanoparticles as carriers of active agents for smart anticorrosive organic coatings: A critical review," *Nanoscale*, vol. 13, No. 20, pp. 9091–9111, 2021. doi: 10.1039/d1nr01899j

[10] Frank, L. A., Onzi, G. R., Morawski, A. S., Pohlmann, A. R., Guterres, S. S., and Contri, R. V., "Chitosan as a coating material for nanoparticles intended for biomedical applications," *Reactive and Functional Polymers*, vol. 147, pp. 104–459, February 2020. doi: 10.1016/j.reactfunctpolym.2019.104459

[11] Lamprakou, Z., Bi, H., Weinell, C. E., Tortajada, S., and Dam-Johansen, K., "Smart epoxy coating with mesoporous silica nanoparticles loaded with calcium phosphate for corrosion protection," *Progress in Organic Coatings*, vol. 165, p. 106740, 2022. doi: 10.1016/j.porgcoat.2022.106740

[12] Tavandashti, N. P., and dan Sanjabi, S., "Corrosion study of hybrid sol–gel coatings containing boehmite nanoparticles loaded with cerium nitrate corrosion inhibitor," *Progress in Organic Coatings*, vol. 69, pp. 384–391, December 2010. doi: 10.1016/j.porgcoat.2010.07.012

[13] Sehrish, H., Fayyed, E., Shakoor, R. A., Kahraman, R., and Abdullah, A., "Improved self-healing performance of polymeric nanocomposites reinforced with talc nanoparticles (TNPs) and urea-formaldehyde microcapsules (UFMCs)," vol. 14, 2021. doi: 10.1016/j.arabjc.2020.102926

[14] Karekar, S. E., Bagale, U. D., Sonawane, S. H., Bhanvase, B. A., and Pinjari, D. V., "A smart coating established with encapsulation of zinc molybdate centred nanocontainer for active corrosion protection of mild steel: Release kinetics of corrosion inhibitor," *Composite Interfaces*, vol. 25, pp. 785–808, February 2018. doi: 10.1080/09276440.2018.1439631

[15] Suba Kannaiyan, Easwaramoorthi, and Andal Gopal, *Corrosion Inhbition Using Nanomaterials-An Overview*, Department of Chemistry, B.S Abdur Rehman University, Chennai, Tamilnadu, pp. 167–170, March 2016.

[16] Zandi-Zand, R., Ershad-Langroudi, A., and Rahimi, A., "Silica based organic–inorganic hybrid nanocomposite coatings for corrosion protection," *Progress in Organic Coatings*, vol. 53, No. 4, pp. 286–291, 2005. doi:10.1016/j.porgcoat.2005.03.009

[17] Zea, C., Barranco-García, R., Chico, B., Díaz, I., Morcillo, M., and De La Fuente, D., "Smart mesoporous silica nanocapsules as environmentally friendly anticorrosive pigments," *International Journal of Corrosion*, vol. 2015, 2015, pp. 1–15. doi: 10.1155/2015/426397

[18] Chen, Z., Yang, W., Chen, Y., Yin, X., and Liu, Y., "Smart coatings embedded with polydopamine-decorated layer-by-layer assembled SnO_2 nanocontainers for the corrosion protection of 304 stainless steels," *Journal of Colloid and Interface Science*, vol. 579, pp. 741–753, 2020. doi: 10.1016/j.jcis.2020.06.118

[19] Zahidahm, K. A., Kakooei, Saeid, Ismail, M. C., and Raja, P. B., "Halloysite nanotubes as nanocontainer for smart coating application: A review," *Progress in Organic Coatings*, vol. 111, pp. 175–185, June 2017. doi: 10.1016/j.porgcoat.2017.05.018

[20] Zheludkevich, M. L., Tedim, J., and Ferreira, M. G. S., "Smart coatings for active corrosion protection based on multi-functional micro and nanocontainers," *Electrochimica Acta*, vol. 82, pp. 314–323, November 2012. doi: 10.1016/j.electacta.2012.04.095

[21] Ahmed A. Farag, *Applications of Nanomaterials in Corrosion Protection Coatings and Inhibitors*. De Gruyter, October 2019. doi: 10.1515/corrrev-2019-0011

[22] Aleksandra Gavrilovic-Wohlmuther, Andreas Laskos, and Erich Kny, "Corrosion inhibitoreloaded smart nanocontainers," In *Corrosion Protection at the Nanoscale*, pp. 203–223, Elsevier, 2020. doi: 10.1016/B978-0-12-819359-4.00012-X

[23] Avila-Gonzalez, C., Cruz-Silva, R., Menchaca, C., Sepulveda-Guzman, S., and Uruchurtu, J., "Use of silica tubes as nanocontainers for corrosion inhibitor storage," *Journal of Nanotechnology*, vol. 2011, pp. 23–48, 2011. doi:10.1155/2011/461313

[24] Sengupta, S., Murmu, M., Mandal, S., Hirani, H. and Banerjee, P., 2021. "Competitive corrosion inhibition performance of alkyl/acyl substituted 2-(2-hydroxybenzylideneamino) phenol protecting mild steel used in adverse acidic medium: A dual approach analysis using FMOs/molecular dynamics simulation corroborated experimental findings". *Colloids and Surfaces A: Physicochemical and Engineering Aspects*, vol. 617, p.126314, 2021.

[25] De Luna, M. S., Buonocore, G., Di Carlo, G., Giuliani, C., Ingo, G. M., and Lavorgna, M., "Protection of bronze artefacts through polymeric coatings based on nanocarriers filled with corrosion inhibitors," In *AIP Conference Proceedings,* vol. 1736, No. 1, p. 020048. AIP Publishing LLC, May 2016. doi: 10.1063/1.4949623

[26] Elshad Abdullayev and Yuri M Lvov, "Clay nanotubes for corrosion inhibitor encapsulation: Release control with end stoppers," *Journal of Materials Chemistry*, vol. 20, pp. 6681–6687, May 2010. doi: 10.1039/c0jm00810a

[26] Lingwei Ma, Jinke Wang, Dawei Zhang, Yao Huang, Luyao Huang, Panjun Wang, Hongchang Qian, Xiaogang Li, H. A. Terryn, and J. M.C. Mol., "Dual-action self-healing protective coatings with photothermal responsive corrosion inhibitor nanocontainers," *Chemical Engineering Journal*, vol. 404, pp. 1–11, January 2021. doi: 10.1016/j.cej.2020.127118

[27] Parijat Burhagohain and Gitalee Shrma, "A descriptive study on the use of nanomaterials as corrosion inhibitors in oil and gas industry," *International Journal of Scientific and Technology*, vol. 8, pp. 3999–4002, November 2019.

[28] Qian, B., Michailidis, M., Bilton, M., Hobson, T., Zheng, Z., and Shchukin, D., "Tannic complexes coated nanocontainers for controlled release of corrosion inhibitors in self-healing coatings," *Electrochimica Acta*, vol. 297, pp. 1035–1041, 2019. doi: 10.1016/j.electacta.2018.12.062

[29] Raja, P. B., Assad, M. A., and Ismail, M., "Inhibitor-encapsulated smart nanocontainers for the controlled release of corrosion inhibitors," In *Corrosion Protection at the Nanoscale*, pp. 91–105. Elsevier, 2020. doi: 10.1016/B978-0-12-819359-4.00006-4

[30] Yuanchao Feng and Y. Frank Cheng, "An intelligent coating doped with inhibitor-encapsulated nanocontainers for corrosion protection of pipeline steel," *Chemical Engineering Journal*, vol. 315, pp. 537–551, May 2017. doi: 10.1016/j.cej.2017.01.064

[31] Zea, C., Alcántara, J., Barranco-García, R., Morcillo, M., and De la Fuente, D., "Synthesis and characterization of hollow mesoporous silica nanoparticles for smart corrosion protection," *Nanomaterials*, vol. 8, No. 7, p. 478, 2018. doi: 10.3390/nano8070478

[32] Zheng Liu, Biao Zhang, Hao Yu, Zhicai Zhang, Wenjuan Jiang, and Zengsheng Ma, "A smart anticorrosive epoxy coating based on graphene oxide/functional mesoporous silica nanoparticles for controlled release of corrosion inhibitors," *Coatings*, vol. 12, pp. 1–13, November 2022. doi: 10.3390/coatings12111749

[33] Atta, M., El-Mahdy, G. A., Al-Lohedan, H. A., and Al Hussain, S. A., "Corrosion inhibition of nanocomposite based on acrylamide copolymers /magnetite for steel," *Digest Journal of Nanomaterials and Biostructures*, vol. 9, pp.627–639, May 2014.

[34] Hoseinzadeh, R., and Javadpour, S., "Formulation of a smart nanocomposite coating with pH-responsive loaded halloysite and investigation of its anticorrosion behaviour," *Indian Academy of Sciences*, vol. 230, pp. 220–230, December 2020. doi: 10.1007/s12034-020-02130-6

[35] Razaghi Kashani, O., Amiri, S., and "Hosseini-zori, M., Self-healing and anti-corrosion nanocomposite coatings based on polyurethane nanocapsules containing mercapto benzimidazole," *Journal of Nanostructures*, vol. 12, No. 3, pp. 726–737, 2022. doi: 10.22052/JNS.2022.03.025

[36] Sarah B. Ulaeto, Jerin K. Pancrecious, and T.P.D. Rajan, "Anticorrosive enamels: Evaluation of nanocomposite additives on enamel efficiency on cold rolled steel," *ASM International*, vol. 30, pp. 4103–4116, June 2021. doi: 10.1007/s11665-021-05881-3

[37] Khoirudin, W.Y., "External corrosion of pipe riser API 5L X52 on tidal zone of offshore poduction platform," *Senatik*, vol. 5, December 2019. doi: 10.28989/senatik.v5i0.363

[38] Bratovcic, A., and Suljagic, J., "Micro-and nano-encapsulation in food industry," *Croatian Journal of Food Science and Technology*, vol. 11, No. 1, pp. 113–121, 2019. doi: 10.17508/CJFST.2019.11.1.17

[39] Cai, H., Wang, P., and Zhang, D., "Smart anticorrosion coating based on stimuli-responsive micro/nanocontainer: A review," *Journal of Oceanology and Limnology*, vol. 38, No. 4, pp. 1045–1063, 2020. doi: 10.1007/s00343-020-0058-x

[40] Dennis, R. V., Patil, V., Andrews, J. L., Aldinger, J. P., Yadav, G. D., and Banerjee, S., "Hybrid nanostructured coatings for corrosion protection of base metals: a sustainability perspective," *Materials Research Express*, vol. 2, No. 3, p. 032001, 2015. doi: 10.1088/2053-1591/2/3/032001

[41] Grigoriev, D., Shchukina, E., and Shchukin, D. G., "Nanocontainers for self-healing coatings," *Advanced Materials Interfaces*, vol. 4, No. 1, 1600318, 2017. doi: 10.1002/admi.201600318

[42] Huige Wei, Yiran Wang, Jiang Guo, Nancy Z. Shen, Dawei Jiang, Xi Zhang, Xingru Yan, Jiahua Zhu, Qiang Wang, Lu Shao, Hongfei Lin, Suying Wei, and Zhanhu Guo, "Advanced micro/nanocapsules for self-healing smart anticorrosion coatings," *The Royal Society of Chemistry*, vol. 3, pp. 469–480, October 2014. doi: 10.1039/c4ta04791e

[43] Feng, Yuanchao. "Intelligent nanocoatings for corrosion protection of steels," In *PRISM Repository*, University of Calgary, 2017. doi:10.11575/PRISM/26592

[44] Nguyen, T. A., and Assadi, A. A., "Smart nanocontainers: Preparation, loading/release processes and applications," *Kenkyu Journal of Nanotechnology Nanoscience*, vol. 4, No. 4, p. 1, 2018.

[45] Pierre Loison, *Development of a Smart Coating Based on Hollow Nanoparticles for Corrosion Detection and Protection*, Materials Université de La Rochelle, April 2019.

[46] Salaluk, S., Auepattana-Aumrung, K., Thongchaivetcharat, K., Lv, L. P., Wang, Y., Viyanit, E., and Crespy, D., "Nanonetwork composite coating for sensing and corrosion inhibition," *Advanced Materials Interfaces*, vol. 7, No. 20, p. 2001073, 2020. doi: 10.1002/admi.202001073

[47] Pearman, B. P., Calle ,L. M., and Zhang, X., "Characterization of encapsulated corrosion inhibitors for environmentally friendly smart coatings," Paper No. 77, September 2015.

[48] Khashayar Saleh and Pierre Guigon, *Coating and Encapsulation Processes in Powder Technology*, Chemical Engineering Department, University of Technology, chapter 7, pp. 323–375, 2007.

[49] Zoi Lamprakou, Huichao Bi, C.E. Weinell, and K. D. Johansen, "Encapsulated corrosion inhibitive pigment for smart epoxy coating development: An investigation of leaching behavior of inhibitive ions," *ACS Omega*, vol. 8, pp. 14420–14429, April 2023. doi: 10.1021/acsomega.2c07853

[50] Heakal, F. E. T., and Elkholy, A. E., "Smart coatings on magnesium alloys in transportation industries," In *Advances in Smart Coatings and Thin Films for Future Industrial and Biomedical Engineering Applications*, pp. 263–287, Elsevier, 2020. doi: 10.1016/B978-0-12-849870-5.00003-3

[51] Li, W., Hintze, P., Calle, L. M., Buhrow, J., Curran, J., Muehlberg, A. J., and Taylor, S. R., "Smart coating for corrosion indication and prevention: Recent progress," *CORROSION 2009,* 2009.

[52] Makhlouf, Abdel Salam Hamdy, ed. *Handbook of Smart Coatings for Materials Protection*, vol. 64. Elsevier, 2014.

[53] Samadzadeh, M., Hatami Boura, S., Peikari, M., Kasiriha, S. M., and Ashrafi, A., "A review on self healing coatings based on micro/nanocapsules," *Progress in Organic Coatings*, 2010, vol. 68, pp. 159–164.

[54] Mohammadloo, H. E., Mirabedini, S. M., and Pezeshk-Fallah, H., "Microencapsulation of quinoline and cerium based inhibitors for smart coating application: Anti-corrosion, morphology". *Progress in Organic Coatings*, vol. 137, p.105339, 2019.

[55] Richards, C. A. J., McMurray, H. N., and Williams, G., "Smart-release inhibition of corrosion driven organic coating failure on zinc by cationic benzotriazole based pigments," *Corrosion Science*, vol. 154, pp. 101–110, 2019. doi: 10.1016/j.corsci.2019.04.005

[56] Montemor, M.F., "Functional and smart coatings for corrosion protection: A review of recent advances," *Surface & Coatings Technology*, vol. 258, pp. 17–37, June 2014. doi: 10.1016/j.surfcoat.2014.06.031

[57] Jing Li, Zhenglin Tao, Jincan Cui, Shuling Shen, and Hanxun Qiu, "Facile fabrication of dual functional graphene oxide microcapsules carrying corrosion inhibitor and encapsulating self-healing agent," *Polymers*, vol. 14, pp. 1–9, September 2022. doi: 10.3390/polym14194067

[58] Zhang, F., Ju, P., Pan, M., Zhang, D., Huang, Y., Li, G., and Li, X., "Self-healing mechanisms in smart protective coatings: A review," *Corrosion Science*, vol. 144, pp. 74–88, 2018. doi: 10.1016/j.corsci.2018.08.005

[59] Alicja Stankiewicz and M. B. Barker, "Development of self-healing coatings for corrosion protection on metallic structures," *Smart Materials and Structures*, vol. 25, pp. 1–27, July 2016. doi: 10.1088/0964-1726/25/8/084013

[60] Andreeva, D. V., and Shchukin, D. G., "Smart self-repairing protective coatings," *Materials Today*, vol. 11, No. 10, pp. 24–30, 2008. doi: 10.1016/S1369-7021(08)70204-9

[61] Nagappan, S., Moorthy, M.S., Rao, K.M., and Ha, C.S., Stimuli-responsive smart polymeric coatings: an overview," *Industrial Applications for Intelligent Polymers and Coatings*, pp. 27–49, 2016. doi: 10.1007/978-3-319-26893-4_2

[62] Latnikova, A., *Polymeric Capsules for Self-Healing Anticorrosion Coatings.* (Doctoral dissertation, Universität Potsdam), 2012.

[63] Li, J., Feng, Q., Cui, J., Yuan, Q., Qiu, H., Gao, S., and Yang, J., "Self-assembled graphene oxide microcapsules in Pickering emulsions for self-healing waterborne polyurethane coatings," *Composites Science and Technology*, vol. 151, pp. 282–290, 2017. doi: 10.1016/j.compscitech.2017.07.031

[64] Manoj, A., Ramachandran, R., and Menezes, P. L. "Self-healing and superhydrophobic coatings for corrosion inhibition and protection," *The International Journal of Advanced Manufacturing Technology*, vol. 106, pp. 2119–2131, 2020. doi: 10.1007/s00170-019-04758-z

[65] Sehrish Habib, R. A. Shakoor, and Ramazan Kahraman, "A focused review on smart carriers tailored for corrosion protection: Developments, applications, and challenges," *Progress in Organic Coatings*, vol. 154, pp. 106–218, May 2021. doi: 10.1016/j.porgcoat.2021.106218

[66] Zadeh, M. A., Van Der Zwaag, S., and Garcia, S. J. "Routes to extrinsic and intrinsic self-healing corrosion protective sol-gel coatings: A review," *Self-Healing Materials*, vol. 1, 2013. doi: 10.2478/shm-2013-0001

[67] Maia, K. A. Yasakau, J. Carneiro, S. Kallip, J. Tedim, T. Henriques, A. Cabral, J. Venancio, M. L. Zheludkevich, and M. G. S. Ferreira, "Corrosion protection of AA2024 by sol–gel coatings modified with MBT-loaded polyurea microcapsules," *Chemical Engineering Journal*, vol. 283, pp. 1108–1117, January 2016. doi: 10.1016/j.cej.2015.07.087

[68] Khramov, A. N., Voevodin, N. N., Balbyshev, V. N., and Donley, M. S., "Hybrid organo-ceramic corrosion protection coatings with encapsulated organic corrosion inhibitors", *Thin Solid Films*, vol. 447, pp. 549–557, 2004. doi: 10.1016/j.tsf.2003.07.016

[69] Dariva, G., and Galio, A. F., *Corrosion Inhibitors — Principles, Mechanisms and Applications*, Chapter 16, February 2014. doi: 10.5772/57255

[70] Gece, G., "The use of quantum chemical methods in corrosion inhibitor studies," *Corrosion Science*, vol. 50, No. 11, pp. 2981–2992, 2008. doi:10.1016/j.corsci.2008.08.043

[71] Emran, K. M., Ali, S. M., and Al Lehaibi, H. A., "Green Methods for Corrosion Control," In *Corrosion Inhibitors, Principles and Recent Applications*, chapter 3, pp. 61–78, Elsevier, December 2017. doi: 10.5772/intechopen.72762

[72] Mohammadloo, H. Eivaz, Mirabedini, S. M., and Pezeshk-Fallah, H. "Microencapsulation of quinoline and cerium based inhibitors for smart coating application: Anti-corrosion, morphology and adhesion study," *Progress in Organic Coatings*, vol. 137, 105339, 2019. doi: 10.1016/j.porgcoat.2019.105339

[73] Nazeer, A. A., and Madkour, M., "Potential use of smart coatings for corrosion protection of metals and alloys: A review," *Journal of Molecular Liquids*, vol. 253, pp. 11–22, 2018. doi: 10.1016/j.molliq.2018.01.027

[74] Palumbo, G., "Smart coatings for corrosion protection by adopting microcapsules," *Physical Sciences Reviews*, vol. 1, No. 5, 2016. doi: 10.1515/psr-2015-0006

[75] Nóvoa, X. R., and Pérez, C., "The use of smart coatings for metal corrosion control," *Current Opinion in Electrochemistry*, 101324, 2023. doi: 10.1016/j.coelec.2023.101324

[76] Khelifa, F., Habibi, Y., Benard, F., and Dubois, P., "Smart acrylic coatings containing silica particles for corrosion protection of aluminum and other metals," In *Handbook of Smart Coatings for Materials Protection*, pp. 423–458. Woodhead Publishing, 2014.

[77] Calle, L., Hintze, P., Li, W., Buhrow, J., and Curran, J., "Launch pad coatings for smart corrosion control," In *SpaceOps 2010 Conference Delivering on the Dream Hosted by NASA Marshall Space Flight Center and Organized by AIAA* (p. 2017). April 2010. doi: 10.2514/6.2010-2017

[78] Frederico Maia, Joao Tedim, Aleksey D. Lisenkov, Andrei N. Salak, Mikhail L. Zheludkevich, and Mário G. S. Ferreira. "Silica nanocontainers for active corrosion protection," *Nanoscale,* vol. 4, No. 4, pp. 1287–1298, 2012.

17 Conclusion and Future Outlooks

Abhinay Thakur, Ashish Kumar, Harpreet Kaur, Ambrish Singh, Elyor Berdimurodov, and Akshay Kumar

17.1 INTRODUCTION

In recent years, the development of eco-benign smart coatings has gained significant attention in the field of corrosion protection. These coatings offer a sustainable and environmentally friendly solution for inhibiting corrosion in various industries, ranging from transportation to infrastructure. One of the key advancements in this area is the incorporation of encapsulated corrosion inhibitors within the coatings, providing enhanced performance and long-term protection [1–3]. Companies such as AkzoNobel, PPG Industries, Hempel, Jotun, Nippon Paint, BASF, and Sherwin-Williams are global leaders in the coatings industry and have been actively involved in the development and application of eco-benign smart coatings with encapsulated corrosion inhibitors [4, 5]. For instance, AkzoNobel's Interpon Cr range of coatings incorporates encapsulated corrosion inhibitors to provide long-lasting corrosion protection for various metal substrates. PPG Industries has also developed eco-friendly coatings with encapsulated corrosion inhibitors that offer enhanced durability and corrosion resistance. Hempel and Jotun have invested in the development of eco-benign smart coatings utilizing encapsulated corrosion inhibitors, specifically targeting industries such as offshore, marine, and infrastructure. Nippon Paint, BASF, and Sherwin-Williams have also embraced the development of eco-benign smart coatings with encapsulated corrosion inhibitors [6–9]. These companies have introduced sustainable coatings that provide efficient and long-term corrosion protection in applications ranging from automotive and electronics to construction and industrial equipment. The integration of encapsulated corrosion inhibitors within eco-benign smart coatings addresses the limitations of traditional corrosion prevention methods, such as barrier coatings and sacrificial anodes. By utilizing controlled release mechanisms, these coatings ensure sustained and efficient delivery of the corrosion inhibitors, optimizing their inhibitive properties and minimizing waste.

Encapsulated corrosion inhibitors have emerged as a promising solution to address the limitations of conventional corrosion inhibitors. These inhibitors are encapsulated within micro- or nano-sized containers, such as microcapsules or nanoparticles, which release the active corrosion inhibitor compounds in a controlled manner [8, 10, 11]. This controlled release mechanism offers several advantages, including prolonged protection, reduced toxicity, and minimized environmental impact. The concept of encapsulated corrosion inhibitors for smart coatings is rooted in the field of self-healing materials, which aim to repair damage autonomously. In the case of corrosion protection, the encapsulated inhibitors are designed to release their active ingredients when triggered by specific stimuli, such as the presence of corrosion products or changes in pH or temperature [12–14]. This targeted release ensures that the inhibitors are activated precisely where and when they are needed, optimizing their effectiveness and reducing wastage. One of the key benefits of encapsulated corrosion inhibitors is their ability to provide long-term protection. Traditional corrosion inhibitors

DOI: 10.1201/9781032677255-17

often suffer from issues such as leaching, volatilization, or depletion over time, leading to a decrease in their performance. In contrast, the encapsulation of corrosion inhibitors helps to preserve their integrity and control their release rate, ensuring a sustained supply of inhibitors throughout the coating's lifetime. This extended protection significantly enhances the durability and reliability of the coatings, reducing the need for frequent reapplications and maintenance [15–17]. Moreover, encapsulated corrosion inhibitors offer the advantage of reduced toxicity and environmental impact compared to their traditional counterparts. Conventional corrosion inhibitors can be harmful to human health and the environment, especially when they leach into surrounding ecosystems or water sources. Figure 17.1 shows the pathway of the smart anticorrosive epoxy coating focused on graphene oxide/functional mesoporous silica nanoparticles. The encapsulation technology provides a protective barrier that prevents the direct contact of the inhibitors with the environment until they are released at the corrosion site. This containment minimizes the potential for toxic exposure, making encapsulated corrosion inhibitors a safer and more sustainable alternative.

Furthermore, the encapsulation process allows for the customization and tailoring of corrosion inhibitors based on specific coating requirements. Different types of encapsulation materials and techniques can be employed to achieve desired release profiles, such as delayed or triggered release.

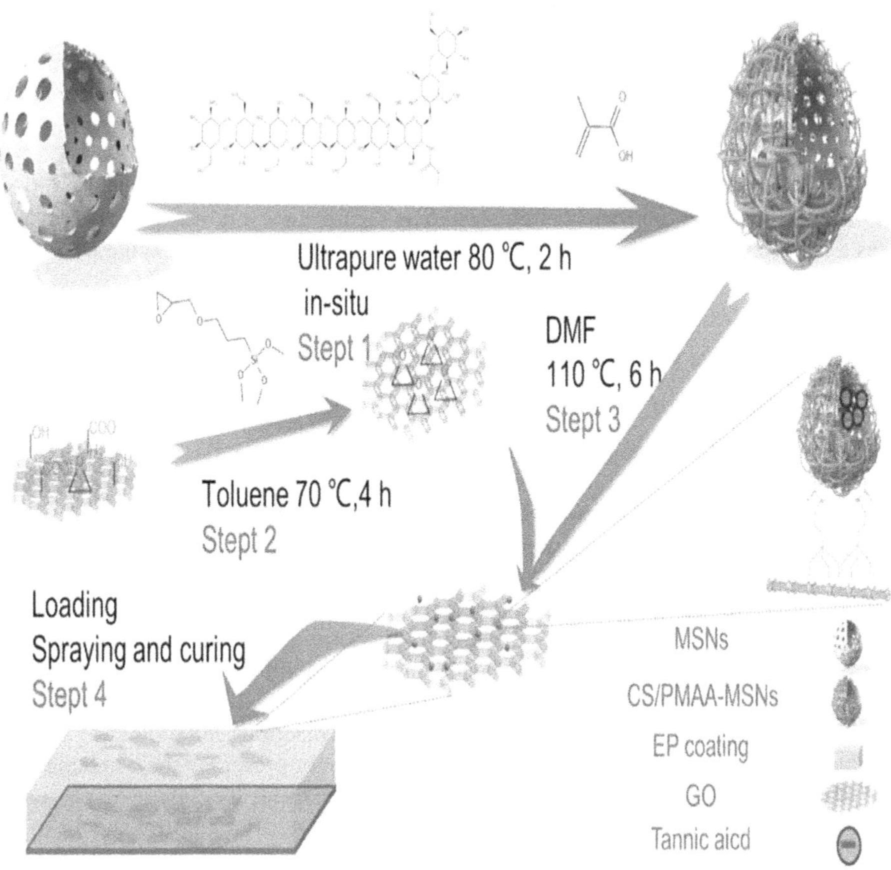

FIGURE 17.1 The schematic pathway of the smart epoxy coating using graphene oxide/functional mesoporous silica nanoparticles for effective corrosion protection. (Adapted with permission from Ref. [18] MDPI. Distributed under CCBY 4.0.)

This flexibility enables the design of smart coatings that are responsive to varying environmental conditions, extending their applicability to diverse industries and settings [19, 20]. The field of encapsulated corrosion inhibitors for eco-benign smart coatings is rapidly evolving, driven by ongoing research and development efforts. Various encapsulation techniques, such as layer-by-layer assembly, core-shell structures, and polymeric micro/nanocapsules, have been explored to achieve efficient and controlled release of corrosion inhibitors. Researchers are also investigating the use of different corrosion inhibitors, including organic compounds, inorganic salts, and nanostructured materials, to enhance the performance and broaden the application range of these coatings. Additionally, advancements in nanotechnology and materials science have contributed to the development of innovative encapsulation materials with improved properties, such as enhanced stability, mechanical strength, and chemical compatibility. These materials play a vital role in determining the release kinetics, encapsulation efficiency, and overall performance of the corrosion inhibitors within the coatings. The continuous progress in material synthesis and characterization techniques is expected to further optimize the encapsulation process and enhance the properties of eco-benign smart coatings [21–23].

The purpose of this chapter is to provide a comprehensive understanding of encapsulated corrosion inhibitors and their role in enhancing the performance and sustainability of protective coatings. It serves as a guide for researchers, industry professionals, and policymakers interested in corrosion protection strategies and the development of eco-benign smart coatings. The chapter aims to summarize the key findings presented in the subsequent sections, discuss potential future directions for research and development, and emphasize the need for standardization, scalability, and cost-effectiveness in the production and application of encapsulated corrosion inhibitors. Furthermore, the chapter encourages collaboration among stakeholders to accelerate the adoption of these innovative solutions and promote sustainable corrosion protection practices. By setting the stage for the subsequent sections, the introduction lays the foundation for a comprehensive exploration of encapsulated corrosion inhibitors, their significance, and the purpose of the chapter. It provides a roadmap for readers to delve into the topic and gain insights into the potential benefits and future outlooks of this emerging field.

17.2 INTEGRATION OF ECO-BENIGN SMART COATINGS

17.2.1 Application in the Automotive Industry

The automotive industry is a major consumer of protective coatings due to the high exposure of vehicles to environmental conditions that promote corrosion [24, 25]. Factors such as high humidity, temperature fluctuations, road salt, and chemicals encountered during regular use pose significant challenges to maintaining the appearance and structural integrity of automotive components. The integration of eco-benign smart coatings, incorporating encapsulated corrosion inhibitors, offers a promising solution to combat these challenges effectively. One of the key advantages of eco-benign smart coatings in the automotive industry is their ability to provide an additional layer of protection against corrosion. By incorporating encapsulated corrosion inhibitors, these coatings ensure a controlled and sustained release of the inhibitors, extending the lifespan of automotive components and structures. The inhibitors act as a barrier, inhibiting corrosion reactions and protecting the underlying materials from degradation. This enhanced protection not only preserves the aesthetic appearance of vehicles but also safeguards their structural integrity, ensuring the safety of passengers [26]. In addition to corrosion protection, eco-benign smart coatings offer improvements in durability and performance. Traditional coatings may require frequent repainting or touch-ups to maintain their protective properties, resulting in increased costs and environmental impacts associated with coating application and disposal. In contrast, the use of encapsulated corrosion inhibitors in smart coatings reduces the need for frequent maintenance and repainting, resulting in cost savings for both manufacturers and vehicle owners. This also translates to reduced environmental impacts, as fewer coating materials need to be manufactured and disposed of over the lifespan of a vehicle. Here are some examples of the application of eco-benign smart coatings, incorporating encapsulated corrosion inhibitors, in the automotive industry:

- **Corrosion Protection for Vehicle Frames:** The frames of vehicles are particularly susceptible to corrosion due to exposure to moisture, road salt, and other environmental factors. Eco-benign smart coatings can be applied to the vehicle frames, incorporating encapsulated corrosion inhibitors that provide long-lasting protection against corrosion [27–35]. This helps to maintain the structural integrity of the vehicle and extend its lifespan.
- **Enhanced Durability for Exterior Coatings:** The exterior coatings of vehicles are exposed to various elements, such as sunlight, rain, and pollutants, which can degrade their performance over time. By incorporating encapsulated corrosion inhibitors, eco-benign smart coatings offer improved durability, reducing the need for frequent repainting and touch-ups. This not only saves costs for vehicle owners but also reduces the environmental impact associated with coating maintenance.
- **Self-Cleaning Properties for Windows and Windshields:** Eco-benign smart coatings can be designed to have self-cleaning properties when applied to vehicle windows and windshields. These coatings incorporate hydrophobic or superhydrophobic properties, preventing water and dirt from adhering to the surface. As a result, the windows and windshields stay cleaner for longer periods, reducing the need for frequent cleaning and the use of chemical detergents.
- **Thermal Management for Vehicle Interiors:** Eco-benign smart coatings can be developed with thermal management properties, which help regulate the temperature inside the vehicle. These coatings can reflect a portion of the sun's heat, reducing the reliance on air-conditioning and improving energy efficiency. By maintaining a more comfortable interior temperature, eco-benign smart coatings contribute to the overall comfort and energy efficiency of the vehicle.
- **Anti-Fouling Properties for Exterior Surfaces:** Vehicles traveling on roads or parked outdoors are often exposed to contaminants such as dust, dirt, and pollutants, which can accumulate on the exterior surfaces. Eco-benign smart coatings with anti-fouling properties prevent the adhesion of these contaminants, making it easier to keep the vehicle clean. This not only improves the appearance but also helps maintain the aerodynamic efficiency of the vehicle, resulting in fuel savings [36].
- **Anti-Icing Properties for External Mirrors and Sensors:** In regions with cold climates, ice formation on external mirrors and sensors can impair visibility and functionality. Eco-benign smart coatings can incorporate anti-icing properties, preventing the formation of ice on these surfaces. By reducing the need for manual de-icing or the use of chemical de-icers, these coatings enhance safety, reduce maintenance requirements, and minimize the environmental impact of de-icing practices.

Furthermore, eco-benign smart coatings contribute to energy efficiency and sustainability in the automotive industry [37–46]. These coatings can incorporate self-cleaning properties, reducing the need for water-intensive cleaning processes and the use of chemical detergents. Self-cleaning coatings prevent the adhesion of dirt, dust, and pollutants, making it easier to maintain the cleanliness of vehicles with minimal effort. By reducing the frequency of cleaning, water consumption can be reduced, leading to the conservation of resources and a more sustainable approach to vehicle maintenance. Moreover, eco-benign smart coatings can integrate various functionalities that enhance the overall performance and efficiency of vehicles. For example, these coatings can incorporate thermal management properties, helping to regulate the temperature of vehicle surfaces and reducing the energy consumption of air-conditioning systems [47, 48]. Anti-fouling properties can also be integrated, preventing the accumulation of contaminants on vehicle surfaces and improving aerodynamic performance, leading to fuel efficiency gains. Additionally, anti-icing properties in smart coatings can prevent the formation of ice on vehicle surfaces, enhancing safety during winter conditions and reducing the reliance on de-icing chemicals. The application of eco-benign smart coatings in the automotive industry aligns with the growing demand for sustainable transportation solutions [49–58]. As the automotive sector seeks to reduce its environmental footprint and comply with stricter regulations, the integration of eco-friendly coatings becomes increasingly important. These coatings offer a combination of corrosion protection, durability, improved performance, and energy efficiency, making them an attractive choice for manufacturers and consumers who prioritize sustainability.

17.2.2 Application in the Aerospace Industry

The aerospace industry is renowned for its stringent safety requirements, rigorous standards, and extreme operating conditions. The integration of eco-benign smart coatings, incorporating encapsulated corrosion inhibitors, offers numerous applications and advantages in this industry, addressing corrosion challenges, improving operational efficiency, reducing maintenance costs, and promoting sustainable practices.

- **Corrosion Protection for Aircraft Structures**

Aircraft structures are exposed to a wide range of environmental factors during flight, including high humidity, temperature fluctuations, moisture, and corrosive agents. Corrosion can pose significant risks to the structural integrity of aircraft components, leading to safety concerns and increased maintenance needs [23, 59]. Eco-benign smart coatings with encapsulated corrosion inhibitors provide an effective solution for corrosion protection in the aerospace industry. These coatings can be applied to various aircraft surfaces, including the fuselage, wings, landing gear, and engine components. By releasing corrosion inhibitors over time, they form a protective barrier that inhibits corrosion reactions, preventing the degradation of structural materials. The controlled release of inhibitors ensures sustained effectiveness, even in harsh aerospace environments. This extends the service life of aircraft structures, reduces maintenance requirements, and enhances safety and reliability.

- **Anti-Icing and De-Icing Capabilities**

Icing is a critical issue in the aerospace industry, as it can affect the performance of aircraft and compromise flight safety. Eco-benign smart coatings can incorporate anti-icing and de-icing capabilities, providing a solution to mitigate the formation and accumulation of ice on aircraft surfaces [60]. These coatings can be designed to have superhydrophobic properties, preventing the adhesion of water droplets and ice. By reducing ice formation, they enhance aerodynamic efficiency, reduce fuel consumption, and improve aircraft performance. Additionally, these coatings can be engineered to have low ice adhesion strength, enabling easier removal of ice and reducing the need for extensive de-icing procedures.

- **Thermal Management and Heat Dissipation**

In the aerospace industry, thermal management is crucial to ensure the optimal performance and reliability of aircraft systems. Eco-benign smart coatings can be tailored to have thermal management properties, facilitating heat dissipation and regulation. These coatings can have high thermal emissivity, allowing efficient radiation of heat from aircraft surfaces. By dissipating heat effectively, they help prevent overheating of critical components and contribute to the overall thermal balance of the aircraft. This reduces the strain on cooling systems, enhances energy efficiency, and extends the lifespan of sensitive equipment.

- **Drag Reduction and Fuel Efficiency**

Drag is a significant factor affecting the fuel efficiency of aircraft. Eco-benign smart coatings can be engineered to have anti-fouling and low-friction properties, reducing drag and improving fuel economy. These coatings can incorporate self-cleaning properties, preventing the adhesion of contaminants such as dirt, insects, and pollutants. By maintaining a clean and smooth surface, they minimize surface roughness and airflow disruption, reducing drag and optimizing aerodynamic performance. This leads to fuel savings, increased range, and reduced emissions, aligning with the aerospace industry's sustainability goals.

- **Enhanced Inspection and Maintenance**

Eco-benign smart coatings can also integrate inspection and maintenance functionalities, facilitating proactive maintenance practices in the aerospace industry. These coatings can incorporate sensors or indicators that detect corrosion, cracks, or other structural anomalies in real time. By providing early warning signs of potential issues, these coatings enable timely inspections and maintenance interventions. This improves safety, reduces downtime, and minimizes the risk of costly repairs or component failures. Additionally, these coatings can self-heal small-scale damage, reducing the need for immediate repair and extending the lifespan of aircraft components.

- **Environmental Sustainability**

The aerospace industry is increasingly focused on environmental sustainability and reducing its carbon footprint. Eco-benign smart coatings align with this objective by promoting sustainable practices. These coatings offer several environmental benefits, such as reducing the use of toxic or harmful chemicals in corrosion protection, minimizing the release of pollutants during maintenance activities, and optimizing fuel efficiency. By incorporating eco-benign materials and techniques, the aerospace industry can reduce its environmental impact while maintaining high standards of safety and performance [61, 62].

17.2.3 Application in the Marine Industry

The marine industry operates in one of the harshest environments, where corrosion poses a significant challenge to the longevity and performance of marine structures and equipment. The integration of eco-benign smart coatings, incorporating encapsulated corrosion inhibitors, offers tremendous potential and benefits in the marine industry. These coatings provide effective corrosion protection, enhance operational efficiency, reduce maintenance costs, and contribute to sustainable practices.

17.2.3.1 Corrosion Protection for Marine Structures

Marine structures, such as offshore platforms, ships, and underwater infrastructure, are constantly exposed to corrosive elements like saltwater, humidity, and marine organisms. Corrosion can cause severe damage to these structures, compromising their integrity and durability. Eco-benign smart coatings with encapsulated corrosion inhibitors provide an innovative solution for corrosion protection in the marine industry. These coatings can be applied to various marine surfaces, including the hulls of ships, offshore pipelines, and underwater equipment. By releasing corrosion inhibitors over time, they form a protective barrier that prevents corrosion initiation and propagation. This extends the service life of marine structures, reduces maintenance requirements, and enhances safety in the marine environment.

17.2.3.2 Biofouling Resistance

Biofouling, the accumulation of marine organisms on submerged surfaces, is a prevalent issue in the marine industry. It can lead to increased fuel consumption, reduced maneuverability, and damage to marine equipment. Eco-benign smart coatings can be engineered to possess anti-fouling properties, preventing the attachment and growth of marine organisms [63–66]. These coatings can incorporate bio-repellent substances or surface textures that deter fouling organisms, reducing the need for toxic antifouling paints and minimizing the environmental impact. By maintaining clean and smooth surfaces, eco-benign smart coatings improve hydrodynamic efficiency, enhance fuel efficiency, and reduce maintenance costs.

17.2.3.3 Erosion and Abrasion Resistance

In the marine industry, structures and equipment are subjected to erosion and abrasion caused by water currents, wave action, and abrasive particles in the marine environment. Eco-benign smart coatings can be formulated to have high erosion and abrasion resistance, protecting vulnerable surfaces from damage. These coatings can incorporate durable materials and protective additives that withstand harsh conditions, minimizing wear and extending the lifespan of marine equipment. By reducing the need for frequent repairs and replacements, these coatings contribute to cost savings and sustainable resource utilization.

17.2.3.4 Chemical Resistance

Marine structures and equipment often encounter aggressive chemicals, such as acids, alkalis, and pollutants, which can cause corrosion and deterioration. Eco-benign smart coatings can offer excellent chemical resistance, acting as a barrier against chemical attacks. These coatings can be designed to withstand specific chemicals present in the marine environment, ensuring the long-term protection of marine structures and equipment. By mitigating chemical-induced corrosion, eco-benign smart coatings reduce maintenance needs, enhance operational efficiency, and promote environmental stewardship.

17.2.3.5 Environmental Sustainability

The marine industry is increasingly focused on environmental sustainability, aiming to reduce its ecological footprint and preserve marine ecosystems. Eco-benign smart coatings align with these objectives by promoting sustainable practices. These coatings minimize the use of toxic or hazardous chemicals in corrosion protection, reducing the release of pollutants into the marine environment [67, 68]. They also contribute to fuel efficiency, as the smooth and clean surfaces provided by the coatings reduce drag, leading to reduced fuel consumption and greenhouse gas emissions. By adopting eco-benign smart coatings, the marine industry can enhance its sustainability profile while ensuring the longevity and reliability of marine structures and equipment.

17.2.4 Application in the Infrastructure Sector

The infrastructure sector plays a critical role in a society's functioning, encompassing a wide range of structures such as bridges, buildings, pipelines, and transportation networks. However, these structures face the constant threat of corrosion, which poses significant challenges to their durability, safety, and maintenance. To address these challenges, the integration of eco-benign smart coatings, incorporating encapsulated corrosion inhibitors, offers immense potential in the infrastructure sector, providing effective corrosion protection, extending service life, reducing maintenance costs, and promoting sustainable infrastructure practices. One of the key advantages of eco-benign smart coatings in the infrastructure sector is their ability to protect structural components from corrosion. Structural elements like steel beams, concrete reinforcement, and pipelines are particularly vulnerable to corrosion due to their exposure to environmental elements and aggressive substances [69, 70]. By applying eco-benign smart coatings with encapsulated corrosion inhibitors to these components, a protective layer is formed, inhibiting corrosion reactions. The controlled release of corrosion inhibitors over time ensures sustained effectiveness, even in harsh conditions. This extended protection helps to significantly prolong the service life of infrastructure components, reducing the need for frequent repairs or replacements and enhancing the overall safety and functionality of bridges, buildings, and other structures.

In addition to corrosion protection, eco-benign smart coatings also offer excellent resistance against chemical attacks. Infrastructure structures often come into contact with aggressive chemicals such as acids, alkalis, salts, and pollutants, which can accelerate corrosion processes. Eco-benign smart coatings can be tailored to withstand specific chemicals present in the environment, providing long-term protection. By mitigating chemical-induced corrosion, these coatings minimize structural degradation, reduce maintenance needs, and contribute to the longevity of infrastructure assets. Water intrusion is another common cause of corrosion in infrastructure, particularly in structures exposed to moisture such as bridges, tunnels, and parking garages. Eco-benign smart coatings excel in providing superior water and moisture resistance, acting as a barrier to prevent the penetration of water into structural components. By forming a waterproof barrier, these coatings protect infrastructure surfaces from water-related damage and corrosion [71, 72]. This enhanced water resistance preserves the structural integrity of infrastructure, reduces maintenance costs, and supports sustainable infrastructure practices. Fire protection is a critical consideration in the infrastructure sector for the safety and resilience of buildings, tunnels, and transportation systems. Eco-benign smart coatings can be engineered to possess fire-resistant properties, providing passive fire protection. These coatings can incorporate flame-retardant additives or intumescent formulations that expand when exposed to high temperatures, forming an insulating layer. By delaying the spread of fire and reducing the damage caused by heat, eco-benign smart coatings enhance the fire safety of infrastructure structures and contribute to the overall resilience of the built environment.

In addition to the technical advantages, the use of eco-benign smart coatings aligns with sustainable maintenance practices in the infrastructure sector. These coatings offer several benefits that contribute to sustainability objectives. By extending the service life of infrastructure components, they reduce the need for frequent repairs or replacements, resulting in cost savings and reduced resource consumption. The incorporation of encapsulated corrosion inhibitors in the coatings minimizes the use of toxic or harmful chemicals, improving environmental stewardship. Furthermore, eco-benign smart coatings can be designed to have self-healing capabilities, where small-scale damage automatically repairs, further reducing maintenance requirements and extending the lifespan of infrastructure. Beyond the functional and sustainable aspects, eco-benign smart coatings also enhance the aesthetics of infrastructure structures. These coatings can be formulated to provide various finishes, colors, and textures, allowing for customization and architectural integration [73–75]. By improving the visual appeal of infrastructure, eco-benign smart coatings contribute to the overall aesthetics of cities and urban landscapes, enhancing public perception and satisfaction.

17.2.5 Benefits and Advantages of Eco-Benign Smart Coatings

Eco-benign smart coatings, incorporating encapsulated corrosion inhibitors, offer numerous benefits and advantages across various industries and applications. These coatings provide effective corrosion protection, promote sustainability, enhance performance, reduce maintenance costs, and offer additional functionalities. In this section, we will delve into the benefits and advantages of eco-benign smart coatings in more detail.

17.2.5.1 Corrosion Protection

One of the primary advantages of eco-benign smart coatings is their ability to provide efficient corrosion protection. Corrosion is a pervasive issue that affects the durability and integrity of structures, equipment, and surfaces. By incorporating encapsulated corrosion inhibitors, these coatings act as a barrier against corrosive elements, preventing corrosion initiation and progression. The controlled release of inhibitors ensures long-term protection, extending the service life of assets and reducing the need for frequent repairs or replacements. Eco-benign smart coatings offer an effective and sustainable solution to combat corrosion in diverse applications.

17.2.5.2 Sustainability

Eco-benign smart coatings align with sustainable practices and environmental stewardship. These coatings are formulated using environmentally friendly materials and technologies, reducing the use of hazardous substances [76]. The incorporation of encapsulated corrosion inhibitors minimizes the release of toxic compounds into the environment, promoting eco-benign corrosion protection. Furthermore, the extended durability provided by these coatings reduces the consumption of resources and energy associated with maintenance, repairs, and replacements. By promoting sustainable corrosion protection, eco-benign smart coatings contribute to a greener and more sustainable future.

17.2.5.3 Performance Enhancement

In addition to corrosion protection, eco-benign smart coatings offer performance-enhancing properties. These coatings can be tailored to meet specific performance requirements, such as abrasion resistance, chemical resistance, ultraviolet (UV) resistance, and thermal stability. By enhancing the performance of surfaces and structures, eco-benign smart coatings improve the overall functionality and reliability of various applications. For example, in the automotive industry, these coatings can provide scratch resistance and self-cleaning properties, reducing the need for frequent washing and maintenance. The integration of advanced functionalities into these coatings enhances their value and versatility across industries.

17.2.5.4 Cost Savings

Eco-benign smart coatings contribute to cost savings in several ways. Firstly, the extended service life provided by these coatings reduces the frequency of maintenance, repairs, and replacements. This translates into significant cost reductions for industries and asset owners, minimizing downtime and operational disruptions. Additionally, eco-benign smart coatings eliminate or reduce the need for costly surface preparation, such as sandblasting or chemical stripping, before coating application. The ability of these coatings to self-heal small-scale damage also reduces maintenance costs over time. By optimizing maintenance cycles and reducing the total cost of ownership, eco-benign smart coatings deliver economic benefits to industries and end-users.

17.2.5.5 Additional Functionalities

Eco-benign smart coatings can be designed to incorporate additional functionalities beyond corrosion protection. These coatings have the potential to offer self-cleaning properties, anti-fouling capabilities, anti-icing properties, thermal management, and antimicrobial effects, among others. For example, in the marine industry, these coatings can prevent the attachment of marine organisms, reducing drag and fuel consumption. The integration of these functionalities into a single coating system simplifies the overall coating process, reduces the number of layers required, and enhances the overall performance of the coated surface.

17.2.5.6 Aesthetics and Versatility

Eco-benign smart coatings provide aesthetic benefits by offering a wide range of colors, finishes, and textures. This versatility allows for customization and architectural integration, enhancing the visual appeal of surfaces and structures [77–79]. These coatings can be used in various industries, including automotive, architecture, infrastructure, and consumer goods, to achieve desired aesthetic effects. The ability to combine aesthetic appeal with performance attributes makes eco-benign smart coatings a desirable choice for both functional and decorative applications.

17.2.5.7 Ease of Application

Eco-benign smart coatings are designed for ease of application, making them suitable for various coating processes and application methods. They can be applied using conventional methods such as spraying, brushing, or dipping, and can be easily integrated into existing coating systems. The compatibility of these coatings with different substrates, including metals, concrete, plastics, and composites, further expands their applicability across industries. The ease of application facilitates the adoption of eco-benign smart coatings in various manufacturing and maintenance processes, ensuring seamless integration into existing operations.

17.2.5.8 Innovation and Research Opportunities

The development and implementation of eco-benign smart coatings present a significant opportunity for innovation and research. This field is continuously evolving, and ongoing research aims to improve the performance, durability, and sustainability of these coatings. Advances in encapsulation technologies, corrosion inhibitor selection, self-healing mechanisms, and nanomaterials contribute to the development of next-generation eco-benign smart coatings. Researchers, scientists, and industry professionals are actively exploring new materials, fabrication techniques, and application methods to further enhance the benefits and advantages of these coatings.

17.3 FUTURE DIRECTIONS FOR RESEARCH AND DEVELOPMENT

17.3.1 Utilization of Bio-Based Encapsulation Materials

The utilization of bio-based encapsulation materials represents a significant advancement in the development of eco-benign smart coatings. Bio-based materials are derived from renewable sources, such as plant-based polymers, and offer several advantages over conventional synthetic materials. One key advantage is their inherent biodegradability, which reduces the environmental impact associated with the end-of-life disposal of coatings. Bio-based materials can undergo natural degradation processes, minimizing the accumulation of non-biodegradable waste in the environment. Moreover, bio-based encapsulation materials exhibit low toxicity, making them safer for human health and the environment compared to synthetic materials that may contain harmful chemicals. This is particularly important in applications where coatings come into direct contact with sensitive environments or are used in industries that prioritize worker safety. The compatibility of bio-based materials with corrosion inhibitors is another crucial aspect. Corrosion inhibitors are chemical compounds that prevent or slow down the corrosion process. It is essential for the encapsulation material to effectively retain and release the corrosion inhibitors over an extended period to provide continuous protection [80, 81]. Bio-based encapsulation materials have been extensively studied to ensure compatibility with different types of corrosion inhibitors, ensuring their sustained performance and effectiveness in eco-benign smart coatings.

Research in the utilization of bio-based encapsulation materials focuses on the synthesis and characterization of these materials. Various techniques are employed to extract or synthesize bio-based polymers, such as cellulose, chitosan, starch, and lignin, from renewable sources. These materials are then modified and tailored to meet the specific requirements of encapsulation, such as controlling the release kinetics of corrosion inhibitors and optimizing the mechanical and barrier properties of the coatings. The development of efficient encapsulation processes, including coating formulation, deposition techniques, and curing methods, is also a focus of research to ensure the successful integration of bio-based materials into coating systems. The performance of bio-based encapsulation materials is evaluated through rigorous testing and characterization methods. This includes assessing their mechanical properties, such as tensile strength and flexibility, to ensure that they can withstand the application and service conditions of the coatings. The barrier properties of the encapsulation materials are also examined to determine their ability to prevent the penetration of corrosive agents and protect the underlying substrate. Compatibility testing is conducted to verify the stability and release characteristics of the encapsulated corrosion inhibitors, ensuring their sustained effectiveness over time.

The utilization of bio-based encapsulation materials in eco-benign smart coatings offers several benefits and advantages. Firstly, it enhances the sustainability profile of the coatings by utilizing renewable resources and reducing reliance on nonrenewable fossil-based materials. This aligns with the growing demand for environmentally friendly solutions and supports the transition toward a more sustainable and circular economy. Secondly, the use of bio-based materials promotes a safer working environment for coating applicators and users. The low toxicity of these materials reduces the potential health risks associated with exposure to harmful chemicals during coating application and maintenance. Furthermore, the compatibility of bio-based encapsulation materials with corrosion inhibitors ensures the efficient and controlled release of these inhibitors, providing long-lasting protection against corrosion. The sustained release of corrosion inhibitors is crucial in maintaining the integrity and performance of coatings, particularly in harsh and aggressive environments [82–85]. The use of bio-based encapsulation materials can enhance the corrosion resistance of coatings and extend the service life of protected structures, resulting in cost savings and reduced maintenance requirements. The utilization of bio-based encapsulation materials also contributes to the development of a bioeconomy, where renewable resources are harnessed for various industrial applications. This promotes economic growth, job creation, and a reduced dependency on finite resources. Additionally, it stimulates research and innovation in the field of bio-based materials, leading to further advancements and discoveries in sustainable coating technologies.

17.3.2 Incorporation of Self-Healing Mechanisms into Coatings

The incorporation of self-healing mechanisms into coatings represents a significant advancement in the field of corrosion protection. Self-healing coatings have the ability to autonomously repair damage and restore their protective properties, thereby extending the lifespan of the coating and enhancing its performance in corrosive environments. This innovative approach has gained considerable attention and interest in recent years, and ongoing research and development efforts are focused on exploring various self-healing mechanisms and their integration into eco-benign smart coatings [86, 87]. Self-healing mechanisms can be broadly categorized into intrinsic and extrinsic approaches. Intrinsic self-healing involves incorporating healing agents directly into the coating formulation, while extrinsic self-healing relies on the presence of external stimuli or triggers to initiate the healing process. Both approaches aim to repair and restore the coating's integrity, preventing the propagation of corrosion and maintaining the barrier properties of the coating. One of the commonly studied self-healing mechanisms is based on the incorporation of microcapsules containing healing agents within the coating matrix. These microcapsules rupture upon the occurrence of damage, releasing the healing agents into the damaged region. The healing agents, which can be in the form of liquid monomers, polymers, or reactive particles, react and polymerize in situ, filling the cracks or voids and restoring the coating's continuity. This autonomous healing process occurs without human intervention and is driven by the inherent properties of the coating.

Another approach to self-healing involves the use of microvascular networks within the coating. These networks consist of hollow channels or capillaries filled with healing agents. When the coating is damaged, the release of the healing agents is triggered, allowing them to flow into the damaged region and initiate the healing process [88]. This approach mimics the vascular system found in living organisms and offers a unique and efficient method of self-repair. In addition to microcapsules and microvascular networks, other self-healing mechanisms explored in research include the use of reversible bonds, shape memory polymers, and embedded nanoparticles. Reversible bonds allow the damaged coating to revert to its original state through reversible chemical reactions. Shape memory polymers can recover their original shape and properties when subjected to specific external stimuli, such as heat or light. Embedded nanoparticles can act as catalysts to initiate repair reactions and facilitate the healing process. The integration of self-healing mechanisms into eco-benign smart coatings offers several advantages. Firstly, it significantly improves the durability and performance of coatings by effectively repairing damage and preventing the propagation of corrosion. This leads to extended service life, reduced maintenance needs, and cost savings for industries and infrastructure owners. Self-healing coatings can also contribute to improved safety and reliability, as they help maintain the structural integrity of protected components and structures.

Furthermore, the incorporation of self-healing mechanisms enhances sustainability by reducing the consumption of resources and minimizing waste. The ability of coatings to repair themselves reduces the need for frequent reapplication or replacement, thereby reducing material consumption and waste generation. This aligns with the principles of a circular economy, where resources are utilized efficiently and waste is minimized. The development of self-healing coatings faces several challenges that require further research and development. One of the challenges is optimizing the healing efficiency and effectiveness of the coatings. This includes improving the encapsulation or delivery systems for the healing agents, enhancing their reactivity, and ensuring their compatibility with the coating matrix [89]. The development of advanced characterization techniques and testing methods is also necessary to evaluate the performance and healing capabilities of self-healing coatings under various environmental conditions. Another challenge is the scalability and cost-effectiveness of self-healing technologies. While self-healing coatings show great potential, the implementation on a large scale requires cost-effective manufacturing processes and materials. Researchers are exploring cost-effective methods of encapsulating healing agents and developing scalable production techniques to make self-healing coatings commercially viable.

17.3.3 Nanotechnology for Enhanced Corrosion Protection

Nanotechnology has emerged as a promising field for enhancing corrosion protection in various industries. By leveraging the unique properties and behavior of nanoscale materials, researchers and engineers are developing innovative solutions to combat corrosion and extend the lifespan of protective coatings. Nanotechnology-based approaches offer numerous advantages, including improved barrier properties, increased adhesion, enhanced mechanical strength, and tailored functionality, making them an attractive option for achieving enhanced corrosion protection. One key advantage of nanotechnology in corrosion protection is the ability to engineer nanoscale coatings with superior barrier properties [90–92]. Nanomaterials, such as nanoparticles and nanocomposites, can be incorporated into coating formulations to create densely packed, uniform, and defect-free layers. This results in coatings with reduced porosity and enhanced impermeability, effectively preventing the penetration of corrosive agents and minimizing the degradation of underlying substrates. The nanoscale dimensions of the materials allow for improved coating coverage and conformability, ensuring comprehensive protection of complex geometries and hard-to-reach areas.

Nanotechnology also enables the enhancement of coating adhesion to the substrate. Traditional coatings can suffer from poor adhesion, leading to premature delamination and reduced corrosion resistance. By incorporating nanoscale materials, such as nanotubes or nanoparticles, into the coating matrix, interfacial interactions can be improved, resulting in enhanced adhesion strength. The high surface area-to-volume ratio of nanomaterials facilitates stronger bonding with the substrate, leading to coatings with improved adhesion and reduced susceptibility to delamination. In addition to barrier properties and adhesion, nanotechnology offers the opportunity to enhance the mechanical strength of coatings. Nanoscale reinforcements, such as nanoparticles or nanofibers, can be dispersed within the coating matrix to reinforce its structure. These reinforcements act as strengthening agents, improving the mechanical properties of the coating, including hardness, toughness, and resistance to wear and abrasion. The incorporation of nanoscale reinforcements results in coatings that are more resistant to mechanical stress, thereby increasing their durability and longevity in corrosive environments. Furthermore, nanotechnology enables the tailoring of functional properties within coatings to address specific corrosion challenges. For instance, nanoscale particles can be loaded with corrosion inhibitors and incorporated into the coating formulation to provide a controlled release of these inhibitors over time. This ensures continuous protection against corrosion and extends the effectiveness of the coatings. Additionally, nanomaterials can be engineered to possess other functionalities, such as self-cleaning, antimicrobial, or photovoltaic properties, further enhancing the overall performance and value of the coatings.

Nanotechnology-based coatings also offer the advantage of versatility and compatibility with various coating systems and substrates. Nanomaterials can be integrated into a wide range of coating formulations, including solvent-based, water-based, and powder coatings, without compromising their processability or performance. Moreover, they can be applied to different substrates, such as metals, polymers, or ceramics, allowing for the protection of diverse materials and structures in various industries. The development and application of nanotechnology for enhanced corrosion protection involve ongoing research and development efforts. One focus of research is the synthesis and characterization of nanomaterials with tailored properties suitable for corrosion protection [93]. This includes the development of scalable and cost-effective synthesis methods to produce nanomaterials in large quantities while maintaining their desired characteristics. Characterization techniques, such as electron microscopy, spectroscopy, and surface analysis, are employed to study the morphology, structure, and composition of nanomaterials and evaluate their performance in corrosion protection applications. Another area of research is the optimization of nanomaterial dispersion and incorporation into coating formulations. Achieving uniform dispersion and stability of nanomaterials within the coating matrix is essential to ensure their effective performance and long-term stability. Researchers are exploring various techniques, such as sonication, surfactant-assisted dispersion, and surface modification, to enhance the dispersion and prevent the agglomeration of nanoparticles or nanofillers. This ensures their homogeneous distribution throughout the coating, maximizing their corrosion protection properties.

The evaluation and testing of nanotechnology-based coatings for corrosion protection require the development of specialized testing methodologies. These methods assess key performance parameters, such as corrosion resistance, adhesion strength, mechanical properties, and durability, under relevant environmental conditions. Accelerated corrosion testing, electrochemical techniques, salt spray testing, and mechanical testing are among the commonly employed methods to evaluate the performance and reliability of nanotechnology-based coatings. Although nanotechnology offers significant potential for enhanced corrosion protection, there are challenges that need to be addressed for its widespread adoption. One challenge is the potential toxicity and environmental impact of nanomaterials. It is crucial to understand and mitigate the potential risks associated with the release of nanoparticles into the environment during coating application, service, and end-of-life stages. Research efforts are focused on developing safe handling practices, improving material characterization techniques, and evaluating the long-term environmental impact of nanotechnology-based coatings [94, 95]. Another challenge is the scalability and cost-effectiveness of nanotechnology-based coatings. The synthesis and production of nanomaterials on a large scale can be complex and costly. Researchers are exploring scalable manufacturing processes and cost-effective approaches to enable the practical application of nanotechnology in corrosion protection. This includes optimizing the production parameters, such as reaction conditions, energy consumption, and material utilization, to reduce costs and increase efficiency.

17.3.4 Exploration of New Encapsulation Techniques

The exploration of new encapsulation techniques holds significant promise for the advancement of encapsulated corrosion inhibitors in eco-benign smart coatings. Encapsulation plays a crucial role in protecting and delivering corrosion inhibitors effectively to the coating system, enhancing their performance and longevity. Traditional encapsulation methods, such as core-shell structures and microencapsulation, have been widely employed. However, ongoing research focuses on developing innovative encapsulation techniques that offer improved efficiency, stability, controlled release, and compatibility with eco-friendly materials. One area of exploration is the development of nanoencapsulation techniques. Nanoencapsulation involves the encapsulation of corrosion inhibitors at the nanoscale, providing several advantages over traditional encapsulation methods. Nanoencapsulation techniques offer precise control over the size, shape, and composition of encapsulated particles, enabling enhanced stability and controlled release of corrosion inhibitors. These techniques include electrostatic assembly, layer-by-layer deposition, nanoprecipitation, and emulsion-based methods. By utilizing nanoencapsulation, corrosion inhibitors can be uniformly distributed within the coating matrix, resulting in improved corrosion protection and prolonged inhibitor release. Another area of exploration is the utilization of stimuli-responsive encapsulation systems. Stimuli-responsive encapsulation allows for the triggered release of corrosion inhibitors in response to specific environmental conditions, such as changes in pH, temperature, or humidity. By incorporating stimuli-responsive materials, such as polymers or microgels, into the encapsulation system, corrosion inhibitors can be released selectively when needed. This approach ensures the efficient utilization of inhibitors and enhances their performance in dynamic corrosion environments. Stimuli-responsive encapsulation techniques offer potential applications in self-healing coatings, where corrosion inhibitors are released upon detection of localized damage or exposure to corrosive species.

Additionally, the exploration of bio-based encapsulation techniques is gaining attention in the development of eco-friendly smart coatings. Bio-based encapsulation utilizes natural polymers, such as cellulose, chitosan, or alginate, derived from renewable resources. These materials offer biodegradability, low toxicity, and compatibility with corrosion inhibitors, making them suitable for eco-benign coating systems [96, 97]. Bio-based encapsulation techniques can involve solvent-free processes, such as electrospraying or supercritical fluid techniques, which reduce environmental impact and energy consumption. The utilization of bio-based encapsulation techniques aligns with sustainable practices and contributes to the development of environmentally friendly corrosion protection solutions. Furthermore, the exploration of advanced encapsulation techniques aims to overcome existing limitations and challenges in traditional encapsulation methods. One challenge is the release kinetics of corrosion inhibitors, as sustained and controlled release is crucial for long-term corrosion protection. Researchers are investigating novel encapsulation methods that allow for tunable release profiles, such as encapsulation within porous structures or the use of functional coatings with permeable barriers. These approaches enable precise control over the release rate and duration of corrosion inhibitors, ensuring their optimal utilization and longevity [95, 98].

Moreover, the combination of multiple encapsulation techniques, known as hybrid encapsulation, is being explored to synergistically enhance the performance of encapsulated corrosion inhibitors. Hybrid encapsulation involves the integration of different encapsulation methods, such as core-shell structures with nanoencapsulation or microencapsulation with stimuli-responsive coatings. This approach provides enhanced stability, prolonged release, and improved compatibility with diverse coating systems. By combining the strengths of different encapsulation techniques, hybrid encapsulation offers versatile and tailored solutions for corrosion protection in various applications. The exploration of new encapsulation techniques also involves the development of advanced characterization and evaluation methods. Researchers are utilizing advanced analytical techniques, such as microscopy, spectroscopy, and thermal analysis, to study the morphology, structure, and release behavior of encapsulated corrosion inhibitors. These characterization methods provide valuable insights into the encapsulation process, allowing for the optimization and improvement of encapsulation techniques. Additionally, researchers are developing robust testing methodologies to evaluate the performance and durability of encapsulated corrosion inhibitors under relevant environmental conditions.

17.3.5 Advances in Performance Evaluation Methods

Advances in performance evaluation methods play a critical role in the development and implementation of encapsulated corrosion inhibitors for eco-benign smart coatings. Accurate and reliable evaluation techniques are essential for assessing the effectiveness, durability, and long-term performance of these coatings in real-world corrosion environments. Over the years, significant progress has been made in the development of advanced performance evaluation methods that provide valuable insights into the behavior of encapsulated corrosion inhibitors. These advancements encompass various aspects, including accelerated testing, characterization techniques, and predictive modeling. Accelerated testing methods have been widely used to evaluate the performance of coatings and corrosion inhibitors within a shorter time frame than natural exposure conditions. These methods simulate aggressive environmental conditions, such as high humidity, temperature fluctuations, salt spray, or corrosive chemicals, to accelerate the corrosion processes. One commonly used accelerated testing method is salt fog testing, where coated samples are exposed to a controlled saltwater mist to evaluate the resistance of the coating to corrosion. Other accelerated tests include cyclic testing, electrochemical methods, and exposure to corrosive gases [99, 100]. These methods enable researchers to assess the corrosion resistance and durability of encapsulated corrosion inhibitors in a relatively short time, providing valuable data for performance evaluation.

Characterization techniques have also advanced significantly, allowing for a more detailed understanding of the behavior and performance of encapsulated corrosion inhibitors. Here are a few examples of advances in performance evaluation methods:

- **Electrochemical Impedance Spectroscopy (EIS):** EIS is a widely used technique for evaluating the corrosion resistance of coatings and the effectiveness of encapsulated corrosion inhibitors. It measures the impedance response of a coating system to an applied alternating current, providing information about the electrochemical processes occurring at the coating–metal interface. By analyzing the impedance spectra, researchers can assess parameters such as coating resistance, charge transfer resistance, and capacitance, which are indicative of the coating's barrier properties and corrosion protection efficiency.
- **Salt Spray Testing:** Salt spray testing is an accelerated corrosion test that exposes coated samples to a highly corrosive saltwater mist. This test simulates harsh environmental conditions, such as coastal or industrial atmospheres, where corrosion is more severe. By subjecting coated samples to prolonged exposure to salt spray, researchers can evaluate the resistance of coatings and encapsulated corrosion inhibitors to corrosion and assess their durability and performance over time.
- **Scanning Electron Microscopy (SEM):** SEM is a powerful imaging technique that provides high-resolution images of coating surfaces. It enables researchers to examine the morphology, uniformity, and adhesion of coatings, as well as the distribution and behavior of encapsulated corrosion inhibitors. SEM can reveal details about the coating–substrate interface, the presence of defects or delamination, and the formation of corrosion products. This information is crucial for understanding the performance and effectiveness of encapsulated corrosion inhibitors.
- **Fourier-Transform Infrared Spectroscopy (FTIR):** FTIR is a spectroscopic technique used to analyze the chemical composition of coatings and identify functional groups present in the coating matrix and encapsulated particles. By analyzing the infrared absorption spectra, researchers can determine the presence of specific corrosion inhibitors, assess their

compatibility with the coating matrix, and study the interactions between the coating and encapsulated particles. FTIR provides valuable information for optimizing the formulation and performance of encapsulated corrosion inhibitors [101].

- **Computational Modeling:** Computational modeling has emerged as a valuable tool for predicting the performance of coatings and encapsulated corrosion inhibitors. These models utilize principles of corrosion science, such as electrochemical kinetics and diffusion, to simulate the behavior of coatings under different environmental conditions. By integrating experimental data, such as coating thickness, environmental parameters, and inhibitor release kinetics, researchers can predict the long-term performance of encapsulated corrosion inhibitors, optimize coating formulations, and evaluate their effectiveness in various corrosion scenarios. These advanced characterization techniques provide researchers with a comprehensive understanding of the performance and mechanisms of encapsulated corrosion inhibitors, aiding in the optimization and development of more effective coatings.

Predictive modeling has emerged as a powerful tool in performance evaluation, allowing researchers to simulate and predict the behavior of coatings and encapsulated corrosion inhibitors under different environmental conditions. Computational models based on principles of corrosion science, such as electrochemical kinetics and diffusion, enable the prediction of coating performance, corrosion rates, and inhibitor release kinetics. These models take into account factors such as coating thickness, corrosion product formation, environmental parameters, and the behavior of encapsulated particles. By integrating experimental data with predictive modeling, researchers can optimize the formulation of coatings, evaluate different encapsulation strategies, and predict the long-term performance of encapsulated corrosion inhibitors. Predictive modeling provides insights into the durability and performance of coatings, facilitating informed decision-making in the development and selection of eco-benign smart coatings. Furthermore, advancements in performance evaluation methods have led to the development of standardized testing protocols and guidelines. Standardization plays a crucial role in ensuring consistent and reliable performance evaluation across different laboratories and industries. Organizations such as ASTM International, NACE International, and the International Organization for Standardization (ISO) have established guidelines and standards for testing the corrosion resistance of coatings and the performance of corrosion inhibitors [102, 103]. These standards provide a framework for conducting tests, specifying parameters, and reporting results, facilitating comparability and reproducibility of performance evaluation data. Standardized testing protocols promote quality control, enhance credibility, and enable effective communication among researchers, manufacturers, and end-users.

17.4 STANDARDIZATION, SCALABILITY, AND COST-EFFECTIVENESS

17.4.1 Need for Standardization in Encapsulation Processes

Standardization is a critical aspect of encapsulation processes for the successful development, implementation, and commercialization of encapsulated corrosion inhibitors in eco-benign smart coatings. It ensures the reproducibility, comparability, and reliability of research findings, facilitates technology transfer, and promotes the adoption of these coatings in various industries. The need for standardization arises in various aspects of encapsulation processes, including the synthesis and characterization of encapsulation materials, evaluation and testing of encapsulated corrosion inhibitors, and characterization and quantification of corrosion inhibition efficiency. Firstly, standardization is essential for the synthesis and characterization of encapsulation materials. Encapsulation materials, such as polymers, ceramics, and hybrid materials, are critical components that provide protection and controlled release of corrosion inhibitors. Standardized methods and protocols for the selection and preparation of coating matrices and encapsulant materials ensure consistent and reproducible encapsulation of corrosion inhibitors. This includes the characterization of material properties, such as particle size distribution, surface morphology, chemical composition, and encapsulation efficiency [104, 105]. Standardization in material synthesis and characterization facilitates the comparison of different encapsulation systems and their performance in corrosion protection. Secondly, standardization is required for the evaluation and testing of encapsulated corrosion inhibitors. Standardized testing protocols enable researchers and industry professionals to compare the performance of different coatings, assess corrosion resistance, and validate the effectiveness of encapsulation techniques. This includes standardized testing methods for corrosion protection, durability, adhesion, mechanical properties, and release kinetics of encapsulated corrosion inhibitors. Standardization ensures that the test conditions, parameters, and measurement techniques are consistent, enabling reliable and meaningful comparisons between different coatings and encapsulation approaches.

Furthermore, standardization is necessary for the characterization and quantification of corrosion inhibition efficiency. Standardized measurement techniques, such as electrochemical methods (e.g., electrochemical impedance spectroscopy, and potentiodynamic polarization), weight loss analysis, and salt spray testing, provide consistent and comparable data on corrosion rates, inhibitor release kinetics, and long-term performance of coatings. Harmonized reporting formats and criteria for assessing corrosion protection efficiency facilitate data interpretation, benchmarking, and decision-making processes. Standardization in performance evaluation methods ensures that the corrosion inhibition efficiency of encapsulated corrosion inhibitors can be reliably assessed and compared. Standardization in encapsulation processes also contributes to the scalability and commercialization of encapsulated corrosion inhibitors. It enables the transfer of technology from the laboratory to industrial settings by providing consistent and reproducible methods and protocols. Standardized processes facilitate the integration of encapsulated corrosion inhibitors into existing coating manufacturing processes, ensuring compatibility with different coating systems, substrates, and application methods [106]. This scalability allows for the large-scale production of eco-benign smart coatings, making them economically viable and accessible to industries such as automotive, aerospace, marine, and infrastructure. In addition to the technical aspects, standardization also promotes safety and environmental considerations. Standardized processes help identify and mitigate potential risks associated with the use of encapsulation materials, corrosion inhibitors, and coating application methods. It ensures that the coatings meet regulatory and safety requirements, reducing the potential impact on human health and the environment. Standardization encourages the adoption of sustainable practices, such as the use of nontoxic and environmentally friendly encapsulation materials, waste reduction strategies, and efficient manufacturing processes.

Collaboration among researchers, industry professionals, and standardization organizations is crucial for the development and implementation of standardized protocols in encapsulation processes. The establishment of international standards and guidelines specific to encapsulated corrosion inhibitors and eco-benign smart coatings can provide a common framework for researchers, manufacturers, and end-users. This collaboration ensures that standardization efforts consider the diverse needs and requirements of different industries and applications, fostering the widespread adoption of encapsulated corrosion inhibitors and promoting sustainable corrosion protection practices.

17.4.2 Scalability Challenges and Opportunities

Scalability is a critical consideration for the successful implementation of encapsulated corrosion inhibitors in large-scale production and industrial applications. While encapsulation techniques have shown promising results at the laboratory scale, their translation to industrial settings presents several challenges and opportunities. This discussion will explore the scalability challenges and opportunities associated with the production and application of encapsulated corrosion inhibitors.

- **Optimization and Control of Encapsulation Processes**

One of the key challenges in scalability is the optimization and control of encapsulation processes. Laboratory-scale encapsulation methods often involve batch processes with limited throughput, which may not be suitable for large-scale production. Scaling up these processes requires the development of continuous and high-throughput encapsulation techniques that can meet the demand for eco-benign smart coatings in industries such as automotive, aerospace, marine, and infrastructure. The optimization of process parameters, such as flow rates, temperature, and pressure, is crucial to ensure consistent and reliable encapsulation results on a larger scale [107, 108].

- **Material Availability and Cost Considerations**

Scalability also requires consideration of material availability and cost-effectiveness. The choice of encapsulation materials, corrosion inhibitors, and coating matrices should take into account their availability, cost, and sustainability aspects. Large-scale production often demands significant quantities of materials, and ensuring a stable supply chain is essential. Additionally, cost-effective manufacturing processes should be developed to make the production of encapsulated corrosion inhibitors economically viable. Efficient material utilization, waste reduction strategies, and recycling or reuse of materials can contribute to cost-effectiveness.

- **Compatibility with Coating Systems and Application Methods**

To achieve scalability, the compatibility of encapsulated corrosion inhibitors with different coating systems and application methods is crucial. Encapsulation techniques should be adaptable to various coating formulations, substrates, and application processes, such as spray coating, dip coating, and electrodeposition [109]. Compatibility with existing coating equipment and application protocols is essential to facilitate the integration of encapsulated corrosion inhibitors into existing manufacturing processes without significant modifications. The scalability of encapsulation processes relies on the ability to incorporate encapsulated corrosion inhibitors seamlessly into established coating production lines.

- **Performance and Long-Term Stability**

Scalability requires ensuring the performance and long-term stability of encapsulated corrosion inhibitors. As the production volume increases, maintaining the desired corrosion protection efficacy becomes crucial. Proper characterization and quality control processes should be in place to monitor the performance of encapsulated corrosion inhibitors and ensure consistent quality across large-scale production batches. Long-term stability studies should be conducted to assess the durability and reliability of the coatings over extended periods. Understanding the degradation mechanisms and potential limitations of encapsulated corrosion inhibitors on a larger scale is essential for successful scalability.

- **Innovation and Process Optimization**

Scalability challenges also present opportunities for innovation and process optimization. Research and development efforts should focus on developing scalable encapsulation techniques that maintain the quality, performance, and reliability of encapsulated corrosion inhibitors. Process optimization, automation, and integration of advanced manufacturing technologies, such as microfluidics and nanofabrication, can enhance the scalability of encapsulation processes. Continuous improvement in production efficiency, yield, and product consistency can contribute to the scalability of eco-benign smart coatings.

- **Collaboration and Knowledge Sharing**

Scalability efforts require collaboration and knowledge sharing among researchers, industry professionals, and stakeholders. Collaboration allows for the exchange of ideas, expertise, and experiences related to the production and application of encapsulated corrosion inhibitors. It fosters innovation, enables the identification of best practices, and promotes the development of standardized protocols and guidelines for large-scale production [110, 111]. Collaborative efforts can also address common scalability challenges and collectively work toward overcoming them.

17.4.3 Cost-Effectiveness Considerations

Cost-effectiveness is a critical factor in the successful implementation of encapsulated corrosion inhibitors in eco-benign smart coatings. While the development of effective corrosion protection is important, it must also be economically viable to ensure widespread adoption in various industries. This discussion will explore the cost-effectiveness considerations associated with the production, application, and maintenance of encapsulated corrosion inhibitors.

- **Material Costs**

One of the primary cost considerations is the materials used in the encapsulation process. Encapsulation materials, corrosion inhibitors, coating matrices, and other additives all contribute to the overall cost of the coating system. It is essential to evaluate the availability, cost, and performance of these materials to ensure cost-effectiveness. Researchers and manufacturers should explore sustainable sourcing options, such as bio-based or recycled materials, to mitigate the impact on costs. Optimizing material utilization and reducing waste can also help minimize material costs.

- **Manufacturing Processes**

The cost-effectiveness of encapsulated corrosion inhibitors is closely tied to the manufacturing processes employed. The choice of encapsulation technique, process parameters, and equipment can significantly impact production costs. Developing efficient and scalable manufacturing processes is crucial to achieving cost-effectiveness. Automation, continuous processing, and optimization of production parameters can help reduce labor costs and increase production efficiency. Investing in advanced manufacturing technologies, such as microfluidics or nanofabrication, may offer opportunities for cost reduction and process improvement.

- **Application Methods**

The application of encapsulated corrosion inhibitors also influences cost-effectiveness. The choice of application methods, such as spray coating, dip coating, or electrodeposition, can impact material usage, labor requirements, and overall coating performance. Selecting an application method that ensures uniform and efficient coverage while minimizing waste is important. Compatibility with existing coating equipment and processes can help reduce implementation costs and facilitate the integration of encapsulated corrosion inhibitors into existing manufacturing workflows.

- **Performance and Durability**

Cost-effectiveness considerations extend beyond the initial production and application stages to the long-term performance and durability of the coatings. High-quality encapsulation and coating systems that demonstrate superior corrosion protection and durability can offer significant cost savings over time. Coatings with longer service life reduce the need for frequent reapplication or maintenance, resulting in lower overall costs. Evaluating the long-term performance of encapsulated corrosion inhibitors and conducting lifecycle cost analyses can provide insights into their cost-effectiveness.

- **Maintenance and Repair**

The cost-effectiveness of encapsulated corrosion inhibitors also depends on the ease of maintenance and repair. In the event of damage or corrosion, the ability to repair or recoat the affected areas without extensive surface preparation can lead to cost savings. Coatings that require minimal surface preparation and allow for spot repairs can help reduce maintenance costs. Additionally, considering the compatibility of encapsulated corrosion inhibitors with common repair methods and materials is important for ensuring cost-effective maintenance practices.

- **Return on Investment (ROI)**

Assessing the ROI is crucial when evaluating the cost-effectiveness of encapsulated corrosion inhibitors [71, 112–114]. The initial investment in research, development, and implementation must be justified by the expected benefits and cost savings over time. ROI calculations should consider factors such as reduced maintenance costs, extended asset lifespan, improved operational efficiency, and enhanced product performance. Conducting cost–benefit analyses and comparing the lifetime costs of encapsulated corrosion inhibitors with traditional corrosion protection methods can provide insights into their cost-effectiveness.

- **Industry-Specific Considerations**

Cost-effectiveness considerations may vary across different industries and applications. Industries with high asset values, such as aerospace or infrastructure, may prioritize long-term durability and reduced maintenance costs, even if the upfront investment is higher. On the other hand, industries with large-scale production requirements, such as automotive, may prioritize cost optimization and scalability [115, 116]. Understanding the specific cost drivers and requirements of each industry can help tailor cost-effective solutions for different applications.

17.5 PROMISING PROSPECTS OF ENCAPSULATED CORROSION INHIBITORS AND ECO-BENIGN SMART COATINGS

17.5.1 Benefits and Applications

Encapsulated corrosion inhibitors and eco-benign smart coatings offer a range of benefits and applications that hold promising prospects for various industries. These innovative solutions provide enhanced corrosion protection, improved durability, controlled release of inhibitors, and environmental sustainability. This section will discuss the benefits and applications of encapsulated corrosion inhibitors and eco-benign smart coatings.

17.5.1.1 Enhanced Corrosion Protection

Encapsulated corrosion inhibitors offer several advantages over traditional methods, providing improved corrosion protection and enhancing the durability of coating systems. The encapsulation process involves incorporating the corrosion inhibitor within a protective barrier, which prevents its premature release and degradation. This controlled release mechanism is key to the effectiveness of encapsulated corrosion inhibitors. By encapsulating the corrosion inhibitor, its release is regulated, ensuring a more efficient and prolonged delivery of the inhibitor to the substrate. This controlled release allows for a continuous and sustained inhibition of corrosion, even in harsh and corrosive environments [117, 118]. As a result, the corrosion inhibitor can effectively mitigate the electrochemical reactions that lead to corrosion, reducing corrosion rates and preventing the degradation of the coated surface. The encapsulation process also offers protection to the corrosion inhibitor itself. The protective barrier surrounding the inhibitor shields it from external factors such as moisture, oxygen, and contaminants that could cause its premature degradation. This protection extends the lifespan of the corrosion inhibitor, allowing it to remain active and effective for a longer period. One of the key benefits of encapsulated corrosion inhibitors is their ability to provide targeted corrosion protection. The encapsulation process allows for the selective release of the inhibitor in response to specific environmental conditions, such as changes in pH, temperature, or the presence of corrosive agents. This targeted release ensures that the inhibitor is deployed precisely when and where it is needed, maximizing its inhibitive properties and optimizing its effectiveness.

Encapsulated corrosion inhibitors also contribute to the overall durability of the coating system. By reducing corrosion rates and protecting the substrate from degradation, these inhibitors help to maintain the structural integrity of assets and infrastructure. This, in turn, extends the service life of coated surfaces, reducing the need for frequent maintenance and replacement. The cost savings associated with improved durability make encapsulated corrosion inhibitors a cost-effective solution in the long run. Furthermore, the use of encapsulated corrosion inhibitors promotes environmental sustainability. The controlled release mechanism ensures the efficient utilization of the inhibitor, minimizing waste and reducing the environmental impact. The prolonged effect of the inhibitor also reduces the frequency of reapplication, leading to lower chemical consumption and waste generation over time. Additionally, encapsulated corrosion inhibitors often utilize eco-benign materials that are biodegradable and have low toxicity, aligning with sustainable practices and minimizing environmental risks [119–121].

Encapsulated corrosion inhibitors find extensive applications across various industries. They are particularly valuable in sectors such as transportation (including automotive, aerospace, and marine), infrastructure (bridges, pipelines, and buildings), energy (oil and gas, and renewable energy), and manufacturing. In these industries, assets and structures are exposed to corrosive environments, and effective corrosion protection is critical for their performance, safety, and longevity.

17.5.1.2 Improved Durability

Eco-benign smart coatings incorporating encapsulated corrosion inhibitors offer significant improvements in durability compared to conventional coatings. The integration of corrosion inhibitors within the encapsulation matrix provides a controlled release mechanism that sustains the protection of the coating system over an extended period. This controlled release allows for continuous and effective delivery of the inhibitors, resulting in enhanced durability and prolonged corrosion resistance. One of the key advantages of eco-benign smart coatings is their ability to provide long-term protection against corrosion. The encapsulation process ensures that the corrosion inhibitors are released gradually and in a controlled manner, continuously inhibiting the electrochemical reactions that lead to corrosion. This sustained protection mechanism allows the coating to maintain its corrosion resistance properties over an extended lifespan, even in harsh and corrosive environments. As a result, eco-benign smart coatings exhibit superior durability, minimizing the need for frequent recoating or maintenance. The extended durability of eco-benign smart coatings brings several benefits to various industries [122–124]. In the automotive sector, for example, where vehicles are exposed to diverse environmental conditions, the durability of coatings is crucial for protecting the vehicle's exterior from corrosion. Eco-benign smart coatings can withstand exposure to moisture, salt, UV radiation, and other corrosive agents, ensuring the longevity of the coating and preserving the aesthetics of the vehicle. In the aerospace industry, eco-benign smart coatings play a vital role in protecting aircraft components and structures from corrosion. Aircraft are exposed to high-altitude conditions, temperature variations, and atmospheric pollutants, making corrosion protection a critical concern. The extended durability of eco-benign smart coatings ensures that aircraft surfaces remain protected for prolonged periods, reducing the frequency of maintenance and improving operational efficiency.

Similarly, in the marine industry, eco-benign smart coatings provide enhanced durability for ship hulls, offshore structures, and marine equipment. These coatings withstand the harsh conditions of saltwater, wave action, and fouling organisms, ensuring the longevity of the coatings and minimizing the need for dry-docking and maintenance. The oil and gas industry also benefits from the extended durability of eco-benign smart coatings. Pipelines, storage tanks, and equipment used in the oil and gas sector are exposed to aggressive environments, including corrosive fluids and high temperatures. The ability of eco-benign smart coatings to maintain their corrosion resistance over a prolonged period ensures the integrity of these critical assets and reduces the risk of leaks, failures, and costly repairs. In the infrastructure sector, eco-benign smart coatings offer improved durability for bridges, buildings, and other infrastructure components. These coatings provide long-lasting corrosion protection, preserving structural integrity and reducing the need for frequent maintenance and rehabilitation. The extended durability of eco-benign smart coatings enhances the lifespan of infrastructure assets, resulting in significant cost savings over time. Furthermore, the improved durability of eco-benign smart coatings contributes to environmental sustainability. By reducing the need for frequent recoating or maintenance, these coatings minimize the consumption of resources, energy, and chemicals. They also help to reduce waste generation and the environmental impact associated with coating application and disposal.

17.5.1.3 Controlled Release of Inhibitors

One of the key advantages of encapsulated corrosion inhibitors is their ability to control the release of the inhibitor within the coating system. This controlled release mechanism offers several benefits in terms of optimizing the inhibitive properties of the corrosion inhibitor and minimizing waste. The encapsulation matrix acts as a reservoir for the corrosion inhibitor, holding it securely within the coating system. The matrix is designed to release the inhibitor gradually over time, based on specific triggers or environmental conditions. These triggers can include factors such as pH levels, temperature variations, or the presence of corrosive agents. By tailoring the encapsulation materials and techniques, the release kinetics of the corrosion inhibitor can be precisely controlled, providing a sustained and efficient delivery of the inhibitor. The controlled release of the inhibitor ensures continuous and targeted protection against corrosion. Instead of a rapid and immediate release, which may result in the depletion of the inhibitor and reduced effectiveness over time, encapsulated corrosion inhibitors provide a controlled and prolonged release. This sustained release maintains an optimum concentration of the inhibitor at the coating-substrate interface, where it is most needed to inhibit corrosion reactions. The ability to control the release kinetics of encapsulated corrosion inhibitors allows for customization and adaptation to different applications and environmental conditions [22, 125]. Each application may have unique requirements in terms of the duration and intensity of corrosion protection. For example, in a marine environment with high levels of saltwater exposure, a slower and continuous release of the inhibitor may be preferred to provide long-term protection against corrosion. On the other hand, in an environment with periodic exposure to corrosive agents, a more rapid release of the inhibitor during those specific periods may be desired.

By tailoring the encapsulation materials, coating formulations, and encapsulation techniques, it is possible to achieve a wide range of release profiles for the corrosion inhibitor. This flexibility enables the development of customized corrosion protection solutions for different substrates, industries, and environmental conditions. Another advantage of the controlled release mechanism is the minimization of waste. Traditional corrosion protection methods often involve the addition of large amounts of corrosion inhibitors, which can result in an excess of unused or unreacted inhibitors. With encapsulated corrosion inhibitors, the release kinetics can be optimized to match the rate of inhibitor consumption, reducing waste and improving the overall efficiency of the corrosion protection system. Furthermore, the controlled release of encapsulated corrosion inhibitors contributes to cost savings. The prolonged and sustained release of the inhibitor ensures its effectiveness over an extended period, reducing the frequency of reapplication or maintenance. This results in cost savings associated with the purchase of additional corrosion inhibitors and the labor required for frequent coating maintenance.

17.5.1.4 Versatile Applications

Encapsulated corrosion inhibitors and eco-benign smart coatings find applications across various industries and sectors. They can be applied to diverse substrates, including metals, concrete, polymers, and composites. The versatility of these coatings enables their use in industries such as automotive, aerospace, marine, infrastructure, oil and gas, and electronics [126, 127]. Encapsulated corrosion inhibitors can be incorporated into coatings for applications such as corrosion protection of pipelines, offshore structures, bridges, storage tanks, automotive components, and electronic devices.

17.5.1.5 Compatibility with Existing Coating Systems

Encapsulated corrosion inhibitors and eco-benign smart coatings are designed to be compatible with existing coating systems and application methods. They can be seamlessly integrated into conventional coating processes, such as spray coating, dip coating, or electrodeposition, without significant modifications to equipment or procedures. This compatibility facilitates the adoption of encapsulated corrosion inhibitors in industries where established coating systems are already in place, ensuring a smooth transition to eco-benign smart coatings.

17.5.2 Environmental and Sustainability Implications

The environmental and sustainability implications of encapsulated corrosion inhibitors and eco-benign smart coatings contribute to their promising prospects. These coatings offer several environmental benefits compared to traditional corrosion protection methods, including reduced environmental impact, waste reduction, and improved resource efficiency. This section will discuss the environmental and sustainability implications of encapsulated corrosion inhibitors and eco-benign smart coatings.

17.5.2.1 Reduced Environmental Impact

Encapsulated corrosion inhibitors and eco-benign smart coatings have the potential to reduce the environmental impact associated with corrosion protection. Traditional corrosion protection methods often involve the use of toxic or hazardous chemicals, such as chromates or heavy metals, which can pose risks to human health and the environment. In contrast, encapsulated corrosion inhibitors can employ eco-friendly and nontoxic inhibitor materials, minimizing environmental hazards. By replacing or reducing the use of harmful chemicals, these coatings contribute to a cleaner and safer working environment.

17.5.2.2 Waste Reduction

Encapsulated corrosion inhibitors offer waste reduction benefits by optimizing the utilization of inhibitors and reducing their premature release or degradation. The controlled release mechanism of encapsulated corrosion inhibitors ensures more efficient use of the inhibitor, minimizing overuse and waste [128, 129]. Additionally, the encapsulation process allows for the use of smaller quantities of inhibitors while achieving comparable or even superior corrosion protection compared to traditional methods. This reduction in waste generation translates into cost savings and a more sustainable corrosion protection approach.

17.5.2.3 Resource Efficiency

Eco-benign smart coatings promote resource efficiency by enhancing the performance and durability of corrosion protection systems. The extended lifespan of these coatings reduces the frequency of maintenance, recoating, and replacement, thereby conserving resources. Furthermore, the ability to customize and control the release kinetics of encapsulated corrosion inhibitors allows for targeted and efficient use of inhibitors, optimizing their utilization. By maximizing the efficiency of corrosion protection resources, eco-benign smart coatings contribute to sustainable practices and minimize resource consumption.

17.5.2.4 Compatibility with Green Initiatives

Encapsulated corrosion inhibitors and eco-benign smart coatings align with green initiatives and sustainability goals pursued by various industries and regulatory bodies. These coatings can contribute to achieving environmental targets, such as reducing greenhouse gas emissions, improving energy efficiency, and conserving natural resources. The adoption of eco-benign smart coatings demonstrates a commitment to sustainable corrosion protection practices, positioning industries as environmentally responsible and socially conscious.

17.5.3 Role in Promoting Sustainable Corrosion Protection Practices

Encapsulated corrosion inhibitors and eco-benign smart coatings play a crucial role in promoting sustainable corrosion protection practices. They offer a range of advantages that align with the principles of sustainability, including improved efficiency, reduced environmental impact, and enhanced durability. This section will discuss the role of encapsulated corrosion inhibitors and eco-benign smart coatings in promoting sustainable corrosion protection practices.

17.5.3.1 Minimizing Environmental Footprint

Encapsulated corrosion inhibitors and eco-benign smart coatings contribute to minimizing the environmental footprint of corrosion protection practices. By reducing the use of toxic or hazardous chemicals, these coatings mitigate environmental risks and promote a safer working environment. The controlled release of inhibitors ensures their optimal use, minimizing waste generation. The extended durability of eco-benign smart coatings reduces the need for frequent maintenance or recoating, resulting in resource conservation and reduced environmental impact.

17.5.3.2 Enhancing Energy Efficiency

Eco-benign smart coatings can enhance energy efficiency by reducing the energy consumption associated with corrosion protection practices. The improved durability of these coatings minimizes the need for frequent maintenance or recoating, leading to energy savings over the lifecycle of the coating system. Additionally, the compatibility of eco-benign smart coatings with existing coating processes allows for efficient integration into manufacturing workflows, optimizing energy use during the coating application and curing processes.

17.5.3.3 Long-Term Cost Savings

Encapsulated corrosion inhibitors and eco-benign smart coatings offer long-term cost savings, aligning with the principles of sustainability. The extended durability and reduced maintenance requirements of these coatings minimize the lifecycle costs associated with corrosion protection. The controlled release of inhibitors ensures their efficient use, reducing the overall consumption of inhibitors. As a result, industries adopting encapsulated corrosion inhibitors can benefit from reduced operating costs, improved cost-effectiveness, and enhanced ROI.

17.5.3.4 Compliance with Regulations and Standards

Encapsulated corrosion inhibitors and eco-benign smart coatings comply with stringent environmental regulations and sustainability standards. As environmental regulations become increasingly stringent, industries are compelled to adopt more sustainable corrosion protection practices. The use of eco-friendly and nontoxic corrosion inhibitors addresses regulatory requirements and ensures compliance with environmental standards [36, 47, 130]. By embracing encapsulated corrosion inhibitors and eco-benign smart coatings, industries can meet regulatory obligations while also demonstrating their commitment to sustainable practices.

17.5.3.5 Facilitating Asset Management and Sustainability Reporting

The implementation of encapsulated corrosion inhibitors and eco-benign smart coatings facilitates effective asset management and sustainability reporting. The extended durability and enhanced corrosion protection offered by these coatings contribute to the preservation of critical assets and infrastructure. This, in turn, supports long-term asset management strategies and enables industries to optimize their maintenance schedules and budgets. The environmental benefits of encapsulated corrosion inhibitors and eco-benign smart coatings can be incorporated into sustainability reports, allowing industries to demonstrate their environmental performance and commitment to sustainability.

17.6 INDUSTRIES AND COMPANIES GLOBALLY WORKING IN THE ENCAPSULATION OF SMART INHIBITORS FOR CORROSION INHIBITION OF METALS AND ALLOYS

The encapsulation of smart inhibitors for corrosion inhibition of metals and alloys is a rapidly evolving field with significant implications for various industries. Numerous industries and companies worldwide are actively engaged in research, development, and commercialization efforts to advance the use of encapsulated corrosion inhibitors and enhance corrosion protection.

17.6.1 Coatings and Paint Industry

The coatings and paint industry plays a pivotal role in the development and application of encapsulated corrosion inhibitors. These companies specialize in formulating protective coatings that incorporate corrosion inhibitors to provide long-lasting corrosion protection for metal substrates. Notable companies in this industry include AkzoNobel, PPG Industries, and Sherwin-Williams [23, 60]. These companies have dedicated research and development divisions focused on developing innovative coating technologies, including those that utilize encapsulated corrosion inhibitors. By leveraging their expertise in coatings technology, these companies contribute to the development of advanced corrosion protection solutions.

17.6.2 Materials Science and Engineering Companies

Materials science and engineering companies are instrumental in the development and optimization of encapsulation materials and techniques. These companies play a vital role in advancing corrosion protection technologies through their expertise in materials research, synthesis, and characterization. Prominent companies in this sector include 3M, Corning Incorporated, and BASF. These companies are involved in the development of advanced materials, including encapsulation technologies, for various applications, including corrosion protection. Their research and development activities aim to enhance the performance, durability, and sustainability of encapsulated corrosion inhibitors.

17.6.3 Research Institutions and Universities

Research institutions and universities contribute significantly to the advancement of encapsulated corrosion inhibitors through their extensive research and development activities. These institutions focus on understanding the fundamental principles underlying corrosion processes, developing innovative encapsulation techniques, and optimizing the performance of corrosion inhibitors. Notable research institutions and universities in this field include the National Association of Corrosion Engineers (NACE), the Massachusetts Institute of Technology (MIT), and the University of Manchester. NACE conducts research, publishes technical standards, and provides educational resources related to corrosion inhibition, including encapsulation techniques [64, 65, 67]. MIT and the University of Manchester, renowned for their expertise in materials science and engineering, contribute to the development of advanced coatings and encapsulation techniques for the corrosion protection of metals and alloys.

17.6.4 Startups and Technology Companies

The emergence of startups and technology companies in the field of encapsulated corrosion inhibitors brings innovative solutions and technologies to the market. These companies often focus on niche applications or employ novel encapsulation techniques to enhance corrosion protection. Noteworthy startups and technology companies include Greenkote, Saratech, and Green Theme Technologies [64, 65, 67, 70–73]. Greenkote specializes in metal coating solutions for corrosion protection and offers a proprietary encapsulation technology that provides long-lasting corrosion resistance for various metals and alloys. Saratech focuses on developing advanced coatings and surface treatments using encapsulation techniques to enhance the performance and durability of metal surfaces in corrosive environments. Green Theme Technologies develops eco-friendly and sustainable encapsulation materials for corrosion protection, addressing the environmental impact while delivering effective corrosion inhibition for metals and alloys.

The involvement of these industries and companies highlights the widespread interest and investment in encapsulated corrosion inhibitors for enhancing corrosion protection. The collective efforts of these stakeholders are driving advancements in materials, technologies, and applications related to corrosion protection. Their collaborative efforts span various aspects, including research and development, innovation, manufacturing, and commercialization.

17.7 CONCLUSION

17.7.1 Summary of Key Findings

Encapsulated corrosion inhibitors and eco-benign smart coatings have emerged as promising solutions for enhancing corrosion protection, improving durability, and promoting sustainability across various industries. Throughout this chapter, we have explored the benefits, applications, and prospects of these innovative coatings. The encapsulation process, which involves incorporating corrosion inhibitors into protective coatings, offers controlled release and prolonged effectiveness of the inhibitors. This ensures enhanced corrosion protection for a wide range of substrates. It is essential to select suitable encapsulation materials and techniques to achieve optimal performance and compatibility with different corrosion inhibitors. The applications of encapsulated corrosion inhibitors and eco-benign smart coatings are diverse and span industries such as automotive, aerospace, marine, and infrastructure. These coatings provide a protective layer against corrosion, extending component lifespan, enhancing structural integrity, and reducing maintenance costs. Additionally, eco-benign smart coatings offer additional functionalities, including self-cleaning, thermal management, and fire protection, which further enhance their value in various applications.

The environmental and sustainability implications of these coatings are significant. By minimizing the use of toxic or hazardous chemicals, optimizing resource utilization, and reducing waste generation, encapsulated corrosion inhibitors and eco-benign smart coatings contribute to reducing environmental risks. Their extended durability and energy efficiency benefits align with sustainable practices and support the achievement of environmental targets.

17.7.2 Implications for the Future

Based on the findings discussed, several implications for the future of encapsulated corrosion inhibitors and eco-benign smart coatings can be identified.

17.7.2.1 Advancements in Materials and Technologies

Continued research and development in materials science, nanotechnology, and encapsulation techniques will drive the advancement of these coatings. The utilization of bio-based encapsulation materials, incorporation of self-healing mechanisms, and exploration of new encapsulation techniques will contribute to improving their performance, sustainability, and cost-effectiveness.

17.7.2.2 Integration of Emerging Technologies

Emerging technologies such as nanotechnology, artificial intelligence, and the Internet of Things (IoT) can further enhance the capabilities and functionalities of eco-benign smart coatings. Nanotechnology-based approaches can develop nanocomposite coatings with enhanced barrier properties and self-cleaning functionalities for superior corrosion protection. IoT-enabled smart coatings can enable real-time monitoring of corrosion processes and predictive maintenance, optimizing asset management strategies.

17.7.2.3 Standardization and Certification

Standardization of encapsulation processes, testing methods, and performance evaluation criteria is crucial for ensuring consistent quality and reliability of these coatings. Collaboration among researchers, industry professionals, and regulatory bodies is essential to develop comprehensive standards and certifications that address the specific requirements of different industries and applications.

17.7.2.4 Economic Viability and Cost-Efficiency

The economic viability and cost-effectiveness of these coatings play a significant role in their widespread adoption. Further research and development efforts should focus on optimizing production processes, scaling up manufacturing capabilities, and reducing costs associated with raw materials and coating applications. Emphasizing cost-effectiveness considerations, such as reduced maintenance costs and increased asset lifespan, will showcase the long-term economic benefits of these coatings.

17.7.3 Call for Collaboration among Researchers, Industry Professionals, and Policymakers

Realizing the full potential of encapsulated corrosion inhibitors and eco-benign smart coatings requires collaboration among researchers, industry professionals, and policymakers. The following areas of collaboration are crucial.

17.7.3.1 Research and Development Collaboration

Collaboration among researchers from academia, industry, and government institutions is essential to advance the scientific knowledge and technological capabilities of these coatings. Joint research projects can focus on developing new encapsulation materials, exploring innovative coating formulations, and investigating the long-term performance and durability of these coatings in different environments.

17.7.3.2 Industry–Research Partnerships

Partnerships between industry professionals and researchers facilitate the transfer of technology and knowledge from the research stage to practical applications. Industry can provide valuable insights into the specific challenges and requirements of different sectors, while researchers can contribute their expertise in materials science, corrosion protection, and coating technologies. Collaborative efforts can accelerate the commercialization and adoption of these coatings.

17.7.3.3 Policy and Regulatory Collaboration

Policymakers and regulatory bodies play a crucial role in promoting the adoption of sustainable corrosion protection practices. Collaboration among researchers, industry professionals, and policymakers can help develop regulations and standards that encourage the use of encapsulated corrosion inhibitors and eco-benign smart coatings. Policymakers should create incentives and supportive frameworks to promote the adoption of these coatings and encourage industries to prioritize sustainability in their corrosion protection strategies. In conclusion, the integration of encapsulated corrosion inhibitors and eco-benign smart coatings offers promising prospects for enhancing corrosion protection, improving durability, and promoting sustainability. Advancements in materials, technologies, and standardization, coupled with collaboration among researchers, industry professionals, and policymakers, will drive the future development and adoption of these coatings. By embracing these innovative solutions, industries can achieve cost savings, regulatory compliance, and long-term asset protection while fulfilling their environmental and sustainability objectives.

REFERENCES

1. Noor EA. Temperature effects on the corrosion inhibition of mild steel in acidic solutions by aqueous extract of fenugreek leaves. *Int J Electrochem Sci*. 2007;2(12):996–1017.
2. Shahryari Z, Gheisari K, Yeganeh M. Designing a dual barrier-self-healable functional epoxy nanocomposite using 2D-carbon based nano-flakes functionalized with active corrosion inhibitors. *J Mater Res Technol* [Internet]. 2022;22:2746–67. Available from: https://doi.org/10.1016/j.jmrt.2022.12.138
3. Wang J, Wu S, Ma L, Zhao B, Xu H, Ding X, et al. Corrosion resistant coating with passive protection and self-healing property based on Fe3O4-MBT nanoparticles. *Corros Commun* [Internet]. 2022;7:1–11. Available from: https://doi.org/10.1016/j.corcom.2021.12.005
4. Dehghani A, Bahlakeh G, Ramezanzadeh B. Construction of a sustainable/controlled-release nanocontainer of non-toxic corrosion inhibitors for the water-based siliconized film: Estimating the host-guest interactions/desorption of inclusion complexes of cerium acetylacetonate (CeA) with beta-cycl. *J Hazard Mater* [Internet]. 2020;399(May):123046. Available from: https://doi.org/10.1016/j.jhazmat.2020.123046
5. Zeng W, Li W, Tan B, Liu J, Chen J. A research combined theory with experiment of 2-amino-6-(Methylsulfonyl)benzothiazole as an excellent corrosion inhibitor for copper in H2SO4 medium. *J Taiwan Inst Chem Eng* [Internet]. 2021;128:417–29. Available from: https://doi.org/10.1016/j.jtice.2021.08.032
6. Asaldoust S, Ramezanzadeh B. Synthesis and characterization of a high-quality nanocontainer based on benzimidazole-zinc phosphate (ZP-BIM) tailored graphene oxides; A facile approach to fabricating a smart self-healing anti-corrosion system. *J Colloid Interface Sci* [Internet]. 2020;564:230–44. Available from: https://doi.org/10.1016/j.jcis.2019.12.122

7. Akbarzadeh S, Ramezanzadeh M, Ramezanzadeh B, Bahlakeh G. A green assisted route for the fabrication of a high-efficiency self-healing anti-corrosion coating through graphene oxide nanoplatform reduction by Tamarindus indiaca extract. *J Hazard Mater* [Internet]. 2020;390(January):122147. Available from: https://doi.org/10.1016/j.jhazmat.2020.122147
8. Sanaei Z, Shahrabi T, Ramezanzadeh B. Synthesis and characterization of an effective green corrosion inhibitive hybrid pigment based on zinc acetate-Cichorium intybus L leaves extract (ZnA-CIL.L): Electrochemical investigations on the synergistic corrosion inhibition of mild steel in aqueous . *Dye Pigment* [Internet]. 2017;139:218–32. Available from: http://dx.doi.org/10.1016/j.dyepig.2016.12.002
9. Gnedenkov AS, Sinebryukhov SL, Filonina VS, Ustinov AY, Sukhoverkhov S V, Gnedenkov SV. New polycaprolactone-containing self-healing coating design for enhance corrosion resistance of the magnesium and its alloys. *Polymers* (Basel). 2023;15(1):202.
10. Izadi M, Shahrabi T, Ramezanzadeh B. Electrochemical investigations of the corrosion resistance of a hybrid sol–gel film containing green corrosion inhibitor-encapsulated nanocontainers. *J Taiwan Inst Chem Eng* [Internet]. 2017;81:356–72. Available from: https://doi.org/10.1016/j.jtice.2017.10.039
11. Ji X, Wang W, Li W, Zhao X, Liu A, Wang X, et al. pH-responsible self-healing performance of coating with dual-action core-shell electrospun fibers. *J Taiwan Inst Chem Eng*. 2019;104:227–39.
12. Zheludkevich ML, Tedim J, Freire CSR, Fernandes SCM, Kallip S, Lisenkov A, et al. Self-healing protective coatings with "green" chitosan based pre-layer reservoir of corrosion inhibitor. *J Mater Chem*. 2011;21(13):4805–12.
13. Samiee R, Ramezanzadeh B, Mahdavian M, Alibakhshi E. Assessment of the smart self-healing corrosion protection properties of a water-base hybrid organo-silane film combined with non-toxic organic/inorganic environmentally friendly corrosion inhibitors on mild steel. *J Clean Prod* [Internet]. 2019;220:340–56. Available from: https://doi.org/10.1016/j.jclepro.2019.02.149
14. Hu J, Zhu Y, Hang J, Zhang Z, Ma Y, Huang H, et al. The effect of organic core–shell corrosion inhibitors on corrosion performance of the reinforcement in simulated concrete pore solution. *Constr Build Mater* [Internet]. 2021;267:121011. Available from: https://doi.org/10.1016/j.conbuildmat.2020.121011
15. Hoai Vu NS, Hien P Van, Mathesh M, Hanh Thu VT, Nam ND. Improved corrosion resistance of steel in ethanol fuel blend by titania nanoparticles and Aganonerion polymorphum leaf extract. *ACS Omega*. 2019;4(1):146–58.
16. Sanaei Z, Bahlakeh G, Ramezanzadeh B. Active corrosion protection of mild steel by an epoxy ester coating reinforced with hybrid organic/inorganic green inhibitive pigment. *J Alloys Compd* [Internet]. 2017;728:1289–1304. Available from: http://dx.doi.org/10.1016/j.jallcom.2017.09.095
17. Nardeli JV, Fugivara CS, Pinto ERP, Polito WL, Messaddeq Y, Ribeiro SJL, et al. Preparation of polyurethane monolithic resins and modification with a condensed tannin-yielding self-healing property. *Polymers (Basel)*. 2019;11(11):1890.
18. Cao Y, He J, Wu J, Wang X, Lu W, Lin J, et al. A smart anticorrosive epoxy coating based on environmental-stimuli-responsive copolymer assemblies for controlled release of corrosion inhibitors. *Macromol Mater Eng*. 2022;307(7):2100983.
19. Cabello Mendez JA, Pérez Bueno J de J, Meas Vong Y, Portales Martínez B. Cerium compounds coating as a single self-healing layer for corrosion inhibition on aluminum 3003. *Sustainability* 2022;14(22):15056.
20. Kardogan B, Sekercioglu K, Erşan YÇ. Compatibility and biomineralization oriented optimization of nutrient content in nitrate-reducing-biogranules-based microbial self-healing concrete. *Sustainability* 2021;13(16):8990.
21. Xie P, He Y, Zhong F, Zhang C, Chen C, Li H, et al. Cu-BTA complexes coated layered double hydroxide for controlled release of corrosion inhibitors in dual self-healing waterborne epoxy coatings. *Prog Org Coatings*. 2021;153(February):106164.
22. Kaya S, Tüzün B, Kaya C, Obot IB. Determination of corrosion inhibition effects of amino acids: Quantum chemical and molecular dynamic simulation study. *J Taiwan Inst Chem Eng*. 2016;58(June):528–35.
23. Yadav M, Gope L, Sarkar TK. Synthesized amino acid compounds as eco-friendly corrosion inhibitors for mild steel in hydrochloric acid solution: Electrochemical and quantum studies. *Res Chem Intermed*. 2016;42(3):2641–60.

24. El Ibrahimi B, Jmiai A, Bazzi L, El Issami S. Amino acids and their derivatives as corrosion inhibitors for metals and alloys. *Arab J Chem* [Internet]. 2020;13(1):740–71. Available from: https://doi.org/10.1016/j.arabjc.2017.07.013
25. Satpati S, Suhasaria A, Ghosal S, Saha A, Dey S, Sukul D. Amino acid and cinnamaldehyde conjugated Schiff bases as proficient corrosion inhibitors for mild steel in 1 M HCl at higher temperature and prolonged exposure: Detailed electrochemical, adsorption and theoretical study. *J Mol Liq* [Internet]. 2021;324:115077. Available from: https://doi.org/10.1016/j.molliq.2020.115077
26. Fawzy A, Abdallah M, Zaafarany IA, Ahmed SA, Althagafi II. Thermodynamic, kinetic and mechanistic approach to the corrosion inhibition of carbon steel by new synthesized amino acids-based surfactants as green inhibitors in neutral and alkaline aqueous media. *J Mol Liq*. 2018;265:276–91.
27. Thakur A, SAVAŞ K, Kumar A. Recent trends in the characterization and application progress of nano-modified coatings in corrosion mitigation of metals and alloys. *Appl Sci*. 2023;13:730.
28. Verma C, Thakur A, Ganjoo R, Sharma S, Assad H. Coordination bonding and corrosion inhibition potential of nitrogen-rich heterocycles: Azoles and triazines as specific examples. *Coord Chem Rev* [Internet]. 2023;488(November 2022):215177. Available from: https://doi.org/10.1016/j.ccr.2023.215177
29. Sharma D, Thakur A, Sharma MK, Jakhar K, Kumar A, Sharma AK, Hari OM. Synthesis, electrochemical, morphological, computational and corrosion inhibition studies of 3-(5-Naphthalen-2-yl-[1,3,4] oxadiazol-2-yl)-pyridine against mild steel in 1 M HCl. *Asian J Chem*. 2014;35(5):1079–88.
30. Bashir S, Thakur A, Lgaz H, Chung I-M, Kumar A. Computational and experimental studies on Phenylephrine as anti-corrosion substance of mild steel in acidic medium. *J Mol Liq*. 2019;293:111539.
31. Thakur A, Kaya S, Kumar A. Recent innovations in nano container-based self-healing coatings in the construction industry. *Curr Nanosci*. 2021;18(2):203–16.
32. Dhonchak C, Agnihotri N. Computational insights in the spectrophotometrically 4H-chromen-4-one complex using DFT method. *Biointerface Res Appl Chem*. 2023;13(4):357.
33. Thakur A, Kumar A, Zhang R. Alcoholic beverage purification applications of activated carbon. In: Verma C, Quraishi MA, editors. *Activated Carbon: Progress and Applications* [Internet]. The Royal Society of Chemistry; 2023. p. 0. Available from: https://doi.org/10.1039/BK9781839169861-00152
34. Thakur A, Kumar A, Sharma S, Ganjoo R, Assad H. Materials today: Proceedings computational and experimental studies on the efficiency of Sonchus arvensis as green corrosion inhibitor for mild steel in 0.5 M HCl solution. *Mater Today Proc* [Internet]. 2022;66:609–21. Available from: https://doi.org/10.1016/j.matpr.2022.06.479
35. Sharma S, Ganjoo R, Thakur A, Kumar A. Investigation of corrosion performance of expired Irnocam on the mild steel in acidic medium. *Mater Today Proc* [Internet]. 2022;66:543–540. Available from: https://doi.org/10.1016/j.matpr.2022.05.595
36. Florez-Frias EA, Barba V, Lopez-Sesenes R, Landeros-Martínez LL, Los Ríos JPF De, Casales M, et al. Use of a metallic complex derived from Curcuma Longa as green corrosion inhibitor for carbon steel in sulfuric acid. *Int J Corros*. 2021;2021:1–13.
37. Thakur A, Kaya S, Abousalem AS, Sharma S, Ganjoo R, Assad H, et al. Computational and experimental studies on the corrosion inhibition performance of an aerial extract of Cnicus Benedictus weed on the acidic corrosion of mild steel. *Process Saf Environ Prot* [Internet]. 2022;161:801–18. Available from: https://doi.org/10.1016/j.psep.2022.03.082
38. Sharma S, Ganjoo R, Kr. Saha S, Kang N, Thakur A, Assad H, et al. Experimental and theoretical analysis of baclofen as a potential corrosion inhibitor for mild steel surface in HCl medium. *J Adhes Sci Technol* [Internet]. 2021;0(0):1–26. Available from: https://doi.org/10.1080/01694243.2021.2000230
39. Kaya S, Thakur A, Kumar A. The role of in silico / DFT investigations in analyzing dye molecules for enhanced solar cell efficiency and reduced toxicity. *J Mol Graph Model*. 2023;124(June):108536.
40. Bashir S, Lgaz H, Chung IM, Kumar A. Effective green corrosion inhibition of aluminium using analgin in acidic medium: An experimental and theoretical study. *Chem Eng Commun* [Internet]. 2020;208(8):1–10. Available from: https://doi.org/10.1080/00986445.2020.1752680
41. Bashir S, Thakur A, Lgaz H, Chung I-M, Kumar A. Corrosion inhibition performance of acarbose on mild steel corrosion in acidic medium: An experimental and computational study. *Arab J Sci Eng* [Internet]. 2020;45(6):4773–83. Available from: https://doi.org/10.1007/s13369-020-04514-6

42. Thakur A, Kumar A, Kaya S, Marzouki R, Zhang F, Guo L. Recent advancements in surface modification, characterization and functionalization for enhancing the biocompatibility and corrosion resistance of biomedical implants. *Coatings*. 2022;12:1459.
43. Parveen G, Bashir S, Thakur A, Saha SK, Banerjee P, Kumar A. Experimental and computational studies of imidazolium based ionic liquid 1-methyl- 3-propylimidazolium iodide on mild steel corrosion in acidic solution experimental and computational studies of imidazolium based ionic liquid 1-methyl- 3-propylimidazolium. *Mater Res Express*. 2020;7(1):016510.
44. Bashir S, Thakur A, Lgaz H, Chung IM, Kumar A. Corrosion inhibition efficiency of bronopol on aluminium in 0.5 M HCl solution: Insights from experimental and quantum chemical studies. *Surfaces Interfaces* [Internet]. 2020;20(April):100542. Available from: https://doi.org/10.1016/j.surfin.2020.100542
45. Thakur A, Kumar A. Recent trends in nanostructured carbon-based electrochemical sensors for the detection and remediation of persistent toxic substances in real- time analysis. *Mater Res Express*. 2023;10:034001.
46. Thakur A, Kumar A. A review on ammonia derivatives as corrosion inhibitors for metals and alloys. *Eur J Mol Clin Med*. 2020;07(07):49–67.
47. Su H, Liu Y, Gao X, Qian Y, Li W, Ren T, et al. Corrosion inhibition of magnesium alloy in NaCl solution by ionic liquid: Synthesis, electrochemical and theoretical studies]. *J Alloys Compd* [Internet]. 2019;791:681–89. Available from: https://doi.org/10.1016/j.jallcom.2019.03.318
48. Alamiery AA, Isahak WNRW, Takriff MS. Inhibition of mild steel corrosion by 4-benzyl-1-(4-oxo-4-phenylbutanoyl)thiosemicarbazide: Gravimetrical, adsorption and theoretical studies. *Lubricants*. 2021;9(9):1–10.
49. Thakur A, Kumar A. Recent advances on rapid detection and remediation of environmental pollutants utilizing nanomaterials-based (bio)sensors. *Sci Total Environ* [Internet]. 2022;834(January):155219. Available from: https://doi.org/10.1016/j.scitotenv.2022.155219
50. Ganjoo R, Bharmal A, Sharma S, Thakur A, Assad H, Kumar A. Imidazolium based ionic liquids as green corrosion inhibitors against corrosion of mild steel in acidic media. *J Phys Conf Ser*. 2022;2267(1):012023.
51. Thakur A, Kaya S, Abousalem AS, Kumar A. Experimental, DFT and MC simulation analysis of Vicia Sativa weed aerial extract as sustainable and eco-benign corrosion inhibitor for mild steel in acidic environment. *Sustain Chem Pharm* [Internet]. 2022;29(July):100785. Available from: https://doi.org/10.1016/j.scp.2022.100785
52. Thakur A, Sharma S, Ganjoo R, Assad H, Kumar A. Anti-corrosive potential of the sustainable corrosion inhibitors based on biomass waste: A review on preceding and perspective research. *J Phys Conf Ser*. 2022;2267(1):012079.
53. Ganjoo R, Sharma S, Thakur A, Kumar A. Thermodynamic study of corrosion inhibition of dioctylsulfosuccinate sodium salt as corrosion inhibitor against mild steel in 1 M HCl. *Mater Today Proc* [Internet]. 2022;66(1xxxx):1–5. Available from: https://doi.org/10.1016/j.matpr.2022.05.594
54. Kumar A, Thakur A. Overview of the properties, applicability, and recent advancements of some natural products used as potential inhibitors in various corrosive systems. In: *Handbook of Research on Corrosion Sciences and Engineering*. 2003. p. 275–310. IGI Global, 2023.
55. Kumar A, Thakur A. Encapsulated nanoparticles in organic polymers for corrosion inhibition [Internet]. In: *Corrosion Protection at the Nanoscale*. Elsevier Inc.; 2020. 345–62 p. Available from: http://dx.doi.org/10.1016/B978-0-12-819359-4.00018-0
56. Thakur A, Kumar A, Kaya S, Vo DVN, Sharma A. Suppressing inhibitory compounds by nanomaterials for highly efficient biofuel production: A review. *Fuel* [Internet]. 2022;312(September 2021):122934. Available from: https://doi.org/10.1016/j.fuel.2021.122934
57. Thakur A, Kumar A. Sustainable inhibitors for corrosion mitigation in aggressive corrosive media: A comprehensive study. *J Bio-Tribo-Corrosion* [Internet]. 2021;7(2):1–48. Available from: https://doi.org/10.1007/s40735-021-00501-y
58. Sharma S, Ganjoo R, Kr. Saha S, Kang N, Thakur A, Assad H, et al. Investigation of inhibitive performance of betahistine dihydrochloride on mild steel in 1M HCl solution. *J Mol Liq* [Internet]. 2021;347:118383. Available from: https://doi.org/10.1016/j.molliq.2021.118383
59. Hu K, Zhuang J, Zheng C, Ma Z, Yan L, Gu H, et al. Effect of novel cytosine-L-alanine derivative based corrosion inhibitor on steel surface in acidic solution. *J Mol Liq*. 2016;222:109–17.

60. Keramatinia M, Ramezanzadeh B, Mahdavian M. Green production of bioactive components from herbal origins through one-pot oxidation/polymerization reactions and application as a corrosion inhibitor for mild steel in HCl solution. *J Taiwan Inst Chem Eng*. 2019;105(1):134–49.
61. Go LC, Depan D, Holmes WE, Gallo A, Knierim K, Bertrand T, et al. Kinetic and thermodynamic analyses of the corrosion inhibition of synthetic extracellular polymeric substances. *PeerJ Mater Sci*. 2020;2:e4.
62. Wang X, Jiang H, Zhang DX, Hou L, Zhou WJ. Solanum lasiocarpum L. extract as green corrosion inhibitor for A3 steel in 1 M HCl solution. *Int J Electrochem Sci. 2019*;14(2):1178–96.
63. Valdez-Salas B, Vazquez-Delgado R, Salvador-Carlos J, Beltran-Partida E, Salinas-Martinez R, Cheng N, et al. Azadirachta indica leaf extract as green corrosion inhibitor for reinforced concrete structures: Corrosion effectiveness against commercial corrosion inhibitors and concrete integrity. *Materials (Basel)*. 2021;14(12).
64. Casaletto MP, Figà V, Privitera A, Bruno M, Napolitano A, Piacente S. PT Graphical abstract. *Eval Program Plann* [Internet]. 2018; 136:91–105. Available from: https://doi.org/10.1016/j.corsci.2018.02.059
65. Okeniyi JO, Popoola API, Okeniyi ET. Cymbopogon citratus and NaNO2 behaviours in 3.5% NaCl-immersed steel-reinforced concrete: Implications for eco-friendly corrosion inhibitor applications for steel in concrete. *Int J Corros*. 2018;2018:1–11.
66. Martins GR, Guedes D, Marques de Paula UL, de Oliveira MDSP, Lutterbach MTS, Reznik LY, et al. Açaí (Euterpe oleracea mart.) seed extracts from different varieties: A source of proanthocyanidins and eco-friendly corrosion inhibition activity. *Molecules*. 2021;26(11):1–13.
67. Bahlakeh G, Ramezanzadeh B, Dehghani A, Ramezanzadeh M. Novel cost-effective and high-performance green inhibitor based on aqueous Peganum harmala seed extract for mild steel corrosion in HCl solution: Detailed experimental and electronic/atomic level computational explorations. *J Mol Liq* [Internet]. 2019;283:174–95. Available from: https://doi.org/10.1016/j.molliq.2019.03.086
68. Cui M, Yu Y, Zheng Y. Effective corrosion inhibition of carbon steel in hydrochloric acid by dopamine-produced carbon dots. *Polymers (Basel)*. 2021;13(12):1–16.
69. El Ibrahimi B, Jmiai A, El Mouaden K, Oukhrib R, Soumoue A, El Issami S, et al. Theoretical evaluation of some α-amino acids for corrosion inhibition of copper in acidic medium: DFT calculations, Monte Carlo simulations and QSPR studies. *J King Saud Univ – Sci* [Internet]. 2020;32(1):163–71. Available from: https://doi.org/10.1016/j.jksus.2018.04.004
70. Tang J, Wang H, Jiang X, Zhu Z, Xie J, Tang J, et al. Electrochemical behavior of jasmine tea extract as corrosion inhibitor for carbon steel in hydrochloric acid solution. *Int J Electrochem Sci*. 2018;13(4):3625–42.
71. Zhao A, Sun H, Chen L, Huang Y, Lu X, Mu B, et al. Electrochemical studies of bitter gourd (Momordica charantia) fruits as ecofriendly corrosion inhibitor for mild steel in 1 M HCl solution. *Int J Electrochem Sci*. 2019;14(7):6814–25.
72. Fu J, Chen T, Wang M, Yang N, Li S, Wang Y, et al. Acid and alkaline dual stimuli-responsive mechanized hollow mesoporous silica nanoparticles as smart nanocontainers for intelligent anticorrosion coatings. *ACS Nano*. 2013;7(12):11397–408.
73. Stimpfling T, Vialat P, Hintze-Bruening H, Keil P, Shkirskiy V, Volovitch P, et al. Amino acid interleaved layered double hydroxides as promising hybrid materials for AA2024 corrosion inhibition. *Eur J Inorg Chem*. 2016;2016(13–14):2006–16.
74. Awad MI, Saad AF, Shaaban MR, Al Jahdaly BA, Hazazi OA. New insight into the mechanism of the inhibition of corrosion of mild steel by some amino acids. *Int J Electrochem Sci*. 2017;12(2):1657–69.
75. Bouraoui MM, Chettouh S, Chouchane T, Khellaf N. Inhibition efficiency of cinnamon oil as a green corrosion inhibitor. *J Bio- Tribo-Corrosion* [Internet]. 2019;5(1):1–9. Available from: http://dx.doi.org/10.1007/s40735-019-0221-0
76. Kowsari E, Arman SY, Shahini MH, Zandi H, Ehsani A, Naderi R, et al. In situ synthesis, electrochemical and quantum chemical analysis of an amino acid-derived ionic liquid inhibitor for corrosion protection of mild steel in 1M HCl solution. *Corros Sci* [Internet]. 2016;112:73–85. Available from: http://dx.doi.org/10.1016/j.corsci.2016.07.015
77. Nadi I, Belattmania Z, Sabour B, Reani A, Sahibed-dine A, Jama C, et al. Sargassum muticum extract based on alginate biopolymer as a new efficient biological corrosion inhibitor for carbon steel in

hydrochloric acid pickling environment: Gravimetric, electrochemical and surface studies. *Int J Biol Macromol*. 2019;141:137–49.

78. Ma X, Dang R, Kang Y, Gong Y, Luo J, Zhang Y, et al. Electrochemical studies of expired drug (Formoterol) as oilfield corrosion inhibitor for mild steel in H2SO4 media. *Int J Electrochem Sci*. 2020;15(3):1964–81.
79. Feng L, Zhang S, Lu Y, Tan B, Chen S, Guo L. Synergistic corrosion inhibition effect of thiazolyl-based ionic liquids between anions and cations for copper in HCl solution. *Appl Surf Sci* [Internet]. 2019;483:901–11. Available from: https://doi.org/10.1016/j.apsusc.2019.03.299
80. Sair S, Oushabi A, Nehhale K, Abboud Y, Tanane O, El Bouari A. Date palm waste extract as corrosion inhibitor for 304 stainless steel in 1 M HCl solution. *Int J Electrochem Sci*. 2018;13(11):10642–53.
81. Alamri AH, Obot IB. Highly efficient corrosion inhibitor for C1020 carbon steel during acid cleaning in multistage flash (MSF) desalination plant. *Desalination* [Internet]. 2019;470(August):114100. Available from: https://doi.org/10.1016/j.desal.2019.114100
82. Zhang QH, Hou BS, Li YY, Zhu GY, Lei Y, Wang X, et al. Dextran derivatives as highly efficient green corrosion inhibitors for carbon steel in CO2-saturated oilfield produced water: Experimental and theoretical approaches. *Chem Eng J*. 2021;424(April):130519.
83. Junaedi S, Al-Amiery AA, Kadihum A, Kadhum AAH, Mohamad AB. Inhibition effects of a synthesized novel 4-aminoantipyrine derivative on the corrosion of mild steel in hydrochloric acid solution together with quantum chemical studies. *Int J Mol Sci*. 2013;14(6):11915–28.
84. Atta AM, El-Mahdy GA, Al-Lohedan HA, Al-Hussain SA. Synthesis of environmentally friendly highly dispersed magnetite nanoparticles based on rosin cationic surfactants as thin film coatings of steel. *Int J Mol Sci*. 2014;15(4):6974–89.
85. Benabbouha T, Siniti M, El Attari H, Chefira K, Chibi F, Nmila R, et al. Red algae halopitys incurvus extract as a green corrosion inhibitor of carbon steel in hydrochloric acid. *J Bio- Tribo-Corrosion* [Internet]. 2018;4(3):0. Available from: http://dx.doi.org/10.1007/s40735-018-0161-0
86. Gao Y, Fan L, Ward L, Liu Z. Synthesis of polyaspartic acid derivative and evaluation of its corrosion and scale inhibition performance in seawater utilization. *Desalination* [Internet]. 2015;365:220–6. Available from: http://dx.doi.org/10.1016/j.desal.2015.03.006
87. Al-Amiery AA, Mohamad AB, Kadhum AAH, Shaker LM, Isahak WNRW, Takriff MS. Experimental and theoretical study on the corrosion inhibition of mild steel by nonanedioic acid derivative in hydrochloric acid solution. *Sci Rep* [Internet]. 2022;12(1):1–21. Available from: https://doi.org/10.1038/s41598-022-08146-8
88. Ebenso EE, Isabirye DA, Eddy NO. Adsorption and quantum chemical studies on the inhibition potentials of some thiosemicarbazides for the corrosion of mild steel in acidic medium. *Int J Mol Sci*. 2010;11(6):2473–98.
89. Umoren SA, Solomon MM, Obot IB, Suleiman RK. Comparative studies on the corrosion inhibition efficacy of ethanolic extracts of date palm leaves and seeds on carbon steel corrosion in 15% HCl solution. *J Adhes Sci Technol* [Internet]. 2018;32(17):1934–51. Available from: https://doi.org/10.1080/01694243.2018.1455797
90. Kumari P, Lavanya M. Optimization of inhibition efficiency of a schiff base on mild steel in acid medium: Electrochemical and RSM approach. *J Bio- Tribo-Corrosion* [Internet]. 2021;7(3):1–15. Available from: https://doi.org/10.1007/s40735-021-00542-3
91. Oubaaqa M, Ouakki M, Rbaa M, Abousalem AS, Maatallah M, Benhiba F, et al. Insight into the corrosion inhibition of new amino-acids as efficient inhibitors for mild steel in HCl solution: Experimental studies and theoretical calculations. *J Mol Liq* [Internet]. 2021;334:116520. Available from: https://doi.org/10.1016/j.molliq.2021.116520
92. Al-Amiery AA, Kadhum AAH, Mohamad AB, Musa AY, Li CJ. Electrochemical study on newly synthesized chlorocurcumin as an inhibitor for mild steel corrosion in hydrochloric acid. *Materials* (Basel). 2013;6(12):5466–77.
93. Palaniappan N, Cole I, Caballero-Briones F, Manickam S, Justin Thomas KR, Santos D. Experimental and DFT studies on the ultrasonic energy-assisted extraction of the phytochemicals of: Catharanthus roseus as green corrosion inhibitors for mild steel in NaCl medium. *RSC Adv*. 2020;10(9):5399–411.
94. Mehmeti V, Podvorica FI. Experimental and theoretical studies on corrosion inhibition of niobium and tantalum surfaces by carboxylated graphene oxide. *Materials (Basel)*. 2018;11(6):893.

95. Thoume A, Elmakssoudi A, Left DB, Benzbiria N, Benhiba F, Dakir M, et al. Amino acid structure analog as a corrosion inhibitor of carbon steel in 0.5 M H2SO4: Electrochemical, synergistic effect and theoretical studies. *Chem Data Collect* [Internet]. 2020;30:100586. Available from: https://doi.org/10.1016/j.cdc.2020.100586
96. Hao L, Lv G, Zhou Y, Zhu K, Dong M, Liu Y, et al. High performance anti-corrosion coatings of poly (Vinyl Butyral) composites with poly N-(vinyl)pyrrole and carbon black nanoparticles. *Materials (Basel)*. 2018;11(11).
97. Haque J, Srivastava V, Verma C, Quraishi MA. Experimental and quantum chemical analysis of 2-amino-3-((4-((S)-2-amino-2-carboxyethyl)-1H-imidazol-2-yl)thio) propionic acid as new and green corrosion inhibitor for mild steel in 1 M hydrochloric acid solution. *J Mol Liq*. 2017;225:848–55.
98. Berdimurodov E, Kholikov A, Akbarov K, Obot IB, Guo L. Thioglycoluril derivative as a new and effective corrosion inhibitor for low carbon steel in a 1 M HCl medium: Experimental and theoretical investigation. *J Mol Struct* [Internet]. 2021;1234:130165. Available from: https://doi.org/10.1016/j.molstruc.2021.130165
99. Xu XT, Xu HW, Li W, Wang Y, Zhang XY. A combined quantum chemical, molcular dynamics and Monto Carlo study of three amino acids as corroison inhibitors for aluminum in NaCl solution. *J Mol Liq*. 2022;345:117010.
100. Berrissoul A, Ouarhach A, Benhiba F, Romane A, Zarrouk A, Guenbour A, et al. Evaluation of Lavandula mairei extract as green inhibitor for mild steel corrosion in 1 M HCl solution. Experimental and theoretical approach. *J Mol Liq* [Internet]. 2020;313:113493. Available from: https://doi.org/10.1016/j.molliq.2020.113493
101. Ardakani EK, Kowsari E, Ehsani A, Ramakrishna S. Performance of all ionic liquids as the eco-friendly and sustainable compounds in inhibiting corrosion in various media: A comprehensive review. *Microchem J* [Internet]. 2021;165(December 2020):106049. Available from: https://doi.org/10.1016/j.microc.2021.106049
102. Fouda AS, Abdel Haleem E. Berry leaves extract as green effective corrosion inhibitor for cu in nitric acid solutions. *Surf Eng Appl Electrochem*. 2018;54(5):498–507.
103. Arellanes-Lozada P, Olivares-Xometl O, Guzmán-Lucero D, Likhanova NV., Domínguez-Aguilar MA, Lijanova IV., et al. The inhibition of aluminum corrosion in sulfuric acid by poly(1-vinyl-3-alkyl-imidazolium hexafluorophosphate). *Materials (Basel)*. 2014;7(8):5711–34.
104. Rbaa M, Benhiba F, Hssisou R, Lakhrissi Y, Lakhrissi B, Touhami ME, et al. Green synthesis of novel carbohydrate polymer chitosan oligosaccharide grafted on D-glucose derivative as bio-based corrosion inhibitor. *J Mol Liq* [Internet]. 2021;322:114549. Available from: https://doi.org/10.1016/j.molliq.2020.114549
105. Messina E, Giuliani C, Pascucci M, Riccucci C, Staccioli MP, Albini M, et al. Synergistic inhibition effect of chitosan and l-cysteine for the protection of copper-based alloys against atmospheric chloride-induced indoor corrosion. *Int J Mol Sci*. 2021;22(19):10321.
106. Anadebe VC, Okafor CS, Onukwuli OD. Electrochemical, molecular dynamics, adsorption studies and anti-corrosion activities of Moringa Leaf biomolecules on carbon steel surface in alkaline and acid environment. *Chem Data Collect* [Internet]. 2020;28:100437. Available from: https://doi.org/10.1016/j.cdc.2020.100437
107. Fawzy A, Zaafarany IA, Ali HM, Abdallah M. New synthesized amino acids-based surfactants as efficient inhibitors for corrosion of mild steel in hydrochloric acid medium: Kinetics and thermodynamic approach. *Int J Electrochem Sci*. 2018;13(5):4575–4600.
108. Al-Sabagh AM, Nasser NM, El-Azabawy OE, El-Tabey AE. Corrosion inhibition behavior of new synthesized nonionic surfactants based on amino acid on carbon steel in acid media. *J Mol Liq* [Internet]. 2016;219:1078–88. Available from: http://dx.doi.org/10.1016/j.molliq.2016.03.048
109. El-Katori EE, Al-Mhyawi S. Assessment of the Bassia muricata extract as a green corrosion inhibitor for aluminum in acidic solution. *Green Chem Lett Rev* [Internet]. 2019;12(1):31–48. Available from: https://doi.org/10.1080/17518253.2019.1569728
110. Fouda AEAS, Shahba RMA, El-Shenawy AE, Seyam TJA. Evaluation of Cleome Droserifolia (Samwah) as green corrosion inhibitor for mild steel in 1 M HCl solution. *Int J Electrochem Sci*. 2018;13(7):7057–75.

111. Zhao R, Xu W, Yu Q, Niu L. Synergistic effect of SAMs of S-containing amino acids and surfactant on corrosion inhibition of 316L stainless steel in 0.5 M NaCl solution. *J Mol Liq* [Internet]. 2020;318:114322. Available from: https://doi.org/10.1016/j.molliq.2020.114322
112. Akbarzade K, Danaee I. Nyquist plots prediction using neural networks in corrosion inhibition of steel by schiff base. *Iran J Chem Chem Eng*. 2018;37(3):135–43.
113. Akbarzadeh S, Ramezanzadeh M, Ramezanzadeh B, Bahlakeh G. Detailed atomic/molecular-level/electronic-scale computer modeling and electrochemical explorations of the adsorption and anti-corrosion effectiveness of the green nitrogen-based phytochemicals on the mild steel surface in the saline solution. *J Mol Liq* [Internet]. 2020;319:114312. Available from: https://doi.org/10.1016/j.molliq.2020.114312
114. Haris NIN, Sobri S, Yusof YA, Kassim NK. An overview of molecular dynamic simulation for corrosion inhibition of ferrous metals. *Metals (Basel)*. 2021;11(1):1–22.
115. Lebrini M, Suedile F, Salvin P, Roos C, Zarrouk A, Jama C, et al. Bagassa guianensis ethanol extract used as sustainable eco-friendly inhibitor for zinc corrosion in 3% NaCl: Electrochemical and XPS studies. *Surfaces Interfaces*. 2020;20:1–31.
116. Duca DA, Dan ML, Vaszilcsin N. Recycling of expired ceftamil drug as additive in the copper and nickel electrodeposition from acid baths. *Int J Environ Res Public Health*. 2021;18(18):9476.
117. Mouaden KEL, Chauhan DS, Quraishi MA, Bazzi L, Hilali M. Cinnamaldehyde-modified chitosan as a bio-derived corrosion inhibitor for acid pickling of copper: Microwave synthesis, experimental and computational study. *Int J Biol Macromol* [Internet]. 2020;164:3709–17. Available from: https://doi.org/10.1016/j.ijbiomac.2020.08.137
118. Okafor CS, Anadebe VC, Onukwuli OD. Experimental, Statistical modelling and molecular dynamics simulation concept of sapium ellipticum leaf extract as corrosion inhibitor for carbon steel in acid environment. *South Afr J Chem*. 2019;72:164–75.
119. Thakur RC, Kaur R, Kaur H. Transport properties of thiaminehydrochloridein binary aqueous mixtures of galactose. *Eur J Mol Clin Med*. 2020;7(07):2020.
120. Thakur RC, Sharma R, Kaur V, Kaur H. Molecular interactions of Niacin in binary aqueous solutions of KCl and FeCl 3 at different temperatures: Volumetric approach. *Think India J*. 2019;22(37):1604–13.
121. Kaur H, Thakur RC, Kumar H. Effect of proteinogenic amino acids L-serine/L-threonine on volumetric and acoustic behavior of aqueous 1-butyl-3-propyl imidazolium bromide at T = (288.15, 298.15, 308.15, 318.15) K. *J Chem Thermodyn* [Internet]. 2020;150:106211. Available from: https://doi.org/10.1016/j.jct.2020.106211
122. Espinosa T, Sanes J, Bermúdez MD. Halogen-free phosphonate ionic liquids as precursors of abrasion resistant surface layers on AZ31B magnesium alloy. *Coatings*. 2015;5(1):39–53.
123. Kasprzhitskii A, Lazorenko G, Nazdracheva T, Yavna V. Comparative computational study of l-amino acids as green corrosion inhibitors for mild steel. *Computation*. 2021;9(1):1–11.
124. Asaad MA, Ismail M, Khalid NHA, Huseien GF, Raja PB. Elaeis guineensis leaves extracts as eco-friendly corrosion inhibitor for mild steel in hydrochloric acid. *J Teknol*. 2018;80(6):53–9.
125. Haldhar R, Prasad D, Bahadur I, Dagdag O, Berisha A. Evaluation of Gloriosa superba seeds extract as corrosion inhibition for low carbon steel in sulfuric acidic medium: A combined experimental and computational studies. *J Mol Liq* [Internet]. 2021;323:114958. Available from: https://doi.org/10.1016/j.molliq.2020.114958
126. Matos LAC, Taborda MC, Alves GJT, da Cunha MT, Banczek E do P, Oliveira M de F, et al. Application of an acid extract of Barley agro-industrial waste as a corrosion inhibitor for stainless steel AISI 304 in H2SO4. *Int J Electrochem Sci*. 2018;13(2):1577–93.
127. Salleh SZ, Yusoff AH, Zakaria SK, Taib MAA, Abu Seman A, Masri MN, et al. Plant extracts as green corrosion inhibitor for ferrous metal alloys: A review. *J Clean Prod*. 2021;304:127030.
128. Goyal M, Kumar S, ChandrabhanVerma, Bahadur I, Ebenso EE, Lgaz H, et al. Interfacial adsorption behavior of quaternary phosphonium based ionic liquids on metal-electrolyte interface: Electrochemical, surface characterization and computational approaches. *J Mol Liq* [Internet]. 2020;298(1):111995. Available from: https://doi.org/10.1016/j.molliq.2019.111995

129. Abdallah M, Altass HM, Al-Gorair AS, Al-Fahemi JH, Jahdaly BAAL, Soliman KA. Natural nutmeg oil as a green corrosion inhibitor for carbon steel in 1.0 M HCl solution: Chemical, electrochemical, and computational methods. *J Mol Liq* [Internet]. 2021;323:115036. Available from: https://doi.org/10.1016/j.molliq.2020.115036
130. Ahanotu CC, Onyeachu IB, Solomon MM, Chikwe IS, Chikwe OB, Eziukwu CA. Pterocarpus santalinoides leaves extract as a sustainable and potent inhibitor for low carbon steel in a simulated pickling medium. *Sustain Chem Pharm* [Internet]. 2020;15(October 2019):100196. Available from: https://doi.org/10.1016/j.scp.2019.100196

Index

Q

R

S

T

U

V

W

Y

Z

For Product Safety Concerns and Information please contact our EU
representative GPSR@taylorandfrancis.com
Taylor & Francis Verlag GmbH, Kaufingerstraße 24, 80331 München, Germany

www.ingramcontent.com/pod-product-compliance
Lightning Source LLC
Chambersburg PA
CBHW080712220326
41598CB00033B/5387

* 9 7 8 1 0 3 2 6 7 7 2 3 1 *